Numerical Methods Using MATLAB®

Numerical Methods Using MATLAB®

Third Edition

G.R. Lindfield

J.E.T. Penny

AMSTERDAM • BOSTON • HEIDELBERG • LONDON
NEW YORK • OXFORD • PARIS • SAN DIEGO
SAN FRANCISCO • SINGAPORE • SYDNEY • TOKYO
Academic Press is an imprint of Elsevier

Academic Press is an imprint of Elsevier
225 Wyman Street, Waltham, MA 02451,USA
The Boulevard, Langford Lane, Kidlington, Oxford, OX5 1GB, UK

Notices

Knowledge and best practice in this field are constantly changing. As new research and experience broaden our understanding, changes in research methods, professional practices, or medical treatment may become necessary.

Practitioners and researchers must always rely on their own experience and knowledge in evaluating and using any information, methods, compounds, or experiments described herein. In using such information or methods they should be mindful of their own safety and the safety of others, including parties for whom they have a professional responsibility.

To the fullest extent of the law, neither the Publisher nor the authors, contributors, or editors, assume any liability for any injury and/or damage to persons or property as a matter of products liability, negligence or otherwise, or from any use or operation of any methods, products, instructions, or ideas contained in the material herein.

MATLAB® is a trademark of The MathWorks, Inc., and is used with permission. The MathWorks does not warrant the accuracy of the text or exercises in this book. This book's use or discussion of MATLAB® software or related products does not constitute endorsement or sponsorship by The MathWorks of a particular pedagogical approach or particular use of the MATLAB® software.

MATLAB® and Handle Graphics® are registered trademarks of The MathWorks, Inc.

Library of Congress Cataloging-in-Publication Data

Lindfield, G.R. (George R.) Numerical methods using MATLAB® / G.R. Lindfield, J.E.T. Penny. — 3rd ed.
p. cm.
Penny's name appears first on the earlier edition.
Includes bibliographical references and index.
ISBN 978-0-12-386942-5 (pbk.)
1. Numerical analysis—Data processing. 2. MATLAB. I. Penny, J.E.T. (John E.T.) II. Title.
QA297.P45 2012
518.0285'53–dc23 2012015199

British Library Cataloguing-in-Publication Data

A catalogue record for this book is available from the British Library.

For information on all Academic Press publications
visit our website at *http://store.elsevier.com*

Transferred to Digital Printing in 2013

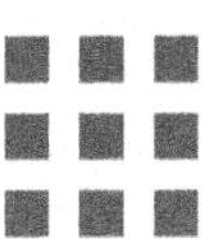

To our wives, Zena Lindfield and Wendy Penny, and our now adult children, Helen and Katy, and Debra, Mark and Joanne, for their patience and support. Also to our various cats, who have walked over, and even slept on, the computer keyboard!

Contents

Preface

The third edition of *Numerical Methods Using MATLAB®* is an extensive development of the first and second editions of this book. All MATLAB scripts and functions have been checked and revised to ensure that they are executable in the current version of MATLAB, version 7.13.

Our primary aim in this text is unchanged from previous editions; it is to introduce the reader to a wide range of numerical algorithms, explain their fundamental principles, and illustrate their application. The algorithms are implemented in the software package MATLAB, which is constantly being enhanced and provides a powerful tool to help with these studies.

Many important theoretical results are discussed, but it is not intended that a detailed and rigorous theoretical development in every area be provided. Rather, we wish to show how numerical procedures can be applied to solve problems from many fields of application, and that the numerical procedures give the expected theoretical performance when used to solve specific problems.

When used with care, MATLAB provides a natural and succinct way of describing numerical algorithms and a powerful means of experimenting with them. However, no tool, irrespective of its power, should be used carelessly or uncritically.

This text allows the reader to study numerical methods by encouraging systematic experimentation with some of the many fascinating problems of numerical analysis. Although MATLAB provides many useful functions, this text also introduces the reader to numerous useful and important algorithms and develops MATLAB functions to implement them. The reader is encouraged to use these functions to produce results in numerical and graphical form. MATLAB provides powerful and varied graphics facilities to give a clearer understanding of the nature of the results produced by the numerical procedures. Particular examples are given throughout the text to illustrate how numerical methods are used to study problems, including applications in the biosciences, chaos, neural networks, engineering, and science.

It should be noted that this introduction to MATLAB is relatively brief and is meant as an aid to the reader. It can in no way be expected to replace the standard MATLAB manual or textbooks devoted to MATLAB software. We provide a broad introduction to the topics, develop algorithms in the form of MATLAB functions, and encourage the reader to experiment with these functions, which have been kept as simple as possible for reasons of clarity. These functions can be improved, and we urge readers to develop those that are of particular interest to them.

In addition to a general introduction to MATLAB, the text covers the solution of linear equations and eigenvalue problems; methods for solving nonlinear equations; numerical integration and differentiation; the solution of initial value and boundary value problems; curve fitting, including splines, least squares, and Fourier analysis; and topics in optimization such as interior point methods, nonlinear programming, and genetic algorithms. Finally, we show how symbolic computing can be integrated with numerical algorithms. Specifically in this third edition, in Chapter 1 we have added descriptions and given examples of some functions recently added to MATLAB and have included a dicussion of handle graphics with examples. Chapter 4 now includes a section on Lobatto's method for integration and the Kronrod extension. Chapter 8 has been extensively revised and includes a description of the continuous genetic algorithm, Moller's scaled conjugate gradient method, and methods for solving constrained optimization problems.

The text contains many worked examples, practice problems (many of which are new to this edition), and solutions. We hope we have provided an interesting range of problems.

The text is suitable for undergraduate and postgraduate students and for those working in industry and education. We hope readers will share our enthusiasm for this area of study. For those who do not currently have access to MATLAB, this text provides a general introduction to a wide range of numerical algorithms and many useful and interesting examples and problems.

For readers of this book, additional materials, including all .m file scripts and functions listed in the text, are available on the book's companion site: *www.elsevierdirect.com/9780123869425*. For instructors using this book as a text for their courses, a solutions manual is available by registering at the textbook site: *www.textbooks.elsevier.com*.

We would like to thank the many readers from all over the world who provided helpful comments, which have enhanced this edition. We also acknowledge the valuable assistance given to us by our colleague, David Wilson, in guiding us in the restructuring of Sections 7.5, 7.6, and 7.7.

We would be pleased to hear from readers who note errors or have suggestions for improvements. Also, we would like to thank key Elsevier staff, including Patricia Osborn, Acquisitions Editor; Kathryn Morrissey, Editorial Project Manager; Joe Hayton, Publisher; Fiona Geraghty, Editorial Project Manager; Kristen Davis, Designer; and Marilyn Rash, Project Manager.

George Lindfield and John Penny
Aston University
Birmingham

List of Figures

1 An Introduction to MATLAB®

MATLAB® is a software package produced by The MathWorks, Inc. (www.mathworks.com) and is available on systems ranging from personal computers to supercomputers, including parallel computing. In this chapter we aim to provide a useful introduction to MATLAB, giving sufficient background for the numerical methods we consider. The reader is referred to the MATLAB manual for a full description of the package.

1.1 The MATLAB Software Package

MATLAB is probably the world's most successful commercial numerical analyis software package, and its name is derived from the term "matrix laboratory." It provides an interactive development tool for scientific and engineering problems and more generally for those areas where significant numeric computations have to be performed. The package can be used to evaluate single statements directly or a list of statements called a script can be prepared. Once named and saved, a script can be executed as an entity. The package was originally based on software produced by the LINPACK and EISPACK projects but currently includes LAPACK and BLAS libraries which represent the current "state-of-the-art" numerical software for matrix computations. MATLAB provides the user with

1. Easy manipulation of matrix structures
2. A vast number of powerful built-in routines that are constantly growing and developing
3. Powerful two- and three-dimensional graphing facilities
4. A scripting system that allows users to develop and modify the software for their own needs
5. Collections of functions, called toolboxes, that may be added to the facilities of the core MATLAB. These are designed for specific applications, for example, neural networks, optimization, digital signal processing, and higher-order spectral analysis.

It is not difficult to use MATLAB, although to use it with maximum efficiency for complex tasks requires experience. Generally MATLAB works with rectangular or square arrays of data (matrices), the elements of which may be real or complex. A scalar quantity is thus a matrix containing a single element. This is an elegant and powerful notion but it can present the user with an initial conceptual difficulty. A user schooled in such languages as C++ or Python is familiar with a pseudo-statement of the form A = B + C and can immediately interpret it as an instruction that A is assigned the sum of values of the numbers stored in

B and C. In MATLAB the variables B and C may represent arrays so that *each element* of the array A will become the sum of the values of corresponding elements of B and C.

There are several languages or software packages that have some similarities to MATLAB. These packages include

APL. The letters stand for A Programming Language. This language was designed mainly for manipulating arrays. It contains many powerful facilities but it used nonstandard symbols and the syntax was unusual. The keyboard had to be remapped for these special characters. This language had an important influence on other languages and was replaced by APL 2, which is still available today.

The NAg Library. This is a very extensive, high-quality collection of subroutines for numerical analysis. There is a MATLAB toolbox for the NAg Library.

Mathematica and Maple. These packages are known for their ability to carry out complicated symbolic mathematical manipulation, but they are also able to undertake high precision numerical computation. In contrast MATLAB is known for its powerful numerical computational and matrix manipulation facilties. However, MATLAB also provides an optional symbolic toolbox. This is discussed in Chapter 9.

Other packages. Packages such as Scilab,[1] Octave[2] (on UNIX platforms only), and Freemat[3] are somewhat similar to MATLAB in that they implement a wide range of numerical methods. A commercial alternative to MATLAB is O-Matrix.[4]

The current MATLAB release, version 7.13.0.564 (R2011b), is available on a wide variety of platforms. Generally Mathworks releases an upgraded version of MATLAB every six months. When MATLAB is invoked it opens a command window; graphics, editing, and help windows may also be opened if required. Users can design their MATLAB working environment as they see fit. MATLAB scripts and functions are generally platform independent and they can be readily ported from one system to another. To install and start MATLAB, readers should consult the manual appropriate to their particular working environment.

The scripts and functions given in this book have been tested under MATLAB release, version 7.13.0.564 (R2011b). However, most of them will work directly using earlier versions of MATLAB although some may require modification.

The remainder of this chapter is devoted to introducing some of the statements and syntax of MATLAB. The intention is to give the reader a sound but brief introduction to the power of MATLAB. Some details of structure and syntax are omitted and must be obtained from the MATLAB manual. A detailed description of MATLAB is given by Higham and Higham (2005). Other sources of information are the Mathworks website and Wikipedia. Wikipedia should be used with some care.

[1] www.scilab.org
[2] www.gnu.orgsoftwareoctave
[3] freemat.sourceforge.net
[4] www.omatrix.com

1.2 Matrices and Matrix Operations in MATLAB

The matrix is fundamental to MATLAB and we have provided a broad and simple introduction to matrices in Appendix A. In MATLAB the names used for matrices must start with a letter and may be followed by any combination of letters or digits. The letters may be upper- or lower- case. Note that throughout this text a distinctive font is used to denote MATLAB statements and output, for example `disp`.

In MATLAB the arithmetic operations of addition, subtraction, multiplication, and division can be performed in the usual way on scalar quanties, but they can also be used directly with matrices or arrays of data. To use these arithmetic operators on matrices, the matrices must first be created. There are several ways of doing this in MATLAB and the simplest method, which is suitable for small matrices, is as follows. We assign an array of values to `A` by opening the **command** window and then typing

```
>> A = [1 3 5;1 0 1;5 0 9]
```

after the prompt >>. Notice that the elements of the matrix are placed in square brackets, each row element separated by at least one space or comma. A semicolon (`;`) indicates the end of a row and the beginning of another. When the return key is pressed the matrix will be displayed:

```
A =

     1     3     5
     1     0     1
     5     0     9
```

All statements are executed by pressing the return or enter key. Thus, for example, by typing `B = [1 3 51;2 6 12;10 7 28]` after the >> prompt, and pressing the return key, we assign values to `B`. To add the matrices in the **command** window and assign the result to `C` we type `C = A+B` and similarly if we type `C = A-B` the matrices are subtracted. In both cases the results are displayed row by row in the **command** window. Note that terminating a MATLAB statement with a semicolon suppresses any output.

For simple problems we can use the **command** window. By simple we mean MATLAB statements of limited complexity; even MATLAB statements of limited complexity can provide some powerful numerical computation. However, if we require the execution of an ordered sequence of MATLAB statements (commands) then it is sensible for these statements to be typed in the MATLAB **editor** window to create a script, which must be saved under a suitable name for future use as required. There will be no execution or output until the name of this script is typed into the **command** window and it is executed by pressing return.

A matrix that has only one row or column is called a vector. A row vector consists of one row of elements and a column vector consists of one column of elements. Conventionally in mathematics, engineering, and science an emboldened uppercase letter is usually used to represent a matrix, for example **A**. An emboldened lowercase letter usually represents a *column* vector, that is, **x**. The transpose operator converts a row to a column and vice versa so that we can represent a row vector as a column vector transposed. Using the superscript ⊤ in mathematics to indicate a transpose, we can write a row vector as $\mathbf{x}^\top$. In MATLAB it is often convenient to ignore the convention that the initial form of a vector is a column; the user can define the initial form of a vector as a row or a column.

The implementation of vector and matrix multiplication in MATLAB is straightforward. Beginning with vector multiplication, we assume that row vectors having the same number of elements have been assigned to `d` and `p`. To multiply them together we write `x = d*p'`. Note that the symbol `'` transposes the row `p` into a column so that the multiplication is valid. The result, `x`, is a scalar. Many practitioners use `.'` to indicate a transpose. The reason for this is discussed in Section 1.4.

Assuming the two matrices `A` and `B` have been assigned, for matrix multiplication the user simply types `C = A*B`. This computes `A` postmultiplied by `B`, assigns the result to `C`, and displays it, providing the multiplication is valid. Otherwise MATLAB gives an appropriate error indication. The conditions for matrix multiplication to be valid are given in Appendix A. Notice that the symbol `*` must be used for multiplication because in MATLAB multiplication is not implied.

A very useful MATLAB function is `whos` (and the similar function, `who`). These functions tell us the current content of the workspace. For example, provided `A`, `B`, and `C` described previously have not been cleared from the memory, then

```
>> whos
  Name          Size                          Bytes  Class
  A             3x3                              72  double array
  B             3x3                              72  double array
  C             3x3                              72  double array

Grand total is 27 elements using 216 bytes
```

This tells us that `A`, `B`, and `C` are all 3×3 matrices. They are stored as double precision arrays. A double precison number requires 8 bytes to store it, so each array of 9 elements requires 72 bytes. Consider now the following operations:

```
>> clear A
>> B = [ ];
>> C = zeros(4,4);

>> whos
  Name          Size                          Bytes  Class
  B             0x0                               0  double array
  C             4x4                             128  double array

Grand total is 16 elements using 128 bytes
```

Here we see that we have cleared (i.e., deleted) A from memory, and assigned an empty matrix to B and a 4×4 array of zeros to C.

Note that the size of matrices can also be determined using the size and length functions:

```
>> A = zeros(4,8);
>> B = ones(7,3);
>> [p q] = size(A)

p =
     4

q =
     8

>> length(A)

ans =
     8

>> L = length(B)

L =
     7
```

size gives the size of the matrix whereas length gives the number of elements in the largest dimension.

1.3 Manipulating the Elements of a Matrix

In MATLAB, matrix elements can be manipulated individually or in blocks. For example,

```
>> X(1,3) = C(4,5)+V(9,1)
>> A(1) = B(1)+D(1)
>> C(i,j+1) = D(i,j+1)+E(i,j)
```

are valid statements relating elements of matrices. Rows and columns can be manipulated as complete entities. Thus A(:,3), B(5,:) refer respectively to the third column of A and fifth row of B. If B has 10 rows and 10 columns—that is, it is a 10×10 matrix—then B(:,4:9) refers to columns 4 through 9 of the matrix. The : by itself indicates all the rows, and hence all elements of columns 4 through 9. Note that in MATLAB, by default, the *lowest matrix index starts at 1*. This can be a source of confusion when implementing some algorithms.

The following examples illustrate some of the ways subscripts can be used in MATLAB. First we assign a matrix:

```
>> A = [2 3 4 5 6;-4 -5 -6 -7 -8; 3 5 7 9 1; ...
            4 6 8 10 12;-2 -3 -4 -5 -6]

A =

     2     3     4     5     6
    -4    -5    -6    -7    -8
     3     5     7     9     1
     4     6     8    10    12
    -2    -3    -4    -5    -6
```

Note the use of ... (an ellipsis) to indicate that the MATLAB statement continues on the next line. Executing the statements

```
>> v = [1 3 5];
>> b = A(v,2)
```

gives

```
b =

     3
     5
    -3
```

Thus b is composed of the elements of the first, third, and fifth rows in the second column of A. Executing

```
>> C = A(v,:)
```

gives

```
C =

     2     3     4     5     6
     3     5     7     9     1
    -2    -3    -4    -5    -6
```

Thus C is composed of the first, third, and fifth rows of A. Executing

```
>> D = zeros(3);
>> D(:,1) = A(v,2)
```

gives

```
D =

     3     0     0
     5     0     0
    -3     0     0
```

Here D is a 3×3 matrix of zeros with column 1 replaced by the first, third, and fifth elements of column 2 of A. Executing

```
>> E = A(1:2,4:5)
```

gives

```
E =
     5     6
    -7    -8
```

Note that if we index an existing square or rectangular array with a single index, then the elements of the array are identified as follows. Index 1 gives the top left element of the array, and the index is incremented down the columns in sequence, from left to right. For example, with reference to the preceding array C

```
C1 = C;
C1(1:4:15) = 10

C1 =
    10     3     4     5    10
     3    10     7     9     1
    -2    -3    10    -5    -6
```

Note that in this example the index is incremented by 4.

When manipulating very large matrices it is easy to become unsure of the size of the matrix. Thus, if we want to find the value of the element in the penultimate row and last column of A defined previously we could write

```
>> size(A)

ans =
     5     5

>> A(4,5)

ans =
    12
```

but it is easier is to use end:

```
>> A(end-1,end)

ans =
    12
```

The reshape function may be used to manipulate a complete matrix. As the name implies, the function reshapes a given matrix into a new matrix of any specified size provided it has an identical number of elements. For example, a 3×4 matrix can be reshaped into a 6×2 matrix but a 3×3 matrix cannot be reshaped into a 5×2 matrix. It is important to note that this function takes each column of the original matrix in turn until the new required column size is achieved and then repeats the process for the next column. For example, consider the matrix P:

```
>> P = C(:,1:4)

P =
     2     3     4     5
     3     5     7     9
    -2    -3    -4    -5

>> reshape(P,6,2)

ans =
     2     4
     3     7
    -2    -4
     3     5
     5     9
    -3    -5

>> s = reshape(P,1,12);
>> s(1:10)

ans =
     2     3    -2     3     5    -3     4     7    -4     5
```

1.4 Transposing Matrices

A simple operation that may be performed on a matrix is transposition, which interchanges rows and columns. Transposition of a vector is briefly discussed in Section 1.2. In MATLAB transposition is denoted by the symbol `'`. For example, consider the matrix A, where

```
>> A = [1 2 3;4 5 6;7 8 9]

A =
     1     2     3
     4     5     6
     7     8     9
```

To assign the transpose of A to B we write

```
>> B = A'

B =
     1     4     7
     2     5     8
     3     6     9
```

Had we used .' to obtain the transform we would have obtained the same result. However, if A is complex then the MATLAB operator ' gives the complex conjugate transpose. For example,

```
>> A = [1+2i 3+5i;4+2i 3+4i]

A =
   1.0000 + 2.0000i   3.0000 + 5.0000i
   4.0000 + 2.0000i   3.0000 + 4.0000i

>> B = A'

B =
   1.0000 - 2.0000i   4.0000 - 2.0000i
   3.0000 - 5.0000i   3.0000 - 4.0000i
```

To provide the transpose without conjugation we execute

```
>> C = A.'

C =
   1.0000 + 2.0000i   4.0000 + 2.0000i
   3.0000 + 5.0000i   3.0000 + 4.0000i
```

1.5 Special Matrices

Certain matrices occur frequently in matrix manipulations and MATLAB ensures that these are generated easily. Some of the most common are `ones(m,n)`, `zeros(m,n)`, `rand(m,n)`, `randn(m,n)`, and `randi(p,m,n)`. These MATLAB functions generate $m \times n$ matrices composed of ones, zeros, uniformly distributed random numbers, normally distributed random numbers, and uniformly distributed random integers, respectively. In the case of `randi(p,m,n)`, `p` is the maximum integer. If only a single scalar parameter is given, then these statements generate a square matrix of the size given by the parameter. The MATLAB function `eye(n)` generates the $n \times n$ unit matrix. The function `eye(m,n)` generates a matrix of m rows and n columns with a diagonal of ones:

```
>> A = eye(3,4), B = eye(4,3)

A =

     1     0     0     0
     0     1     0     0
     0     0     1     0

B =

     1     0     0
     0     1     0
     0     0     1
     0     0     0
```

If we wish to generate a random matrix `C` of the same size as an already existing matrix `A`, then the statement `C = rand(size(A))` can be used. Similarly `D = zeros(size(A))` and `E = ones(size(A))` generate a matrix `D` of zeros and a matrix `E` of ones, both of which are the same size as matrix `A`.

Some special matrices with more complex features are introduced in Chapter 2.

1.6 Generating Matrices and Vectors with Specified Element Values

Here we confine ourselves to some relatively simple examples:

`x = -8:1:8` (or `x = -8:8`) sets x to a vector having the elements $-8, -7, \ldots, 7, 8$

`y = -2:.2:2` sets y to a vector having the elements $-2, -1.8, -1.6, \ldots, 1.8, 2$

`z = [1:3 4:2:8 10:0.5:11]` sets z to a vector having the elements

$$[1 \quad 2 \quad 3 \quad 4 \quad 6 \quad 8 \quad 10 \quad 10.5 \quad 11]$$

The MATLAB function `linspace` also generates a vector. However, in this function the user defines the begining and end values of the vector and the number of elements in the vector. For example,

```
>> w = linspace(-2,2,5)

w =

    -2    -1     0     1     2
```

This is simple and could just as well have been created by `w = -2:1:2` or even `w = -2:2`. However,

```
>> w = linspace(0.2598,0.3024,5)

w =
    0.2598    0.2704    0.2811    0.2918    0.3024
```

Generating this sequence of values by other means would be more difficult. If we require logarithmic spacing then we can use

```
>> w = logspace(1,2,5)

w =
   10.0000   17.7828   31.6228   56.2341  100.0000
```

Note that the values produced are between 10^1 and 10^2, not 1 and 2. Again, generating these values by any other means would require some thought! The user of `logspace` should be warned that if the second parameter is `pi` the values run to π, not 10^π. Consider the following:

```
>> w = logspace(1,pi,5)

w =
   10.0000    7.4866    5.6050    4.1963    3.1416
```

More complicated matrices can be generated by combining other matrices. For example, consider the two statements

```
>> C = [2.3 4.9; 0.9 3.1];
>> D = [C ones(size(C)); eye(size(C)) zeros(size(C))]
```

These two statements generate a new matrix `D` the size of which is double that of the original `C`; thus

```
D =
    2.3000    4.9000    1.0000    1.0000
    0.9000    3.1000    1.0000    1.0000
    1.0000         0         0         0
         0    1.0000         0         0
```

The MATLAB function `repmat` replicates a given matrix a required number of times. For example, assuming the matrix `C` is defined in the preceding, then

```
>> E = repmat(C,2,3)
```

replicates `C` as a block to give a matrix with twice as many rows and three times as many columns. Thus we have a matrix `E` of 4 rows and 6 columns:

```
E =
    2.3000    4.9000    2.3000    4.9000    2.3000    4.9000
    0.9000    3.1000    0.9000    3.1000    0.9000    3.1000
    2.3000    4.9000    2.3000    4.9000    2.3000    4.9000
    0.9000    3.1000    0.9000    3.1000    0.9000    3.1000
```

The MATLAB function `diag` allows us to generate a diagonal matrix from a specified vector of diagonal elements. Thus

```
>> H = diag([2 3 4])
```

generates

```
H =
     2     0     0
     0     3     0
     0     0     4
```

There is a second use of the function `diag`, which is to obtain the elements on the leading diagonal of a given matrix. Consider

```
>> P = rand(3,4)

P =
    0.3825    0.9379    0.2935    0.8548
    0.4658    0.8146    0.2502    0.3160
    0.1030    0.0296    0.5830    0.6325
```

then

```
>> diag(P)

ans =
    0.3825
    0.8146
    0.5830
```

A more complicated form of diagonal matrix is the block diagonal matrix. This type of matrix can be generated using the MATLAB function `blkdiag`. We set matrices `A1` and `A2` as follows:

```
>> A1 = [1 2 5;3 4 6;3 4 5];
>> A2 = [1.2 3.5,8;0.6 0.9,56];
```

Then,

```
>> blkdiag(A1,A2,78)

ans =
   1.0000   2.0000   5.0000        0        0        0        0
   3.0000   4.0000   6.0000        0        0        0        0
   3.0000   4.0000   5.0000        0        0        0        0
        0        0        0   1.2000   3.5000   8.0000        0
        0        0        0   0.6000   0.9000  56.0000        0
        0        0        0        0        0        0  78.0000
```

The preceding functions can be very useful in allowing the user to create matrices with complicated structures, without detailed programming.

1.7 Matrix Functions

Some arithmetic operations are simple to evaluate for single scalar values but involve a great deal of computation for matrices. For large matrices such operations may take a significant amount of time. An example of this is where a matrix is raised to a power. We can write this in MATLAB as `A^p` where `p` is a scalar value and `A` is a square matrix. This produces the power of the matrix for any value of `p`. For the case where the power equals 0.5 it is better to use `sqrtm(A)`, which gives the principal square root of the matrix `A` (see Appendix A, Section A.13). Similarly, for the case where the power equals -1 it is better to use `inv(A)`. Another special operation directly available in MATLAB is `expm(A)`, which gives the exponential of the matrix `A`. The MATLAB function `logm(A)` provides the principal logarithm to the base e of `A`. If `B = logm(A)`, then the principal logarithm `B` is the unique logarithm for which every eigenvalue has an imaginary part lying strictly betweeen $-\pi$ and π.

For example,

```
>> A = [61 45;60 76]

A =
    61    45
    60    76

>> B = sqrtm(A)

B =
    7.0000    3.0000
    4.0000    8.0000
```

```
>> B^2

ans =
    61.0000    45.0000
    60.0000    76.0000
```

1.8 Using the MATLAB \ Operator for Matrix Division

As an example of the power of MATLAB we consider the solution of a system of linear equations. It is easy to solve the problem $ax = b$ where a and b are simple scalar constants and x is the unknown. Given a and b then $x = b/a$. However, consider the corresponding matrix equation

$$\mathbf{Ax} = \mathbf{b} \tag{1.1}$$

where **A** is a square matrix and **x** and **b** are column vectors. We wish to find **x**. Computationally this is a much more difficult problem and in MATLAB it is solved by executing the statement

```
x = A\b
```

This statement uses the important MATLAB division operator \ and solves the linear equation system (1.1).

Solving linear equation systems is an important problem and the computational efficiency and other aspects of this type of problem are discussed in considerable detail in Chapter 2.

1.9 Element-by-Element Operations

Element-by-element operations differ from the standard matrix operations but they can be very useful. They are achieved by using a period or dot (`.`) to precede the operator. If `X` and `Y` are matrices (or vectors), then `X.^Y` raises each *element* of `X` to the power of the corresponding element of `Y`. Similarly `X.*Y` and `Y.\X` multiply or divide each element of `X` by the corresponding element in `Y` respectively. The form `X./Y` gives the same result as `Y.\X`. For these operations to be executed the matrices and vectors used must be the same size. Note that a period is *not* used in the operations + and - because ordinary matrix addition and subtraction *are* element-by-element operations. Examples of element-by-element operations are given as follows:

```
>> A = [1 2;3 4]

A =
     1     2
     3     4
```

```
>> B = [5 6;7 8]

B =
     5     6
     7     8
```

First we use normal matrix multiplication:

```
>> A*B

ans =
    19    22
    43    50
```

However, using the dot operator (.) we have

```
>> A.*B

ans =
     5    12
    21    32
```

which is element-by-element multiplication. Now consider the statement

```
>> A.^B

ans =
           1          64
        2187       65536
```

In the preceding, each element of A is raised to the corresponding power in B.

Element-by-element operations have many applications. An important use is in plotting graphs (see Section 1.13). For example,

```
>> x = -1:0.1:1;
>> y = x.*cos(x);
>> y1 = x.^3.*(x.^2+3*x+sin(x));
```

Notice here that using a vector of many values, x, allows a vector of corresponding values for y and y1 to be computed simultaneously from single statements. Element-by-element operations are in effect processes on scalar quatities performed simultaneously.

1.10 Scalar Operations and Functions

In MATLAB we can define and manipulate scalar quantities, as in most other computer languages, but no distinction is made in the naming of matrices and scalars. Thus A could represent a scalar or matrix quantity. The process of assignment makes the distinction.

For example,

```
>> x = 2;
>> y = x^2+3*x-7

y =

     3

>> x = [1 2;3 4]

x =

     1     2
     3     4

>> y = x.^2+3*x-7

y =

    -3     3
    11    21
```

Note in the preceding examples that when vectors are used the dot must be placed before the operator. This is not required for scalar operations, but does not cause errors if used.

If we multiply a square matrix by itself, for example, in the form `x^2`, then we get full matrix multiplication as shown in the following, rather than element-by-element multiplication as given by `x.^2`.

```
>> y = x^2+3*x-7

y =

     3     9
    17    27
```

A very large number of mathematical functions are directly built into MATLAB. They act on scalar quanties, arrays, or vectors on an element-by-element basis. They may be called by using the function name together with the parameters that define the function. These functions may return one or more values. A small selection of MATLAB functions is given in Table 1.1, which lists the function name, the function use, and an example function call. Note that all function names must be in lowercase letters.

All MATLAB functions were not listed previously, but MATLAB provides a complete range of trigonometric and inverse trigonometric functions, hyperbolic and inverse hyperbolic functions, and logarithmic functions. The following examples illustrate the use of some of the functions listed before.

Table 1.1 Selected MATLAB Mathematical Functions

Function	Function gives	Example
`sqrt(x)`	square root of x	`y = sqrt(x+2.5);`
`abs(x)`	if x is real, gives positive value of x	
	If x is complex, gives scalar measure of x	`d = abs(x)*y;`
`real(x)`	real part of x when x is complex	`d = real(x)*y;`
`imag(x)`	imaginary part of x when x is complex	`d = imag(x)*y;`
`conj(x)`	complex conjugate of x	`x = conj(y);`
`sin(x)`	sine of x in radians	`t = x+sin(x);`
`asin(x)`	inverse sine of x returned in radians	`t = x+sin(x);`
`sind(x)`	sine of x in degrees	`t = x+sind(x);`
`log(x)`	log to base e of x	`z = log(1+x);`
`log10(x)`	log to base 10 of x	`z = log10(1-2*x);`
`cosh(x)`	hyperbolic cosine of x	`u = cosh(pi*x);`
`exp(x)`	exponential of x, i.e., e^x	`p = .7*exp(x);`
`gamma(x)`	gamma function of x	`f = gamma(y);`
`bessel(n,x)`	nth-order Bessel function of x	`f = bessel(2,y);`

```
>> x = [-4 3];
>> abs(x)

ans =
     4     3

>> x = 3+4i;
>> abs(x)

ans =
     5

>> imag(x)

ans =
     4

>> y = sin(pi/4)

y =
    0.7071

>> x = linspace(0,pi,5)
```

```
x =

         0    0.7854    1.5708    2.3562    3.1416

>> sin(x)

ans =

         0    0.7071    1.0000    0.7071    0.0000

>> x = [0 pi/2;pi 3*pi/2]

x =

         0    1.5708
    3.1416    4.7124

>> y = sin(x)

y =

         0    1.0000
    0.0000   -1.0000
```

Some functions perform special calculations for important and general mathematical processes. These functions often require more than one input parameter and may provide several outputs. For example, `bessel(n,x)` gives the *n*th-order Bessel function of *x*. The statement `y = fzero('fun',x0)` determines the root of the function `fun` near `x0` where `fun` is a function defined by the user that provides the equation for which we are finding the root. For examples of the use of `fzero`, see Section 3.1. The statement `[Y,I] = sort(X)` is an example of a function that can return two output values. `Y` is the sorted matrix and `I` is a matrix containing the indices of the sort.

In addition to a large number of mathematical functions, MATLAB provides several utility functions that may be used for examining the operation of scripts. These follow:

- `pause` causes the execution of the script to pause until the user presses a key. Note that the cursor is turned into the symbol P, warning the script is in pause mode. This is often used when the script is operating with `echo on`.
- `echo on` displays each line of script in the **command** window before execution. This is useful for demonstrations. To turn it off, use the statement `echo off`.
- `who` lists the variables in the current workspace.
- `whos` lists all the variables in the current workspace, together with information about their size and class, and so on.

MATLAB also provides functions related to time:

- `clock` returns the current date and time in the form <year month day hour min sec>.
- `etime(t2,t1)` calculates elapsed time between `t1` and `t2`. Note that `t1` and `t2` are output from the clock function.

- `tic ... toc` provides a way of finding the time taken to execute a segment of script. The statement `tic` starts the timing and `toc` gives the elapsed time since the last `tic`.
- `cputime` returns the total time in seconds since MATLAB was launched.

The following script uses the timing functions described in the preceding to estimate the time taken to solve a 1000×1000 system of linear equations:

```
% e3s107.m  Solves a 1000x1000 linear equation system
A = rand(1000); b = rand(1000,1);
T_before = clock;
tic
t0 = cputime;
y = A\b;
timetaken = etime(clock,T_before);
tend = toc;
t1 = cputime-t0;
disp('etime     tic-toc   cputime')
fprintf('%5.2f %10.2f %10.2f\n\n', timetaken,tend,t1);
```

Running this script on a particular computer gave the following results:

```
etime     tic-toc   cputime
 0.30       0.31       0.30
```

The output shows that the three alternative methods of timing give essentially the same value. When measuring computing times the displayed times vary from run to run and the shorter the run time, the greater the percentage variation.

1.11 String Variables

We have found that MATLAB makes no distinction in naming matrices and scalar quantities. This is also true of string variables or strings. For example, `A = [1 2; 3 4]`, `A = 17.23`, or `A = 'help'` are each valid statements and assign an array, a scalar, or a text string respectively to `A`.

Characters and strings of characters can be assigned to variables directly in MATLAB by placing the string in quotes and then assigning it to a variable name. Strings can then be manipulated by specific MATLAB string functions, which we list in this section. Some examples showing the manipulation of strings using standard MATLAB assignments are given in the following.

```
>> s1 = 'Matlab ', s2 = 'is ', s3 = 'useful'

s1 =
Matlab

s2 =
is
```

```
s3 =
useful
```

Strings in MATLAB are represented as vectors of the equivalent ASCII code numbers; it is only the way that we assign and access them that makes them strings. For example, the string `'is '` is actually saved as the vector [105 115 32]. Hence we can see that the ASCII codes for the letters `i` and `s` and a space are 105, 115, and 32, respectively. This vector structure has important implications when we manipulate strings. For example, we can concatenate strings, because of their vector nature, by using the square brackets as follows:

```
>> sc = [s1 s2 s3]

sc =
Matlab is useful
```

Note the spaces are recognized. To identify any item in the string array we can write

```
>> sc(2)

ans =
a
```

To identify a subset of the elements of this string we can write

```
>> sc(3:10)

ans =
tlab is
```

We can display a string vertically by transposing the string vector:

```
>> sc(1:3)'

ans =
M
a
t
```

We can also reverse the order of a substring and assign it to another string as follows:

```
>>a = sc(6:-1:1)

a =
baltaM
```

We can define string arrays as well. For example, using the string `sc` as defined previously we have

```
>> sd = 'Numerical method'
>> s = [sc; sd]

s =
Matlab is useful
Numerical method
```

To obtain the 12th column of this string we use

```
>> s(:,12)

ans =
s
e
```

Note that the string lengths must be the same in order to form a rectangular array of ASCII code numbers. In this case the array is 2×16.

We now show how MATLAB string functions can be used to manipulate strings. To replace one string by another we use `strrep` as follows:

```
>> strrep(sc,'useful','super')

ans =
Matlab is super
```

Notice that this statement causes `useful` in `sc` to be replaced by `super`.

We can determine if a particular character or string is present in another string by using `findstr`. For example,

```
>> findstr(sd,'e')

ans =
     4    12
```

This tells us that the 4th and 12th characters in the string are `'e'`. We can also use this function to find the location of a substring of this string as follows:

```
>> findstr(sd, 'meth')

ans =
    11
```

The string `'meth'` begins at the 11th character in the string. If the substring or character is not in the original string we obtain the result illustrated by the example that follows:

```
>> findstr(sd,'E')

ans =
     [ ]
```

We can convert a string to its ASCII code equivalent by either using the function `double` or invoking any arithmetic operation. Thus, operating on the existing string `sd` we have

```
>> p = double(sd(1:9))

p =
     78   117   109   101   114   105    99    97   108

>> q = 1*sd(1:9)

q =
     78   117   109   101   114   105    99    97   108
```

Note that in the case where we are multiplying the string by 1, MATLAB treats the string as a vector of ASCII equivalent numbers and multiplies it by 1. Recalling that `sd(1:9) = 'Numerical'` we can deduce that the ASCII code for N is 78, for u it is 117, and so on.

We convert a vector of ASCII code to a string using the MATLAB `char` function. For example,

```
>> char(q)

ans =
Numerical
```

To increase each ASCII code number by 3 and then convert to the character equivalent we have

```
>> char(q+3)

ans =
Qxphulfdo

>> char((q+3)/2)

ans =
(<84:6327

>> double(ans)

ans =
    40    60    56    52    58    54    51    50    55
```

As seen in the preceding, `char(q)` converts the ASCII string back to characters. Here we have shown that it is possible to do arithmetic on the ASCII code numbers and, if we wish, convert back to characters. If after manipulation the ASCII code values are noninteger, they are rounded down.

It is important to appreciate that the string `'123'` and the number 123 are not the same. Thus

```
>> a = 123

a =
   123

>> s1 = '123'

s1 =
123
```

Using `whos` shows the class of the variables `a` and `s1` as follows:

```
>> whos
  Name      Size                    Bytes  Class
  a         1x1                         8  double array
  s1        1x3                         6  char array

Grand total is 4 elements using 14 bytes
```

A character requires 2 bytes, while a double precision number requires 8 bytes. We can convert strings to their numeric equivalent using the functions `str2num` and `str2double` as follows:

```
>> x=str2num('123.56')

x =
  123.5600
```

Appropriate strings can be converted to complex numbers but the user should take care, as we illustrate in the following:

```
>> x = str2num('1+2j')

x =
1.0 + 2.0000i
```

but

```
>> x = str2num('1+2 j')

x =
   3.0000                  0 + 1.0000i
```

Note that `str2double` can be used to convert to complex numbers and is more tolerant of spaces.

```
>> x = str2double('1+2 j')

x =
1.0 + 2.0000i
```

There are many MATLAB functions that are available to manipulate strings; for more information see the MATLAB manuals. Here we illustrate the use of some functions.

- `bin2dec('111001')` or `bin2dec('111 001')` returns 57
- `dec2bin(57)` returns the string `'111001'`
- `int2str([3.9 6.2])` returns the string `'4 6'`
- `num2str([3.9 6.2])` returns the string `'3.9 6.2'`
- `str2num('3.9 6.2')` returns 3.9000 6.2000
- `strcat('how ','why ','when')` returns the string `'howwhywhen'`
- `strcmp('whitehouse','whitepaint')` returns 0 because the strings are not identical
- `strncmp('whitehouse','whitepaint',5)` returns 1 because the first 5 characters of the strings are identical
- `date` returns the current date, in the form `24-Aug-2011`

A useful and common application of the function `num2str` is in the `disp` and `title` functions; see Sections 1.12 and 1.13, respectively.

1.12 Input and Output in MATLAB

To output the names and values of variables the semicolon can be omitted from assignment statements. However, this does not produce clear scripts or well-organized and tidy output. It is often better practice to use the function `disp` since this leads to clearer scripts. The `disp` function allows the display of text and values on the screen. To output the contents of the matrix `A` on the screen we write `disp(A)`. Text output must be placed in single quotes, for example,

```
>> disp('This will display this test')
This will display this test
```

Combinations of strings can be printed using square brackets `[ ]`, and numerical values can be placed in text strings if they are converted to strings using the `num2str` function. For example,

```
>> x = 2.678;
>> disp(['Value of iterate is ', num2str(x), ' at this stage'])
```

will place on the screen

```
Value of iterate is 2.678 at this stage
```

The more flexible `fprintf` function allows formatted output to the screen or to a file. It takes the form

```
fprintf('filename','format_string',list);
```

Here `list` is a list of variable names separated by commas. The filename parameter is optional; if not present, output is to the screen rather than to the filename. The format string formats the output. The basic elements that may be used in the format string are

- `%P.Qe` for exponential notation
- `%P.Qf` for fixed point
- `%P.Qg` becomes `%P.Qe` or `%P.Qf`, whichever is shorter
- `\n` gives a new line

Note that `P` and `Q` in the preceding are integers. The integer string characters, including a period (`.`), must follow the `%` symbol and precede the letter `e`, `f`, or `g`. The integer before the period (`P`) sets the field width; the integer after the period (`Q`) sets the number of decimal places after the decimal point. For example, `%8.4f` and `%10.3f` give field width 8 with four decimal places and 10 with three decimal places, respectively. Note that one space is allocated to the decimal point. For example,

```
>> x = 1007.461; y = 2.1278; k = 17;
>> fprintf('\n x = %8.2f y = %8.6f k = %2.0f \n',x,y,k)
```

outputs

```
 x =  1007.46 y = 2.127800 k = 17
```

whereas

```
>> p = sprintf('\n x = %8.2f y = %8.6f k = %2.0f \n',x,y,k)
```

gives

```
p =

 x =  1007.46 y = 2.127800 k = 17
```

Note that `p` is a string vector, and can be manipulated if required.

The degree to which the MATLAB user will want to improve the style of MATLAB output will depend on the circumstances. Is the output for other persons to read, perhaps requiring a clearly structured ouput, or is it just for the user alone, therefore requiring only a simple output? Will the output be filed away for future use, or is it a result that is rapidly discarded? In this text we have given examples of very simple output and sometimes quite elaborate output.

We now consider the input of text and data via the keyboard. An interactive way of obtaining input is to use the function `input`. One form of this function is

```
>> variable = input('Enter data: ');
Enter data: 67.3
```

The `input` function displays the text as a prompt and then waits for a numeric entry from the keyboard, `67.3` in this example. This is assigned to `variable` when return is pressed. Scalar values or arrays can be entered in this way. The alternative form of the `input` function allows string input:

```
>> variable = input('Enter text: ','s');
Enter text: Male
```

This assigns the string `Male` to `variable`.

For large amounts of data, perhaps saved in a previous MATLAB session, the function `load` allows the loading of files from disk using

```
load filename
```

The filename normally ends in `.mat` or `.dat`. A file of sunspot data already exists in the MATLAB package and can be loaded into memory using the command

```
>> load sunspot.dat
```

In the following example, we save the values of `x`, `y`, and `z` in file `test001`, clear the workspace, and then reload `x`, `y`, and `z` into the workspace:

```
>> x = 1:5; y = sin(x); z = cos(x);
>> whos
  Name        Size          Bytes  Class
  x           1x5              40  double array
  y           1x5              40  double array
  z           1x5              40  double array
>> save test001
>> clear all, whos        Nothing listed
>> load test001
>> whos
  Name        Size          Bytes  Class
  x           1x5              40  double array
  y           1x5              40  double array
  z           1x5              40  double array
>> x = 1:5; y = sin(x); z = cos(x);
```

Here we only save x, y in file test002 and then we clear the workspace and reload x, y:

```
>> save test002 x y
>> clear all,  whos  Nothing listed
>> load test002 x y, whos
  Name       Size             Bytes  Class
  x          1x5                 40  double array
  y          1x5                 40  double array
```

Note that the statement `load test002` has the same effect as `load test002 x y`. Finally we clear the workspace and reload x into the workspace:

```
>> clear all, whos   Nothing listed
>> load test002 x, whos
  Name       Size             Bytes  Class
  x          1x5                 40  double array
```

Files composed of Comma Separated Values (CSV) are commonly used to exchange large amounts of tabular data between software applications. The data is stored in plain-text and the fields are separated by commas. The files are easily editable using common spreadsheet applications (e.g., Microsoft Excel). If data has been generated elsewhere and saved as a CSV file it can be imported into MATLAB using `csvread`. We use `csvwrite` to generate a CSV file from MATLAB. In the following MATLAB statements we save the vector p, clear the workspace, and then reload p, but now call it the vector g:

```
>> p = 1:6;
>> whos
  Name       Size             Bytes  Class
  p          1x6                 48  double array

>> csvwrite('test003',p)
>> clear
>> g = csvread('test003')
g =
     1     2     3     4     5
```

1.13 MATLAB Graphics

MATLAB provides a wide range of graphics facilities that may be called from within a script or used simply in command mode for direct execution. We begin by considering the `plot` function. This function takes several forms. For example,

Table 1.2 Symbols and Characters Used in Plotting

Line	Symbol	Point	Symbol	Color	Character
solid	-	point	.	yellow	y
dashed	--	plus	+	red	r
dotted	:	star	*	green	g
dashdot	-.	circle	o	blue	b
		x mark	x	black	k

- `plot(x,y)` plots the vector `x` against `y`. If `x` and `y` are matrices, the first column of `x` is plotted against the first column of `y`. This is then repeated for each pair of columns of `x` and `y`.
- `plot(x1,y1,'type1',x2,y2,'type2')` plots the vector `x1` against `y1` using the the line or point type given by `type1`, and the vector `x2` against `y2` using the line or point type given by `type2`.

The `type` is selected by using the required symbol from Table 1.2. This symbol may be preceded by a character indicating a color.

Semilog and log-log graphs can be obtained by replacing `plot` by `semilogx`, `semilogy`, or `loglog` functions and various other replacements for `plot` are available to give special plots. Titles, axis labels, and other features can be added to a given graph using the functions `xlabel`, `ylabel`, `title`, `grid`, and `text`. These functions have the following forms:

- `title('title')` displays the title that is enclosed between quotes at the top of the graph.
- `xlabel('x_axis_name')` displays the name that is enclosed between quotes for the *x*-axis.
- `ylabel('y_axis_name')` displays the name that is enclosed between quotes for the *y*-axis.
- `grid` superimposes a grid on the graph.
- `text(x,y,'text-at-x,y')` displays text at position (x, y) in the graphics window where x and y are measured in the units of the current plotting axes. There may be one point or many at which text is placed depending on whether or not `x` and `y` are vectors.
- `gtext('text')` allows the placement of text using the mouse by positioning it where the text is required and then pressing the mouse button.
- `ginput` allows information to be taken from a graphics window.

The `ginput` function takes two main forms. The simplest is

```
[x,y] = ginput
```

This inputs an unlimited number of points into the vectors `x` and `y` by positioning the mouse crosshairs at the points required and then pressing the mouse button. To exit

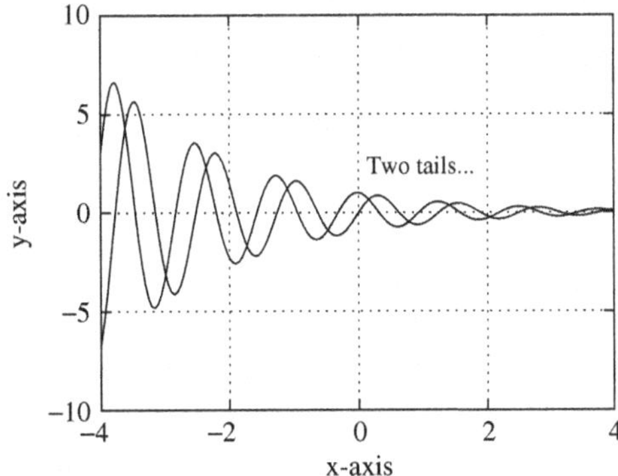

FIGURE 1.1 Superimposed graphs obtained using `plot(x,y)` and `hold` statements.

`ginput` the return key must be pressed. If a specific number of points `n` are required, then we write

```
[x,y] = ginput(n)
```

In addition, the function `axis` allows the user to set the limits of the axes for a particular plot. This takes the form `axis(p)` where `p` is a four-element row vector specifying the lower and upper limits of the axes in the x and y directions. The axis statement must be placed *after* the plot statement to which it refers. Similarly the functions `xlabel`, `ylabel`, `title`, `grid`, `text`, `gtext`, and `axis` must *follow* the plot to which they refer.

The following script gives the plot that is output as Figure. 1.1. The function `hold` is used to ensure that the two graphs are superimposed.

```
% e3s101.m
x = -4:0.05:4;
y = exp(-0.5*x).*sin(5*x);
figure(1), plot(x,y)
xlabel('x-axis'), ylabel('y-axis')
hold on
y = exp(-0.5*x).*cos(5*x);
plot(x,y), grid
gtext('Two tails...')
hold off
```

Script `e3s101.m` illustates how few MATLAB statements are required to generate a graph.

The function `fplot` allows the user to plot a previously defined function between given limits. The important difference between `fplot` and `plot` is that `fplot` chooses the plotting points in the given range adaptively depending on the rate of change of the function at that point. Thus more points are chosen when the function is changing more rapidly. This is illustrated by executing the following MATLAB script:

```
% e3s102.m
y = @(x) sin(x.^3);
x = 2:.04:4;
figure(1)
plot(x,y(x),'o-')
xlabel('x'), ylabel('y')
figure(2)
fplot(y,[2 4])
xlabel('x'), ylabel('y')
```

Note `figure(1)` and `figure(2)` direct the graphic output to separate windows. The interpretation of the anonymous function `@(x) sin(x.^3)` is explained in Section 1.17.

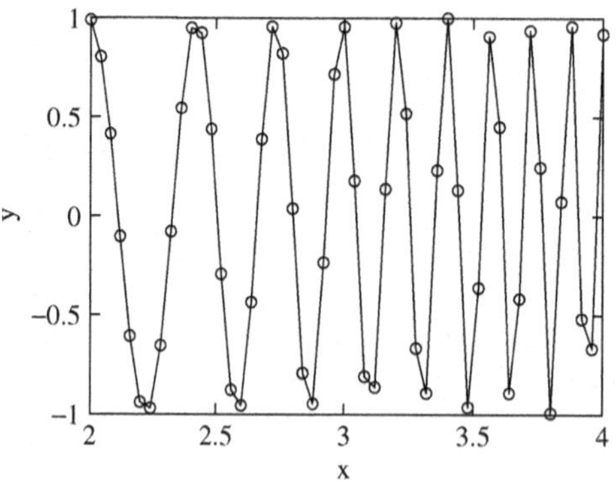

FIGURE 1.2 Plot of $y = \sin(x^3)$ using 51 equispaced plotting points.

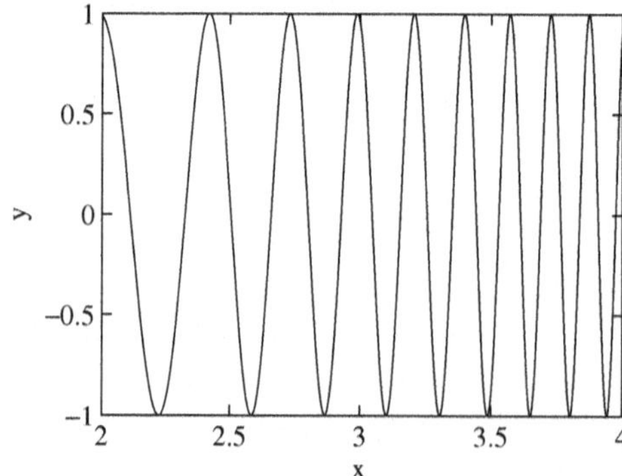

FIGURE 1.3 Plot of $y = \sin(x^3)$ using the function `fplot` to choose plotting points adaptively.

Running the script just given produces Figures 1.2 and 1.3. In the plot example we have deliberately chosen an inadequate number of plotting points and this is reflected in the quality of Figure 1.2. The function `fplot` produces a smoother and more accurate curve. Note that `fplot` only allows a function or functions to be plotted against an independent variable. Parametric plots cannot be created by `fplot`.

The MATLAB function `ezplot` is similar to `fplot` in the sense that we only have to specify the function, but has the disadvantage that the step size is fixed. However, `ezplot` does allow parametric plots and three-dimensional plots. For example,

```
>> ezplot(@(t) (cos(3*t)), @(t) (sin(1.6*t)), [0 50])
```

is a parametric plot but the plot is rather coarse.

We have seen how `fplot` helps in plotting difficult functions. Other functions that help to clarify when the plot of a function is unclear or unpredictable are `ylim` and `xlim`. The function `ylim` allows the user to easily limit the range of the y-axis in the plot and `xlim` performs similarly for the x-axis. Their use is illustrated by the following example.

Figure 1.4 (without the use of `xlim` and `ylim`) is unsatisfactory since it gives little understanding about how the function behaves except at the specific points $x = -2.5$, $x = 1$, and $x = 3.5$.

```
>> x = -4:0.0011:4;
>> y =1./(((x+2.5).^2).*((x-3.5).^2))+1./((x-1).^2);
>> plot(x,y)
>> ylim([0,10])
```

Figure 1.5 shows how the MATLAB statement `ylim([0,10])` restricts the y-axis to a maximum value of 10. This gives a clear picture of the behavior of the graph.

There are a number of special features available in MATLAB for the presentation and manipulation of graphs and some of these will now be discussed. The `subplot` function

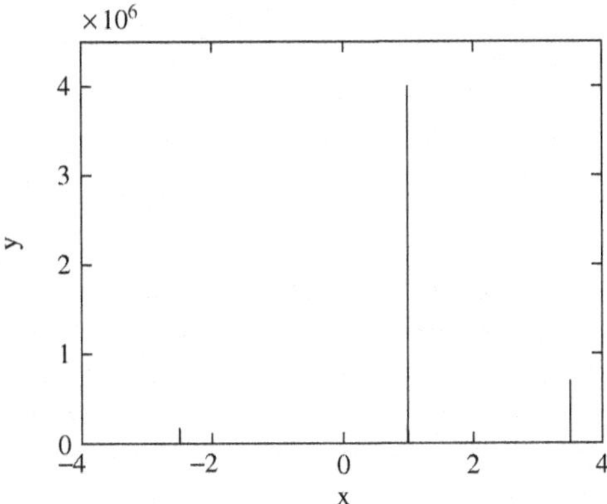

FIGURE 1.4 Function plotted over the range −4 to 4. It has a maximum value of 4×10^6.

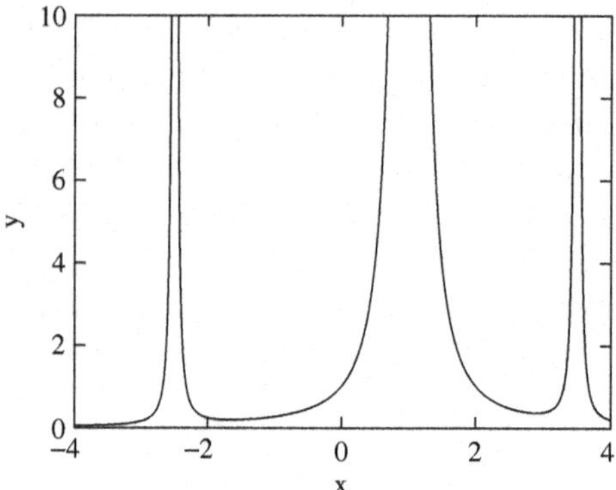

FIGURE 1.5 The same function as plotted in Figure 1.4 but with a limit on the range of the *y*-axis.

takes the form `subplot(p,q,r)` where `p`, `q` splits the figure window into a $p \times q$ grid of cells and places the plot in the *r*th cell of the grid, numbered consecutively along the rows. This is illustrated by running the following script, which generates six different plots, one in each of the six cells. These plots are given in Figure 1.6.

```
% e3s103.m
x = 0.1:.1:5;
subplot(2,3,1), plot(x,x)
title('plot of x'), xlabel('x'), ylabel('y')
subplot(2,3,2), plot(x,x.^2)
title('plot of x^2'), xlabel('x'), ylabel('y')
subplot(2,3,3), plot(x,x.^3)
title('plot of x^3'), xlabel('x'), ylabel('y')
subplot(2,3,4), plot(x,cos(x))
title('plot of cos(x)'), xlabel('x'), ylabel('y')
subplot(2,3,5), plot(x,cos(2*x))
title('plot of cos(2x)'), xlabel('x'), ylabel('y')
subplot(2,3,6), plot(x,cos(3*x))
title('plot of cos(3x)'), xlabel('x'), ylabel('y')
```

The current plot can be held on screen by using the function `hold` and subsequent plots are drawn over it. The function `hold on` switches the hold facility on while `hold off` switches it off. The figure window can be cleared using the function `clf`.

MATLAB provides many other plot functions and styles. To illustrate two of these, `polar` and `compass` plots, we display the roots of $x^5 - 1 = 0$, which have been determined using the MATLAB function `roots`. This function is descibed in detail in Section 3.11. Having determined the five roots of this equation we plot them using both `polar` and `compass`. The function `polar` requires the absolute values and phase angles of the roots, whereas the function `compass` plots the real parts of the roots against their imaginary parts.

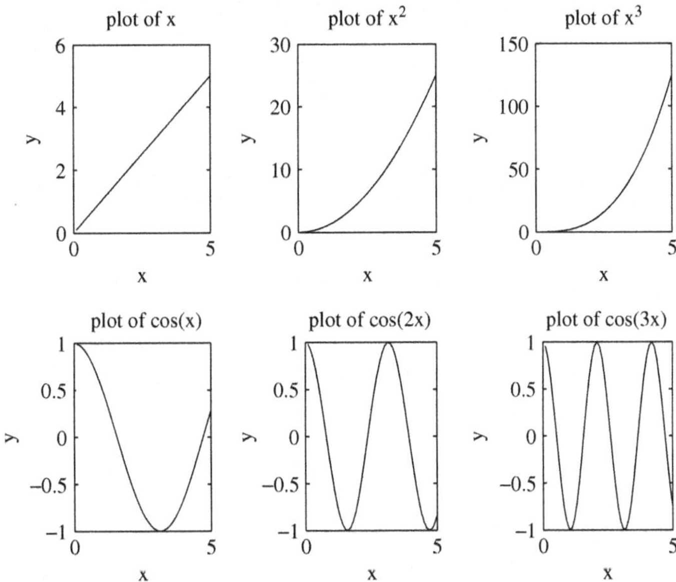

FIGURE 1.6 An example of the use of the `subplot` function.

```
>> p=roots([1 0 0 0 0 1])

p =
  -1.0000
  -0.3090 + 0.9511i
  -0.3090 - 0.9511i
   0.8090 + 0.5878i
   0.8090 - 0.5878i

>> pm = abs(p.')

pm =
    1.0000    1.0000    1.0000    1.0000    1.0000

>> pa = angle(p.')

pa =
    3.1416    1.8850   -1.8850    0.6283   -0.6283

>> subplot(1,2,1), polar(pa,pm,'ok')
>> subplot(1,2,2), compass(real(p),imag(p),'k')
```

Figure 1.7 shows these subplots.

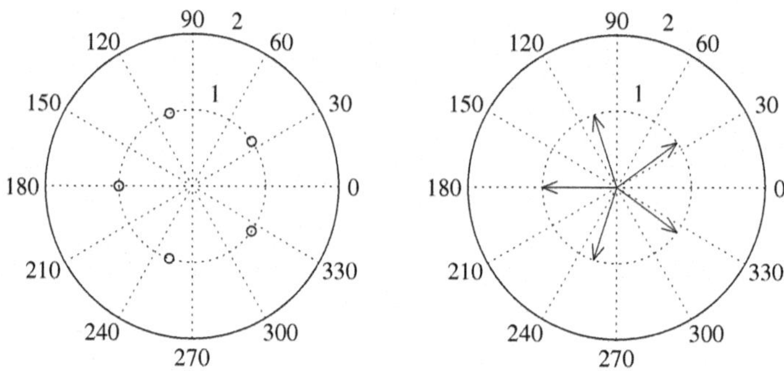

FIGURE 1.7 `polar` and `compass` plots showing the roots of $x^5 - 1 = 0$.

1.14 Three-Dimensional Graphics

It is often convenient to draw a three-dimensional graph of a function or set of data to gain a deeper insight into the nature of the function or data. MATLAB provides powerful and extensive facilities to allow the user to draw a wide range of three-dimensional graphs. Here we only briefly introduce a small selection of these functions. These are the functions `meshgrid`, `mesh`, `surfl`, `contour`, and `contour3`. It should be noted that the more complex graphs of this type may take a significant time to draw on the screen, depending on the algebraic complexity of the function, the amount of detail required, and the power of the computer being used.

Usually three-dimensional functions are plotted to illustrate particular features of the function such as regions where maxima or minima lie. Plotting surfaces to illustrate these features can be difficult and some careful analysis of the function may be needed before the graph is drawn successfully. In addition, even when the region of interest is successfully located and plotted, the feature of interest may be hidden and it is then necessary to choose a different viewpoint. Discontinuities may also be present and cause plotting problems.

For the function $z = f(x, y)$ the MATLAB function `meshgrid` is used to *generate a complete set of points in the x-y plane for the three-dimensional plotting functions.* We can then compute the values of z and these are finally plotted by using one of the functions `mesh`, `surf`, `surfl`, or `surfc` . For example, to plot the function

$$z = (-20x^2 + x)/2 + (-15y^2 + 5y)/2 \quad \text{for} \quad x = -4:0.2:4 \quad \text{and}$$
$$y = -4:0.2:4$$

we first set up the values of the x-y domain and then compute z corresponding to these x and y values using the given function. Finally we plot the three-dimensional graph using the function `surfl`. This is achieved by using the following script. Note how the function `figure` is used to direct the output to a graphics window so that the first plot is not overwritten by the second.

```
% e3s105.m
[x,y] = meshgrid(-4.0:0.2:4.0,-4.0:0.2:4.0);
z = 0.5*(-20*x.^2+x)+0.5*(-15*y.^2+5*y);
figure(1)
surfl(x,y,z); axis([-4 4 -4 4 -400 0])
xlabel('x-axis'), ylabel('y-axis'), zlabel('z-axis')
figure(2)
contour3(x,y,z,15); axis([-4 4 -4 4 -400 0])
xlabel('x-axis'), ylabel('y-axis'), zlabel('z-axis')
figure(3)
contourf(x,y,z,10)
xlabel('x-axis'), ylabel('y-axis')
```

Running this script generates the plots shown in Figures 1.8 through 1.10. The first plot is created using `surfl` and shows the function as a surface; the second is created by `contour3` and is a three-dimensional contour plot of the surface; and the third, created using `contourf`, provides a two-dimensional filled contour plot.

When plotting surfaces a very useful function is `view`. This function allows the surface or mesh to be viewed from different positions. The function has the form `view(az,el)` where `az` is the azimuth and `el` is the elevation of the viewpoint required. Azimuth may be interpreted as the viewpoint rotation about the z-axis and elevation as the rotation of the viewpoint about the x-y plane. A positive value of the elevation gives a view from above the object and a negative value a view from below. Similarly a positive value of azimuth gives a counterclockwise rotation of the viewpoint about the z-axis while a negative value gives a clockwise rotation. If the view function is not used, the default values are -37.5° for the azimuth and 30° for the elevation.

There are many other three-dimensional plotting facilities that are outside the scope of the text; please see the MATLAB manual for more information.

1.15 Manipulating Graphics—Handle Graphics

Handle Graphics allow the user to choose the font type, line thickness, symbol type and size, axes form, and many other features for a particular plot. It introduces more complexity into MATLAB but has considerable benefits. Here we give a very brief introduction to some of the main features. There are two key functions, `get` and `set`. The `get` function allows the user to obtain detailed information about a particular graphics function such as `plot`, `title`, `xlabel`, `ylabel`, and others. The function `set` allows the user to modify the

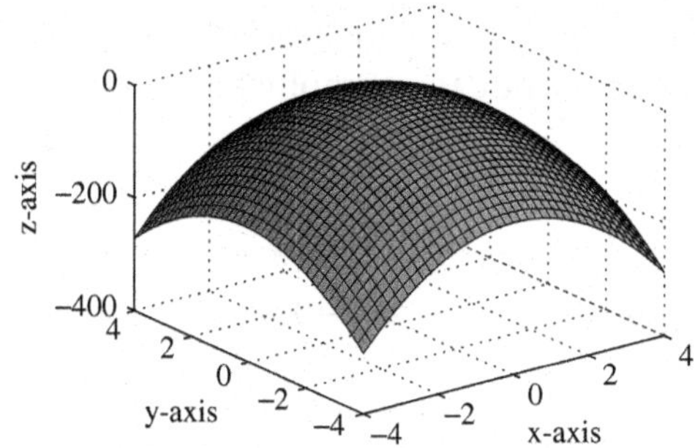

FIGURE 1.8 Three-dimensional surface using default view.

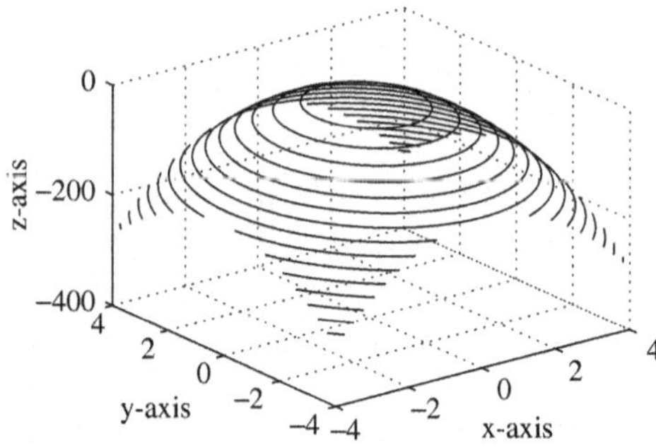

FIGURE 1.9 Three-dimensional contour plot.

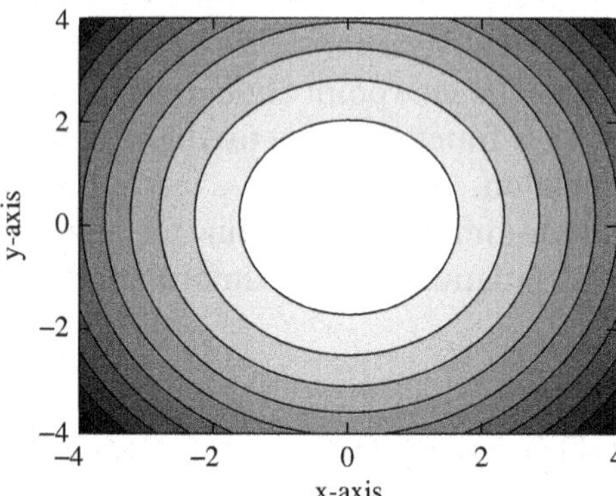

FIGURE 1.10 Filled contour plot.

standard setting for a particular graphics element such as `xlabel` or `plot`. In addition, `gca` can be used with `set` to retrieve the handles of the axes of the current figure and with `get` to manipulate the properties of the axes of that figure.

To illustrate the details involved in a simple graphics statement, consider the following statements, where handles `h` and `h1` have been introduced for the plot and title functions:

```
>> x = -4:.1:4;
>> y = cos(x);
>> h = plot(x,y);
>> h1 = title('cos graph')
```

To obtain information about the detailed structure of the `plot` and `title` functions, we use `get` and the appropriate handle as follows. Note that only a selection of the properties produced by `get` are shown.

```
>> get(h)
                   Color: [0 0 1]
               EraseMode: 'normal'
               LineStyle: '-'
               LineWidth: 0.5000
                  Marker: 'none'
              MarkerSize: 6
         MarkerEdgeColor: 'auto'
         ..........................[etc]
```

However, for the `title` function we have

```
>> get(h1)
FontName = Helvetica
FontSize = [10]
FontUnits = points
  HorizontalAlignment = center
LineStyle = -
LineWidth = [0.5]
Margin = [2]
Position = [-0.00921659 1.03801 1.00011]
Rotation = [0]
String = cos graph
..........................[etc]
```

Notice there are different properties for `plot` and `title`. The follwing example illustrates the use of Handle Graphics:

```
% e3s121.m
% Example for Handle Graphics
x = -5:0.1:5;
subplot(1,3,1)
e1 = plot(x,sin(x)); title('sin x')
```

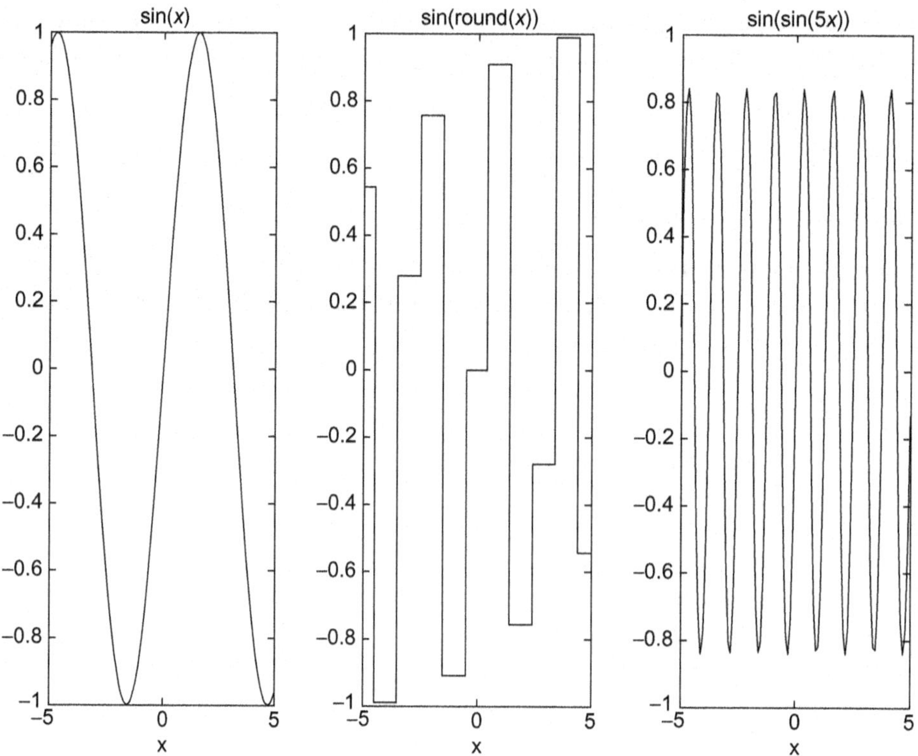

FIGURE 1.11 Plots illustrating aspects of Handle Graphics.

```
subplot(1,3,2)
e2 = plot(x,sin(2*round(x))); title('sin round x')
subplot(1,3,3)
e3 = plot(x,sin(sin(5*x))); title('sin sin 5x')
```

Running this script gives Figure 1.11.

We now modify the preceding script using a sequence of `set` statements as follows:

```
% e3s122.m
% Example for Handle Graphics
x = -5:0.1:5;
s1 = subplot(1,3,1);
e1 = plot(x,sin(x)); t1 = title('sin(x)');
s2 = subplot(1,3,2);
e2 = plot(x,sin(2*round(x))); t2 = title('sin(round(x))');
s3 = subplot(1,3,3);
e3 = plot(x,sin(sin(5*x))); t3 = title('sin(sin(5x))');
% change dimensions of first subplot
```

```
set(s1,'Position',[0.1 0.1 0.2 0.5]);
%change thickness of line of first graph
set(e1,'LineWidth',6)
set(s1,'XTick',[-5 -2  0 2  5])
%Change all titles to italics
set(t1,'FontAngle','italic'), set(t1,'FontWeight','bold')
set(t1,'FontSize',16)
set(t2,'FontAngle','italic')
set(t3,'FontAngle','italic')
%change dimensions of last subplot
set(s3,'Position',[0.7 0.1 0.2 0.5]);
```

The `Position` statement has the values

```
[shift from left, shift from bottom, width, height].
```

The size of the plotting area is taken as a unit square. Thus

```
set(s3,'Position',[0.7 0.1 0.2 0.5]);
```

shifts the figure 0.7 from the left and 0.1 from the bottom; its width is 0.2 and its height is 0.5. It may take some experimentation to get the required effect. Executing this script gives Figure 1.12. Notice the differing sizes of the boxes, thicker line in the first graph, bold title, different ticks on the x-axis, and that all the titles are in italics; many other aspects could have been changed.

The following example shows how we can manipulate the various properties of the axes in Figure 1.13 using `gca` with `get`, which gets the properties of the current axes. The examples that follow show the use of `gca` in altering various properties of the axes:

```
>> x = -1:0.1:2; h = plot(x,cos(2*x));
```

These statements produce the left plot in Figure 1.13.

```
>> get(gca,'FontWeight')

ans =
normal

>> set(gca,'FontWeight','bold')
>> set(gca,'FontSize',16)
>> set(gca,'XTick',[-1 0 1 2])
```

These additional statements provide the right plot of Figure 1.13. Note that the differences produced a larger bold font and fewer x-axis tickmarks.

An alternative approach to manipulating font styles and other features is illustrated in the script that follows Figure 1.14.

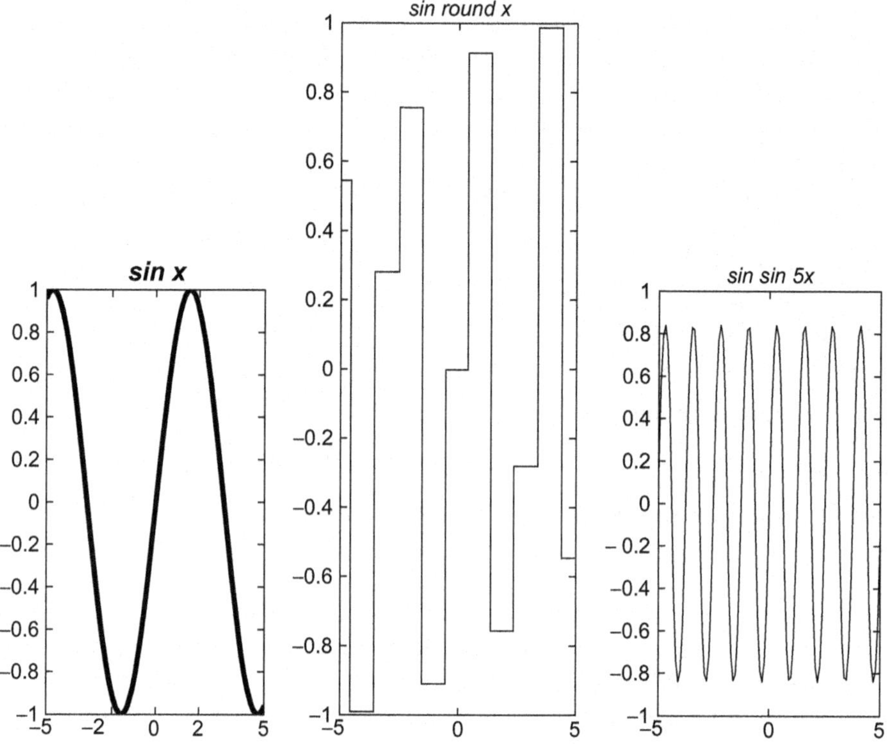

FIGURE 1.12 Plot of functions shown in Figure 1.11 illustrating further Handle Graphics features.

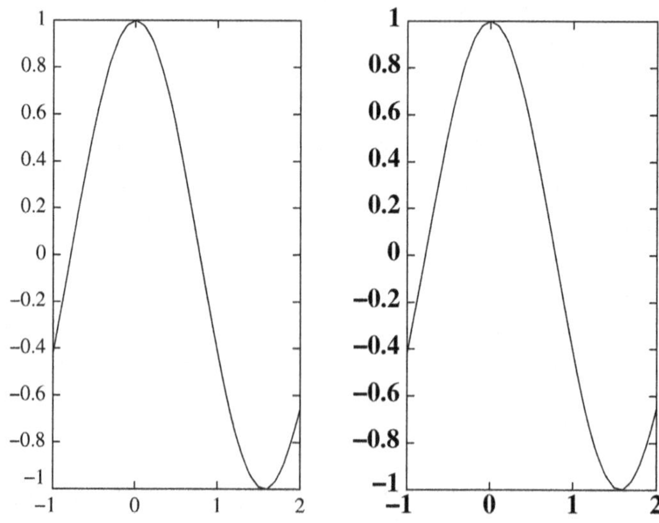

FIGURE 1.13 Plot of $\cos(2x)$. The axes of the right plot are enhanced using Handle Graphics.

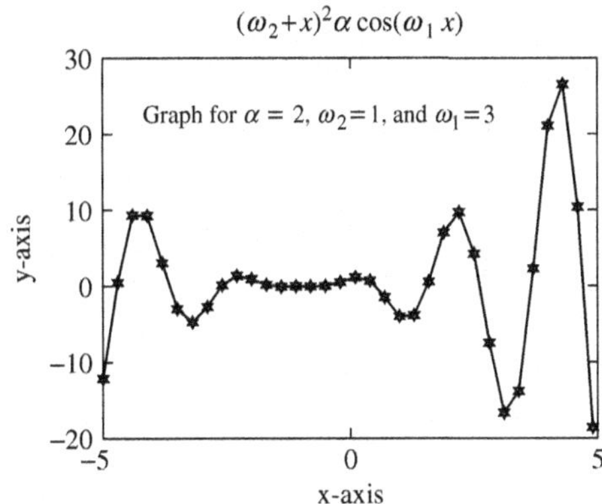

FIGURE 1.14 Plot of $(\omega_2 + x)^2\alpha\cos(\omega_1 x)$.

```
% e3s104.m
% Example of the use of special graphics parameters in MATLAB
% illustrates the use of superscripts, subscripts,
% fontsize and special characters
x = -5:.3:5;
plot(x,(1+x).^2.*cos(3*x),...
'linewidth',1,'marker','hexagram','markersize',12)
title('(\omega_2+x)^2\alpha cos(\omega_1x)','fontsize',14)
xlabel('x-axis'), ylabel('y-axis','rotation',0)
gtext('graph for \alpha = 2,\omega_2 = 1, and \omega_1 = 3')
```

Executing this script provides the graph shown in Figure 1.14.

We now describe the features that were used in this script. We have used Greek characters from an extensive range of symbols that can be introduced using the backslash character "\". The following gives examples of how these characters may be introduced:

- `alpha` gives α
- `beta` gives β
- `gamma` gives γ

Any of the Greek symbols may be obtained by typing the backslash followed by the standard English name of the Greek letter. Titles and axis labels may include superscripts and subscripts by preceding the subscript character by " _ " and the superscript by "∧ ". Font sizes may be specified by placing the additional parameter `'fontsize'` in the `xlabel`, `ylabel`, or `title` statements, followed by and separated by a comma from the actual font size required. For example,

```
title('(\omega_2+x)^2\alpha*cos(\omega_1*x)','fontsize',14)
```

gives

$$(\omega_2 + x)^2\alpha * \cos(\omega_1 * x)$$

in 14-point font. In the `plot` function itself, additional markers for the graph points are available and may be indicated by using the additional parameter `'marker'` followed by the name of the marker. For example,

```
'marker','hexagram'
```

The size of the marker may also be specified using the additional parameter `'markersize'` followed by the required marker size:

```
'markersize',12
```

The line thickness may also be adjusted using the parameter `'linewidth'`, for example,

```
'linewidth',1
```

Finally, the orientation of any label may be changed using `'rotation'`. For example,

```
'rotation',0
```

This additional parameter with the setting zero makes the label horizontal to the *y*-axis rather than the usual vertical orientation; the value of the parameter gives the angle in degrees.

A further more complex example involving reference to a partial differential equation in a MATLAB `text` statement is as follows:

```
gtext('Solution of \partial^2V/\partialx^2+\partial^2V/\partialy^2 = 0')
```

This leads to the text

$$\texttt{Solution of } \partial^2 V/\partial x^2 + \partial^2 V/\partial y^2 = 0$$

being placed in the current graphics window at a point selected using the crosshairs cursor and clicking the mouse button.

In addition, features may be included in conjunction with the \ followed by a font name parameter, which allows the specification of any available font. Examples are `\bf` that gives a bold style and `\it` that gives an italic style.

An important issue in placing figures in a manuscript is that they must have a consistent position and size and must be easy to read. The listed graphics scripts work satisfactorily but would not provide the quality required if directly imported into this manuscript. An example of this is shown in Figure 1.11. To ensure that the size and position of the figures generated by the MATLAB scripts are generally consistent and their fonts easily read, the following statements are added to all scripts producing graphical output except for Figures 1.11, 1.12, and 1.13.

```
set(0,'defaultaxesfontsize',16)
set(0,'defaultaxesfontname','Times New Roman')
set(0,'defaulttextfontsize',12)
set(0,'defaulttextfontname','Times New Roman')
axes('position',[0.30 0.30 0.50 0.50])
```

These statements are examples of Handle Graphics. The first and second statements set the fonts used for the axes to 16-point Times New Roman. The third and fourth statement set the fonts used in the plot to 12-point Times New Roman and the fifth statement controls the size of the graph within the graphs window. Finally, we add the statement `print -deps Fig101.eps` at the end of each script that generates a graph. This statement saves the plot as an extended postscript (eps) format file for inclusion in the manuscript.

This was used in the creation and placement of the MATLAB graphs in this book. These are not shown in the text listings because they are the same for each script.

1.16 Scripting in MATLAB

In some of the previous sections we have created some simple MATLAB scripts that have allowed a series of commands to be executed sequentially. However, many of the features usually found in programming languages are also provided in MATLAB to allow the user to create versatile scripts. The more important of these features are described in this section. It must be noted that scripting is done in the **edit** window using a text editor appropriate to the system, not in the **command** window, which only allows the execution of statements one at a time or several statements provided that they are on the same line.

MATLAB *does not require the declaration of variable types*, but for the sake of clarity the role and nature of key variables may be indicated by using comments. Any text following the symbol % is considered a comment. In addition, there are certain variable names that have predefined special values for the convenience of the user. They can, however, be redefined if required. These are

`pi`	equals π
`inf`	the result of dividing by zero
`eps`	set to the particular machine accuracy
`realmax`	largest positive floating-point number
`realmin`	smallest positive floating-point number
`NaN`	"Not-a-Number"; result of operations with undefined numerical results, such as dividing zero by zero
`i,j`	both equal $\sqrt{-1}$

Assignment statements in a MATLAB script take the form

```
variable = <expression>;
```

The expression is calculated and the value assigned to the variable on the left-hand side. If the semicolon is omitted from the end of these statements, the names of the variable(s) and the assigned value(s) are displayed on the screen. If an expression is not assigned explicitly to a variable then the value of the expression is calculated, assigned to the variable `ans`, and displayed.

In previous sections it was stated that generally a variable in MATLAB is assumed to be a matrix of some kind; its name must start with a letter and may be followed by any combination of digits and letters; a maximum of 32 characters is recognized. It is good practice to use a meaningful variable name. The variable name must not include spaces or hyphens. However, the underscore character is a useful replacement for a space. For example, `test_run` is acceptable; `test run` and `test-run` are not. It is very important to avoid the use of existing MATLAB commands, function names, or even the word MATLAB itself! MATLAB does not prevent their use but using them can lead to problems and inconsistencies. An expression in MATLAB is a valid combination of variables, constants, operators, and functions. Brackets can be used to alter or clarify the precedence of operations. The precedence of operation for simple operators is first \, second *, third /, and finally + and - where

^	raises to a power
*	multiplies
/	divides
+	adds
-	subtracts

The effects of these operators in MATLAB have already been discussed.

Unless there are instructions to the contrary, a set of MATLAB statements in a script is executed in sequence. This is the case in the following example.

```
% e3s106.m
% Matrix calculations for two matrices A and B
A = [1 2 3;4 5 6;7 8 9];
B = [5 -6 -9;1 1 0;24 1 0];
% Addition. Result assigned to C
C = A+B; disp(C)
% Multiplication. Result assigned to D
D = A*B; disp(D)
% Division. Result assigned to E
E = A\B; disp(E)
```

To allow the repeated execution of one or more statements, a `for` loop is used. This takes the form

```
for <loop_variable> = <loop_expression>
    <statements>
end
```

The `<loop_variable>` is a suitably named variable and `<loop_expression>` is usually of the form `n:m` or `m:i:n` where `n`, `i`, and `m` are the initial, incremental, and final values of `<loop_variable>`. They may be constants, variables, or expressions; they can be negative or positive but clearly they should be chosen to take values consistent with the logic of the script. This structure should be used when the loop is to be repeated a predetermined number of times.

Examples

```
for i = 1:n
    for j = 1:m
        C(i,j) = A(i,j)+cos((i+j)*pi/(n+m))*B(i,j);
    end
end

for k = n+2:-1:n/2
    a(k) = sin(pi*k);
    b(k) = cos(pi*k);
end

p = 1;
for a = [2 13 5 11 7 3]
    p = p*a;
end
p

p = 1;
prime_numbs = [2 13 5 11 7 3];
for a = prime_numbs
    p = p*a;
end
p
```

The first example illustrates the use of nested `for` loops, while the second illustrates that upper and lower limits can be expressions and the step value can be negative. The third example shows that the loop does not have to use a uniform step and the fourth example, which gives an identical result to the third, illustrates that the `<loop_expression>` can be any previously defined vector.

When assigning values to a vector in a `for` loop, the reader should note that the vector generated is a *row* vector. For example,

```
for i = 1:4
    d(i) = i^3;
end
```

gives the row vector `d = 1 8 27 64.`

The `while` statement is used when the repetition is subject to a condition being satisfied that is dependent on values generated within the loop. This has the form

```
while <while_expression>
    <statements>
end
```

The `<while_expression>` is a *relational expression* of the form e1 ∘ e2 where e1 and e2 are ordinary arithmetic expressions as described before and ∘ is a relational operator defined as follows:

`==`	equals
`<=`	less than or equals
`>=`	greater than or equals
`~=`	not equals
`<`	less than
`>`	greater than

Relational expressions may be combined using the following *logical operators*:

`&`	the and operator
`|`	the or operator
`~`	the not operator
`&&`	the scalar and operator (if the first condition is false then the second is not evaluated)
`||`	the scalar or operator (if the first condition is true then the second is not evaluated)

Note that false is zero and true is nonzero. Relational operators have a higher order of precedence than logical operators.

Examples of `while` *Loops*

```
dif = 1;
x2 = 1;
while dif>0.0005
    x1 = x2-cos(x2)/(1+x2);
    dif = abs(x2-x1);
    x2 = x1;
end

x = [1 2 3];
y = [4 5 8];
while sum(x) ~= max(y)
    x = x.^2;
    y = y+x;
end
```

Note also that `break` stops the execution (and hence allows exit from) a `while` or `for` loop and that `break` cannot be used outside of a `while` or `for` loop. The statement `return` must be used in these circumstances.

A vital feature of all programming languages is the ability to change the sequence in which instructions are executed within the program. In MATLAB the `if` statement is used to achieve this and has the general form

```
if < if_expression1>
    <statements>
elseif < if_expression2>
    <statements>
elseif < if_expression3>
    <statements>
...
...
else
    <statements>
end
```

Here `<if_expression1>` and so on are relational expressions of the form e1 $\circ$ e2 where e1 and e2 are ordinary arithmetic expressions and $\circ$ is a relational operator as described before. Relational expressions may be combined using logical operators.

Examples

```
for k = 1:n
    for p = 1:m
        if k == p
            z(k,p) = 1;
            total = total+z(k,p);
        elseif k<p
            z(k,p) = -1;
            total = total+z(k,p);
        else
            z(k,p) = 0;
        end
    end
end

if (x~=0) & (x<y)
   b = sqrt(y-x)/x;
   disp(b)
end
```

The MATLAB function `switch` provides an alternative to the `if` structure and is particularly useful when many options must be considered. This has the form

```
switch <condition>
   case
      statements
   case ref2
      statements
   case ref3
      statements
   otherwise
      statements
end
```

The following fragment of code allows the user to choose a particular plot, dependent on the value of *n*. In the following script, the second plot has been chosen by setting `n = 2`.

```
 x = 1:.01:10; n = 2
switch n
    case 1
        plot(x,log(x));
    case 2
        plot(x,x.*log(x));
    case 3
        plot(x,x./(1+log(x)));
    otherwise
        disp('That was an invalid selection.')
end
```

As a further example of the `switch` function, the following fragment of code allows the user to convert an astronomical distance `x` given in `AU` (an astronomical unit), `LY` (a light year), or `pc` (a parsec) to km by setting the string variable `units` to `AU`, `LY`, or `pc`, respectively.

```
x = 2;
units = 'LY'
switch units
    case {'AU' 'Astronomical  Units'}
        km = 149597871*x
    case  {'LY','lightyear'}
        km = 149597871*63241*x
    case  {'pc' 'parsec'}
        km = 149597871*63241*3.26156*x
    otherwise
        disp('That was an invalid selection.')
end
```

Note that if any statement in a MATLAB script is longer than one line then it must be continued by using an ellipsis (...) at the end of the line.

The `menu` function creates a menu window with buttons to allow the user to select options. For example,

```
frequency = 123;
units = menu('Select units for output data', 'rad/s','Hz', 'rev/min')
switch units
    case 1
        disp(frequency)
    case 2
        disp(frequency/(2*pi))
    case 3
        disp(frequency*60/(2*pi))
end
```

creates a small window (called **MENU**) with three buttons, labeled `'rad/s'`, `'Hz'`, and `'rev/min'`. "Clicking" a particular button with the mouse provides a frequency converted to the chosen units.

1.17 User-Defined Functions in MATLAB

MATLAB allows users to define their own functions, but a specific form of definition must be followed. The first form of function is the *m*-file function and is described as follows:

```
function <output_params> = func_name(<input_params>)
<func body>
```

`<input_params>` is a set of variable names separated by commas and `<output_params>` is either a single variable or a list of variables separated by commas or spaces and placed in square brackets. The function body consists of the statements defining the user's function. These statements will utilize the values of the input arguments and *must include statements assigning values to the output parameters.* Once the function is defined *it must be saved as an m-file* under the same name as the given `func_name`. Then the function can be used as required. It is good practice to put some comments describing the nature of the function immediately after the function heading. Writing `help` followed by the function name in the **command** window will access these comments.

To execute the function for specific parameters we write

```
<specific_out_params> = <func_name>(<assigned_input_params>)
```

where the `<assigned_input_params>` term is either a single parameter, or it is a list of parameters separated by commas. The `<assigned_input_params>` must match the `<input_params>` in the function definition.

We now provide two examples of named functions.

Example 1.1

The Fourier series for a sawtooth wave is

$$y(t) = \frac{1}{2} - \sum_{n=1}^{\infty} \sin\left(\frac{2\pi nt}{T}\right)$$

where T is the period of the waveform. We can create a function to evaluate this for given values of t and T. Since we can't sum to infinity, we will sum to m terms, where m is a relatively large value. Thus we can define the MATLAB function `sawblade` as follows. Note that this function has three input arguments and one output.

```
function y = sawblade(t,T,n_trms)
% Evaluates, at instant t, the Fourier approximation of a sawtooth wave of
% period T using the first n_trms terms in the infinte series.
y = 1/2;
for n = 1:n_trms
    y = y - (1/(n*pi))*sin(2*n*pi*t/T);
end
```

We can now use this function for a specific purpose. For example, if we wish to plot this waveform over the range $t = 0$ to 4 and a period of $T = 2$, using only 50 terms in the series, we have

```
c = 1;
for t = 0:0.01:4, y(c) = sawblade(t,2,50); c = c+1; end
plot([0:0.01:4],y)
```

Further valid function calls are

```
y = sawblade(0.2*period,period,terms)
```

where `period` and `terms` have previously assigned values, or

```
y = sawblade(2,5.7,60)
```

or, using the function `feval`

```
y = feval('sawblade',2,5.7,60)
```

A more important application of `feval`, which is widely used in this text, is in the process of defining functions that themselves have functions as parameters. These m-file functions can be evaluated internally in the body of the calling function by using `feval`.

Example 1.2

We now consider a further example that involves the generation of a matrix within a function. The essential features of the finite element method applied to the static and/or dynamic analysis of structures is to express the stiffness and inertia properties of a small section or element of the structure in matrix form. These *element* matrices are then assembled to obtain matrices that describe the overall stiffness and inertia for the whole structure. Knowing the forces acting on the structure, we can obtain the static or dynamic response of the structure. One such element is a uniform circular shaft. For this element the inertia matrix and the stiffness matrix, relating angular accelerations and displacements, repectively, to applied torques are given by

$$\mathbf{K} = \frac{GJ}{L}\begin{bmatrix} 1 & -1 \\ -1 & 1 \end{bmatrix}$$

and

$$\mathbf{M} = \frac{\rho JL}{6}\begin{bmatrix} 2 & 1 \\ 1 & 2 \end{bmatrix}$$

where L is the length of the shaft, G and ρ are material properties, and d is the diameter of the shaft. If we intend to create a finite element package in MATLAB that includes torsional elements, then these matrices are required. The following function generates these matrices from the shaft properties as follows:

```
function [K,M] = tors_el(L,d,rho,G)
J = pi*d^4/32;
K = (G*J/L)*[1 -1;-1 1];
M = (rho*J*L/6)*[2 1;1 2];
```

Note that this function has four input arguments and ouputs two arrays.

Functions can be nested inside other functions. This is only useful if the nested function is only required by the main function. An example of this is shown in Section 3.11.1. Here the function `solveq` is not a generally useful function but it is required by the function `bairstow`. Therefore it is nested in `bairstow`. This arrangment has the advantage that `bairstow` is a complete entity; it does not require `solveq` to be stored and available. It has the minor disadvantage that since it is not stored separately it cannot be used independently of `bairstow`.

A second, simpler form of the MATLAB user-defined function is the anonymous function. This function is not saved as an *m*-file; it is either entered into the workspace from the command window or from a script. For example, suppose we wish to define the function

$$\left(\frac{x}{2.4}\right)^3 - \frac{2x}{2.4} + \cos\left(\frac{\pi x}{2.4}\right)$$

The MATLAB function definition is as follows:

```
>> f = @(x) (x/2.4).^3-2*x/2.4+cos(pi*x/2.4);
```

Example calls of this function are `f([1 2])`, which produces two values corresponding to $x = 1$ and $x = 2$. Another way of using this function is as an input parameter to another function. For example,

```
>> solution = fzero(f,2.9)
solution =
    3.4825
```

This gives the zero of `f` closest to 2.9. Another example of its use is

```
x = 0:0.1:5; plot(x,f(x))
```

Here we must call `f(x)` because the `plot` function needs all the values of the function over the range of `x`. Another form is

```
>> solution = fzero(@(x) (x/2.4).^3-2*x/2.4+cos(pi*x/2.4), 2.9)
```

Here we have used the anonymous function definition directly, rather than assigning it to a handle and then using the handle.

If an m-file function has an anonymous function as one of its input arguments, then this anonymous function can be evaluated directly without the use of the MATLAB function `feval`. If, however, the function in the parameter list may require a multistatement definition, an m-file function must be used and in this case `feval` must be used. In this text, to allow flexibility, when defining m-file functions we have used `feval` so that the user can input a function as an m-file function or as an anonymous function.

For example, we define the m-file functions `sp_cubic` and `minandmax` as follows:

```
function y = sp_cubic(x)
y = x.^3-2*x.^2-6;

function [minimum maximum] = minandmax(f,v)
% v is a vector with the start, increment and end value
y = feval(f,v); minimum = min(y); maximum = max(y);
```

Using this definition of `minandmax` means that `f` can be an anonymous or m-file function. Thus, using the anonymous function's definition for `f` given before, we have

```
>> [lo hi] = minandmax(f,[-5:0.1:5]);
>> fprintf('lo = %8.4f hi = %8.4f\n',lo,hi)

lo = -181.0000 hi =  69.0000
```

Alternatively, using the other form of the function

```
>> [lo hi] = minandmax('sp_cubic',[-5:0.1:5]);
>> fprintf('lo = %8.4f hi = %8.4f\n',lo,hi)

lo = -181.0000 hi =  69.0000
```

gives identical answers. However, suppose we define *m*-file function `minandmax` without the use of `feval` as follows:

```
function [minimum maximum] = minandmax(f,v)
% v is a vector with the start, increment and end value
y = f(v); minimum = min(y); maximum = max(y);
```

Then, if `f` is an anonymous function, we obtain the preceding results, but if `f` is the *m*-file function `sp_cubic`, the function `minandmax` fails as shown in the following:

```
>> [lo hi] = minandmax('sp_cubic',[-5:0.1:5]);
>> fprintf('lo = %8.4f hi = %8.4f\n',lo,hi)
??? Subscript indices must either be real positive integers or logicals.

Error in ==> minandmax at 4
y = f(v);
```

There is another MATLAB user-defined function called the in-line function. However, the anonymous function has made the need for this function limited and it is not discussed further here.

1.18 Data Structures in MATLAB

Previous sections discussed the use of numerical and nonnumerical data. We now introduce the cell array structure, which allows a more complex data structure. The cell data structure is indicated by curly brackets, that is `{ }`. As an example,

```
>> A = cell(4,1);
>> A = {'maths'; 'physics'; 'history'; 'IT'}

A =
    'maths'
    'physics'
    'history'
    'IT'
```

We may refer to the individual components:

```
>> p = A(2)

p =
    'physics'

>> A(3:4)

ans =
    'history'
    'IT'
```

To access the contents of the cell we use curly brackets:

```
>> cont = A{3}

cont =
history
```

Note that `history` is no longer in quotes and thus we can reference individual characters as follows:

```
>> cont(4)

ans =
t
```

A cell array can include both numeric and string data and can also be generated using the `cell` function. For example, to generate a cell with 2 rows and 2 columns we have

```
>> F = cell(2,2)

F =
     [ ]     [ ]
     [ ]     [ ]
```

To assign a scalar, an array, or a character string to a cell we write

```
>> F{1,1} = 2;
>> F{1,2} = 'test';
>> F{2,1} = ones(3);
>> F

F =
    [          2]    'test'
    [3x3 double]        [ ]
```

An equivalent way of generating `F` is

```
>> F = {[2] 'test'; [ones(3)] [ ]}
```

Because we cannot see the detailed content of `F{2,1}` in the preceding, we use the function `celldisp`:

```
>> celldisp(F)

F{1,1} =
     2

F{2,1} =
     1     1     1
     1     1     1
     1     1     1

F{1,2} =
test

F{2,2} =
     [ ]
```

The cell array allows us to group data of different sizes and types together in the form of an array and access its elements by using subscripts.

The last form of data we consider is the structure, implemented in MATLAB using `struct`. This is similar to a cell array but individual cells are indexed by name. A structure combines a number of fields, each of which may be a different type. There is a general name for the field, for example, `'name'` or `'phone number'`. Each of these fields can have specific values such as `'George Brown'` or `'12719'`. To illustrate these points consider the following example, which sets up a structure called `StudentRecords` containing three fields: `NameField`, `FeesField`, and `SubjectField`.

Note that we begin by setting up the information for three students as specific values held in the cell arrays: names, fees, and subjects.

```
>> names = {'A Best', 'D Good', 'S Green', 'J Jones'}

names =
    'A Best'     'D Good'     'S Green'     'J Jones'

>> fees = {333 450 200 800}

fees =
    [333]     [450]     [200]     [800]
```

```
>> subjects = {'cs','cs','maths','eng'}

subjects =
    'cs'    'cs'    'maths'    'eng'

>> StudentRecords = struct('NameField',names,'FeesField',fees,...
                           'SubjectField',subjects)

StudentRecords =
1x4 struct array with fields:
    NameField
    FeesField
    SubjectField
```

Now, having set up our structure, we can refer to each individual record using a subscript:

```
>> StudentRecords(1)

ans =
       NameField: 'A Best'
       FeesField: 333
    SubjectField: 'cs'
```

Further we can examine the contents of the components of each record:

```
>> StudentRecords(1).NameField

ans =
A Best

>> StudentRecords(2).SubjectField

ans =
cs
```

We can change or update the values of the components of the records as follows:

```
>> StudentRecords(3).FeeField = 1000;
```

Now we check the contents of this student's `FeesField`:

```
>> StudentRecords(3).FeeField

ans =
        1000
```

MATLAB provides functions that allow us to convert from one data structure to another, and some of these are listed here.

```
cell2struct
struct2cell
num2cell
str2num
num2str
int2str
double
single
```

Most of these conversions are self-explanatory. For example, `num2str` converts a double precision number to an equivalent string. The function `double` converts to double precision and examples of its usage are given in Chapter 9.

The use of cells and structures is not usually essential in the development of numerical algorithms although they can be used to enhance an algorithm's ease of use. There is an example of the use of structures in Chapter 9.

1.19 Editing MATLAB Scripts

To help the user develop scripts MATLAB provides a comprehensive selection of debugging tools. These can be listed using the command `help debug`.

When typing a script into the MATLAB editor the user should note the small colored square displayed at the top right of the text window. This square is colored red if the script contains one or more fatal syntactic errors, orange warns of possible nonfatal problems, and green indicates no syntactic errors. Each error or warning is also indicated by an appropriately colored dash beneath the square. Touching these dashes will provide a description of the error or warning and the line in which it occurs.

Errors can be found by using `checkcode`. The `mlint` function can also be used but is now obselete and has been replaced by `checkcode`. The following script contains numerous errors and is provided to illustrate the use of `checkcode`:

```
% e3s125.m A script full of errors!!!
A = [1 2 3; 4 5 6
B = [2 3; 7 6 5]
c(1) = 1; c(1) = 2;
for k = 3:9
    c(k) = c(k-1)+c(k-2)
    if k = 3
        displ('k = 3, working well)
end
c
```

Running and checking this script gives the following output:

```
>> e3s125
Error: File: e3s125.m Line: 3 Column: 3
The expression to the left of the equals sign is not a valid target
                                                  for an assignment.
```

Applying `checkcode` to script `e3s125` gives

```
>> checkcode e3s125
L 3 (C 3): Invalid syntax at '='. Possibly, a ), }, or ] is missing.
L 3 (C 16): Parse error at ']': usage might be invalid MATLAB syntax.
L 5 (C 1-3): Invalid use of a reserved word.
L 7 (C 5-6): IF might not be aligned with its matching END (line 9).
L 7 (C 10): Parse error at '=': usage might be invalid MATLAB syntax.
L 8 (C 15-35): A quoted string is unterminated.
L 11 (C 0): Program might end prematurely (or an earlier error
                                            confused Code Analyzer).
```

Note how the line (`L`), character position (`C`), and nature of the error are given so the errors are clearly identified. Of course, some errors cannot be detected at this stage. For example, the following script is a partially corrected version of the preceding script.

```
% e3s125c.m A script less full of errors!!!
A = [1 2 3; 4 5 6];
B = [2 3; 7 6];
c(1) = 1; c(1) = 2;
for k = 3:9
    c(k) = c(k-1)+c(k-2)
    if k == 3
        disp('k = 3, working well')
    end
end
c
```

Running this script gives

```
>> e3s125c
Attempted to access c(2); index out of bounds because numel(c)=1.

Error in e3s125c (line 6)
    c(k) = c(k-1)+c(k-2)
```

```
>> checkcode e3s125c
L 6 (C 5): The variable 'c' appears to change size on every loop
        iteration (within a script). Consider preallocating for speed.
L 6 (C 10): Terminate statement with semicolon to suppress output
                                                    (within a script).
L 11 (C 1): Terminate statement with semicolon to suppress output
                                                    (within a script).
```

Now that the script can be run as far as line 6 other possible errors are detected.

In addition, a menu option "Debug" is provided in the MATLAB text editor.

1.20 Some Pitfalls in MATLAB

We now list five important points that if observed enable the MATLAB user to avoid some significant difficulties. This list is not exhaustive.

- It is important to take care when naming files and functions. Filenames and function names follow the rules for variable names; that is, they must start with a letter followed by a combination of letters or digits and names of existing functions must not be used.
- Do not use MATLAB function names or commands for variable names. For example, if we were so foolish as to assign a number to a variable that we called `sin`, access to the sine function would be lost. For example,

```
>> sin = 4

sin =
     4

>> 3*sin

ans =
    12

>> sin(1)

ans =
     4

>> sin(2)
??? Index exceeds matrix dimensions.

>> sin(1.1)
??? Subscript indices must either be real positive integers or logicals.
```

- Matrix sizes are set by assignment so it is vital to ensure that matrix sizes are compatible. Often it is a good idea initially to assign a matrix to an appropriately sized matrix of zeros; this also makes execution more efficient. For example, consider the following simple script:

```
for i = 1:2
    b(i) = i*i;
end
A = [4 5; 6 7];
A*b'
```

  We assign two elements to `b` in the `for` loop and define `A` to be a 2×2 array, so we would expect this script to succeed. However, if `b` had in the same session been previously set to be a different size matrix, then this script would have failed. To ensure that it works correctly we must either assign `b` to be a null matrix using `b = [ ]`, or make `b` a column vector of two elements by using `b = zeros(2,1)` or by using the `clear` statement to clear all variables from the system.
- Take care with dot products. For example, when creating a user-defined function where any of the input parameters may be vectors, dot products must be used. Note also that `2.^x` and `2. ^x` are different because the space is important. The first example gives the dot power while the second gives 2.0 to the power `x`, not the dot power. Similar care with spaces must be taken when using complex numbers. For example, `A = [1 2-4i]` assigns two elements: 1 and the complex number $2 - \iota 4$. In contrast `B = [1 2 -4i]` assigns three elements: 1, 2, and the imaginary number -4ι.
- At the beginning of a script, it is often good practice to clear variables or set arrays equal to the empty matrix (e.g., `A = [ ]`). This avoids incompatibility in matrix operations.

1.21 Faster Calculations in MATLAB

Calculations can be greatly sped up by using vector operations rather than using a loop to repeat a calculation. To illustrate this consider the following simple examples.

Example 1.3

This script fills the vector **b** using a `for` loop.

```
% e3s108.m
% Fill b with square roots of 1 to 100000 using a for loop
tic;
for i = 1:100000
    b(i) = sqrt(i);
end
```

```
t = toc;
disp(['Time taken for loop method is ', num2str(t)]);
```

Example 1.4

This script fills the vector **b** using a vector operation.

```
% e3s109.m
% Fill b with square roots of 1 to 100000 using a vector
tic
a = 1:100000; b = sqrt(a);
t = toc;
disp(['Time taken for vector method is ',num2str(t)]);
```

If the reader runs these two scripts and compares the time taken they will notice the vector method is substantially faster than the loop method. One of our experiments produced a 400 to 1 ratio for the times taken. There is a need to think very carefully about the way algorithms are implemented in MATLAB, particularly with regard to the use of vectors and arrays.

Problems

1.1. **(a)** Start up MATLAB. In the **command** window type `x = -1:0.1:1` and then execute each of the following statements by typing them in and pressing return:

```
sqrt(x)                 cos(x)
sin(x)                  2./x
x.\ 3                   plot(x, sin(x.^3))
plot(x, cos(x.^4))
```

Examine the effects of each statement carefully.

(b) Execute the following and explain the results:

```
x = [2 3 4 5]
y = -1:1:2
x.^y
x.*y
x./y
```

1.2. **(a)** Set up the matrix `A = [1 5 8;84 81 7;12 34 71]` in the **command** window and examine the contents of `A(1,1)`, `A(2,1)`, `A(1,2)`, `A(3,3)`, `A(1:2,:)`, `A(:,1)`, `A(3,:)`, and `A(:,2:3)`.

(b) What do the following MATLAB statements produce?

```
x = 1:1:10
z = rand(10)
y = [z;x]
c = rand(4)
e = [c eye(size(c)); eye(size(c)) ones(size(c))]
d = sqrt(c)
t1 = d*d
t2 = d.*d
```

1.3. Set up a 4×4 matrix. Given that the function `sum(x)` gives the sum of the elements of the vector `x`, use the function `sum` to find the sums of the first row and second column of the matrix.

1.4. Solve the following system of equations using the MATLAB function `inv` and also using the operators \ and / in the **command** window:

$$2x + y + 5z = 5$$

$$2x + 2y + 3z = 7$$

$$x + 3y + 3z = 6$$

Verify the solution is correct using matrix multiplication.

1.5. Write a simple script to input two square matrices **A** and **B**; then add, subtract, and multiply them. Comment the script and use `disp` to output suitable titles.

1.6. Write a MATLAB script to set up a 4×4 random matrix `A` and a four-element column vector `b`. Calculate `x = A\b` and display the result. Calculate `A*x` and compare it with `b`.

1.7. Write a simple script to plot the two functions $y_1 = x^2 \cos x$ and $y_2 = x^2 \sin x$ on the same graph. Use comments in your script and take $x = -2 : 0.1 : 2$.

1.8. Write a MATLAB script to produce graphs of the functions $y = \cos x$ and $y = \cos(x^3)$ in the range $x = -4 : 0.02 : 4$ using the same axes. Use the MATLAB functions `xlabel`, `ylabel`, and `title` to annotate your graphs clearly.

1.9. Draw the function $y = \exp(-x^2)\cos(20x)$ in the range $x = -2 : 0.1 : 2$. All axes should be labeled and a title included. Compare the results of using the functions `fplot` and `plot` to plot this function.

1.10. Write a MATLAB script to draw the functions $y = 3\sin(\pi x)$ and $y = \exp(-0.2x)$ on the same graph for $x = 0 : 0.02 : 4$. All axes should be labeled. Use `gtext` to label one of the several points of intersection of the graphs.

1.11. Use the functions `meshgrid` and `mesh` to obtain a three-dimensional plot of the function

$$z = 2xy/(x^2 + y^2) \quad \text{for } x = 1:0.1:3 \quad \text{and} \quad y = 1:0.1:3$$

Redraw the surface using the functions `surf`, `surfl`, and `contour`.

1.12. An iterative equation for solving the equation $x^2 - x - 1 = 0$ is given by

$$x_{r+1} = 1 + (1/x_r) \quad \text{for} \quad r = 0, 1, 2, \ldots$$

Given x_0 is 2, write a MATLAB script to solve the equation. Sufficient accuracy is obtained when $|x_{r+1} - x_r| < 0.0005$. Include a check on the answer.

1.13. Given a 4×5 matrix **A**, write a script to find the sums of each of the columns using
(a) The `for ... end` construction
(b) The function `sum`

1.14. Given a vector **x** with n elements, write a MATLAB script to form the products

$$p_k = x_1 x_2 \ldots x_{k-1} x_{k+1} \cdots x_n$$

for $k = 1, 2, \ldots, n$. That is, p_k contains the products of all the vector elements except the kth. Run your script with specific values of x and n.

1.15. The series for $\log_e(1 + x)$ is given by

$$\log_e(1 + x) = x - x^2/2 + x^3/3 - \cdots + (-1)^{k+1} x^k/k \cdots$$

Write a MATLAB script to input a value for x and sum the series while the value of the current term is greater than or equal to the variable *tol*. Use values of $x = 0.5$ and 0.82 and $tol = 0.005$ and 0.0005. The result should be checked by using the MATLAB function `log`. The script should display the value of x and *tol* and the value of $\log_e(1 + x)$ obtained. Use the `input` and `disp` functions to obtain clear output and prompts.

1.16. Write a MATLAB script to generate a matrix that has the values d along the main diagonal and the values c on the diagonals above and below the main diagonal and zero elsewhere. Your script should allow the user to input any values for c and d and work for any size of matrix n. The script should give clear prompts for input and display the results with a suitable heading.

1.17. Write a MATLAB function to solve the quadratic equation

$$ax^2 + bx + c = 0$$

The function will use three input parameters a, b, c and output the values of the two roots. You should take account of the three cases:

(a) No real roots
(b) Real and different roots
(c) Equal roots

1.18. Adjust the function of Problem 1.17 to deal with the case when $a = 0$. That is, when the equation is nonquadratic. In this case include a third output parameter that will have the value 1 if the equation is quadratic and 0 otherwise.

1.19. Write a simple function to define $f(x) = x^2 - \cos(x) - x$ and plot the graph of the function in the range 0 to 2. Use this graph to find an initial approximation to the root and then apply the function `fzero` to find the root to tolerance 0.0005.

1.20. Write a script to generate the sequence of values given by

$$x_{r+1} = \begin{cases} x_r/2 & \text{if } x_r \text{ is even} \\ 3x_r + 1 & \text{if } x_r \text{ is odd} \end{cases} \quad \text{where } r = 0, 1, 2, \ldots$$

where x_0 is any positive integer. The sequence terminates when $x_r = 1$. Show after a sufficient number of steps that the sequence terminates for any value of x_0 you choose. It is interesting to plot the values of x_r against r.

1.21. Write a MATLAB script to plot the surface $z = f(x, y)$ over the ranges $x = -4:0.1:4$ and $y = -4:0.1:4$ where z is given by

$$z = f(x, y) = (1 - x^2)e^{-p} - pe^{-p} - e^{-(x+1)^2 - y^2}$$

and $p = x^2 + y^2$. The script should provide `mesh`, `contour`, and `surf` plots and use the function `subplot` to lay out the three plots one above the other.

1.22. The following three functions are presented in parametric form:

$$x = a(t - \sin(t)) \quad \text{and} \quad y = a(1 - \cos(t))$$

$$x = 2at \quad \text{and} \quad y = 2a/(1 + t^2)$$

$$x = a\,\cos(t) - b\,\cos(at/b) \quad \text{and} \quad y = a\,\sin(t) - b\,\sin(at/b)$$

Write a MATLAB script to plot each of these functions one above the other using the MATLAB `subplot` function given $a = 2$ and $b = 3$; t is assigned the range of values $-10:0.1:10$.

1.23. The Riemann ζ function may be defined as the sum of an infinite series:

$$\zeta(s) = 1 + \frac{1}{2^s} + \frac{1}{3^s} + \frac{1}{4^s} + \cdots + \frac{1}{n^s} \cdots$$

Write a MATLAB script `zetainf(s,acc)` to sum terms of this series until a term is less than `acc`, where `s` is an integer.

1.24. Write a MATLAB function to sum the series

$$s = 1 + 2^2/2! + 3^2/3! + \cdots + n^2/n!$$

to n terms. The function should take the form `sumfac(n)` where `n` is the number of terms used. You may use the MATLAB function `factorial` to evaluate the factorial terms. Write MATLAB statements using this function to sum the series to 5 and 10 terms.

Rewrite your script avoiding the use of the function `factorial` by noting that the $k+1$th term T_{k+1} is given by $T_k \times (k+1)/k$.

1.25. Given the matrix `D = [1 -1; 3 2]`, give the values that will be assigned to `A`, `B`, `C`, and `E` by executing the following MATLAB statements:

(a) `A = D*(D*inv(D))`
(b) `B = D.*D`
(c) `C = [D,ones(2);eye(2),zeros(2)]`
(d) `E = D'*ones(2)*eye(2)`

1.26. The following matrices, called the Dirac matrices, are defined by

$$\mathbf{P}_1 = \begin{bmatrix} \mathbf{0} & \mathbf{I}_2 \\ \mathbf{I}_2 & \mathbf{0} \end{bmatrix}, \ \mathbf{P}_2 = \begin{bmatrix} \mathbf{0} & -\imath\mathbf{I}_2 \\ \imath\mathbf{I}_2 & \mathbf{0} \end{bmatrix}, \ \mathbf{P}_3 = \begin{bmatrix} \mathbf{I}_2 & \mathbf{0} \\ \mathbf{0} & -\mathbf{I}_2 \end{bmatrix}$$

where the $\mathbf{0}$ represents a 2×2 matrix of zeros, $\mathbf{I}_2$ represents a 2×2 unit matrix, and $\imath = \sqrt{(-1)}$. A related set of matrices is given by

$$\mathbf{Q}_k = \begin{bmatrix} \mathbf{0} & \mathbf{P}_k \\ -\mathbf{P}_k & \mathbf{0} \end{bmatrix} \quad \text{for} \quad k = 1,2,3$$

Write MATLAB statements to generate the matrices $\mathbf{P}_1$, $\mathbf{P}_2$, $\mathbf{P}_3$ and the matrices $\mathbf{Q}_k$ for $k = 1, 2, 3$. Note that in $\mathbf{Q}_k$ the $\mathbf{0}$ represents a 4×4 matrix of zeros.

1.27. Plot the function

$$y = \frac{1}{((x+2.5)^2)((x-3.5)^2)}$$

for values of $x = -4 : 0.001 : 4$. Then use the MATLAB functions `xlim` and `ylim` in the form `ylim([0,20])` and `xlim([-3,-2])` to illustrate how this allows considerable clarification of the nature of the function.

1.28. Write user-defined functions for the following functions:

$$\textbf{(a)} \quad y = x^2 \cos(1 + x^2)$$

$$\textbf{(b)} \quad y = \frac{1 + e^x}{\cos(x) + \sin(x)}$$

$$\textbf{(c)} \quad z = \cos(x^2 + y^2)$$

Rewrite each of the previous functions as anonymous functions and illustrate the use of these anonymous functions by using the MATLAB function `subplot` to plot graphs of functions (a) and (b) in the range $x = 0$ to 2.

1.29. Consider the following MATLAB script, which contains some errors. Use the MATLAB function `checkcode` to find these errors.

```
function sol = solvepoly(x0, acc)
%poly solver
d = 1+acc;
whil abs(d)>acc
    x1 = (2*x0^2-1))/x0^2;
    d = x1-x0;
    x0 = x1/x2
end
sol = x0;
```

1.30. The symmetric hyperbolic Fibonacci sine and cosine functions are defined as follows:

$$\mathrm{sFs}(x) = \frac{\gamma^x - \gamma^{-x}}{\sqrt{5}} \quad \text{and} \quad \mathrm{cFs}(x) = \frac{\gamma^x + \gamma^{-x}}{\sqrt{5}}$$

where $\gamma = (1+\sqrt{5})/2$. Also, the complex quasi-sine Fibonacci function is defined as

$$\mathrm{cqsF}(x,n) = \frac{\gamma^x - \cos(n\pi x)\gamma^{-x}}{\sqrt{5}} + \imath\frac{\sin(n\pi x)\gamma^{-x}}{\sqrt{5}}$$

where γ is defined as before.

Write a MATLAB script that begins by defining these three functions as anonymous functions. Then, using these anonymous functions, carry out the following operations within the script:

(a) In a single figure, plot the graphs of sFs(x) and cFs(x) against x over the range −5 to 5.

(b) Plot the real and imaginary parts of the function cqsF(x,5) in 3D space. Plot the real part of the function in the y direction, the imaginary part in the z direction. Plot the function over the range −5 to 5. Use the MATLAB function `plot3`.

Stakhov and Rozin (2005, 2007) provide more information on these functions.

2 Linear Equations and Eigensystems

When physical systems are modeled mathematically, they are sometimes described by linear equation systems or eigensystems, and in this chapter we will examine how such equation systems are solved. Linear equation systems can be expressed in terms of matrices and vectors, and we introduce some of the more important properties of vectors and matrices in Appendix A.

MATLAB is an ideal environment for studying linear algebra, including linear equation systems and eigenvalue problems, because MATLAB functions and operators can work directly on vectors and matrices. It is rich in functions and operators, which facilitate the manipulation of matrices. MATLAB originated as a set of linear algebra operators and functions based on the LINPACK (Dongarra et al., 1979) and EISPACK (Smith et al., 1976; Garbow et al., 1977) routines. These routines were developed specifically to solve linear equations and eigenvalue problems, respectively. In 2000, MATLAB began using the LAPACK library of linear algebra subroutines, which is the modern replacement for LINPACK and EISPACK.

2.1 Introduction

We start with a discussion of linear equation systems and defer discussion of eigensystems until Section 2.15. To illustrate how linear equation systems arise in the modeling of certain physical problems we will consider how current flows are calculated in a simple electrical network. The necessary equations can be developed using one of several techniques; here we use the loop-current method together with Ohm's law and Kirchhoff's voltage law. A *loop current* is assumed to circulate around each loop in the network. Thus, in the network given in Figure 2.1, the loop current I_1 circulates around the closed loop *abcd*. Note that the current $I_1 - I_2$ is assumed to flow in the link connecting b to c. Ohm's law states that the voltage across an ideal resistor is proportional to the current flow through the resistor. For example, for the link connecting b to c

$$V_{bc} = R_2(I_1 - I_2)$$

where R_2 is the value of the resistor in the link connecting b to c. Kirchhoff's voltage law states that the algebraic sum of the voltages around a loop is zero. Applying these laws to the circuit *abcd* of Figure 2.1 we have

$$V_{ab} + V_{bc} + V_{cd} = V$$

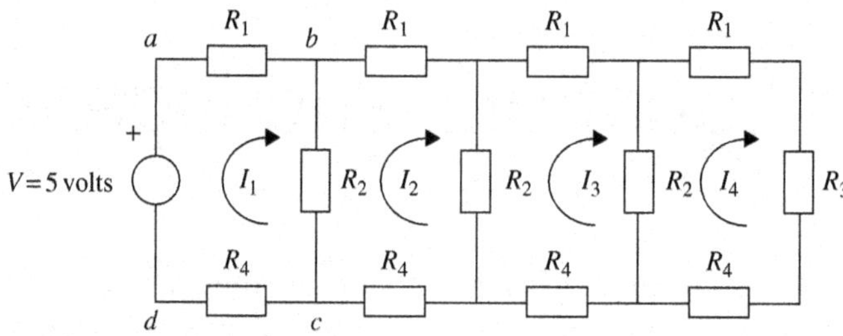

FIGURE 2.1 Electrical network.

Substituting the product of current and resistance for voltage gives

$$R_1I_1 + R_2(I_1 - I_2) + R_4I_1 = V$$

We can repeat this process for each loop to obtain the following four equations:

$$\begin{aligned}(R_1 + R_2 + R_4)I_1 - R_2I_2 &= V\\(R_1 + 2R_2 + R_4)I_2 - R_2I_1 - R_2I_3 &= 0\\(R_1 + 2R_2 + R_4)I_3 - R_2I_2 - R_2I_4 &= 0\\(R_1 + R_2 + R_3 + R_4)I_4 - R_2I_3 &= 0\end{aligned} \tag{2.1}$$

Letting $R_1 = R_4 = 1\Omega$, $R_2 = 2\Omega$, $R_3 = 4\Omega$, and $V = 5$ volts, (2.1) becomes

$$\begin{aligned}4I_1 - 2I_2 &= 5\\-2I_1 + 6I_2 - 2I_3 &= 0\\-2I_2 + 6I_3 - 2I_4 &= 0\\-2I_3 + 8I_4 &= 0\end{aligned}$$

This is a system of linear equations in four variables, $I_1, \ldots, I_4$. In matrix notation it becomes

$$\begin{bmatrix}4 & -2 & 0 & 0\\-2 & 6 & -2 & 0\\0 & -2 & 6 & -2\\0 & 0 & -2 & 8\end{bmatrix}\begin{bmatrix}I_1\\I_2\\I_3\\I_4\end{bmatrix} = \begin{bmatrix}5\\0\\0\\0\end{bmatrix} \tag{2.2}$$

This equation has the form $\mathbf{Ax} = \mathbf{b}$ where $\mathbf{A}$ is a square matrix of known coefficients, in this case relating to the values of the resistors in the circuit. The vector $\mathbf{b}$ is a vector of known coefficients, in this case the voltage applied to each current loop. The vector $\mathbf{x}$ is the vector of unknown currents. Although this set of equations can be solved by hand, the process is time consuming and error prone. Using MATLAB we simply enter matrix `A` and vector `b` and use the command `A\b` as follows:

```
>> A = [4 -2 0 0;-2 6 -2 0;0 -2 6 -2;0 0 -2 8];
>> b = [5 0 0 0].';
>> A\b

ans =
    1.5426
    0.5851
    0.2128
    0.0532
```

The sequence of operations that are invoked by this apparently simple command is examined in Section 2.3.

In many electrical networks the ideal resistors of Figure 2.1 are more accurately represented by electrical impedances. When a harmonic alternating current (AC) supply is connected to the network, electrical engineers represent the impedances by complex quantities. This is to account for the effect of capacitance and/or inductance. To illustrate this we will replace the 5 volt DC supply to the network of Figure 2.1 with a 5 volt AC supply and replace the ideal resistors $R_1,\ldots,R_4$ by impedances $Z_1,\ldots,Z_4$. Thus (2.1) becomes

$$\begin{aligned}(Z_1+Z_2+Z_4)I_1-Z_2I_2&=V\\(Z_1+2Z_2+Z_4)I_2-Z_2I_1-Z_2I_3&=0\\(Z_1+2Z_2+Z_4)I3-Z_2I_2-Z_2I_4&=0\\(Z_1+Z_2+Z_3+Z_4)I_4-Z_2I_3&=0\end{aligned}\tag{2.3}$$

At the frequency of the 5 volt AC supply we will assume that $Z_1=Z_4=(1+0.5\jmath)$, $Z_2=(2+0.5\jmath)$, and $Z_3=(4+1\jmath)$, where $\jmath=\sqrt{-1}$. Electrical engineers prefer to use $\jmath$ rather than $\imath$ for $\sqrt{-1}$. This avoids any possible confusion with I or i, which are normally used to denote the current in a circuit. Thus (2.3) becomes

$$\begin{aligned}(4+1.5\jmath)I_1-(2+0.5\jmath)I_2&=5\\-(2+0.5\jmath)I_1+(6+2.0\jmath)I_2-(2+0.5\jmath)I_3&=0\\-(2+0.5\jmath)I_2+(6+2.0\jmath)I_3-(2+0.5\jmath)I_4&=0\\-(2+0.5\jmath)I_3+(8+2.5\jmath)I_4&=0\end{aligned}$$

This system of linear equations becomes, in matrix notation,

$$\begin{bmatrix}(4+1.5\jmath)&-(2+0.5\jmath)&0&0\\-(2+0.5\jmath)&(6+2.0\jmath)&-(2+0.5\jmath)&0\\0&-(2+0.5\jmath)&(6+2.0\jmath)&-(2+0.5\jmath)\\0&0&-(2+0.5\jmath)&(8+2.5\jmath)\end{bmatrix}\begin{bmatrix}I_1\\I_2\\I_3\\I_4\end{bmatrix}=\begin{bmatrix}5\\0\\0\\0\end{bmatrix}\tag{2.4}$$

Note that the coefficient matrix is now complex. This does not present any difficulty for MATLAB because the operation `A\b` works directly with both real and complex numbers. Thus

```
>> p = 4+1.5i; q = -2-0.5i;
>> r = 6+2i; s = 8+2.5i;
>> A = [p q 0 0;q r q 0;0 q r q;0 0 q s];
>> b = [5 0 0 0].';
>> A\b

ans =
    1.3008 - 0.5560i
    0.4560 - 0.2504i
    0.1530 - 0.1026i
    0.0361 - 0.0274i
```

Note that strictly we have no need to reenter the values in vector `b`, assuming that we have not cleared the memory, reassigned the vector `b`, or quit MATLAB. The answer shows that currents flowing in the network are complex. This means that there is a phase difference between the applied harmonic voltage and the currents flowing.

We will now begin a more detailed examination of linear equation systems.

2.2 Linear Equation Systems

In general, a linear equation system can be written in matrix form as

$$\mathbf{Ax} = \mathbf{b} \tag{2.5}$$

where $\mathbf{A}$ is an $n \times n$ matrix of known coefficients, $\mathbf{b}$ is a column vector of n known coefficients, and $\mathbf{x}$ is the column vector of n unknowns. We have already seen an example of this type of equation system in Section 2.1 where the matrix equation (2.2) is the matrix equivalent of the linear equations (2.1).

The equation system (2.5) is called homogeneous if $\mathbf{b} = \mathbf{0}$ and inhomogeneous if $\mathbf{b} \neq \mathbf{0}$. Before attempting to solve an equation system it is reasonable to ask if it has a solution and if so is it unique? A linear inhomogeneous equation system may be *consistent* and have one or an infinity of solutions or be *inconsistent* and have no solution. This is illustrated in Figure 2.2 for a system of three equations in three variables x_1, x_2, and x_3. Each equation represents a plane surface in the x_1, x_2, x_3 space. In Figure 2.2(a) the three planes have a common point of intersection. The coordinates of the point of intersection give the unique solution for the three equations. In Figure 2.2(b) the three planes intersect in a line. Any point on the line of intersection represents a solution so there is no unique solution but an infinite number of solutions satisfying the three equations. In Figure 2.2(c) two of the surfaces are parallel to each other and therefore they never intersect, while in Figure 2.2(d)

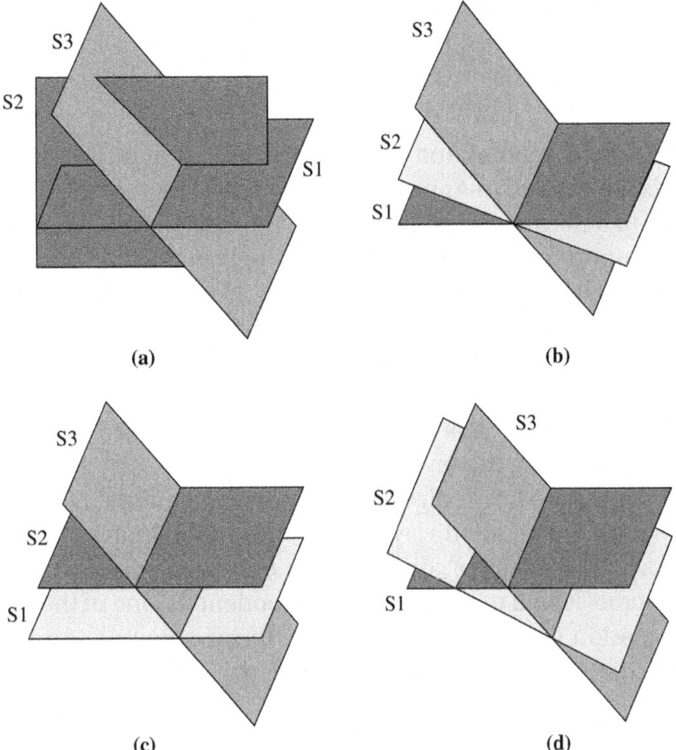

FIGURE 2.2 Three intersecting planes representing three equations in three variables. (a) Three plane surfaces intersecting in a point. (b) Three plane surfaces intersecting in a line. (c) Three plane surfaces, two of which do not intersect. (d) Three plane surfaces intersecting in three lines.

the line of intersection of each pair of surfaces is different. In both of these cases there is no solution and the equations these surfaces represent are inconsistent.

To obtain an algebraic solution to the inhomogeneous equation system (2.5) we multiply both sides of (2.5) by a matrix called the inverse of **A**, denoted by $\mathbf{A}^{-1}$:

$$\mathbf{A}^{-1}\mathbf{A}\mathbf{x} = \mathbf{A}^{-1}\mathbf{b} \tag{2.6}$$

where $\mathbf{A}^{-1}$ is defined by

$$\mathbf{A}^{-1}\mathbf{A} = \mathbf{A}\mathbf{A}^{-1} = \mathbf{I} \tag{2.7}$$

and **I** is the identity matrix. Thus we obtain

$$\mathbf{x} = \mathbf{A}^{-1}\mathbf{b} \tag{2.8}$$

The standard algebraic formula for the inverse of **A** is

$$\mathbf{A}^{-1} = \text{adj}(\mathbf{A})/|\mathbf{A}| \tag{2.9}$$

where $|\mathbf{A}|$ is the determinant of **A** and adj(**A**) is the adjoint of **A**. The determinant and the adjoint of a matrix are defined in Appendix A. Equations (2.8) and (2.9) are algebraic statements allowing us to determine **x** but they do not provide an efficient means of solving the system because computing $\mathbf{A}^{-1}$ using (2.9) is extremely inefficient, involving order $(n+1)!$ multiplications where n is the number of equations. However, (2.9) is theoretically important because it shows that if $|\mathbf{A}| = 0$ then **A** does not have an inverse. The matrix **A** is then said to be singular and a unique solution for **x** does not exist. Thus establishing that $|\mathbf{A}|$ is nonzero is one way of showing that an inhomogeneous equation system is a consistent system with a unique solution. It is shown in Sections 2.6 and 2.7 that (2.5) can be solved without formally determining the inverse of **A**.

An important concept in linear algebra is the rank of a matrix. For a square matrix, the rank is the number of independent rows or columns in the matrix. Independence can be explained as follows. The rows (or columns) of a matrix can clearly be viewed as a set of vectors. A set of vectors is said to be linearly independent if none of them can be expressed as a linear combination of any of the others. By linear combination we mean a sum of scalar multiples of the vectors. For example, the matrix

$$\begin{bmatrix} 1 & 2 & 3 \\ -2 & 1 & 4 \\ -1 & 3 & 4 \end{bmatrix} \text{ or } \begin{bmatrix} [1 & 2 & 3] \\ [-2 & 1 & 4] \\ [-1 & 3 & 7] \end{bmatrix} \text{ or } \begin{bmatrix} \begin{bmatrix} 1 \\ 2 \\ 3 \end{bmatrix} & \begin{bmatrix} -2 \\ 1 \\ 4 \end{bmatrix} & \begin{bmatrix} -1 \\ 3 \\ 7 \end{bmatrix} \end{bmatrix}$$

has linearly *dependent* rows and columns. This is because $\text{row3} - \text{row1} - \text{row2} = 0$ and $\text{column3} - 2(\text{column2}) + \text{column1} = 0$. There is only one equation relating the rows (or columns) and thus there are two independent rows (or columns). Hence this matrix has a rank of 2. Now consider

$$\begin{bmatrix} 1 & 2 & 3 \\ 2 & 4 & 6 \\ 3 & 6 & 9 \end{bmatrix}$$

Here $\text{row2} = 2(\text{row1})$ and $\text{row3} = 3(\text{row1})$. There are two equations relating the rows and hence only one row is independent and the matrix has a rank of 1. Note that the number of independent rows and columns in a square matrix is identical; that is, its row rank and column rank are equal. In general matrices may be nonsquare and the rank of an $m \times n$ matrix **A** is written rank(**A**). Matrix **A** is said to be of full rank if $\text{rank}(\mathbf{A}) = \min(m, n)$; otherwise $\text{rank}(\mathbf{A}) < \min(m, n)$ and **A** is said to be rank deficient. MATLAB provides the function `rank`, which works with both square and nonsquare matrices. In practice, MATLAB determines the rank of a matrix from its singular values; see Section 2.10.

For example, consider the following MATLAB statements:

```
>> D = [1 2 3;3 4 7;4 -3 1;-2 5 3;1 -7 6]

D =
     1     2     3
     3     4     7
     4    -3     1
    -2     5     3
     1    -7     6

>> rank(D)

ans =
     3
```

Thus D is of full rank since the rank equals the minimum size of the matrix.

A useful operation in linear algebra is the conversion of a matrix to its reduced row echelon form (RREF). The RREF is defined in Appendix A. In MATLAB we can use the `rref` function to compute the RREF of a matrix as follows:

```
>> rref(D)

ans =
     1     0     0
     0     1     0
     0     0     1
     0     0     0
     0     0     0
```

It is a property of the RREF of a matrix that the number of rows with at least one nonzero element equals the rank of the matrix. In this example we see that there are three rows in the RREF of the matrix containing a nonzero element, confirming that the matrix rank is 3. The RREF also allows us to determine whether a system has a unique solution or not.

We have discussed a number of important concepts relating to the nature of linear equations and their solutions. We now summarize the equivalencies between these concepts. Let $\mathbf{A}$ be an $n \times n$ matrix. If $\mathbf{Ax} = \mathbf{b}$ is consistent and has a unique solution, then all of the following statements are true:

$\mathbf{Ax} = \mathbf{0}$ has only the trivial solution $\mathbf{x} = \mathbf{0}$.

$\mathbf{A}$ is nonsingular and $\det(\mathbf{A}) \neq 0$.

The RREF of $\mathbf{A}$ is the identity matrix.

A has n linearly independent rows and columns.

A has full rank, i.e., rank(**A**) $= n$.

In contrast, if $\mathbf{Ax} = \mathbf{b}$ is either inconsistent or consistent but with more than one solution, then all of the following statements are true:

$\mathbf{Ax} = \mathbf{0}$ has more than one solution.

A is singular and det(**A**) $= 0$.

The RREF of **A** contains at least one zero row.

A has linearly dependent rows and columns.

A is rank deficient, i.e., rank(**A**) $< n$.

So far we have only considered the case where there are as many equations as unknowns. Now we consider the cases where there are fewer or more equations than unknown variables.

If there are fewer equations than unknowns, then the system is said to be underdetermined. The equation system does not have a unique solution; it is either consistent with an infinity of solutions, or inconsistent with no solution. These conditions are illustrated by Figure 2.3. The diagram shows two plane surfaces in three-dimensional space, representing two equations in three variables. It is seen that the planes either intersect in a line so that the equations are consistent with an infinity of solutions represented by the line of intersection, or the surfaces do not intersect and the equations they represent are inconsistent.

Consider the following system of equations:

$$\begin{bmatrix} 1 & 2 & 3 & 4 \\ -4 & 2 & -3 & 7 \end{bmatrix} \begin{bmatrix} x_1 \\ x_2 \\ x_3 \\ x_4 \end{bmatrix} = \begin{bmatrix} 1 \\ 3 \end{bmatrix}$$

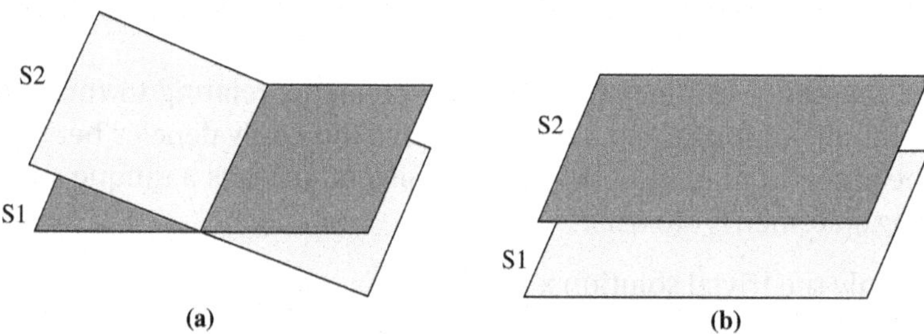

FIGURE 2.3 Planes representing an underdetermined system of equations. (a) Two plane surfaces intersecting in a line. (b) Two plane surfaces which do not intersect.

This underdetermined system can be rearranged as follows:

$$\begin{bmatrix} 1 & 2 \\ -4 & 2 \end{bmatrix} \begin{bmatrix} x_1 \\ x_2 \end{bmatrix} + \begin{bmatrix} 3 & 4 \\ -3 & 7 \end{bmatrix} \begin{bmatrix} x_3 \\ x_4 \end{bmatrix} = \begin{bmatrix} 1 \\ 3 \end{bmatrix}$$

or

$$\begin{bmatrix} 1 & 2 \\ -4 & 2 \end{bmatrix} \begin{bmatrix} x_1 \\ x_2 \end{bmatrix} = \begin{bmatrix} 1 \\ 3 \end{bmatrix} - \begin{bmatrix} 3 & 4 \\ -3 & 7 \end{bmatrix} \begin{bmatrix} x_3 \\ x_4 \end{bmatrix}$$

Thus we have reduced this to a system of two equations in two unknowns, provided values are assumed for x_3 and x_4. Thus the problem has an infinity of solutions, depending on the values chosen for x_3 and x_4.

If a system has more equations than unknowns, then the system is said to be overdetermined. Figure 2.4 shows four plane surfaces in three-dimensional space, representing four equations in three variables. Figure 2.4(a) shows all four planes intersecting in a single point so that the system of equations is consistent with a unique solution. Figure 2.4(b) shows all the planes intersecting in a line and this represents a consistent system with an infinity of solutions. Figure 2.4(d) shows planes that represent an inconsistent system of equations with no solution. In Figure 2.4(c) the planes do not intersect in a single point and so the system of equations is inconsistent. However, in this example the points of intersection of groups of three planes (i.e., (S1, S2, S3), (S1, S2, S4), (S1, S3, S4), and (S2, S3, S4)) are close to each other and a mean point of intersection could be determined and used as an approximate solution. This example of marginal inconsistency often arises because the coefficients in the equations are determined experimentally; if the coefficients were known exactly, it is likely that the equations would be consistent with a unique solution. Rather than accepting that the system is inconsistent we may ask what the best solution is that satisfies the equations approximately. In Sections 2.11 and 2.12 we deal with the problem of overdetermined and underdetermined systems in more detail.

2.3 Operators \ and / for Solving **Ax** = **b**

The purpose of this section is to introduce the reader to the MATLAB operator \. A detailed discussion of the algorithms behind its operation will be given in later sections. This operator is a very powerful one that provides a unified approach to the solution of many categories of linear equation systems. The operators / and \ perform matrix "division" and have identical effects. Thus to solve $\mathbf{Ax} = \mathbf{b}$ we may write either `x=A\b` or `x'=b'/A'`. In the latter case the solution **x** is expressed as a row rather than a column vector. The operator / or \, when solving $\mathbf{Ax} = \mathbf{b}$, selects the appropriate algorithm dependent on the form of the matrix **A**. These cases are outlined next:

- if **A** is a triangular matrix, the system is solved by back or forward substitution alone, described in Section 2.6.

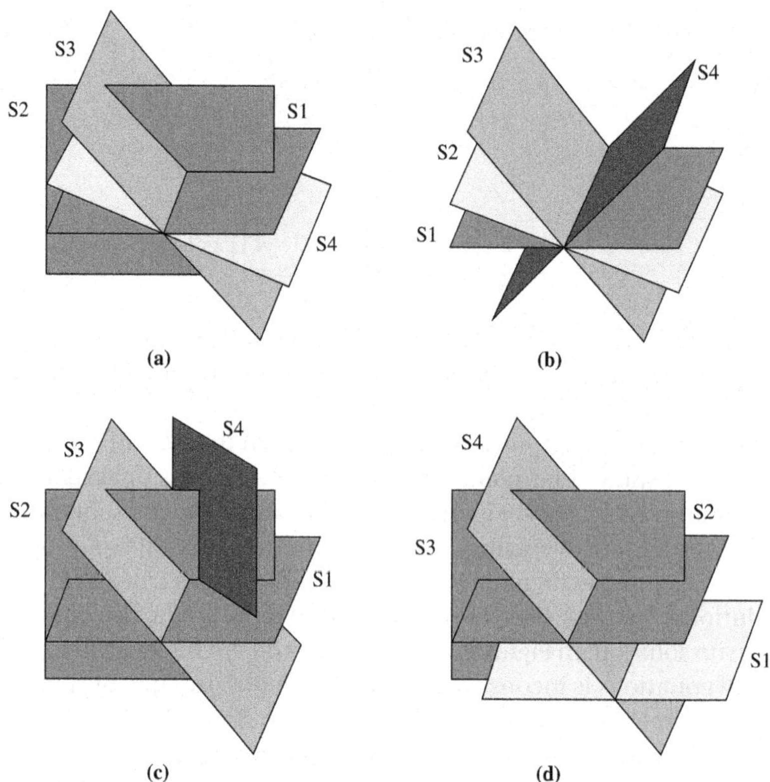

FIGURE 2.4 Planes representing an overdetermined system of equations. (a) Four plane surfaces intersecting in a point. (b) Four plane surfaces intersecting in a line. (c) Four plane surfaces not intersecting at a single point; points of intersection of (S1, S2, S3) and (S1, S2, S4) are visible. (d) Four plane surfaces representing inconsistent equations.

- elseif **A** is a positive definite, square symmetric, or Hermitian matrix, Cholesky decomposition (described in Section 2.8) is applied. When **A** is sparse, Cholesky decomposition is preceded by a symmetric minimum degree preordering (described in Section 2.14).
- elseif **A** is a square matrix, general LU decomposition (described in Section 2.7) is applied. If **A** is sparse, this is preceded by a nonsymmetric minimum degree preordering (described in Section 2.14).
- elseif **A** is a full nonsquare matrix, QR decomposition (described in Section 2.9) is applied.
- elseif **A** is a sparse nonsquare matrix, it is augmented and then a minimum degree preordering is applied, followed by sparse Gaussian elimination (described in Section 2.14).

The MATLAB \ operator can also be used to solve $\mathbf{AX} = \mathbf{B}$ where **B** and the unknown **X** are $m \times n$ matrices. This could provide a simple method of finding the inverse of **A**. If we make

B the identity matrix **I** then we have

$$\mathbf{AX} = \mathbf{I}$$

and **X** must be the inverse of **A** since $\mathbf{AA}^{-1} = \mathbf{I}$. Thus in MATLAB we could determine the inverse of **A** by using the statement `A\eye(size(A))`. However, MATLAB provides the function `inv(A)` to find the inverse of a matrix. It is important to stress that the inverse of a matrix should only be determined if it is specifically required. If we require the solution of a set of linear equations it is more efficient to use the operators \ or /.

We now examine some cases to show how the \ operator works, beginning with the solution of a system where the system matrix is triangular. The experiment in this case examines the time taken by the operator \ to solve a system when it is full and then when the same system is converted to triangular form by zeroing appropriate elements to produce a triangular matrix. The script used for this experiment is

```
% e3s201.m
disp('   n     full-time  full-time/n^3   tri-time  tri-time/n^2');
A = [ ]; b = [ ];
for n = 2000:500:6000
    A = 100*rand(n); b = [1:n].';
    tic, x = A\b; t1 = toc;
    t1n = 5e9*t1/n^3;
    for i = 1:n
        for j = i+1:n
            A(i,j) = 0;
        end
    end
    tic, x = A\b; t2 = toc;
    t2n = 1e9*t2/n^2;
    fprintf('%6.0f %9.4f %12.4f %12.4f %11.4f\n',n,t1,t1n,t2,t2n)
end
```

The results for a series of randomly generated $n \times n$ matrices are as follows:

n	full-time	full-time/n^3	tri-time	tri-time/n^2
2000	1.7552	1.0970	0.0101	2.5203
2500	3.3604	1.0753	0.0151	2.4151
3000	5.4936	1.0173	0.0209	2.3275
3500	8.5735	0.9998	0.0282	2.3001
4000	12.6882	0.9913	0.0358	2.2393
4500	17.5680	0.9639	0.0453	2.2392
5000	24.8408	0.9936	0.0718	2.8703

Column 1 of this table gives the size of the square matrix, n. To demonstrate that the operator \ takes account of the triangular form, columns 2 and 3 contain the time taken and the time taken divided by n^3 and multiplied by the scaling factor 5×10^9 for the full matrix problem. Columns 4 and 5 give the time taken and the time taken divided by n^2 and multiplied by the scaling factor 1×10^9 for the triangular system. These interesting results show that for the full matrix the time taken by \ is approximately proportional to n^3 while for the triangular system the time taken by \ is approximately proportional to n^2. This is the expected result for simple back substitution. In addition we see a considerable reduction in the time taken to solve the system when the operator \ is used with a triangular system.

We now perform experiments to examine the effects of using the operator \ with positive definite symmetric systems. This is a more complex problem than those previously discussed and the script that follows implements this test. It is based on comparing the application of the \ operator to a positive definite system and a nonpositive definite system of equations. We can create a positive definite matrix by letting `A = M*M'`. Where `M` is any matrix, but in this case the matrix will be of random numbers. `A` will then be a positive definite matrix. To generate a nonpositive definite system we add a random matrix to the positive definite matrix and we compare the time required to solve the two forms of matrix. The script takes the form

```
% e3s202.m
disp(' n       time-pos    time-pos/n^3  time-npos    time-b/n^3');
for n = 100:100:1000
    A = [ ]; M = 100*randn(n,n);
    A = M*M'; b = [1:n].';
    tic, x = A\b; t1 = toc*1000;
    t1d = t1/n^3;
    A = A+rand(size(A));
    tic, x = A\b; t2 = toc*1000;
    t2d = t2/n^3;
    fprintf('%4.0f %10.4f %14.4e %11.4f %13.4e\n',n,t1,t1d,t2,t2d)
end
```

The result of running this script is

```
   n       time-pos    time-pos/n^3  time-npos    time-b/n^3
 100        0.9881     9.8811e-007      1.2085   1.2085e-006
 200        3.5946     4.4932e-007      3.0903   3.8629e-007
 300        5.0646     1.8758e-007      9.7878   3.6251e-007
 400       10.3890     1.6233e-007     20.4892   3.2014e-007
 500       18.0235     1.4419e-007     36.5653   2.9252e-007
 600       18.1892     8.4209e-008     37.7766   1.7489e-007
 700       26.5483     7.7400e-008     58.3854   1.7022e-007
 800       39.6402     7.7422e-008     79.4285   1.5513e-007
 900       58.5519     8.0318e-008    110.5409   1.5163e-007
1000       67.9078     6.7908e-008    130.2029   1.3020e-007
```

Column 1 of this table gives n, the size of the matrix. Column 2 gives the time multiplied by 1000 for the positive definite matrix and column 4 gives the time multiplied by 1000 for the nonpositive definite matrix. These results show that the time taken to determine the solution for the system is somewhat faster for the positive definite system. This is because the operator \ checks to see if the matrix is positive definite and if so uses the more efficient Cholesky decomposition. Columns 3 and 5 give the times divide by the size of the matrix cubed to illustrate that the processing time is approximately proportional to n^3.

The next test we perform examines how the operator \ succeeds with the very badly conditioned Hilbert matrix. The test gives the time taken to solve the system and the accuracy of the solution given by the Euclidean norm of the residuals, that is, norm($\mathbf{Ax} - \mathbf{b}$). For the definition of the norm see Appendix A, Section A.10. In addition, the test compares these results for the \ operator with the results obtained using the inverse, that is, $\mathbf{x} = \mathbf{A}^{-1}\mathbf{b}$. The script for this test is

```
% e3s203.m
disp(' n  time-slash acc-slash   time-inv   acc-inv   condition');
for n = 4:2:20
    A = hilb(n); b = [1:n].';
    tic, x = A\b; t1 = toc; t1 = t1*10000;
    nm1 = norm(b-A*x);
    tic, x = inv(A)*b; t2 = toc; t2 = t2*10000;
    nm2 = norm(b-A*x);
    c = cond(A);
    fprintf('%2.0f %10.4f %10.2e %8.4f %11.2e %11.2e \n',n,t1,nm1,t2,nm2,c)
end
```

This produces the following table of results:

```
 n  time-slash acc-slash   time-inv   acc-inv    condition
 4     1.6427  1.39e-013   0.8549   9.85e-014   1.55e+004
 6     0.9415  5.22e-012   0.7710   2.02e-009   1.50e+007
 8     1.1454  5.35e-010   0.8465   3.19e-006   1.53e+010
10     1.2627  3.53e-008   1.5477   2.47e-004   1.60e+013
12     1.9332  1.40e-006   1.5589   9.39e-001   1.74e+016
14     2.1958  3.36e-005   1.5924   3.39e+002   5.13e+017
16     2.3187  5.76e-006   1.6650   1.02e+002   4.52e+017
18     2.4836  5.25e-005   2.0589   2.31e+002   1.57e+018
20     2.4417  1.11e-005   2.0869   3.72e+002   2.57e+018
```

This output has been edited to remove warnings about the ill-conditioning of the matrix for $n >= 10$. Column 1 gives the size of the matrix. Columns 2 and 3 give the time taken multiplied by 10,000 and accuracy when using the \ operator. Columns 4 and 5 give the same information when using the `inv` function. Column 6 gives the condition number of the system matrix. When the condition number is large, the matrix is nearly singular and the equations are ill-conditioned. This is fully described in Section 2.4.

The results in the preceding table demonstrate convincingly the superiority of the \ operator over the `inv` function for solving a system of linear equations. It is considerably more accurate than using matrix inversion. However, it should be noted that the accuracy falls off as the matrix becomes increasingly ill-conditioned.

The MATLAB operator \ can also be used to solve under- and overdetermined systems. In this case the \ operator uses a least squares approximation, discussed in detail in Section 2.12.

2.4 Accuracy of Solutions and Ill-Conditioning

We now consider factors that affect the accuracy of the solution of $\mathbf{Ax} = \mathbf{b}$ and how any inaccuracies can be detected. A further discussion on the accuracy of the solution of this equation system is given in Appendix B, Section B.3. We begin with the following examples.

Example 2.1

Consider the following MATLAB statements:

```
>> A = [3.021 2.714 6.913;1.031 -4.273 1.121;5.084 -5.832 9.155]

A =
    3.0210    2.7140    6.9130
    1.0310   -4.2730    1.1210
    5.0840   -5.8320    9.1550

>> b = [12.648 -2.121 8.407].'

b =
   12.6480
   -2.1210
    8.4070

>> A\b

ans =
    1.0000
    1.0000
    1.0000
```

This result is correct and easily verified by substitution into the original equations.

Example 2.2

Consider Example 2.1 with `A(2,2)` changed from −4.2730 to −4.2750:

```
>> A(2,2) = -4.2750

A =
      3.0210     2.7140     6.9130
      1.0310    -4.2750     1.1210
      5.0840    -5.8320     9.1550

>> A\b

ans =
    -1.7403
     0.6851
     2.3212
```

Here we have a solution that is very different from that of Example 2.1, even though the only change in the equation system is less than 0.1% in coefficient `A(2,2)`.

The two examples just shown have dramatically different solutions because the coefficient matrix **A** is ill-conditioned. Ill-conditioning can be interpreted graphically by representing each of the equation systems by three plane surfaces, in the manner shown earlier in Figure 2.2. In an ill-conditioned system at least two of the surfaces will be almost parallel so that the point of intersection of the surfaces will be very sensitive to small changes in slope, caused by small changes in coefficient values.

A system of equations is said to be ill-conditioned if a relatively small change in the elements of the coefficient matrix **A** causes a relatively large change in the solution. Conversely a system of equations is said to be well-conditioned if a relatively small change in the elements of the coefficient matrix **A** causes a relatively small change in the solution. Clearly we require a measure of the condition of a system of equations. We know that a system of equations without a solution—the very worst condition possible—has a coefficient matrix with a determinant of zero. It is therefore tempting to think that the size of the determinant of **A** can be used as a measure of condition. However, if $\mathbf{Ax} = \mathbf{b}$ and **A** is an $n \times n$ diagonal matrix with each element on the leading diagonal equal to s, then **A** is perfectly conditioned, regardless of the value of s. But the determinant of **A** in this case is s^n. Thus, the size of the determinant of **A** is not a suitable measure of condition because in this example it changes with s even though the condition of the system is constant.

Two of the functions MATLAB provides to estimate the condition of a matrix are `cond` and `rcond`. The function `cond` is a sophisticated function and is based on singular value decomposition, discussed in Section 2.10. For a perfect condition `cond` is unity but gives a large value for a matrix that is ill-conditioned. The function `rcond` is less reliable but usually faster. This function gives a value between zero and one. The smaller the value, the worse the conditioning. The reciprocal of `rcond` is usually of the same order of magnitude as `cond`. We now illustrate these points with two examples.

Example 2.3

Illustration of a perfectly conditioned system:

```
>> A = diag([20 20 20])
A =
    20     0     0
     0    20     0
     0     0    20

>> [det(A) rcond(A) cond(A)]

ans =
        8000            1            1
```

Example 2.4

Illustration of a badly conditioned system:

```
>> A = [1 2 3;4 5 6;7 8 9.000001];
>> format short e
>> [det(A) rcond(A) 1/rcond(A) cond(A)]

ans =
 -3.0000e-006  6.9444e-009  1.4400e+008  1.0109e+008
```

Note that the reciprocal of the `rcond` value is close to the value of `cond`. Using the MATLAB functions `cond` and `rcond` we now investigate the condition number of the Hilbert matrix (defined in Problem 2.1), using the script shown next:

```
% e3s204.m Hilbert matrix test.
disp('    n          cond               rcond       log10(cond)')
for n = 4:2:20
    A = hilb(n);
    fprintf('%5.0f %16.4e',n,cond(A));
    fprintf('%16.4e %10.2f\n',rcond(A),log10(cond(A)));
end
```

Running this script gives

n	cond	rcond	log10(cond)
4	1.5514e+004	3.5242e-005	4.19
6	1.4951e+007	3.4399e-008	7.17
8	1.5258e+010	2.9522e-011	10.18
10	1.6025e+013	2.8286e-014	13.20
12	1.7352e+016	2.6328e-017	16.24
14	5.1317e+017	1.7082e-019	17.71
16	4.5175e+017	4.6391e-019	17.65
18	1.5745e+018	5.8371e-020	18.20
20	2.5710e+018	1.9953e-019	18.41

This shows that the Hilbert matrix is ill-conditioned even for relatively small values of n, the size of the matrix. The last column of the preceding output gives the value of $\log_{10}$ of the condition number of the appropriate Hilbert matrix. This gives a rule of thumb estimate of the number of significant figures lost in solving an equation system with this matrix or inverting the matrix.

The Hilbert matrix of order n was generated in the preceding script using the MATLAB function `hilb(n)`. Other important matrices with interesting structures and properties, such as the Hadamard matrix and the Wilkinson matrix, can be obtained using, in these cases, the MATLAB functions `hadamard(n)` and `wilkinson(n)` where `n` is the required size of the matrix. In addition, many other interesting matrices can be accessed using the `gallery` function. In almost every case we can choose the size of the matrix and in many cases we can also choose other parameters within the matrix. Example calls are

```
gallery('hanowa',6,4)
gallery('cauchy',6)
gallery('forsythe',6,8)
```

The next section begins the detailed examination of one of the algorithms used by the \ operator.

2.5 Elementary Row Operations

We now examine the operations that can usefully be carried out on each equation of a system of equations. Such a system will have the form

$$a_{11}x_1 + a_{12}x_2 + \cdots + a_{1n}x_n = b_1$$

$$a_{21}x_2 + a_{22}x_2 + \cdots + a_{2n}x_n = b_2$$

$$\cdots\cdots\cdots\cdots\cdots$$

$$a_{n1}x_n + a_{n2}x_2 + \cdots + a_{nn}x_n = b_n$$

or in matrix notation

$$\mathbf{Ax} = \mathbf{b}$$

where

$$\mathbf{A} = \begin{bmatrix} a_{11} & a_{12}\dots & a_{1n} \\ a_{21} & a_{22}\dots & a_{2n} \\ \vdots & \vdots & \vdots \\ a_{n1} & a_{n2}\dots & a_{nn} \end{bmatrix} \quad \mathbf{b} = \begin{bmatrix} b_1 \\ b_2 \\ \vdots \\ b_n \end{bmatrix} \quad \mathbf{x} = \begin{bmatrix} x_1 \\ x_2 \\ \vdots \\ x_n \end{bmatrix}$$

A is called the coefficient matrix. Any operation performed on an equation must be applied to both its left and right sides. With this in mind it is helpful to combine the coefficient matrix **A** with the right side vector **b**:

$$\mathbf{A} = \begin{bmatrix} a_{11} & a_{12}\dots & a_{1n} & b_1 \\ a_{21} & a_{22}\dots & a_{2n} & b_2 \\ \vdots & \vdots & \vdots & \vdots \\ a_{n1} & a_{n2}\dots & a_{nn} & b_n \end{bmatrix}$$

This new matrix is called the augmented matrix and we will write it as [**A b**]. We have chosen to adopt this notation because it is consistent with MATLAB notation for combining **A** and **b**. Note that if **A** is an $n \times n$ matrix, then the augmented matrix is an $n \times (n+1)$ matrix. Each row of the augmented matrix holds all the coefficients of an equation and any operation must be applied to every element in the row. The three elementary row operations described in the following can be applied to individual equations in a system without altering the solution of the equation system. They are

1. Interchange the position of any two rows (i.e., equations).
2. Multiply a row (i.e., equation) by a nonzero scalar.
3. Replace a row by the sum of the row and a scalar multiple of another row.

These elementary row operations can be used to solve some important problems in linear algebra and we now discuss an application of them.

2.6 Solution of Ax = b by Gaussian Elimination

Gaussian elimination is an efficient way to solve equation systems, particularly those with a nonsymmetric coefficient matrix having a relatively small number of zero elements. The method depends entirely on using the three elementary row operations described in Section 2.5. Essentially the procedure is to form the augmented matrix for the system and then reduce the coefficient matrix part to an upper triangular form. To illustrate the

Table 2.1 Gaussian Elimination to Transform an Augmented Matrix to Upper Triangular Form

A1	[3]	6	9	3	**Stage 1**: Initial matrix
A2	2	(4+p)	2	4	
A3	−3	−4	−11	−5	
A1	3	6	9	3	**Stage 2**: Reduce col 1 of rows 2 and 3 to zero
B2=A2−2(A1)/3	0	p	−4	2	
B3=A3+3(A1)/3	0	2	−2	−2	
A1	3	6	9	3	**Stage 3**: Interchange rows 2 and 3
B3	0	[2]	−2	−2	
B2	0	p	−4	2	
A1	3	6	9	3	**Stage 4**: Reduce col 2 of row 3 to zero
B3	0	2	−2	−2	
C3=B2−p(B3)/2	0	0	[(−4+p)]	(2+p)	

systematic use of the elementary row operations we consider the application of Gaussian elimination to solve the following equation system:

$$\begin{bmatrix} 3 & 6 & 9 \\ 2 & (4+p) & 2 \\ -3 & -4 & -11 \end{bmatrix} \begin{bmatrix} x_1 \\ x_2 \\ x_3 \end{bmatrix} = \begin{bmatrix} 3 \\ 4 \\ -5 \end{bmatrix} \tag{2.10}$$

where the value of p is known. Table 2.1 shows the sequence of operations, beginning at Stage 1 with the augmented matrix. In Stage 1 the element in the first column of the first row (enclosed in a box in the table) is designated the pivot. We wish to make the elements of column 1 in rows 2 and 3 zero. To achieve this, we divide row 1 by the pivot and then add or subtract a suitable multiple of the modified row 1 to or from rows 2 and 3. The result of this is shown in Stage 2 of the table. We then select the next pivot. This is the element in the second column of the new second row, which in Stage 2 is equal to p. If p is large, this does not present a problem, but if p is small, then numerical problems may arise because we will be dividing all the elements of the new row 2 by this small quantity p. If p is zero, then we have an impossible situation because we cannot divide by zero.

This difficulty is not related to ill-conditioning; indeed this particular equation system is quite well conditioned when p is zero. To circumvent these problems the usual procedure is to interchange the row in question with the row containing the element of largest modulus in the column *below the pivot*. In this way we provide a new and larger pivot. This procedure is called partial pivoting. If we assume in this case that $p < 2$, then we interchange rows 2 and 3 as shown in Stage 3 of the table to replace p by 2 as the pivot. From row 3 we now subtract row 2 divided by the pivot and multiply by a coefficient in order to make the element of column 2, row 3, zero. Thus in Stage 4 of the table the original coefficient matrix has been reduced to an upper triangular matrix. If, for example, $p = 0$, we obtain

$$3x_1 + 6x_2 + 9x_3 = 3 \tag{2.11}$$

$$2x_2 - 2x_3 = -2 \tag{2.12}$$

$$-4x_3 = 2 \tag{2.13}$$

We can now obtain the values of the unknowns x_1, x_2, and x_3 by a process called back substitution. We solve the equations in reverse order. Thus from (2.13), $x_3 = -0.5$. From (2.12), knowing x_3, we have $x_2 = -1.5$. Finally from (2.11), knowing x_2 and x_3, we have $x_1 = 5.5$.

It can be shown that the determinant of a matrix can be evaluated from the product of the elements on the main diagonal provided at Stage 3 in Table 2.1. This product must be multiplied by $(-1)^m$ where m is the number of row interchanges used. For example, in the preceding problem, with $p = 0$, one row interchange is used so that $m = 1$ and the determinant of the coefficient matrix is given by $3 \times 2 \times (-4) \times (-1)^1 = 24$.

A method for solving a linear equation system that is closely related to Gaussian elimination is Gauss–Jordan elimination. The method uses the same elementary row operations but differs from Gaussian elimination because elements both below and above the leading diagonal are reduced to zero. This means that back substitution is avoided. For example, solving system (2.10) with $p = 0$ leads to the following augmented matrix:

$$\begin{bmatrix} 3 & 0 & 0 & 16.5 \\ 0 & 2 & 0 & -3.0 \\ 0 & 0 & -4 & 2.0 \end{bmatrix}$$

Thus $x_1 = 16.5/3 = 5.5$, $x_2 = -3/2 = -1.5$, and $x_3 = 2/-4 = -0.5$.

Gaussian elimination requires order $n^3/3$ multiplications followed by back substitution requiring order n^2 multiplications. Gauss–Jordan elimination requires order $n^3/2$ multiplications. Thus for large systems of equations (say $n > 10$), Gauss–Jordan elimination requires approximately 50% more operations than Gaussian elimination.

2.7 LU Decomposition

LU decomposition (or factorization) is a similar process to Gaussian elimination and is equivalent in terms of elementary row operations. The matrix **A** can be decomposed so that

$$\mathbf{A} = \mathbf{LU} \tag{2.14}$$

where **L** is a lower triangular matrix with a leading diagonal of ones and **U** is an upper triangular matrix. Matrix **A** may be real or complex. Compared with Gaussian elimination, LU decomposition has a particular advantage when the equation system we wish to solve, $\mathbf{Ax}=\mathbf{b}$, has more than one right side or when the right sides are not known in advance. This is because the factors **L** and **U** are obtained explicitly and they can be used for any right sides as they arise without recalculating **L** and **U**. Gaussian elimination does not determine **L** explicitly but rather forms $\mathbf{L}^{-1}\,\mathbf{b}$ so that all right sides must be known when the equation is solved.

The major steps required to solve an equation system by LU decomposition are as follows. Since $\mathbf{A}=\mathbf{LU}$, then $\mathbf{Ax}=\mathbf{B}$ becomes

$$\mathbf{LUx}=\mathbf{b}$$

where **b** is not restricted to a single column. Letting $\mathbf{y}=\mathbf{Ux}$ leads to

$$\mathbf{Ly}=\mathbf{b}$$

Because **L** is a lower triangular matrix this equation is solved efficiently by forward substitution. To find **x** we now solve

$$\mathbf{Ux}=\mathbf{y}$$

Because **U** is an upper triangular matrix, this equation can also be solved efficiently by back substitution.

We now illustrate the LU decomposition process by solving (2.10) with $p=1$. We are not concerned with **b** and we do not form an augmented matrix. We proceed exactly as with Gaussian elimination (see Table 2.1), except that we keep a record of the elementary row operations performed at the ith stage in $\mathbf{T}^{(i)}$ and place the results of these operations in a matrix $\mathbf{U}^{(i)}$ rather than overwriting **A**.

We begin with the matrix

$$\mathbf{A}=\begin{bmatrix} 3 & 6 & 9 \\ 2 & 5 & 2 \\ -3 & -4 & -11 \end{bmatrix}$$

Following the same operations as used in Table 2.1, we will create a matrix $\mathbf{U}^{(1)}$ with zeros below the leading diagonal in the first column using the following elementary row operations:

$$\text{row 2 of }\mathbf{U}^{(1)} = \text{row 2 of }\mathbf{A} - 2(\text{row 1 of}\mathbf{A})/3 \tag{2.15}$$

and

$$\text{row 3 of }\mathbf{U}^{(1)} = \text{row 3 of }\mathbf{A} + 3(\text{row 1 of}\mathbf{A})/3 \tag{2.16}$$

Now **A** can be expressed as the product $\mathbf{T}^{(1)}\ \mathbf{U}^{(1)}$ as follows:

$$\begin{bmatrix} 3 & 6 & 9 \\ 2 & 5 & 2 \\ -3 & -4 & -11 \end{bmatrix} = \begin{bmatrix} 1 & 0 & 0 \\ 2/3 & 1 & 0 \\ -1 & 0 & 1 \end{bmatrix} \begin{bmatrix} 3 & 6 & 9 \\ 0 & 1 & -4 \\ 0 & 2 & -2 \end{bmatrix}$$

Note that row 1 of **A** and row 1 of $\mathbf{U}^{(1)}$ are identical. Thus row 1 of $\mathbf{T}^{(1)}$ has a unit entry in column 1 and zero elsewhere. The remaining rows of $\mathbf{T}^{(1)}$ are determined from (2.15) and (2.16). For example, row 2 of $\mathbf{T}^{(1)}$ is derived by rearranging (2.15); thus

$$\text{row 2 of } \mathbf{A} = \text{ row 2 of } \mathbf{U}^{(1)} + 2(\text{row 1 of} \mathbf{A})/3 \tag{2.17}$$

or

$$\text{row 2 of } \mathbf{A} = 2(\text{row 1 of } \mathbf{U}^{(1)})/3 + \text{row 2 of } \mathbf{U}^{(1)} \tag{2.18}$$

since row 1 of $\mathbf{U}^{(1)}$ is identical to row 1 of **A**. Hence row 2 of $\mathbf{T}^{(1)}$ is [2/3 1 0].

We now move to the next stage of the decomposition process. In order to bring the largest element of column 2 in $\mathbf{U}^{(1)}$ onto the leading diagonal we must interchange rows 2 and 3. Thus $\mathbf{U}^{(1)}$ becomes the product $\mathbf{T}^{(2)}\ \mathbf{U}^{(2)}$ as follows:

$$\begin{bmatrix} 3 & 6 & 9 \\ 0 & 1 & -4 \\ 0 & 2 & -2 \end{bmatrix} = \begin{bmatrix} 1 & 0 & 0 \\ 0 & 0 & 1 \\ 0 & 1 & 0 \end{bmatrix} \begin{bmatrix} 3 & 6 & 9 \\ 0 & 2 & -2 \\ 0 & 1 & -4 \end{bmatrix}$$

Finally, to complete the process of obtaining an upper triangular matrix we make

$$\text{row 3 of } \mathbf{U} = \text{ row 3 of } \mathbf{U}^{(2)} - (\text{row 2 of } \mathbf{U}^{(2)})/2$$

Hence $\mathbf{U}^{(2)}$ becomes the product $\mathbf{T}^{(3)}\ \mathbf{U}$ as follows:

$$\begin{bmatrix} 3 & 6 & 9 \\ 0 & 2 & -2 \\ 0 & 1 & -4 \end{bmatrix} = \begin{bmatrix} 1 & 0 & 0 \\ 0 & 1 & 0 \\ 0 & 1/2 & 1 \end{bmatrix} \begin{bmatrix} 3 & 6 & 9 \\ 0 & 2 & -2 \\ 0 & 0 & -3 \end{bmatrix}$$

Thus $\mathbf{A} = \mathbf{T}^{(1)}\ \mathbf{T}^{(2)}\ \mathbf{T}^{(3)}\ \mathbf{U}$, implying that $\mathbf{L} = \mathbf{T}^{(1)}\ \mathbf{T}^{(2)}\ \mathbf{T}^{(3)}$ as follows:

$$\begin{bmatrix} 1 & 0 & 0 \\ 2/3 & 1 & 0 \\ -1 & 0 & 1 \end{bmatrix} \begin{bmatrix} 1 & 0 & 0 \\ 0 & 0 & 1 \\ 0 & 1 & 0 \end{bmatrix} \begin{bmatrix} 1 & 0 & 0 \\ 0 & 1 & 0 \\ 0 & 1/2 & 1 \end{bmatrix} = \begin{bmatrix} 1 & 0 & 0 \\ 2/3 & 1/2 & 1 \\ -1 & 1 & 0 \end{bmatrix}$$

Note that owing to the row interchanges **L** is not strictly a lower triangular matrix but it can be made so by interchanging rows.

MATLAB implements LU factorization by using the function `lu` and may produce a matrix that is not strictly a lower triangular matrix. However, a permutation matrix **P** may be produced, if required, such that **LU** = **PA** with **L** lower triangular.

We now show how the MATLAB function `lu` deals with the preceding example:

```
>> A = [3 6 9;2 5 2;-3 -4 -11]

A =
     3     6     9
     2     5     2
    -3    -4   -11
```

To obtain the **L** and **U** matrices, we must use that MATLAB facility of assigning two parameters simultaneously as follows:

```
>> [L1 U] = lu(A)

L1 =
    1.0000          0          0
    0.6667     0.5000     1.0000
   -1.0000     1.0000          0

U =
     3     6     9
     0     2    -2
     0     0    -3
```

Note that the `L1` matrix is not in lower triangular form, although its true form can easily be deduced by interchanging rows 2 and 3 to form a triangle. To obtain a true lower triangular matrix we must assign three parameters as follows:

```
>> [L U P] = lu(A)

L =
    1.0000          0          0
   -1.0000     1.0000          0
    0.6667     0.5000     1.0000

U =
     3     6     9
     0     2    -2
     0     0    -3

P =
     1     0     0
     0     0     1
     0     1     0
```

In the preceding output `P` is the permutation matrix such that `L*U = P*A` or `P'*L*U = A`. Thus `P'*L` is equal to `L1`.

The MATLAB operator \ determines the solution of $\mathbf{Ax} = \mathbf{b}$ using LU factorization. As an example of an equation system with multiple right sides we solve $\mathbf{AX} = \mathbf{B}$ where

$$\mathbf{A} = \begin{bmatrix} 3 & 4 & -5 \\ 6 & -3 & 4 \\ 8 & 9 & -2 \end{bmatrix} \quad \text{and} \quad \mathbf{B} = \begin{bmatrix} 1 & 3 \\ 9 & 5 \\ 9 & 4 \end{bmatrix}$$

Performing LU decomposition such that $\mathbf{LU} = \mathbf{A}$ gives

$$\mathbf{L} = \begin{bmatrix} 0.375 & -0.064 & 1 \\ 0.750 & 1 & 0 \\ 1 & 0 & 0 \end{bmatrix} \quad \text{and} \quad \mathbf{U} = \begin{bmatrix} 8 & 9 & -2 \\ 0 & -9.75 & 5.5 \\ 0 & 0 & -3.897 \end{bmatrix}$$

Thus $\mathbf{LY} = \mathbf{B}$ is given by

$$\begin{bmatrix} 0.375 & -0.064 & 1 \\ 0.750 & 1 & 0 \\ 1 & 0 & 0 \end{bmatrix} \begin{bmatrix} y_{11} & y_{12} \\ y_{21} & y_{22} \\ y_{31} & y_{32} \end{bmatrix} = \begin{bmatrix} 1 & 3 \\ 9 & 5 \\ 9 & 4 \end{bmatrix}$$

We note that implicitly we have two systems of equations, which when separated can be written

$$\mathbf{L} \begin{bmatrix} y_{11} \\ y_{21} \\ y_{31} \end{bmatrix} = \begin{bmatrix} 1 \\ 9 \\ 9 \end{bmatrix} \quad \text{and} \quad \mathbf{L} \begin{bmatrix} y_{12} \\ y_{22} \\ y_{32} \end{bmatrix} = \begin{bmatrix} 3 \\ 5 \\ 4 \end{bmatrix}$$

In this example $\mathbf{L}$ is not strictly a lower triangular matrix owing to the reordering of the rows. However, the solution of this equation is still found by forward substitution. For example, $1y_{11} = b_{31} = 9$, so that $y_{11} = 9$. Then $0.75y_{11} + 1y_{21} = b_{21} = 9$. Hence $y_{21} = 2.25$, and so on. The complete $\mathbf{Y}$ matrix is

$$\mathbf{Y} = \begin{bmatrix} 9.000 & 4.000 \\ 2.250 & 2.000 \\ -2.231 & 1.628 \end{bmatrix}$$

Finally, solving $\mathbf{UX} = \mathbf{Y}$ by back substitution gives

$$\mathbf{X} = \begin{bmatrix} 1.165 & 0.891 \\ 0.092 & -0.441 \\ 0.572 & -0.418 \end{bmatrix}$$

The MATLAB function `det` determines the determinant of a matrix using LU factorization as follows. Since $\mathbf{A} = \mathbf{LU}$ then $|\mathbf{A}| = |\mathbf{L}|\,|\mathbf{U}|$. The elements of the leading diagonal of $\mathbf{L}$ are all ones so that $|\mathbf{L}| = 1$. Since **U** is upper triangular, its determinant is the product of the elements of its leading diagonal. Thus, taking account of row interchanges, the appropriately signed product of the diagonal elements of **U** gives the determinant.

2.8 Cholesky Decomposition

Cholesky decomposition or factorization is a form of triangular decomposition that can only be applied to positive definite symmetric or positive definite Hermitian matrices. A symmetric or Hermitian matrix **A** is said to be positive definite if $\mathbf{x}^\top \mathbf{A}\mathbf{x} > 0$ for any nonzero **x**. A more useful definition of a positive definite matrix is one that has all eigenvalues greater than zero. The eigenvalue problem is discussed in Section 2.15. If **A** is symmetric or Hermitian, we can write

$$\mathbf{A} = \mathbf{P}^\top \mathbf{P} \ \ (\text{or } \mathbf{A} = \mathbf{P}^\mathrm{H}\mathbf{P} \text{ when } \mathbf{A} \text{ is Hermitian}) \tag{2.19}$$

where **P** is an upper triangular matrix. The algorithm computes **P** row by row by equating coefficients of each side of (2.19). Thus $p_{11}, p_{12}, p_{13}, \ldots, p_{22}, p_{23}, \ldots$ are determined in sequence, ending with p_{nn}. Coefficients on the leading diagonal of **P** are computed from expressions that involve determining a square root. For example,

$$p_{22} = \sqrt{a_{22} - p_{12}^2}$$

A property of positive definite matrices is that the term under the square root is always positive and so the square root will be real. Furthermore, row interchanges are not required because the dominant coefficients will always be on the main diagonal. The whole process requires only about half as many multiplications as LU decomposition. Cholesky factorization is implemented for positive definite symmetric matrices in MATLAB by the function `chol`. For example, consider the Cholesky factorization of the following positive definite Hermitian matrix:

```
>> A = [2 -i 0;i 2 0;0 0 3]

A =
   2.0000                  0 - 1.0000i        0
        0 + 1.0000i   2.0000                  0
        0                  0             3.0000

>> P = chol(A)

P =
   1.4142                  0 - 0.7071i        0
        0             1.2247                  0
        0                  0             1.7321
```

When the operator \ detects a symmetric positive definite or Hermitian positive definite system matrix, it solves $\mathbf{Ax} = \mathbf{b}$ using the following sequence of operations. $\mathbf{A}$ is factorized into $\mathbf{P}^\top\mathbf{P}$, and $\mathbf{y}$ is set to $\mathbf{Px}$; then $\mathbf{P}^\top\mathbf{y} = \mathbf{b}$. The algorithm solves for $\mathbf{y}$ by forward substitution since $\mathbf{P}^\top$ is a lower triangular matrix. Then $\mathbf{x}$ can be determined from $\mathbf{y}$ by backward substitution since $\mathbf{P}$ is an upper triangular matrix. We can illustrate the steps in this process by the following example:

$$\mathbf{A} = \begin{bmatrix} 2 & 3 & 4 \\ 3 & 6 & 7 \\ 4 & 7 & 10 \end{bmatrix} \text{ and } \mathbf{b} = \begin{bmatrix} 2 \\ 4 \\ 8 \end{bmatrix}$$

Then by Cholesky factorization

$$\mathbf{P} = \begin{bmatrix} 1.414 & 2.121 & 2.828 \\ 0 & 1.225 & 0.817 \\ 0 & 0 & 1.155 \end{bmatrix}$$

Now since $\mathbf{P}^\top\mathbf{y} = \mathbf{b}$, solving for $\mathbf{y}$ by forward substitution gives

$$\mathbf{y} = \begin{bmatrix} 1.414 \\ 0.817 \\ 2.887 \end{bmatrix}$$

Finally, solving $\mathbf{Px} = \mathbf{y}$ by back substitution gives

$$\mathbf{x} = \begin{bmatrix} -2.5 \\ -1.0 \\ 2.5 \end{bmatrix}$$

We now compare the performance of the operator \ with the function `chol`. Clearly their performance should be similar in the case of a positive definite matrix. To generate a symmetric positive define matrix in the following script, we multiply a matrix by its transpose:

```
% e3s205.m
disp('  n        time-backslash  time-chol');
for n = 300:100:1300
    A = [ ]; M = 100*randn(n,n);
    A = M*M'; b = [1:n].';
    tic, x = A\b; t1 = toc;
    tic, R = chol(A);
    v = R.'\b; x = R\b;
    t2 = toc;
    fprintf('%4.0f %14.4f %13.4f \n',n,t1,t2)
end
```

Running this script gives

n	time-backslash	time-chol
300	0.0053	0.0073
400	0.0105	0.0115
500	0.0182	0.0216
600	0.0176	0.0197
700	0.0263	0.0281
800	0.0368	0.0385
900	0.0510	0.0519
1000	0.0666	0.0668
1100	0.0862	0.0869
1200	0.1113	0.1065
1300	0.1449	0.1438

The similarity in performance of the function `chol` and the operator \ is borne out by the preceding table. In this table, column 1 is the size of the matrix and column 2 gives the time taken using the \ operator. Column 3 gives the time taken using Cholesky decomposition to solve the same problem.

Cholesky factorization *can* be applied to a symmetric matrix that is not positive definite but the process does not possess the numerical stability of the positive definite case. Furthermore, one or more rows in **P** may be purely imaginary. For example,

$$\text{If } \mathbf{A} = \begin{bmatrix} 1 & 2 & 3 \\ 2 & -5 & 9 \\ 3 & 9 & 4 \end{bmatrix} \text{ then } \mathbf{P} = \begin{bmatrix} 1 & 2 & 3 \\ 0 & 3\imath & -\imath \\ 0 & 0 & 2\imath \end{bmatrix}$$

This is not implemented in MATLAB.

2.9 QR Decomposition

We have seen how a square matrix can be decomposed or factorized into the product of a lower and an upper triangular matrix by the use of elementary row operations. An alternative decomposition is into an upper triangular matrix and an orthogonal matrix if **A** is real or a unitary matrix if **A** is complex. This is called QR decomposition. Thus

$$\mathbf{A} = \mathbf{Q}\,\mathbf{R}$$

where **R** is the upper triangular matrix and **Q** is the orthogonal, or the unitary matrix. If **Q** is orthogonal, $\mathbf{Q}^{-1} = \mathbf{Q}^{\top}$ and if **Q** is unitary, $\mathbf{Q}^{-1} = \mathbf{Q}^{\mathrm{H}}$. The preceding are very useful properties.

There are several procedures that provide QR decomposition; here we present Householder's method. To decompose a real matrix, Householder's method begins by defining a matrix **P**:

$$\mathbf{P} = \mathbf{I} - 2\mathbf{w}\mathbf{w}^\top \tag{2.20}$$

P is symmetrical and providing $\mathbf{w}^\top\mathbf{w} = 1$, **P** is also orthogonal. The orthogonality can easily be verified by expanding the product $\mathbf{P}^\top\mathbf{P} = \mathbf{P}\mathbf{P}$ as follows:

$$\begin{aligned}\mathbf{P}\mathbf{P} &= \left(\mathbf{I} - 2\mathbf{w}\mathbf{w}^\top\right)\left(\mathbf{I} - 2\mathbf{w}\mathbf{w}^\top\right) \\ &= \mathbf{I} - 4\mathbf{w}\mathbf{w}^\top + 4\mathbf{w}\mathbf{w}^\top\left(\mathbf{w}\mathbf{w}^\top\right) = \mathbf{I}\end{aligned}$$

To decompose **A** into **QR**, we begin by forming the vector $\mathbf{w}_1$ from the coefficients of the first column of **A** as follows:

$$\mathbf{w}_1^\top = \mu_1\left[(a_{11} - s_1)\ a_{21}\ a_{31} \ldots a_{n1}\right]$$

where

$$\mu_1 = \frac{1}{\sqrt{2s_1(s_1 - a_{11})}} \text{ and } s_1 = \pm\left(\sum_{j=1}^{n} a_{j1}^2\right)^{1/2}$$

By substituting for μ_1 and s_1 in $\mathbf{w}_1$ it can be verified that the necessary orthogonality condition, $\mathbf{w}_1^\top\mathbf{w}_1 = 1$, is satisfied. Substituting $\mathbf{w}_1$ into (2.20) we generate an orthogonal matrix $\mathbf{P}^{(1)}$.

The matrix $\mathbf{A}^{(1)}$ is now created from the product $\mathbf{P}^{(1)}\mathbf{A}$. It can easily be verified that all elements in the first column of $\mathbf{A}^{(1)}$ are zero except for the element on the leading diagonal, which is equal to s_1. Thus

$$\mathbf{A}^{(1)} = \mathbf{P}^{(1)}\mathbf{A} = \begin{bmatrix} s_1 & + & \cdots & + \\ 0 & + & \cdots & + \\ \vdots & \vdots & & \vdots \\ 0 & + & \cdots & + \\ 0 & + & \cdots & + \end{bmatrix}$$

In the matrix $\mathbf{A}^{(1)}$, $+$ indicates a nonzero element.

We now begin the second stage of the orthogonalization process by forming $\mathbf{w}_2$ from the coefficients of the second column of $\mathbf{A}^{(1)}$:

$$\mathbf{w}_2^\top = \mu_2\left[0\ \left(a_{22}^{(1)} - s_2\right)\ a_{32}^{(1)}\ a_{42}^{(1)} \cdots a_{n2}^{(1)}\right]$$

where a_{ij} are the coefficients of **A** and

$$\mu_2 = \frac{1}{\sqrt{2s_2(s_2 - a_{22}^{(1)})}} \text{ and } s_2 = \pm\left(\sum_{j=2}^{n} (a_{j2}^{(1)})^2\right)^{1/2}$$

Then the orthogonal matrix $\mathbf{P}^{(2)}$ is generated from

$$\mathbf{P}^{(2)} = \mathbf{I} - 2\mathbf{w}_2\mathbf{w}_2^\top$$

The matrix $\mathbf{A}^{(2)}$ is then created from the product $\mathbf{P}^{(2)}\mathbf{A}^{(1)}$ as follows:

$$\mathbf{A}^{(2)} = \mathbf{P}^{(2)}\mathbf{A}^{(1)} = \mathbf{P}^{(2)}\mathbf{P}^{(1)}\mathbf{A} = \begin{bmatrix} s_1 & + & \cdots & + \\ 0 & s_2 & \cdots & + \\ \vdots & \vdots & & \vdots \\ 0 & 0 & \cdots & + \\ 0 & 0 & \cdots & + \end{bmatrix}$$

Note that $\mathbf{A}^{(2)}$ has zero elements in its first two columns except for the elements on and above the leading diagonal. We can continue this process $n-1$ times until we obtain an upper triangular matrix $\mathbf{R}$. Thus

$$\mathbf{R} = \mathbf{P}^{(n-1)} \ldots \mathbf{P}^{(2)}\mathbf{P}^{(1)}\mathbf{A} \tag{2.21}$$

Note that since $\mathbf{P}^{(i)}$ is orthogonal, the product $\mathbf{P}^{(n-1)} \ldots \mathbf{P}^{(2)}\mathbf{P}^{(1)}$ is also orthogonal.

We wish to determine the orthogonal matrix $\mathbf{Q}$ such that $\mathbf{A} = \mathbf{QR}$. Thus $\mathbf{R} = \mathbf{Q}^{-1}\mathbf{A}$ or $\mathbf{R} = \mathbf{Q}^\top\mathbf{A}$. Hence, from (2.21),

$$\mathbf{Q}^\top = \mathbf{P}^{(n-1)} \ldots \mathbf{P}^{(2)}\mathbf{P}^{(1)}$$

Apart from the signs associated with the columns of $\mathbf{Q}$ and the rows of $\mathbf{R}$, the decomposition is unique. These signs are dependent on whether the positive or negative square root is taken in determining s_1, s_2, and so on. Complete decomposition of the matrix requires $2n^3/3$ multiplications and n square roots. To illustrate this procedure consider the decomposition of the matrix

$$\mathbf{A} = \begin{bmatrix} 4 & -2 & 7 \\ 6 & 2 & -3 \\ 3 & 4 & 4 \end{bmatrix}$$

Thus

$$s_1 = \surd\left(4^2 + 6^2 + 3^2\right) = 7.8102$$

$$\mu_1 = 1/\surd[2 \times 7.8102 \times (7.8102 - 4)] = 0.1296$$

$$\mathbf{w}_1^\top = 0.1296[(4 - 7.8102)\ 6\ 3] = [-0.4939\ 0.7777\ 0.3889]$$

Using (2.20) we generate $\mathbf{P}^{(1)}$ and hence $\mathbf{A}^{(1)}$ as follows:

$$\mathbf{P}^{(1)} = \begin{bmatrix} 0.5121 & 0.7682 & 0.3841 \\ 0.7682 & -0.2097 & -0.6049 \\ 0.3841 & -0.6049 & 0.6976 \end{bmatrix}$$

$$\mathbf{A}^{(1)} = \mathbf{P}^{(1)}\mathbf{A} = \begin{bmatrix} 7.8102 & 2.0486 & 2.8168 \\ 0 & -4.3753 & 3.5873 \\ 0 & 0.8123 & 7.2936 \end{bmatrix}$$

Note that we have reduced the elements of the first column of $\mathbf{A}^{(1)}$ below the leading diagonal to zero. We continue with the second stage:

$$s_2 = \surd\left\{(-4.3753)^2 + 0.8123^2\right\} = 4.4501$$

$$\mu_2 = 1/\surd\{2 \times 4.4501 \times (4.4501 + 4.3753)\} = 0.1128$$

$$\mathbf{w}_2^{\top} = 0.1128\,[0 \;\; (-4.3753 - 4.4501) \;\; 0.8123] = [0 \;\; -0.9958 \;\; 0.0917]$$

$$\mathbf{P}^{(2)} = \begin{bmatrix} 1 & 0 & 0 \\ 0 & -0.9832 & 0.1825 \\ 0 & 0.1825 & 0.9832 \end{bmatrix}$$

$$\mathbf{R} = \mathbf{A}^{(2)} = \mathbf{P}^{(2)}\mathbf{A}^{(1)} = \begin{bmatrix} 7.8102 & 2.0486 & 2.8168 \\ 0 & 4.4501 & -2.1956 \\ 0 & 0 & 7.8259 \end{bmatrix}$$

Note that we have now reduced the first two columns of $\mathbf{A}^{(2)}$ below the leading diagonal to zero. This completes the process to determine the upper triangular matrix $\mathbf{R}$. Finally we determine the orthogonal matrix $\mathbf{Q}$ as follows:

$$\mathbf{Q} = \left(\mathbf{P}^{(2)}\mathbf{P}^{(1)}\right)^{\top} = \begin{bmatrix} 0.5121 & -0.6852 & 0.5179 \\ 0.7682 & 0.0958 & -0.6330 \\ 0.3841 & 0.7220 & 0.5754 \end{bmatrix}$$

It is not necessary for the reader to carry out the preceding calculations since MATLAB provides the function `qr` to carry out this decomposition. For example,

```
>> A = [4 -2 7;6 2 -3;3 4 4]

A =

     4    -2     7
     6     2    -3
     3     4     4
```

```
>> [Q R] = qr(A)

Q =
   -0.5121    0.6852    0.5179
   -0.7682   -0.0958   -0.6330
   -0.3841   -0.7220    0.5754

R =
   -7.8102   -2.0486   -2.8168
         0   -4.4501    2.1956
         0         0    7.8259
```

One advantage of QR decomposition is that it can be applied to nonsquare matrices, decomposing an $m \times n$ matrix into an $m \times m$ orthogonal matrix and an $m \times n$ upper triangular matrix. Note that if $m > n$, the decomposition is not unique.

2.10 Singular Value Decomposition

The singular value decomposition (SVD) of an $m \times n$ matrix **A** is given by

$$\mathbf{A} = \mathbf{USV}^{\top} \text{ (or } \mathbf{A} = \mathbf{USV}^{\mathrm{H}} \text{ if } \mathbf{A} \text{ is complex)}$$

where **U** is an orthogonal $m \times m$ matrix and **V** is an orthogonal $n \times n$ matrix. If **A** is complex then **U** and **V** are unitary matrices. In all cases **S** is a real diagonal $m \times n$ matrix. The elements of the leading diagonal of this matrix are called the singular values of **A**. Normally they are arranged in decreasing value so that $s_1 > s_2 > \cdots > s_n$. Thus

$$\mathbf{S} = \begin{bmatrix} s_1 & 0 & \ldots & 0 \\ 0 & s_2 & \ldots & 0 \\ \vdots & \vdots & & \vdots \\ 0 & 0 & \ldots & s_n \\ 0 & 0 & \ldots & 0 \\ \vdots & \vdots & & \vdots \\ 0 & 0 & \ldots & 0 \end{bmatrix}$$

The singular values are the nonnegative square roots of the eigenvalues of $\mathbf{A}^{\top}\mathbf{A}$. Because $\mathbf{A}^{\top}\mathbf{A}$ is symmetric or Hermitian these eigenvalues are real and nonnegative so that the singular values are also real and nonnegative. Algorithms for computing the SVD of a matrix are given by Golub and Van Loan (1989).

The SVD of a matrix has several important applications. In Section 2.2 we introduced the reduced row echelon form of a matrix and explained how the MATLAB function `rref` gives information from which the rank of a matrix can be deduced. However, rank can be more effectively determined from the SVD of a matrix since its rank is equal to its number of nonzero singular values. Thus for a 5×5 matrix of rank 3, s_4, and s_5 are zero. In practice, rather than counting the nonzero singular values, MATLAB determines rank from the SVD by counting the number of singular values greater than some tolerance value. This is a more realistic approach to determining rank than counting any nonzero value, however small.

To illustrate how singular value decomposition helps us to examine the properties of a matrix we will use the MATLAB function `svd` to carry out a singular value decomposition and compare it with the function `rref`. Consider the following example in which a Vandermonde matrix is created using the MATLAB function `vander`. The Vandermonde matrix is known to be ill-conditioned. SVD allows us to examine the nature of this ill-conditioning. In particular, a zero or a very small singular value indicates rank deficiency and this example shows that the singular values are becoming relatively close to this condition. In addition SVD allows us to compute the condition number of the matrix. In fact, the MATLAB function `cond` uses SVD to compute the condition number and this gives the same values as obtained by dividing the largest singular value by the smallest singular value. Additionally, the Euclidean norm of the matrix is supplied by the first singular value. Comparing the SVD with the RREF process in the following script, we see that using the MATLAB functions `rref` and `rank` give the rank of this special Vandermonde matrix as 5 but tell us nothing else. There is no warning that the matrix is badly conditioned.

```
>> c = [1 1.01 1.02 1.03 1.04];
>> V = vander(c)

V =

    1.0000    1.0000    1.0000    1.0000    1.0000
    1.0406    1.0303    1.0201    1.0100    1.0000
    1.0824    1.0612    1.0404    1.0200    1.0000
    1.1255    1.0927    1.0609    1.0300    1.0000
    1.1699    1.1249    1.0816    1.0400    1.0000

>> format long
>> s = svd(V)

s =

   5.210367051037899
   0.101918335876689
   0.000699698839445
   0.000002352380295
   0.000000003294983
```

```
>> norm(V)

ans =
   5.210367051037899

>> cond(V)

ans =
     1.581303246763933e+009

>> s(1)/s(5)

ans =
     1.581303246763933e+009

>> rank(V)

ans =
       5

>> rref(V)

ans =
       1       0       0       0       0
       0       1       0       0       0
       0       0       1       0       0
       0       0       0       1       0
       0       0       0       0       1
```

The following example is very similar to the preceding one but the Vandermonde matrix has now been generated to be rank deficient. The smallest singular value, although not zero, is zero to machine precision and `rank` returns the value of 4.

```
>> c = [1 1.01 1.02 1.03 1.03];
>> V = vander(c)

V =
    1.0000    1.0000    1.0000    1.0000    1.0000
    1.0406    1.0303    1.0201    1.0100    1.0000
    1.0824    1.0612    1.0404    1.0200    1.0000
    1.1255    1.0927    1.0609    1.0300    1.0000
    1.1255    1.0927    1.0609    1.0300    1.0000
```

```
>> format long e
>> s = svd(V)

s =
     5.187797954424026e+000
     8.336322098941414e-002
     3.997349250042135e-004
     8.462129966456217e-007
                          0

>> format short
>> rank(V)

ans =
     4

>> rref(V)

ans =
    1.0000         0         0         0   -0.9424
         0    1.0000         0         0    3.8262
         0         0    1.0000         0   -5.8251
         0         0         0    1.0000    3.9414
         0         0         0         0         0

>> cond(V)

ans =
   Inf
```

The `rank` function does allow the user to vary the tolerance. However, tolerance should be used with care since the `rank` function counts the number of singular values greater than tolerance and this gives the rank of the matrix. If tolerance is very small (i.e., smaller than the machine precision), the rank may be miscounted.

2.11 The Pseudo-Inverse

Here we discuss the pseudo-inverse and in Section 2.12 apply it to solve over- and under-determined systems.

If $\mathbf{A}$ is an $m \times n$ rectangular matrix, then the system

$$\mathbf{Ax} = \mathbf{b} \tag{2.22}$$

cannot be solved by inverting **A**, since **A** is not a square matrix. Assuming an equation system with more equations than variables (i.e., $m > n$), then by premultiplying (2.22) by $\mathbf{A}^\top$ we can convert the system matrix to a square matrix as follows:

$$\mathbf{A}^\top\mathbf{A}\mathbf{x} = \mathbf{A}^\top\mathbf{b}$$

The product $\mathbf{A}^\top\mathbf{A}$ is square and, provided it is nonsingular, it can be inverted to give the solution to (2.22):

$$\mathbf{x} = \left(\mathbf{A}^\top\mathbf{A}\right)^{-1}\mathbf{A}^\top\mathbf{b} \tag{2.23}$$

Let

$$\mathbf{A}^+ = \left(\mathbf{A}^\top\mathbf{A}\right)^{-1}\mathbf{A}^\top \tag{2.24}$$

The matrix $\mathbf{A}^+$ is called the Moore–Penrose pseudo-inverse of **A** or just the pseudo-inverse. Thus the solution of (2.22) is

$$\mathbf{x} = (\mathbf{A}^+)\,\mathbf{b} \tag{2.25}$$

This definition of the pseudo-inverse, $\mathbf{A}^+$, requires **A** to have full rank. If **A** is full rank and $m > n$, then rank(**A**) $= n$. Now rank($\mathbf{A}^\top\mathbf{A}$) $=$ rank(**A**) and hence rank($\mathbf{A}^\top\mathbf{A}$) $= n$. Since $\mathbf{A}^\top\mathbf{A}$ is an $n \times n$ array, $\mathbf{A}^\top\mathbf{A}$ is automatically full rank and $\mathbf{A}^+$ is then a unique $m \times n$ array. If **A** is rank deficient, then $\mathbf{A}^\top\mathbf{A}$ is rank deficient and cannot be inverted.

If **A** is square and nonsingular, then $\mathbf{A}^+ = \mathbf{A}^{-1}$. If **A** is complex then

$$\mathbf{A}^+ = \left(\mathbf{A}^\mathrm{H}\mathbf{A}\right)^{-1}\mathbf{A}^\mathrm{H} \tag{2.26}$$

where $\mathbf{A}^\mathrm{H}$ is the conjugate transpose, described in Appendix A, Section A.6. The product $\mathbf{A}^\top\mathbf{A}$ has a condition number, which is the square of the condition number of **A**. This has implications for the computations involved in $\mathbf{A}^+$.

It can be shown that the pseudo-inverse has the following properties:

1. $\mathbf{A}(\mathbf{A}^+)\mathbf{A} = \mathbf{A}$
2. $(\mathbf{A}^+)\mathbf{A}(\mathbf{A}^+) = \mathbf{A}^+$
3. $(\mathbf{A}^+)\mathbf{A}$ and $\mathbf{A}(\mathbf{A}^+)$ are symmetrical matrices.

We must now consider the situation that pertains when **A** of (2.22) is $m \times n$ with $m < n$; that is, an equation system with more variables than equations. If **A** is full rank, then rank(**A**) $= m$. Now rank($\mathbf{A}^\top\mathbf{A}$) $=$ rank(**A**) and hence rank($\mathbf{A}^\top\mathbf{A}$) $= m$. Since $\mathbf{A}^\top\mathbf{A}$ is an $n \times n$ matrix, $\mathbf{A}^\top\mathbf{A}$ is rank deficient and cannot be inverted, even though **A** is full rank. We can avoid this problem by recasting (2.22) as follows:

$$\mathbf{A}\mathbf{x} = \left(\mathbf{A}\mathbf{A}^\top\right)\left(\mathbf{A}\mathbf{A}^\top\right)^{-1}\mathbf{b}$$

and hence

$$\mathbf{x} = \mathbf{A}^\top \left(\mathbf{A}\mathbf{A}^\top\right)^{-1} \mathbf{b}$$

Thus

$$\mathbf{x} = (\mathbf{A}^+)\,\mathbf{b}$$

where $\mathbf{A}^+ = \mathbf{A}^\top(\mathbf{A}\mathbf{A}^\top)^{-1}$ and is the pseudo-inverse. Note that $\mathbf{A}\mathbf{A}^\top$ is an $m \times m$ array with rank m and can thus be inverted.

It has been shown that if **A** is rank deficient, (2.24) cannot be used to determine the pseudo-inverse of **A**. This doesn't mean that the pseudo-inverse does not exist; it always exists but we must use a different method to evaluate it. When **A** is rank deficient, or close to rank deficient, $\mathbf{A}^+$ is best calculated from the singular value decomposition (SVD) of **A**. If **A** is real, the SVD of **A** is $\mathbf{USV}^\top$ where **U** is an orthogonal $m \times m$ matrix, **V** is an orthogonal $n \times n$ matrix, and **S** is a $n \times m$ matrix of singular values. Thus the SVD of $\mathbf{A}^\top$ is $\mathbf{VS}^\top\mathbf{U}^\top$ so that

$$\mathbf{A}^\top\mathbf{A} = (\mathbf{VS}^\top\mathbf{U}^\top)(\mathbf{USV}^\top) = \mathbf{VS}^\top\mathbf{SV}^\top \text{ since } \mathbf{U}^\top\mathbf{U} = \mathbf{I}$$

Hence

$$\begin{aligned}\mathbf{A}^+ &= (\mathbf{VS}^\top\mathbf{SV}^\top)^{-1}\mathbf{VS}^\top\mathbf{U}^\top = \mathbf{V}^{-\top}(\mathbf{S}^\top\mathbf{S})^{-1}\mathbf{V}^{-1}\mathbf{VS}^\top\mathbf{U}^\top \\ &= \mathbf{V}(\mathbf{S}^\top\mathbf{S})^{-1}\mathbf{S}^\top\mathbf{U}^\top\end{aligned} \tag{2.27}$$

We note that $\mathbf{V}^{-\top} = (\mathbf{V}^\top)^{-1} = (\mathbf{V}^\top)^\top = \mathbf{V}$ because by orthogonality $\mathbf{VV}^\top = \mathbf{I}$. Since **V** is an $m \times m$ matrix, **U** is an $n \times n$ matrix, and **S** is an $n \times m$ matrix, then (2.27) is conformable (i.e., matrix multiplication is possible); see Appendix A, Section A.5.

Consider now the case when **A** is rank deficient. In this situation $\mathbf{S}^\top\mathbf{S}$ cannot be inverted because of the very small or zero singular values. To deal with this problem we take only the r nonzero singular values of the matrix so that **S** is an $r \times r$ matrix where r is the rank of **A**. To make the multiplications of (2.27) conformable we take the first r columns of **V** and the first r rows of $\mathbf{U}^\top$—that is, the first r columns of **U**. This is illustrated in the second of following examples in which the pseudo-inverse of **A** is determined.

Example 2.5

Consider the following matrix:

$$\mathbf{A} = \begin{bmatrix} 1 & 2 & 3 \\ 4 & 5 & 9 \\ 5 & 6 & 7 \\ -2 & 3 & 1 \end{bmatrix}$$

Computing the pseudo-inverse of **A** using a MATLAB implementation of (2.24) we have

```
>> A = [1 2 3;4 5 9;5 6 7;-2 3 1];
>> rank(A)

ans =
     3
```

We note that A is full rank. Thus

```
>> A_cross = inv(A.'*A)*A.'

A_cross =
   -0.0747   -0.1467    0.2500   -0.2057
   -0.0378   -0.2039    0.2500    0.1983
    0.0858    0.2795   -0.2500   -0.0231
```

The MATLAB function pinv provides this result directly and with greater accuracy.

```
A*A_cross*A

ans =
    1.0000    2.0000    3.0000
    4.0000    5.0000    9.0000
    5.0000    6.0000    7.0000
   -2.0000    3.0000    1.0000

>> A*A_cross

ans =
    0.1070    0.2841    0.0000    0.1218
    0.2841    0.9096    0.0000   -0.0387
    0.0000    0.0000    1.0000   -0.0000
    0.1218   -0.0387   -0.0000    0.9834

>> A_cross*A

ans =
    1.0000    0.0000    0.0000
    0.0000    1.0000    0.0000
   -0.0000   -0.0000    1.0000
```

Note that these calculations verify that A*A_cross*A equals A and that both A*A_cross and A_cross*A are symmetrical.

Example 2.6

Consider the following rank deficient matrix:

$$\mathbf{G} = \begin{bmatrix} 1 & 2 & 3 \\ 4 & 5 & 9 \\ 7 & 11 & 18 \\ -2 & 3 & 1 \\ 7 & 1 & 8 \end{bmatrix}$$

Using MATLAB, we have

```
>> G = [1 2 3;4 5 9;7 11 18;-2 3 1;7 1 8]

G =
     1     2     3
     4     5     9
     7    11    18
    -2     3     1
     7     1     8

>> rank(G)

ans =
     2
```

Note that G has a rank of 2 (i.e., it is rank deficient) and we cannot use (2.24) to determine its pseudo-inverse. We now find the SVD of G:

```
>> [U S V] = svd(G)

U =
   -0.1381    0.0839    0.9724   -0.0044   -0.1681
   -0.4115    0.0215    0.0539   -0.6081    0.6764
   -0.8258    0.2732   -0.2165    0.0607   -0.4392
   -0.0524    0.5650    0.0366    0.6373    0.5201
   -0.3563   -0.7737    0.0572    0.4695    0.2253

S =
   26.8394         0         0
         0    6.1358         0
         0         0    0.0000
         0         0         0
         0         0         0
```

```
V =
    -0.3709   -0.7274   -0.5774
    -0.4445    0.6849   -0.5774
    -0.8154   -0.0425    0.5774
```

We now select the two significant singular values for use in the subsequent computation:

```
>> SS = S(1:2,1:2)

SS =
    26.8394         0
          0    6.1358
```

To make the multiplication conformable we use only the first two columns of `U` and `V`.

```
>> G_cross = V(:,1:2)*inv(SS.'*SS)*SS.'*U(:,1:2).'

G_cross =
    -0.0080    0.0031   -0.0210   -0.0663    0.0966
     0.0117    0.0092    0.0442    0.0639   -0.0805
     0.0036    0.0124    0.0232   -0.0023    0.0162
```

This result can be obtained directly using the `pinv` function, which is based on the singular value decomposition of `G`.

```
>> G*G_cross

ans =
     0.0261    0.0586    0.1369    0.0546   -0.0157
     0.0586    0.1698    0.3457    0.0337    0.1300
     0.1369    0.3457    0.7565    0.1977    0.0829
     0.0546    0.0337    0.1977    0.3220   -0.4185
    -0.0157    0.1300    0.0829   -0.4185    0.7256

>> G_cross*G

ans =
     0.6667   -0.3333    0.3333
    -0.3333    0.6667    0.3333
     0.3333    0.3333    0.6667
```

Note that `G*G_cross` and `G_cross*G` are symmetric.

In the following section we will apply these methods to solve over- and underdetermined systems and discuss the meaning of the solution.

2.12 Over- and Underdetermined Systems

We will begin by examining examples of overdetermined systems, that is, systems of equations in which there are more equations than unknown variables.

Although overdetermined systems may have a unique solution, most often we are concerned with equation systems that are generated from experimental data, which can lead to a relatively small degree of inconsistency between the equations. For example, consider the following overdetermined system of linear equations:

$$\begin{aligned} x_1 + x_2 &= 1.98 \\ 2.05x_1 - x_2 &= 0.95 \\ 3.06x_1 + x_2 &= 3.98 \\ -1.02x_1 + 2x_2 &= 0.92 \\ 4.08x_1 - x_2 &= 2.90 \end{aligned} \tag{2.28}$$

Figure 2.5 shows that (2.28) is such a system; the lines do not intersect in a point, although there is a point that *nearly* satisfies all the equations.

We would like to choose the best point of all in the region defined by the intersections. One criterion for doing this is that the chosen solution should minimize the sum of squares of the residual errors (or residuals) of the equations. For example, consider the equation system (2.28). Letting $r_1, \ldots, r_5$ be the residuals, then

$$\begin{aligned} x_1 + x_2 - 1.98 &= r_1 \\ 2.05x_1 - x_2 - 0.95 &= r_2 \\ 3.06x_1 + x_2 - 3.98 &= r_3 \\ -1.02x_1 + 2x_2 - 0.92 &= r_4 \\ 4.08x_1 - x_2 - 2.90 &= r_5 \end{aligned}$$

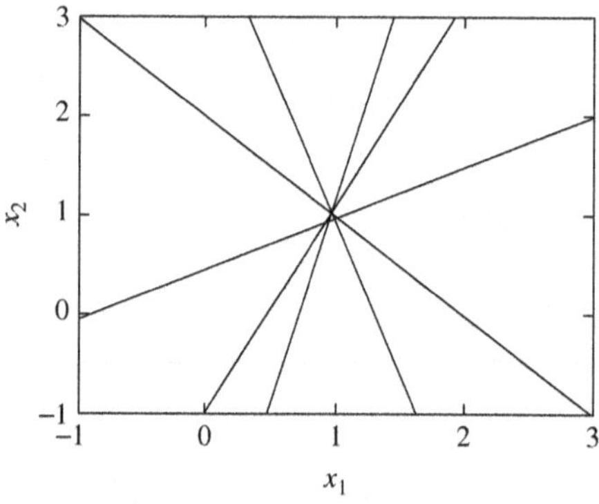

FIGURE 2.5 Plot of inconsistent equation system (2.28).

In this case the sum of the residuals squared is given by

$$S = \sum_{i=1}^{5} r_i^2 \tag{2.29}$$

We wish to minimize S and we can do this by making

$$\frac{\partial S}{\partial x_k} = 0, \quad k = 1, 2$$

Now

$$\frac{\partial S}{\partial x_k} = \sum_{i=1}^{5} 2r_i \frac{\partial r_i}{\partial x_k}, \quad k = 1, 2$$

and thus

$$\sum_{i=1}^{5} r_i \frac{\partial r_i}{\partial x_k} = 0, \quad k = 1, 2 \tag{2.30}$$

It can be shown that minimizing the sum of the squares of the equation residuals using (2.30) gives an identical solution to that given by the pseudo-inverse method of solving the equation system.

When solving a set of overdetermined equations, determining the pseudo-inverse of the system matrix is only part of the process and normally we do not require this interim result. The MATLAB operator \ solves overdetermined systems automatically. Thus the operator may be used to solve any linear equation system.

In the following example we compare the results obtained using the operator \ and using the pseudo-inverse for solving (2.28). The MATLAB script is

```
% e3s206.m
A = [1 1;2.05 -1;3.06 1;-1.02 2;4.08 -1];
b = [1.98;0.95;3.98;0.92;2.90];
x = pinv(A)*b
norm_pinv = norm(A*x-b)
x = A\b
norm_op = norm(A*x-b)
```

Running this script gives the following numeric output:

```
x =

    0.9631
    0.9885
```

```
norm_pinv =
     0.1064

x =
     0.9631
     0.9885

norm_op =
     0.1064
```

Here both the MATLAB operator \ and the function `pinv` have provided the same "best-fit" solution for the inconsistent set of equations. Figure 2.6 shows the region where these equations intersect in greater detail than Figure 2.5. The symbol "+" indicates the MATLAB solution, which lies in this region. The norm of $\mathbf{Ax}-\mathbf{b}$ is the square root of the sum of the squares of the residuals and provides a measure of how well the equations are satisfied.

The MATLAB operator \ does not solve an overdetermined system by using the pseudo-inverse, as given in (2.24). Instead, it solves (2.22) directly by QR decomposition. QR decomposition can be applied to both square and rectangular matrices providing the number of rows is greater than the number of columns. For example, applying the MATLAB function `qr` to solve the overdetermined system (2.28) we have

```
>> A = [1 1;2.05 -1;3.06 1;-1.02 2;4.08 -1];
>> b = [1.98 0.95 3.98 0.92 2.90].';
>> [Q R] = qr(A)

Q =
   -0.1761    0.4123   -0.7157   -0.2339   -0.4818
   -0.3610   -0.2702    0.0998    0.6751   -0.5753
   -0.5388    0.5083    0.5991   -0.2780   -0.1230
    0.1796    0.6839   -0.0615    0.6363    0.3021
   -0.7184   -0.1756   -0.3394    0.0857    0.5749

R =
   -5.6792    0.7237
         0    2.7343
         0         0
         0         0
         0         0
```

In the equation $\mathbf{Ax}=\mathbf{b}$ we have replaced $\mathbf{A}$ by $\mathbf{QR}$ so that $\mathbf{QRx}=\mathbf{b}$. Let $\mathbf{Rx}=\mathbf{y}$. Thus we have $\mathbf{y}=\mathbf{Q}^{-1}\mathbf{b}=\mathbf{Q}^{\top}\mathbf{b}$ since $\mathbf{Q}$ is orthogonal. Once $\mathbf{y}$ is determined we can efficiently determine

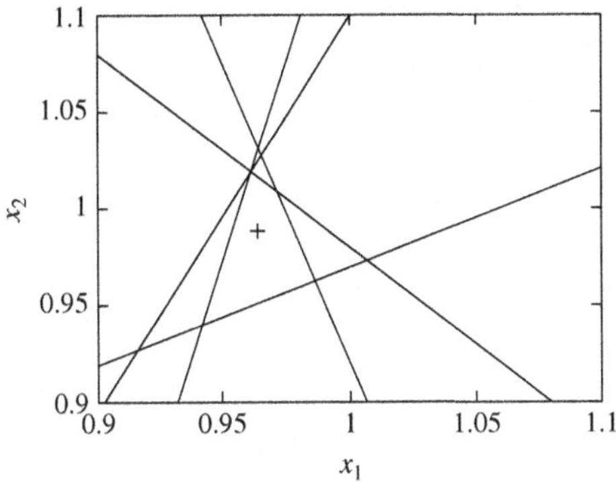

FIGURE 2.6 Plot of inconsistent equation system (2.28) showing the region of intersection of the equations, where + indicates the "best" solution.

x by back substitution since **R** is upper triangular. Thus, continuing the previous example,

```
>> y = Q.'*b

y =
   -4.7542
    2.7029
    0.0212
   -0.0942
   -0.0446
```

Using the second row of **R** and the second row of **y** we can determine x_2. From the first row of **R** and the first row of **y** we can determine x_1 since x_2 is known. Thus

$$-5.6792x_1 + 0.7237x_2 = -4.7542$$

$$2.7343x_2 = 2.7029$$

give $x_1 = 0.9631$ and $x_2 = 0.9885$, as before. The MATLAB operator \ implements this sequence of operations.

We now consider a case where the coefficient matrix of the overdetermined system is rank deficient. The following example is rank deficient and represents a system of parallel lines.

$$x_1 + 2x_2 = 1.00$$

$$x_1 + 2x_2 = 1.03$$

$$x_1 + 2x_2 = 0.97$$

$$x_1 + 2x_2 = 1.01$$

In MATLAB this becomes

```
>> A = [1 2;1 2;1 2;1 2]

A =
         1     2
         1     2
         1     2
         1     2

>> b = [1 1.03 0.97 1.01].'

b =
       1.0000
       1.0300
       0.9700
       1.0100

>> y = A\b
Warning: Rank deficient, rank = 1,  tol =    3.552714e-015.

y =
            0
       0.5012

>> norm(y)

ans =
       0.5012
```

The user is warned that this system is rank deficient. We have solved the system using the \ operator and now solve it using the `pinv` function as follows:

```
>> x = pinv(A)*b

x =
       0.2005
       0.4010

>> norm(x)

ans =
       0.4483
```

We see that when the `pinv` function and the \ operator are applied to rank deficient systems, the `pinv` function gives the solution with the smallest Euclidean norm; see Appendix A, Section A.10. Clearly there is no unique solution to this system since it represents a set of parallel lines.

We now turn to the problem of underdetermined systems. Here there is insufficient information to determine a unique solution. For example, consider the equation system

$$x_1 + 2x_2 + 3x_3 + 4x_4 = 1$$
$$-5x_1 + 3x_2 + 2x_3 + 7x_4 = 2$$

Expressing these equations in MATLAB we have

```
>> A = [1 2 3 4;-5 3 2 7];
>> b = [1 2].';
>> x1 = A\b

x1 =
   -0.0370
         0
         0
    0.2593

>> x2 = pinv(A)*b

x2 =
   -0.0780
    0.0787
    0.0729
    0.1755
```

We calculate the norms:

```
>> norm(x1)

ans =
    0.2619

>> norm(x2)

ans =
    0.2199
```

The first solution, `x1`, is a solution which satisfies the system; the second solution, `x2`, satisfies the system of equations but also gives the solution with the minimum norm.

The definition of the Euclidean or 2-norm of a vector is the square root of the sum of the squares of the elements of the vector; see Appendix A, Section A.10. The shortest distance between a point in space and the origin is given by Pythagoras's theorem as the square root of the sum of squares of the coordinates of the point. Thus the Euclidean norm of a vector, which is a point on a line, surface, or hypersurface, may be interpreted geometrically as the distance between this point and the origin. The vector with the minimum norm must be the point on the line, surface, or hypersurface that is closest to the origin. The line joining this vector to the origin must be perpendicular to the line, surface, or hypersurface. Giving the minimum norm solution has the advantage that, whereas there are an infinite number of solutions to an underdetermined problem, there is only one minimum norm solution. This provides a uniform result.

To complete the discussion of over- and underdetermined systems we consider the use of the `lsqnonneg` function, which solves the nonnegative least squares problem. This solves the problem of finding a solution **x** to the linear equation system

$$\mathbf{Ax} = \mathbf{b} \text{ subject to } \mathbf{x} \geq \mathbf{0}$$

where **A** and **b** must be real. This is equivalent to the problem of finding the vector **x** that minimizes norm$(\mathbf{Ax} - \mathbf{b})$ subject to $\mathbf{x} \geq \mathbf{0}$.

We can call the MATLAB function `lsqnonneg` for a specific problem using the statement

```
x = lsqnonneg(A,b)
```

where `A` corresponds to **A** in our definition and `b` to **b**. The solution is given by `x`. Consider the example which follows. Solve

$$\begin{bmatrix} 1 & 1 & 1 & 1 & 0 \\ 1 & 2 & 3 & 0 & 1 \end{bmatrix} \begin{bmatrix} x_1 \\ x_2 \\ x_3 \\ x_4 \\ x_5 \end{bmatrix} = \begin{bmatrix} 7 \\ 12 \end{bmatrix}$$

subject to $x_i \geq 0$, $i = 1, 2, \ldots, 5$. In MATLAB this becomes

```
>> A = [1 1 1 1 0;1 2 3 0 1];
>> b = [7 12].';
```

Solving this system gives

```
>> x = lsqnonneg(A,b)

x =
     0
     0
     4
     3
     0
```

We can solve this using \ but this will not ensure nonnegative values for x.

```
>> x2 = A\b

x2 =
          0
          0
     4.0000
     3.0000
          0
```

In this case we do obtain a nonnegative solution but this is fortuitous.

The following example illustrates how the `lsqnonneg` function forces a nonnegative solution that best satisfies the equations:

$$\begin{bmatrix} 3.0501 & 4.8913 \\ 3.2311 & -3.2379 \\ 1.6068 & 7.4565 \\ 2.4860 & -0.9815 \end{bmatrix} \begin{bmatrix} x_1 \\ x_2 \end{bmatrix} = \begin{bmatrix} 2.5 \\ 2.5 \\ 0.5 \\ 2.5 \end{bmatrix} \tag{2.31}$$

```
>> A = [3.0501 4.8913;3.2311 -3.2379; 1.6068 7.4565;2.4860 -0.9815];
>> b = [2.5 2.5 0.5 2.5].';
```

We can compute the solution using \ or the `lsqnonneg` function:

```
>> x1 = A\b

x1 =
    0.8307
   -0.0684
```

```
>> x2 = lsqnonneg(A,b)

x2 =
    0.7971
         0

>> norm(A*x1-b)

ans =
    0.7040

>> norm(A*x2-b)

ans =
    0.9428
```

Thus, the best fit is given by using the operator \, but if we require all components of the solution to be nonnegative, then we must use the `lsqnonneg` function.

2.13 Iterative Methods

Except in special circumstances it is unlikely that any function or script developed by the user will outperform a function or operator that is an integral part of MATLAB. Thus we cannot expect to develop a function that will determine the solution of $\mathbf{Ax} = \mathbf{b}$ more efficiently than by using the MATLAB operation `A\b`. However, we describe iterative methods here for the sake of completeness.

Iterative methods of solution are developed as follows. We begin with a system of linear equations

$$\begin{array}{ccccc}
a_{11}x_1+ & a_{12}x_2+ & \ldots & +a_{1n}x_n & = b_1 \\
a_{21}x_1+ & a_{22}x_2+ & \ldots & +a_{2n}x_n & = b_2 \\
\vdots & \vdots & & \vdots & \vdots \\
a_{n1}x_1+ & a_{n2}x_2+ & \ldots & +a_{nn}x_n & = b_n
\end{array}$$

These can be rearranged to give

$$\begin{array}{l}
x_1 = (b_1 - a_{12}x_2 - a_{13}x_3 - \ldots - a_{1n}x_n)\,/\,a_{11} \\
x_2 = (b_2 - a_{21}x_1 - a_{23}x_3 - \ldots - a_{2n}x_n\,/\,a_{22} \\
\vdots \qquad\quad \vdots \qquad\qquad\quad \vdots \qquad \vdots \\
x_n = (b_n - a_{n1}x_1 - a_{n2}x_2 - \ldots - a_{n,n-1}x_{n-1})\,/a_{nn}
\end{array} \tag{2.32}$$

If we assume initial values for x_i, where $i = 1, \ldots, n$, and substitute these values into the right side of the preceding equations, we may determine new values for the x_i from (2.32). The iterative process is continued by substituting these values of x_i into the right side of the equations, and so on. There are several variants of the process. For example, we can use old values of x_i in the right side of the equations to determine *all* the new values of x_i in the left side of the equation. This is called Jacobi or simultaneous iteration. Alternatively, we may use a new value of x_i in the right side of the equation as soon as it is determined, to obtain the other values of x_i in the right side. For example, once a new value of x_1 is determined from the first equation of (2.32), it is used in the second equation, together with the old $x_3, \ldots, x_n$ to determine x_2. This is called Gauss–Seidel or cyclic iteration.

The conditions for convergence for this type of iteration are

$$|a_{ii}| >> \sum_{j=1,\ j\neq i}^{n} |a_{ij}| \text{ for } i = 1, 2, \ldots, n$$

Thus these iterative methods are only guaranteed to work when the coefficient matrix is diagonally dominant. An iterative method based on conjugate gradients for the solution of systems of linear equations is discussed in Chapter 8.

2.14 Sparse Matrices

Sparse matrices arise in many problems of science and engineering—for example, in linear programming and the analysis of structures. Indeed, most large matrices that arise in the analysis of physical systems are sparse and the recognition of this fact makes the solution of linear systems with millions of coefficients feasible. The aim of this section is to give a brief description of the extensive sparse matrix facilities available in MATLAB and to give practical illustrations of their value through examples. For background information on how MATLAB implements the concept of sparsity, see Gilbert et al. (1992).

It is difficult to give a simple quantitative answer to the question: When is a matrix sparse? A matrix is sparse if it contains a high proportion of zero elements. However, this is significant only if the sparsity is of such an extent that we can utilize this feature to reduce the computation time and storage facilities required for operations used on such matrices. One major way in which time can be saved in dealing with sparse matrices is to avoid unnecessary operations on zero elements.

MATLAB *does not automatically treat a matrix as sparse* and the sparsity features of MATLAB are not introduced until invoked. Thus the user determines whether a matrix is in the sparse class or the full class. If the user considers a matrix to be sparse and wants to use this fact to advantage, the matrix must first be converted to sparse form. This is achieved by using the function `sparse`. Thus `b = sparse(a)` converts the matrix `a` to sparse form and subsequent MATLAB operations will take account of this sparsity. If we wish to return this matrix to full form, we simply use `c = full(b)`. However, the sparse function can also be used directly to generate sparse matrices.

It is important to note that binary operators *, +, -, /, and \ produce sparse results if *both* operands are sparse. Thus the property of sparsity may survive a long sequence of matrix operations. In addition, such functions as `chol(A)` and `lu(A)` produce sparse results if the matrix `A` is sparse. However, in mixed cases, where one operand is sparse and the other is full, the result is generally a full matrix. Thus the property of sparsity may be inadvertently lost. Notice in particular that `eye(n)` is not in the sparse class of matrices in MATLAB but a sparse identity matrix can be created using `speye(n)`. Thus the latter should be used in manipulations with sparse matrices.

We will now introduce some of the key MATLAB functions for dealing with sparse matrices, describe their use and, where appropriate, give examples of their application. The simplest MATLAB function that helps in dealing with sparsity is the function `nnz(a)`, which provides the number of nonzero elements in a given matrix `a`, regardless of whether it is sparse or full. A function that enables us to examine whether a given matrix has been defined or has been propagated as sparse is the function `issparse(a)`, which returns the value 1 if the matrix `a` is sparse or 0 if it is not sparse. The function `spy(a)` allows the user to view the structure of a given matrix `a` by displaying symbolically only its nonzero elements; see Figure 2.7 later in the chapter for examples.

Before we can illustrate the action of these and other functions, it is useful to generate some sparse matrices. This is easily done using a different form of the `sparse` function. This time the function is supplied with the location of the nonzero entries in the matrix, the value of these entries, the size of the sparse matrix, and the space allocated for the nonzero entries. This function call takes the form `sparse(i, j, nzvals, m, n, nzmax)`. This generates an `m` $\times$ `n` matrix and allocates the nonzero values in the vector `nzvals` to the positions in the matrix given by the vectors `i` and `j`. The row position is given by `i` and the column position by `j`. Space is allocated for `nzmax` nonzeros. Since all but one parameter is optional, there are many forms of this function. We cannot give examples of all these forms but the following cases illustrate its use.

```
>> colpos = [1 2 1 2 5 3 4 3 4 5];
>> rowpos = [1 1 2 2 2 4 4 5 5 5];
>> value = [12 -4 7 3 -8 -13 11 2 7 -4];
>> A = sparse(rowpos,colpos,value,5,5)
```

These statements give the following output:

```
A =
   (1,1)        12
   (2,1)         7
   (1,2)        -4
   (2,2)         3
```

```
(4,3)        -13
(5,3)          2
(4,4)         11
(5,4)          7
(2,5)         -8
(5,5)         -4
```

We see that a 5×5 sparse matrix with 10 nonzero elements has been generated with the required coefficient values in the required positions. This sparse matrix can be converted to a full matrix as follows:

```
>> B = full(A)

B =
    12    -4     0     0     0
     7     3     0     0    -8
     0     0     0     0     0
     0     0   -13    11     0
     0     0     2     7    -4
```

This is the equivalent full matrix. Now the following statements test to see if the matrices `A` and `B` are in the sparse class and give the number of nonzeros they contain.

```
>> [issparse(A) issparse(B) nnz(A) nnz(B)]

ans =
     1     0    10    10
```

Clearly these functions give the expected results. Since `A` is a member of the class of sparse matrices, the value of `issparse(A)` is 1. However, although `B` looks sparse, it is not *stored* as a sparse matrix and hence is not in the class of sparse matrices within the MATLAB environment. The next example shows how to generate a large 5000×5000 sparse matrix and compares the time required to solve a linear system of equations involving this sparse matrix with the time required for the equivalent full matrix. The script for this is

```
% e3s207.m Generates a sparse triple diagonal matrix
n = 5000;
rowpos = 2:n; colpos = 1:n-1;
values = 2*ones(1,n-1);
Offdiag = sparse(rowpos,colpos,values,n,n);
A = sparse(1:n,1:n,4*ones(1,n),n,n);
A = A+Offdiag+Offdiag.';
%generate full matrix
B = full(A);
```

```
%generate arbitrary right hand side for system of equations
rhs = [1:n].';
tic, x = A\rhs; f1 = toc;
tic, x = B\rhs; f2 = toc;
fprintf('Time to solve sparse matrix = %8.5f\n',f1);
fprintf('Time to solve  full  matrix = %8.5f\n',f2);
```

This provides the following results:

```
Time to solve sparse matrix =  0.00051
Time to solve  full  matrix =  5.74781
```

In this example there is a major reduction in the time taken to solve the system when using the sparse class of matrix. We now perform a similar exercise, this time to determine the `lu` decomposition of a 5000×5000 matrix:

```
% e3s208.m
n = 5000;
offdiag = sparse(2:n,1:n-1,2*ones(1,n-1),n,n);
A = sparse(1:n,1:n,4*ones(1,n),n,n);
A = A+offdiag+offdiag';
%generate full matrix
B = full(A);
%generate arbitrary right hand side for system of equations
rhs = [1:n]';
tic, lu1 = lu(A); f1 = toc;
tic, lu2 = lu(B); f2 = toc;
fprintf('Time for sparse LU = %8.4f\n',f1);
fprintf('Time for  full  LU = %8.4f\n',f2);
```

The time taken to solve the systems is

```
Time for sparse LU =   0.0056
Time for  full  LU =   9.6355
```

Again this provides a considerable reduction in the time taken.

An alternative way to generate sparse matrices is to use the functions `sprandn` and `sprandsym`. These provide random sparse matrices and random sparse symmetric matrices, respectively. The call

```
A = sprandn(m,n,d)
```

produces an $m \times n$ random matrix with normally distributed nonzero entries of density `d`. The density is the proportion of the nonzero entries to the total number of entries in the matrix. Thus `d` must be in the range 0 to 1. To produce a symmetric random matrix with normally distributed nonzero entries of density d, we use

```
A = sprandsys(n,d)
```

Examples of calls of these functions are given by

```
>> A = sprandn(5,5,0.25)

A =
   (2,1)       -0.4326
   (3,3)       -1.6656
   (5,3)       -1.1465
   (4,4)        0.1253
   (5,4)        1.1909
   (4,5)        0.2877

>> B = full(A)

B =
         0         0         0         0         0
   -0.4326         0         0         0         0
         0         0   -1.6656         0         0
         0         0         0    0.1253    0.2877
         0         0   -1.1465    1.1909         0

>> As = sprandsym(5,0.25)

As =
   (3,1)        0.3273
   (1,3)        0.3273
   (5,3)        0.1746
   (5,4)       -0.0376
   (3,5)        0.1746
   (4,5)       -0.0376
   (5,5)        1.1892

>> Bs = full(As)

Bs =
         0         0    0.3273         0         0
         0         0         0         0         0
    0.3273         0         0         0    0.1746
         0         0         0         0   -0.0376
         0         0    0.1746   -0.0376    1.1892
```

An alternative call for `sprandsym` is given by

```
A = sprandsym(n,density,r)
```

If `r` is a scalar, then this produces a random sparse symmetric matrix with a condition number equal to `1/r`. Remarkably, if `r` is a vector of length n, a random sparse matrix with eigenvalues equal to the elements of `r` is produced. Eigenvalues are discussed in Section 2.15. A positive definite matrix has all positive eigenvalues and consequently such a matrix can be generated by choosing each of the n elements of `r` to be positive. An example of this form of call is

```
>> Apd = sprandsym(6,0.4,[1 2.5 6 9 2 4.3])

Apd =
   (1,1)        1.0058
   (2,1)       -0.0294
   (4,1)       -0.0879
   (1,2)       -0.0294
   (2,2)        8.3477
   (4,2)       -1.9540
   (3,3)        5.4937
   (5,3)       -1.3300
   (1,4)       -0.0879
   (2,4)       -1.9540
   (4,4)        3.1465
   (3,5)       -1.3300
   (5,5)        2.5063
   (6,6)        4.3000

>> Bpd = full(Apd)

Bpd =
    1.0058   -0.0294         0   -0.0879         0         0
   -0.0294    8.3477         0   -1.9540         0         0
         0         0    5.4937         0   -1.3300         0
   -0.0879   -1.9540         0    3.1465         0         0
         0         0   -1.3300         0    2.5063         0
         0         0         0         0         0    4.3000
```

This provides an important method for generating test matrices with required properties since, by providing a list of eigenvalues with a range of values, we can produce positive definite matrices that are very badly conditioned.

We now return to examine further the value of using sparsity. The reasons for the very high level of improvement in computing efficiency when using the \ operator, illustrated in the example at the beginning of this section, are complex. The process includes a special preordering of the columns of the matrix. This special preordering, called *minimum degree ordering*, is used in the case of the \ operator. This preordering takes different forms depending on whether the matrix is symmetric or nonsymmetric. The aim of any preordering is to reduce the amount of *fill-in* from any subsequent matrix operations. Fill-in is the introduction of additional nonzero elements.

We can examine this preordering process using the spy function and the function symand, which implements *symmetric minimum degree ordering* in MATLAB. The function is automatically applied when working on matrices that belong to the class of sparse matrices for the standard functions and operators of MATLAB. However, if we are required to use this preordering in nonstandard applications, then we may use the symmmd function. The following examples illustrate the use of this function.

We first consider the simple process of multiplication applied to a full and a sparse matrix. The sparse multiplication uses the minimum degree ordering. The following script generates a sparse matrix, obtains a minimum degree ordering for it, and then examines the result of multiplying the matrix by itself transposed. This is compared with the same operations carried out on the full matrix, and the time required for each operation is compared.

```
% e3s209.m
% generate a sparse matrix
n = 3000;
offdiag = sparse(2:n,1:n-1,2*ones(1,n-1),n,n);
offdiag2 = sparse(4:n,1:n-3,3*ones(1,n-3),n,n);
offdiag3 = sparse(n-5:n,1:6,7*ones(1,6),n,n);
A = sparse(1:n,1:n,4*ones(1,n),n,n);
A = A+offdiag+offdiag'+offdiag2+offdiag2'+offdiag3+offdiag3';
A = A*A.';
% generate full matrix
B = full(A);
m_order = symamd(A);
tic
spmult = A(m_order,m_order)*A(m_order,m_order).';
flsp = toc;
tic, fulmult = B*B.'; flful = toc;
fprintf('Time for sparse mult = %6.4f\n',flsp)
fprintf('Time for  full  mult = %6.4f\n',flful)
```

Running this script results in the following output:

```
Time for sparse mult = 0.0184
Time for  full  mult = 3.8359
```

We now perform a similar experiment to the preceding but for a more complex numerical process than multiplication. In the script that follows we examine LU decomposition. We consider the result of using a minimum degree ordering on the LU decomposition process by comparing the performance of the `lu` function with and without a preordering. The script has the form

```
% e3s210.m
% generate a sparse matrix
n = 100;
offdiag = sparse(2:n,1:n-1,2*ones(1,n-1),n,n);
offdiag2 = sparse(4:n,1:n-3,3*ones(1,n-3),n,n);
offdiag3 = sparse(n-5:n,1:6,7*ones(1,6),n,n);
A = sparse(1:n,1:n,4*ones(1,n),n,n);
A = A+offdiag+offdiag'+offdiag2+offdiag2'+offdiag3+offdiag3';
A = A*A.';
A1 = flipud(A);
A = A+A1;
n1 = nnz(A)
B = full(A); %generate full matrix
m_order = symamd(A);
tic, lud = lu(A(m_order,m_order)); flsp = toc;
n2 = nnz(lud)
tic, fullu = lu(B); flful = toc;
n3 = nnz(fullu)
subplot(2,2,1), spy(A,'k');
title('Original matrix')
subplot(2,2,2), spy(A(m_order,m_order),'k')
title('Ordered matrix')
subplot(2,2,3), spy(fullu,'k')
title('LU decomposition,unordered matrix')
subplot(2,2,4), spy(lud,'k')
title('LU decomposition, ordered matrix')
fprintf('Time for sparse lu = %6.4f\n',flsp)
fprintf('Time for  full  lu = %6.4f\n',flful)
```

Running this script gives

```
n1 =

        2096


n2 =

        1307
```

```
n3 =

        4465

Time for sparse lu = 0.0013
Time for  full  lu = 0.0047
```

As expected, by using a sparse operation we achieve a reduction in the time taken to determine the LU decomposition. Figure 2.7 shows the original matrix with 2096 nonzero elements, the reordered matrix (which has the same number of nonzeros), and the LU decomposition structure both with and without minimum degree ordering. Notice that the number of nonzeros in the LU matrices with preordering is 1307 and without is 4465. Thus there is a large increase in the number of nonzero elements in the LU matrices without preordering. In contrast, LU decomposition of the preordered matrix has produced fewer nonzeros than the original matrix. The reduction of fill-in is an important feature of sparse numerical processes and may ultimately lead to great saving in computational effort. Note that if the size of the matrices is increased from 100×100 to 3000×3000 then the output from the preceding script is

```
n1 =

       65896

n2 =

       34657
```

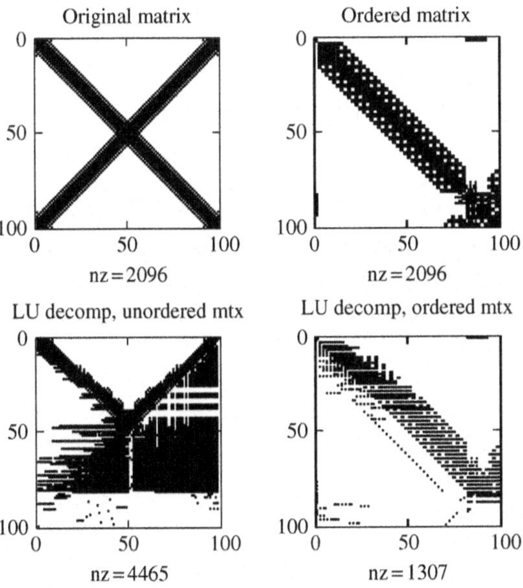

FIGURE 2.7 Effect of minimum degree ordering on LU decomposition.

```
n3 =

       526810

Time for sparse lu = 0.0708
Time for  full  lu = 2.3564
```

Here we obtain a more substantial reduction by using sparse operations.

The MATLAB function symamd provides a minimum degree ordering for symmetric matrices. For nonsymmetric matrices MATLAB provides the function colmmd, which gives the column minimum degree ordering for nonsymmetric matrices. An alternative ordering, which is used to reduce bandwidth, is the reverse Cuthill-McKee ordering. This is implemented in MATLAB by the function symrcm. The execution of the statement `p = symrcm(A)` provides the permutation vector p to produce the required preordering and `A(p,p)` is the reordered matrix.

We have shown that in general taking account of sparsity will provide savings in floating-point operations. However, these savings fall off as the matrices on which we are operating become less sparse, as the following example illustrates.

```
% e3s211.m
n = 1000;  b = 1:n;
disp('   density    time_sparse   time_full');
for density = 0.004:0.003:0.039
    A = sprandsym(n,density)+0.1*speye(n);
    density = density+1/n;
    tic, x = A\b'; f1 = toc;
    B = full(A);
    tic, y = B\b'; f2 = toc;
    fprintf('%10.4f %12.4f %12.4f\n',density,f1,f2);
end
```

In the preceding script a diagonal of elements has been added to the randomly generated sparse matrix. This is done to ensure that each row of the matrix contains a nonzero element; otherwise, the matrix may be singular. Adding this diagonal modifies the density. If the original $n \times n$ matrix has a density of d, then, assuming that this matrix has only zeros on the diagonal, the modified density is $d + 1/n$.

```
   density   time_sparse   time_full
    0.0050        0.0204      0.1907
    0.0080        0.0329      0.1318
    0.0110        0.0508      0.1332
    0.0140        0.0744      0.1399
    0.0170        0.0892      0.1351
    0.0200        0.1064      0.1372
```

```
0.0230        0.1179        0.1348
0.0260        0.1317        0.1381
0.0290        0.1444        0.1372
0.0320        0.1516        0.1369
0.0350        0.1789        0.1404
0.0380        0.1627        0.1450
```

This output shows that the advantage of using a sparse class of matrix diminishes as the density increases.

Another application where sparsity is important is in solving the least squares problem. This problem is known to be ill-conditioned and hence any saving in computational effort is particularly beneficial. This is directly implemented by using `A\b` where `A` is nonsquare and sparse. To illustrate the use of the \ operator with sparse matrices and compare its performance when no account is taken of sparsity, we use the following script:

```
% e3s212.m
% generate a sparse triple diagonal matrix
n = 1000;
rowpos = 2:n;  colpos = 1:n-1;
values = ones(1,n-1);
offdiag = sparse(rowpos,colpos,values,n,n);
A = sparse(1:n,1:n,4*ones(1,n),n,n);
A = A+offdiag+offdiag';
%Now generate a sparse least squares system
Als = A(:,1:n/2);
%generate full matrix
Cfl = full(Als);
rhs = 1:n;
tic, x = Als\rhs'; f1 = toc;
tic, x = Cfl\rhs'; f2 = toc;
fprintf('Time for sparse least squares solve = %8.4f\n',f1)
fprintf('Time for  full  least squares solve = %8.4f\n',f2)
```

This provides the following results:

```
Time for sparse least squares solve =   0.0023
Time for  full  least squares solve =   0.2734
```

Again we see the advantage of using sparsity.

We have not covered all aspects of sparsity or described all the related functions. However, we hope this section has provided a helpful introduction to this difficult but important and valuable development of MATLAB.

2.15 The Eigenvalue Problem

Eigenvalue problems arise in many branches of science and engineering. For example, the vibration characteristics of structures are determined from the solution of an algebraic eigenvalue problem. Here we consider a particular example of a system of masses and springs shown in Figure 2.8. The equations of motion for this system are

$$\begin{aligned}
m_1\ddot{q}_1 + (k_1+k_2+k_4)\,q_1 - k_2q_2 - k_4q_3 &= 0\\
m_2\ddot{q}_2 - k_2q_1 + (k_2+k_3)\,q_2 - k_3q_3 &= 0\\
m_3\ddot{q}_3 - k_4q_1 - k_3q_2 + (k_3+k_4)\,q_3 &= 0
\end{aligned} \tag{2.33}$$

where m_1, m_2, and m_3 are the system masses and $k_1,\ldots,k_4$ are the spring stiffnesses. If we assume an harmonic solution for each coordinate, then $q_i(t) = u_i\exp(\jmath\omega t)$ where $\jmath = \sqrt{-1}$, for $i = 1$, 2, and 3. Hence $d^2q_i/dt^2 = -\omega^2u_i\exp(\jmath\omega t)$. Substituting in (2.33) and canceling the common factor $\exp(j\omega t)$ gives

$$\begin{aligned}
-\omega^2m_1u_1 + (k_1+k_2+k_4)\,u_1 - k_2u_2 - k_4u_3 &= 0\\
-\omega^2m_2u_2 - k_2u_1 + (k_2+k_3)\,u_2 - k_3u_3 &= 0\\
-\omega^2m_3u_3 - k_4u_1 - k_3u_2 + (k_3+k_4)\,u_3 &= 0
\end{aligned} \tag{2.34}$$

If $m_1 = 10$ kg, $m_2 = 20$ kg, $m_3 = 30$ kg, $k_1 = 10$ kN/m, $k_2 = 20$ kN/m, $k_3 = 25$ kN/m, and $k_4 = 15$ kN/m, then (2.34) becomes

$$\begin{aligned}
-\omega^2 10u_1 + 45000u_1 - 20000u_2 - 15000u_3 &= 0\\
-\omega^2 20u_1 - 20000u_1 + 45000u_2 - 25000u_3 &= 0\\
-\omega^2 30u_1 - 15000u_1 - 25000u_2 + 40000u_3 &= 0
\end{aligned}$$

This can be expressed in matrix notation as

$$-\omega^2\mathbf{Mu} + \mathbf{Ku} = \mathbf{0} \tag{2.35}$$

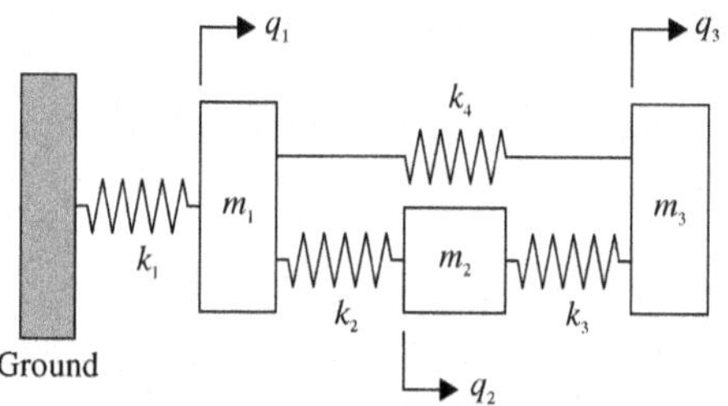

FIGURE 2.8 Mass-spring system with three degrees of freedom.

where

$$\mathbf{M} = \begin{bmatrix} 10 & 0 & 0 \\ 0 & 20 & 0 \\ 0 & 0 & 30 \end{bmatrix} \text{kg} \quad \text{and} \quad \mathbf{K} = \begin{bmatrix} 45 & -20 & -15 \\ -20 & 45 & -25 \\ -15 & -25 & 40 \end{bmatrix} \text{kN/m}$$

Equation (2.35) can be rearranged in a variety of ways. For example, it can be written

$$\mathbf{Mu} = \lambda \mathbf{Ku} \quad \text{where} \quad \lambda = \frac{1}{\omega^2} \tag{2.36}$$

This is an algebraic eigenvalue problem and solving it determines values for $\mathbf{u}$ and λ. MATLAB provides a function `eig` to solve the eigenvalue problem. To illustrate its use we apply it to the solution of (2.35).

```
>> M = [10 0 0;0 20 0;0 0 30];
>> K = 1000*[45 -20 -15;-20 45 -25;-15 -25 40];
>> lambda = eig(M,K).'

lambda =
    0.0002    0.0004    0.0073

>> omega = sqrt(1./lambda)

omega =
   72.2165   52.2551   11.7268
```

This result tells us that the system of Figure 2.8 vibrates with natural frequencies 11.72, 52.25, and 72.21 rad/s. In this example we have chosen not to determine `u`. We will discuss further the use of the function `eig` in Section 2.17.

Having provided an example of an eigenvalue problem, we consider the standard form of this problem:

$$\mathbf{Ax} = \lambda \mathbf{x} \tag{2.37}$$

This equation is an algebraic eigenvalue problem where $\mathbf{A}$ is a given $n \times n$ matrix of coefficients, $\mathbf{x}$ is an unknown column vector of n elements, and λ is an unknown scalar. Equation (2.37) can alternatively be written as

$$(\mathbf{A} - \lambda \mathbf{I})\mathbf{x} = \mathbf{0} \tag{2.38}$$

Our aim is to discover the values of $\mathbf{x}$, called the characteristic vectors or eigenvectors, and the corresponding values of λ, called the characteristic values or eigenvalues. The values of λ that satisfy (2.38) are given by the roots of the equation

$$|\mathbf{A} - \lambda \mathbf{I}| = 0 \tag{2.39}$$

These values of λ are such that $(\mathbf{A} - \lambda\mathbf{I})$ is singular. Since (2.38) is a homogeneous equation, nontrivial solutions exist for these values of λ. Evaluation of the determinant (2.39) leads to an nth-degree polynomial in λ, which is called the characteristic polynomial. This characteristic polynomial has n roots, some of which may be repeated, giving the n values of λ. In MATLAB we can create the coefficients of the characteristic polynomial using the function `poly`, and the roots of the resulting characteristic polynomial can be found using the function `roots`. For example, if

$$\mathbf{A} = \begin{bmatrix} 1 & 2 & 3 \\ 4 & 5 & -6 \\ 7 & -8 & 9 \end{bmatrix}$$

then we have

```
>> A = [1 2 3;4 5 -6;7 -8 9];
>> p = poly(A)

p =
      1.0000   -15.0000   -18.0000   360.0000
```

Hence the characteristic equation is $\lambda^3 - 15\lambda^2 - 18\lambda + 360 = 0$. To find the roots of this we use the statement

```
>> roots(p).'

ans =
    14.5343    -4.7494     5.2152
```

We can verify this result using the function `eig`:

```
>> eig(A).'

ans =
    -4.7494     5.2152    14.5343
```

Having obtained the eigenvalues we can substitute back into (2.38) to obtain linear equations for the characteristic vectors:

$$(\mathbf{A} - \lambda_i\mathbf{I})\mathbf{x} = \mathbf{0} \quad i = 1, 2, \ldots, n \tag{2.40}$$

This homogeneous system provides n nontrivial solutions for $\mathbf{x}$. However, the use of (2.39) and (2.40) is not a practical means of solving eigenvalue problems.

We now consider the properties of eigensolutions where the system matrix is real. If $\mathbf{A}$ is a real symmetric matrix the eigenvalues of $\mathbf{A}$ are real, but not necessarily positive, and the corresponding eigenvectors are also real. In addition, if λ_i, $\mathbf{x}_i$ and λ_j, $\mathbf{x}_j$ satisfy the eigenvalue problem (2.37) and λ_i and λ_j are distinct, then

$$\mathbf{x}_i^\top \mathbf{x}_j = 0 \quad i \neq j \tag{2.41}$$

and

$$\mathbf{x}_i^\top \mathbf{A}\mathbf{x}_j = 0 \quad i \neq j \tag{2.42}$$

Equations (2.41) and (2.42) are called the orthogonality relationships. Note that if $i = j$, then in general $\mathbf{x}_i^\top \mathbf{x}_i$ and $\mathbf{x}_i^\top \mathbf{A}\mathbf{x}_i$ are not zero. The vector $\mathbf{x}_i$ includes an arbitrary scalar multiplier because the vector multiplies both sides of (2.37). Hence the product $\mathbf{x}_i^\top \mathbf{x}_i$ must be arbitrary. However, if the arbitrary scalar multiplier is adjusted so that

$$\mathbf{x}_i^\top \mathbf{x}_i = 1 \tag{2.43}$$

then

$$\mathbf{x}_i^\top \mathbf{A}\mathbf{x}_i = \lambda_i \tag{2.44}$$

and the eigenvectors are then said to be *normalized*. Sometimes the eigenvalues are not distinct and the eigenvectors associated with these equal or repeated eigenvalues are not necessarily orthogonal. If $\lambda_i = \lambda_j$ and the other eigenvalues, λ_k, are distinct, then

$$\left.\begin{array}{l} \mathbf{x}_i^\top \mathbf{x}_k = 0 \\ \mathbf{x}_j^\top \mathbf{x}_k = 0 \end{array}\right\} \quad k = 1, 2, \ldots, n, \quad k \neq i, \quad k \neq j \tag{2.45}$$

For consistency we can choose to make $\mathbf{x}_i^\top \mathbf{x}_j = 0$. When $\lambda_i = \lambda_j$, the eigenvectors $\mathbf{x}_i$ and $\mathbf{x}_j$ are not unique and a linear combination of them (i.e., $\alpha\mathbf{x}_i + \gamma\mathbf{x}_j$, where α and γ are arbitrary constants), also satisfies the eigenvalue problem.

Let us now consider the case where $\mathbf{A}$ is real but not symmetric. A pair of related eigenvalue problems can arise as follows:

$$\mathbf{A}\mathbf{x} = \lambda\mathbf{x} \tag{2.46}$$

$$\mathbf{A}^\top \mathbf{y} = \beta\mathbf{y} \tag{2.47}$$

and (2.47) can be transposed to give

$$\mathbf{y}^\top \mathbf{A} = \beta\mathbf{y}^\top \tag{2.48}$$

The vectors $\mathbf{x}$ and $\mathbf{y}$ are called the right and left vectors of $\mathbf{A}$, respectively. The equations $|\mathbf{A} - \lambda\mathbf{I}| = 0$ and $|\mathbf{A}^\top - \beta\mathbf{I}| = 0$ must have the same solutions for λ and β because the determinant of a matrix and the determinant of its transpose are equal. Thus the eigenvalues of $\mathbf{A}$ and $\mathbf{A}^\top$ are identical but the eigenvectors $\mathbf{x}$ and $\mathbf{y}$ will, in general, differ from each other.

The eigenvalues and eigenvectors of a nonsymmetric real matrix are either real or pairs of complex conjugates. If λ_i, $\mathbf{x}_i$, $\mathbf{y}_i$ and λ_j, $\mathbf{x}_j$, $\mathbf{y}_j$ are solutions that satisfy the eigenvalue problems of (2.46) and (2.47) and λ_i and λ_j are distinct, then

$$\mathbf{x}_i^\top \mathbf{x}_j = 0 \quad i \neq j \tag{2.49}$$

and

$$\mathbf{x}_i^\top \mathbf{A}\mathbf{x}_j = 0 \quad i \neq j \tag{2.50}$$

Equations (2.49) and (2.50) are called the biorthogonal relationships. As with (2.43) and (2.44) if, in these equations, $i = j$, then in general $\mathbf{y}_i^\top \mathbf{x}_i$ and $\mathbf{y}_i^\top \mathbf{A}\mathbf{x}_i$ are not zero. The eigenvectors $\mathbf{x}_i$ and $\mathbf{y}_i$ include arbitrary scaling factors and so the product of these vectors will also be arbitrary. However, if the vectors are adjusted so that

$$\mathbf{y}_i^\top \mathbf{x}_i = 1 \tag{2.51}$$

then

$$\mathbf{y}_i^\top \mathbf{A}\mathbf{x}_i = \lambda_i \tag{2.52}$$

We cannot, in these circumstances, describe either $\mathbf{x}_i$ or $\mathbf{y}_i$ as normalized. The vectors still include an arbitrary scale factor; only their product is uniquely chosen.

2.16 Iterative Methods for Solving the Eigenvalue Problem

The first of two simple iterative procedures described here determines the dominant or largest eigenvalue. The method, which is called the power method or matrix iteration, can be used on both symmetric and nonsymmetric matrices. However, for a nonsymmetric matrix the user must be alert to the possibility that there is not a single real dominant eigenvalue value but a complex conjugate pair. Under these conditions simple iteration does not converge.

Consider the eigenvalue problem defined by (2.37) and let the vector $\mathbf{u}_0$ be an initial trial solution. The vector $\mathbf{u}_0$ is an unknown linear combination of all the eigenvectors of the system provided they are linearly independent. Thus

$$\mathbf{u}_0 = \sum_{i=1}^{n} \alpha_i \mathbf{x}_i \tag{2.53}$$

where α_i are unknown coefficients and $\mathbf{x}_i$ are the unknown eigenvectors. Let the iterative scheme be

$$\mathbf{u}_1 = \mathbf{A}\mathbf{u}_0,\ \mathbf{u}_2 = \mathbf{A}\mathbf{u}_1, \ldots, \mathbf{u}_p = \mathbf{A}\mathbf{u}_{p-1} \tag{2.54}$$

Substituting (2.53) into the sequence (2.54) we have

$$\begin{aligned}
\mathbf{u}_1 &= \sum_{i=1}^{n} \alpha_i \mathbf{A}\mathbf{x}_i = \sum_{i=1}^{n} \alpha_i \lambda_i \mathbf{x}_i \quad \text{since } \mathbf{A}\mathbf{x}_i = \lambda_i \mathbf{x}_i \\
\mathbf{u}_2 &= \sum_{i=1}^{n} \alpha_i \lambda_i \mathbf{A}\mathbf{x}_i = \sum_{i=1}^{n} \alpha_i \lambda_i^2 \mathbf{x}_i \\
&\cdots\cdots\cdots\cdots\cdots\cdots \\
\mathbf{u}_p &= \sum_{i=1}^{n} \alpha_i \lambda_i^{p-1} \mathbf{A}\mathbf{x}_i = \sum_{i=1}^{n} \alpha_i \lambda_i^p \mathbf{x}_i
\end{aligned} \tag{2.55}$$

The final equation can be rearranged as follows:

$$\mathbf{u}_p = \lambda_1^p \left[\alpha_1 \mathbf{x}_1 + \sum_{i=2}^{n} \alpha_i \left(\frac{\lambda_i}{\lambda_1} \right)^p \mathbf{x}_i \right] \tag{2.56}$$

It is the accepted convention that the n eigenvalues of a matrix are numbered such that

$$|\lambda_1| > |\lambda_2| > \cdots > |\lambda_n|$$

Hence

$$\left[\frac{\lambda_i}{\lambda_1} \right]^p$$

tends to zero as p tends to infinity for $i = 2, 3, \ldots, n$. As p becomes large, we have from (2.56):

$$\mathbf{u}_p \Rightarrow \lambda_1^p \alpha_1 \mathbf{x}_1$$

Thus $\mathbf{u}_p$ becomes proportional to $\mathbf{x}_1$ and the ratio between corresponding components of $\mathbf{u}_p$ and $\mathbf{u}_{p-1}$ tends to λ_1.

The algorithm is not usually implemented exactly as described previously because problems could arise due to numeric overflows. Usually, after each iteration, the resulting trial vector is normalized by dividing it by its largest element, thereby reducing the largest element in the vector to unity. This can be expressed mathematically as

$$\left.\begin{aligned}
\mathbf{v}_p &= \mathbf{A}\mathbf{u}_p \\
\mathbf{u}_{p+1} &= \left(\frac{1}{\max(\mathbf{v}_p)} \right) \mathbf{v}_p
\end{aligned}\right\} \quad p = 0, 1, 2, \ldots \tag{2.57}$$

where $\max(\mathbf{v}_p)$ is the element of $\mathbf{v}_p$ with the maximum modulus. The pair of equations (2.57) are iterated until convergence is achieved. This modification to the algorithm does not affect the rate of convergence of the iteration. In addition to preventing the buildup of very large numbers, the modification described before has the added advantage that it is now much easier to decide at what stage the iteration should be terminated.

Post-multiplying the coefficient matrix **A** by one of its eigenvectors gives the eigenvector multiplied by the corresponding eigenvalue. Thus, when we stop the iteration because $\mathbf{u}_{p+1}$ is sufficiently close to $\mathbf{u}_p$ to ensure convergence, $\max(\mathbf{v}_p)$ will be an estimate of the eigenvalue.

The rate of convergence of the iteration is primarily dependent on the distribution of the eigenvalues; the smaller the ratios $|\lambda_i/\lambda_1|$, where $i = 2, 3, \ldots, n$, the faster the convergence. The following MATLAB function `eigit` implements the iterative method to find the dominant eigenvalue and the associated eigenvector.

```
function [lam u iter] = eigit(A,tol)
% Solves EVP to determine dominant eigenvalue and associated vector
% Sample call: [lam u iter] = eigit(A,tol)
% A is a square matrix, tol is the accuracy
% lam is the dominant eigenvalue, u is the associated vector
% iter is the number of iterations required
[n n] = size(A);
err = 100*tol;
u0 = ones(n,1);  iter = 0;
while err>tol
    v = A*u0;
    u1 = (1/max(v))*v;
    err = max(abs(u1-u0));
    u0 = u1;  iter = iter+1;
end
u = u0;  lam = max(v);
```

We now apply this method to find the dominant eigenvalue and corresponding vector for the following eigenvalue problem.

$$\begin{bmatrix} 1 & 2 & 3 \\ 2 & 5 & -6 \\ 3 & -6 & 9 \end{bmatrix} \begin{bmatrix} x_1 \\ x_2 \\ x_3 \end{bmatrix} = \lambda \begin{bmatrix} x_1 \\ x_2 \\ x_3 \end{bmatrix} \tag{2.58}$$

```
>> A = [1 2 3;2 5 -6;3 -6 9];
>> [lam u iterations] = eigit(A,1e-8)

lam =
   13.4627

u =
    0.1319
   -0.6778
    1.0000

iterations =
    18
```

The dominant eigenvalue, to eight decimal places, is 13.46269899.

Iteration can also be used to determine the smallest eigenvalue of a system. The eigenvalue problem $\mathbf{Ax} = \lambda\mathbf{x}$ can be rearranged to give

$$\mathbf{A}^{-1}\mathbf{x} = (1/\lambda)\mathbf{x}$$

Here iteration will converge to the largest value of $1/\lambda$—that is, the smallest value of λ. However, as a general rule, matrix inversion should be avoided, particularly in large systems.

We have seen that direct iteration of $\mathbf{Ax} = \lambda\mathbf{x}$ leads to the largest or dominant eigenvalue. A second iterative procedure, called inverse iteration, provides a powerful method of determining subdominant eigensolutions. Consider again the eigenvalue problem of (2.37). Subtracting $\mu\mathbf{x}$ from both sides of this equation we have

$$(\mathbf{A} - \mu\mathbf{I})\mathbf{x} = (\lambda - \mu)\mathbf{x} \tag{2.59}$$

$$(\mathbf{A} - \mu\mathbf{I})^{-1}\mathbf{x} = \left(\frac{1}{\lambda - \mu}\right)\mathbf{x} \tag{2.60}$$

Consider the iterative scheme that begins with a trial vector $\mathbf{u}_0$. Then, using the equivalent of (2.57), we have

$$\left.\begin{aligned} \mathbf{v}_s &= (\mathbf{A} - \mu\mathbf{I})^{-1}\mathbf{u}_s \\ \mathbf{u}_{s+1} &= \left(\frac{1}{\max(\mathbf{v}_s)}\right)\mathbf{v}_s \end{aligned}\right\} \quad s = 0, 1, 2, \ldots \tag{2.61}$$

Iteration will lead to the largest value of $1/(\lambda - \mu)$—that is, the smallest value of $(\lambda - \mu)$. The smallest value of $(\lambda - \mu)$ implies that the value of λ will be the eigenvalue closest to μ and $\mathbf{u}$ will have converged to the eigenvector $\mathbf{x}$ corresponding to this particular eigenvalue. Thus, by a suitable choice of μ, we have a procedure for finding subdominant eigensolutions.

Iteration is terminated when $\mathbf{u}_{s+1}$ is sufficiently close to $\mathbf{u}_s$. When convergence is complete

$$\frac{1}{\lambda - \mu} = \max(\mathbf{v}_s)$$

Thus the value of λ nearest to μ is given by

$$\lambda = \mu + \frac{1}{\max(\mathbf{v}_s)} \tag{2.62}$$

The rate of convergence is fast, provided the chosen value of μ is close to an eigenvalue. If μ is equal to an eigenvalue, then $(\mathbf{A} - \mu\mathbf{I})$ is singular. In practice this seldom presents difficulties because it is unlikely that μ would be chosen, by chance, to exactly equal an eigenvalue. However, if $(\mathbf{A} - \mu\mathbf{I})$ is singular then we have confirmation that the eigenvalue is known to a very high precision. The corresponding eigenvector can then be obtained by changing μ by a small quantity and iterating to determine the eigenvector.

Although inverse iteration can be used to find the eigensolutions of a system about which we have no previous knowledge, it is more usual to use inverse iteration to refine the approximate eigensolution obtained by some other technique. In practice $(\mathbf{A}-\mu\mathbf{I})^{-1}$ is not formed explicitly; instead, $(\mathbf{A}-\mu\mathbf{I})$ is usually decomposed into the product of a lower and an upper triangular matrix. Explicit matrix inversion is avoided and is replaced by two efficient substitution procedures. In the simple MATLAB implementation of this procedure shown next, the operator \ is used to avoid matrix inversion.

```
function [lam u iter] = eiginv(A,mu,tol)
% Determines eigenvalue of A closest to mu with a tolerance tol.
% Sample call: [lam u] = eiginv(A,mu,tol)
% lam is the eigenvalue and u the corresponding eigenvector.
[n,n] = size(A);
err = 100*tol;
B = A-mu*eye(n,n);
u0 = ones(n,1);
iter = 0;
while err>tol
    v = B\u0; f = 1/max(v);
    u1 = f*v;
    err = max(abs(u1-u0));
    u0 = u1; iter = iter+1;
end
u = u0; lam = mu+f;
```

We now apply this function to find the eigenvalue of (2.58) nearest to 4 and the corresponding eigenvector.

```
>> A = [1 2 3;2 5 -6;3 -6 9];
>> [lam u iterations] = eiginv(A,4,1e-8)


lam =
    4.1283


u =

    1.0000
    0.8737
    0.4603


iterations =
     6
```

The eigenvalue closest to 4 is 4.12827017 to eight decimal places. The functions eigit and eiginv should be used with care when solving large-scale eigenvalue problems since convergence is not always guaranteed and in adverse circumstances may be slow.

We now discuss the MATLAB function eig in some detail.

2.17 The MATLAB Function eig

There are many algorithms available to solve the eigenvalue problem. The method chosen is influenced by many factors such as the form and size of the eigenvalue problem, whether or not it is symmetric, whether it is real or complex, whether or not only the eigenvalues are required, and whether all or only some of the eigenvalues and vectors are required.

We now describe the algorithms that are used in the MATLAB function eig. This MATLAB function can be used in several forms and, in the process, makes use of different algorithms. The different forms are as follows:

1. `lambda = eig(a)`
2. `[u lambda] = eig(a)`
3. `lambda = eig(a,b)`
4. `[u lambda]=eig(a,b)`

where lambda is a vector of eigenvalues in (1) and (3) and a diagonal matrix with the eigenvalues on the diagonal in (2) and (4). In these latter cases u is a matrix, the columns of which are the eigenvectors.

For real matrices the MATLAB function eig(a) proceeds as follows. If **A** is a general matrix, it is first reduced to Hessenberg form using Householder's transformation method. A Hessenberg matrix has zeros everywhere below the diagonal except for the first subdiagonal. If **A** is a symmetric matrix, the transform creates a tridiagonal matrix. Then the eigenvalues and eigenvectors of the real upper Hessenberg matrix are found by the iterative application of the QR procedure. The QR procedure involves decomposing the Hessenberg matrix into an upper triangular and a unitary matrix. The method is as follows:

1. $k = 0$.
2. Decompose $\mathbf{H}_k$ into $\mathbf{Q}_k$ and $\mathbf{R}_k$ such that $\mathbf{H}_k = \mathbf{Q}_k\mathbf{R}_k$ where $\mathbf{H}_k$ is a Hessenberg or tridiagonal matrix.
3. Compute $\mathbf{H}_{k+1} = \mathbf{R}_k\mathbf{Q}_k$. The estimates of the eigenvalues equal diag($\mathbf{H}_{k+1}$).
4. Check the accuracy of the eigenvalues. If the process has not converged, $k = k + 1$; repeat from (2).

The values on the leading diagonal of $\mathbf{H}_k$ tend to the eigenvalues. The following script uses the MATLAB function hess to convert the original matrix to the Hessenberg form, followed by the iterative application of the qr function to determine the eigenvalues of a symmetric matrix. Note that in this script we have iterated 10 times rather than use a formal test for convergence since the purpose of the script is merely to illustrate the functioning of the iterative application of the QR procedure.

```
% e3s213.m
A = [5 4 1 1;4 5 1 1; 1 1 4 2;1 1 2 4];
H1 = hess(A);
for i = 1:10
    [Q R] = qr(H1);
    H2 = R*Q;  H1 = H2;
    p = diag(H1)';
    fprintf('%2.0f %8.4f %8.4f',i,p(1),p(2))
    fprintf('%8.4f %8.4f\n',p(3),p(4))
end
```

Running this script gives

```
 1   1.0000   8.3636  6.2420   2.3944
 2   1.0000   9.4940  5.4433   2.0627
 3   1.0000   9.8646  5.1255   2.0099
 4   1.0000   9.9655  5.0329   2.0016
 5   1.0000   9.9913  5.0084   2.0003
 6   1.0000   9.9978  5.0021   2.0000
 7   1.0000   9.9995  5.0005   2.0000
 8   1.0000   9.9999  5.0001   2.0000
 9   1.0000  10.0000  5.0000   2.0000
10   1.0000  10.0000  5.0000   2.0000
```

The iteration converges to the values 1, 10, 5, and 2, which are the correct values. This QR iteration could be applied directly to the full matrix **A** but in general it would be inefficient. We have not given details of how the eigenvectors are computed.

When there are two real or complex arguments in the MATLAB function `eig`, the QZ algorithm is used instead of the QR algorithm. The QZ algorithm (Golub and Van Loan, 1989) has been modified to deal with the complex case. When `eig` is called using a single complex matrix `A` then the algorithm works by applying the QZ algorithm to `eig(a,eye(size(A)))`. The QZ algorithm begins by noting that there exists a unitary **Q** and **Z** such that $\mathbf{Q}^{\mathrm{H}}\mathbf{AZ} = \mathbf{T}$ and $\mathbf{Q}^{\mathrm{H}}\mathbf{BZ} = \mathbf{S}$ are both upper triangular. This is called generalized Schur decomposition. Providing s_{kk} is not zero then the eigenvalues are computed from the ratio t_{kk}/s_{kk}, where $k = 1,2,\ldots,n$. The following script demonstrates that the ratios of the diagonal elements of the **T** and **S** matrices give the required eigenvalues.

```
% e3s214.m
A = [10+2i 1 2;1-3i 2 -1;1 1 2];
b = [1 2-2i -2;4 5 6;7+3i 9 9];
[T S Q Z V] = qz(A,b);
r1 = diag(T)./diag(S)
r2 = eig(A,b)
```

Running this script gives

```
r1 =
   1.6154 + 2.7252i
  -0.4882 - 1.3680i
   0.1518 + 0.0193i

r2 =
   1.6154 + 2.7252i
  -0.4882 - 1.3680i
   0.1518 + 0.0193i
```

Schur decomposition is closely related to the eigenvalue problem. The MATLAB function `schur(a)` produces an upper triangular matrix **T** with real eigenvalues on its diagonal and complex eigenvalues in 2×2 blocks on the diagonal. Thus **A** can be written

$$\mathbf{A} = \mathbf{U}\mathbf{T}\mathbf{U}^{\mathrm{H}}$$

where **U** is a unitary matrix such that $\mathbf{U}^{\mathrm{H}}\mathbf{U} = \mathbf{I}$. The following script shows the similarity between Schur decomposition and the eigenvalues of a given matrix.

```
% e3s215.m
A = [4 -5 0 3;0 4 -3 -5;5 -3 4 0;3 0 5 4];
T = schur(A), lam = eig(A)
```

Running this script gives

```
T =
   12.0000    0.0000   -0.0000   -0.0000
         0    1.0000   -5.0000   -0.0000
         0    5.0000    1.0000   -0.0000
         0         0         0    2.0000

lam =
  12.0000
   1.0000 + 5.0000i
   1.0000 - 5.0000i
   2.0000
```

We can readily identify the four eigenvalues in the matrix `T`. The following script compares the performance of the `eig` function when solving various classes of problem.

```
% e3s216.m
disp('        real1    realsym1      real2    realsym2      comp1        comp2')
for n = 100:50:500
    A = rand(n); C = rand(n);
    S = A+C*i;
    T = rand(n)+i*rand(n);
    tic, [U,V] = eig(A); f1 = toc;
    B = A+A.'; D = C+C.';
    tic, [U,V] = eig(B); f2 = toc;
    tic, [U,V] = eig(A,C); f3 = toc;
    tic, [U,V] = eig(B,D); f4 = toc;
    tic, [U,V] = eig(S); f5 = toc;
    tic, [U,V] = eig(S,T); f6 = toc;
    fprintf('%12.3f %10.3f %10.3f %10.3f %10.3f %10.3f\n',f1,f2,f3,f4,f5,f6);
end
```

This script gives the time taken (in seconds) to carry out the various operations. The output is as follows:

real1	realsym1	real2	realsym2	comp1	comp2
0.042	0.009	0.063	0.061	0.039	0.037
0.067	0.014	0.086	0.090	0.067	0.106
0.129	0.028	0.228	0.184	0.116	0.200
0.182	0.046	0.430	0.425	0.186	0.432
0.270	0.073	0.729	0.724	0.279	0.782
0.371	0.104	1.277	1.257	0.373	1.232
0.514	0.154	2.006	2.103	0.538	2.104
0.708	0.205	3.055	3.097	0.698	2.919
0.946	0.278	4.403	4.187	0.901	4.344

In some circumstances not all the eigenvalues and eigenvectors are required. For example, in a complex engineering structure, modeled with many hundreds of degrees of freedom, we may only require the first 15 eigenvalues, giving the natural frequencies of the model, and the corresponding eigenvectors. MATLAB provides the function `eigs`, which finds a small number of eigenvalues, such as those with the largest amplitude, the largest or smallest real or imaginary part, and so on. This function is particularly useful when seeking a small number of eigenvalues of very large sparse matrices. Eigenvalue reduction algorithms are used to reduce the size of eigenvalue problem (for example,

Guyan, 1965) but still allow selected eigenvalues to be computed to an acceptable level of accuracy.

MATLAB also includes the facility to find the eigenvalues of a sparse matrix. The following script compares the number of floating-point operations required to find the eigenvalues of a matrix treated as sparse with the corresponding number required to find the eigenvalues of the corresponding full matrix.

```
% e3s217.m
% generate a sparse triple diagonal matrix
n = 2000;
rowpos = 2:n; colpos = 1:n-1;
values = ones(1,n-1);
offdiag = sparse(rowpos,colpos,values,n,n);
A = sparse(1:n,1:n,4*ones(1,n),n,n);
A = A+offdiag+offdiag.';
% generate full matrix
B = full(A);
tic, eig(A); sptim = toc;
tic, eig(B); futim = toc;
fprintf('Time for sparse eigen solve = %8.6f\n',sptim)
fprintf('Time for  full  eigen solve = %8.6f\n',futim)
```

The results from running this script are as follows:

```
Time for sparse eigen solve = 0.349619
Time for  full  eigen solve = 3.000229
```

Clearly there is a significant savings in time.

2.18 Summary

We have described many of the important algorithms related to computational matrix algebra and shown how the power of MATLAB can be used to illustrate the application of these algorithms in a revealing way. We have shown how to solve fully over- and underdetermined systems and eigenvalue problems. We have drawn the attention of the reader to the importance of sparsity in linear systems and demonstrated its significance. The scripts provided should help readers to develop their own applications.

In Chapter 9 we show how the symbolic toolbox can be usefully applied to solve some problems in linear algebra.

Problems

2.1. An $n \times n$ Hilbert matrix, $\mathbf{A}$, is defined by

$$a_{ij} = 1/(i+j-1) \quad \text{for} \quad i,j = 1,2,\ldots, n$$

Find the inverse of $\mathbf{A}$ and the inverse of $\mathbf{A}^\top\mathbf{A}$ for $n = 5$. Then, noting that

$$(\mathbf{A}^\top\mathbf{A})^{-1} = \mathbf{A}^{-1}(\mathbf{A}^{-1})^\top$$

find the inverse of $\mathbf{A}^\top\mathbf{A}$ using this result for $n = 3$, 4, 5, and 6. Compare the accuracy of the two results by using the inverse Hilbert function `invhilb` to find the exact inverse using $(\mathbf{A}^\top\mathbf{A})^{-1} = \mathbf{A}^{-1}(\mathbf{A}^{-1})^\top$. *Hint*: Compute norm$(\mathbf{P}-\mathbf{R})$ and norm$(\mathbf{Q}-\mathbf{R})$ where $\mathbf{P} = (\mathbf{A}^\top\mathbf{A})^{-1}$ and $\mathbf{Q} = \mathbf{A}^{-1}(\mathbf{A}^{-1})^\top$ and $\mathbf{R}$ is the exact inverse of $\mathbf{Q}$ obtained by using the `invhilb` function. .

2.2. Find the condition number of $\mathbf{A}^\top\mathbf{A}$ where $\mathbf{A}$ is an $n \times n$ Hilbert matrix, defined in Problem 2.1, for $n = 3, 4, \ldots, 6$. How do these results relate to the results of Problem 2.1?

2.3. It can be proved that the series $(\mathbf{I}-\mathbf{A}^{-1}) = \mathbf{I}+\mathbf{A}+\mathbf{A}^2+\mathbf{A}^3+\cdots$, where $\mathbf{A}$ is an $n \times n$ matrix, converges if the eigenvalues of $\mathbf{A}$ are all less than unity. The following $n \times n$ matrix satisfies this condition if $a+2b < 1$ and a and b are positive:

$$\begin{bmatrix} a & b & 0 & \ldots & 0 & 0 & 0 \\ b & a & b & \ldots & 0 & 0 & 0 \\ \vdots & \vdots & \vdots & & \vdots & \vdots & \vdots \\ 0 & 0 & 0 & \ldots & b & a & b \\ 0 & 0 & 0 & \ldots & 0 & b & a \end{bmatrix}$$

Experiment with this matrix for various values of n, a, and b to illustrate that the series converges under the condition stated.

2.4. Use the function `eig` to find the eigenvalues of the following matrix:

$$\begin{bmatrix} 2 & 3 & 6 \\ 2 & 3 & -4 \\ 6 & 11 & 4 \end{bmatrix}$$

Then use the `rref` function on the matrix $(\mathbf{A}-\lambda\mathbf{I})$, taking λ equal to any of the eigenvalues. Solve the resulting equations by hand to obtain the eigenvector of the matrix. *Hint*: Note that an eigenvector is the solution of $(\mathbf{A}-\lambda\mathbf{I})\mathbf{x} = \mathbf{0}$ for λ equal to a specific eigenvalue. Assume an arbitrary value for x_3.

2.5. For the system given in Problem 2.3, find the eigenvalues, assuming both full and sparse forms with $n = 10:10:30$. Compare your results with the exact solution given by

$$\lambda_k = a + 2b\cos\{k\pi/(n+1)\}, \quad k = 1, 2, \ldots$$

2.6. Find the solution of the overdetermined system that follows using `pinv`, `qr`, and the \ operator.

$$\begin{bmatrix} 2.0 & -3.0 & 2.0 \\ 1.9 & -3.0 & 2.2 \\ 2.1 & -2.9 & 2.0 \\ 6.1 & 2.1 & -3.0 \\ -3.0 & 5.0 & 2.1 \end{bmatrix} \begin{bmatrix} x_1 \\ x_2 \\ x_3 \end{bmatrix} = \begin{bmatrix} 1.01 \\ 1.01 \\ 0.98 \\ 4.94 \\ 4.10 \end{bmatrix}$$

2.7. Write a script to generate $\mathbf{E} = \{1/(n+1)\}\mathbf{C}$ where

$$\begin{aligned} c_{ij} &= i(n-i+1) && \text{if } i = j \\ &= c_{i,j-1} - i && \text{if } j > i \\ &= c_{ji} && \text{if } j < i \end{aligned}$$

Having generated $\mathbf{E}$ for $n = 5$, solve $\mathbf{Ex} = \mathbf{b}$ where $\mathbf{b} = [1:n]^\top$ by

(a) Using the \ operator

(b) Using the `lu` function and solving $\mathbf{Ux} = \mathbf{y}$ and $\mathbf{Ly} = \mathbf{b}$

2.8. Determine the inverse of $\mathbf{E}$ of Problem 2.7 for $n = 20$ and 50. Compare with the exact inverse, which is a matrix with 2 along the main diagonal and -1 along the upper and lower subdiagonals and zero elsewhere.

2.9. Determine the eigenvalues of $\mathbf{E}$ defined in Problem 2.7 for $n = 20$ and 50. The exact eigenvalues for this system are given by $\lambda_k = 1/[2 - 2\cos\{k\pi/(n+1)\}]$ where $k = 1, \ldots, n$.

2.10. Determine the condition number of $\mathbf{E}$ of Problem 2.7, using the MATLAB function `cond`, for $n = 20$ and 50. Compare your results with the theoretical expression for the condition number, which is $4n^2/\pi^2$.

2.11. Find the eigenvalues and the left and right eigenvectors using the MATLAB function `eig` for the matrix

$$\mathbf{A} = \begin{bmatrix} 8 & -1 & -5 \\ -4 & 4 & -2 \\ 18 & -5 & -7 \end{bmatrix}$$

2.12. For the following matrix $\mathbf{A}$, using `eigit`, `eiginv`, determine

(a) The largest eigenvalue

(b) The eigenvalue nearest 100

(c) The smallest eigenvalue

$$\mathbf{A} = \begin{bmatrix} 122 & 41 & 40 & 26 & 25 \\ 40 & 170 & 25 & 14 & 24 \\ 27 & 26 & 172 & 7 & 3 \\ 32 & 22 & 9 & 106 & 6 \\ 31 & 28 & -2 & -1 & 165 \end{bmatrix}$$

2.13. Given that

$$\mathbf{A} = \begin{bmatrix} 1 & 2 & 2 \\ 5 & 6 & -2 \\ 1 & -1 & 0 \end{bmatrix} \quad \text{and} \quad \mathbf{B} = \begin{bmatrix} 2 & 0 & 1 \\ 4 & -5 & 1 \\ 1 & 0 & 0 \end{bmatrix}$$

and defining **C** by

$$\mathbf{C} = \begin{bmatrix} \mathbf{A} & \mathbf{B} \\ \mathbf{B} & \mathbf{A} \end{bmatrix}$$

verify using `eig` that the eigenvalues of **C** are given by a combination of the eigenvalues of $\mathbf{A}+\mathbf{B}$ and $\mathbf{A}-\mathbf{B}$.

2.14. Write a MATLAB script to generate the matrix

$$\mathbf{A} = \begin{bmatrix} n & n-1 & n-2 & \dots & 2 & 1 \\ n-1 & n-1 & n-2 & \dots & 2 & 1 \\ n-2 & n-2 & n-2 & \dots & 2 & 1 \\ \vdots & \vdots & \vdots & \vdots & \vdots & \vdots \\ 2 & 2 & 2 & \dots & 2 & 1 \\ 1 & 1 & 1 & \dots & 1 & 1 \end{bmatrix}$$

The eigenvalues of this matrix are given by the formula

$$\lambda_i = \frac{1}{2}\left[1 - \cos\frac{(2i-1)\pi}{2n+1}\right], \quad i = 1, 2 \dots, n$$

Taking $n = 5$ and $n = 50$ and using the MATLAB function `eig`, find the largest and smallest eigenvalues. Verify your results are correct using the preceding formula.

2.15. Taking $n = 10$, find the eigenvalues of the matrix

$$\mathbf{A} = \begin{bmatrix} 1 & 0 & 0 & \dots & 0 & 1 \\ 0 & 1 & 0 & \dots & 0 & 2 \\ 0 & 0 & 1 & \dots & 0 & 3 \\ \vdots & \vdots & \vdots & \dots & \vdots & \vdots \\ 0 & 0 & 0 & \dots & 1 & n-1 \\ 1 & 2 & 3 & \dots & n-1 & n \end{bmatrix}$$

using `eig`. As an alternative, find the eigenvalues of **A** by first generating the characteristic polynomial using `poly` for the matrix **A** and then using `roots` to find the roots of the resulting polynomial. What conclusions do you draw from these results?

2.16. For the matrix given in Problem 2.12, use `eig` to find the eigenvalues. Then find the eigenvalues of **A** by first generating the characteristic polynomial for **A** using `poly` and then using `roots` to find the roots of the resulting polynomial. Use `sort` to compare the results of the two approaches. What conclusions do you draw from these results?

2.17. For the matrix given in Problem 2.14, taking $n = 10$, show that the trace is equal to the sum of the eigenvalues and the determinant is equal to the product of the eigenvalues. Use the MATLAB functions `det`, `trace`, and `eig`.

2.18. The matrix **A** is defined as follows:

$$\mathbf{A} = \begin{bmatrix} 2 & -1 & 0 & 0 & \dots & 0 \\ -1 & 2 & -1 & 0 & \dots & 0 \\ 0 & -1 & 2 & -1 & \dots & 0 \\ \vdots & \vdots & \vdots & \vdots & \vdots & \vdots \\ 0 & 0 & \dots & -1 & 2 & -1 \\ 0 & 0 & \dots & 0 & -1 & 2 \end{bmatrix}$$

The condition number for this matrix takes the form $c = pn^q$ where n is the size of the matrix, c is the condition number, and p and q are constants. By computing the condition number for the matrix **A** for $n = 5:5:50$ using the MATLAB function `cond`, fit the function pn^q to the set of results you produce. *Hint*: Take logs of both sides of the equation for c and solve the system of overdetermined equations using the \ operator.

2.19. An approximation for the inverse of $(\mathbf{I} - \mathbf{A})$ where **I** is an $n \times n$ unit matrix and **A** is an $n \times n$ matrix is given by

$$(\mathbf{I} - \mathbf{A})^{-1} = \mathbf{I} + \mathbf{A} + \mathbf{A}^2 + \mathbf{A}^3 + \cdots$$

This series only converges and the approximation is only valid if the maximum eigenvalue of **A** is less than 1. Write a MATLAB function `invapprox(A,k)` that obtains an approximation to $(\mathbf{I}-\mathbf{A})^{-1}$ using k terms of the given series. The function must find all eigenvalues of A using the MATLAB function `eig`. If the largest eigenvalue is greater than one then a message will be output indicating that the method fails. Otherwise, the function will compute an approximation to $(\mathbf{I}-\mathbf{A})^{-1}$ using k terms of the series expansion given. Taking $k=4$, test the function on the matrices:

$$\begin{bmatrix} 0.2 & 0.3 & 0 \\ 0.3 & 0.2 & 0.3 \\ 0 & 0.3 & 0.2 \end{bmatrix} \quad \text{and} \quad \begin{bmatrix} 1.0 & 0.3 & 0 \\ 0.3 & 1.0 & 0.3 \\ 0 & 0.3 & 1.0 \end{bmatrix}$$

Use the `norm` function to compare the accuracy of the inverse of the matrix $(\mathbf{I}-\mathbf{A})$ found using the MATLAB `inv` function and the function `invapprox(A,k)` for $k=4$, 8, 16.

2.20. The system of equations $\mathbf{Ax}=\mathbf{b}$, where **A** is a matrix of m rows and n columns, **x** is an n element column vector, and **b** is an m element column vector, is said to be underdetermined if $n > m$. The direct use of the MATLAB `inv` function to solve this system fails since the matrix **A** is not square. However, multiplying both sides of the equation by $\mathbf{A}^\top$ gives

$$\mathbf{A}^\top\mathbf{Ax}=\mathbf{A}^\top\mathbf{b}$$

$\mathbf{A}^\top\mathbf{A}$ is a square matrix and the MATLAB `inv` function can now be used to solve the system. Write a MATLAB function to use this result to solve underdetermined systems. The function should allow the input of the **b** vector and the **A** matrix, form the necessary matrix products, and use the MATLAB `inv` function to solve the system. The accuracy of the solution should be checked using the MATLAB `norm` function to measure the difference between **Ax** and **b**. The function must also include the direct use of the MATLAB \ symbol to solve the same underdetermined linear system, again with a check on the accuracy of the solution that uses the MATLAB `norm` function to measure the difference between **Ax** and **b**. The function should take the form `udsys(A,b)` and return the solutions given by the different methods and the norms produced by the two methods. Test your program by using it to solve the underdetermined system of linear equations $\mathbf{Ax}=\mathbf{b}$ where

$$\mathbf{A}=\begin{bmatrix} 1 & -2 & -5 & 3 \\ 3 & 4 & 2 & -7 \end{bmatrix} \quad \text{and} \quad \mathbf{b}=\begin{bmatrix} -10 \\ 20 \end{bmatrix}$$

What conclusions do you draw regarding the two methods by comparing the norms that the two methods produce?

2.21. An orthogonal matrix **A** is defined as a square matrix such that the product of the matrix and its transpose equals the unit matrix or

$$\mathbf{A}\mathbf{A}^\top = \mathbf{I}$$

Use MATLAB to verify that the following matrices are orthogonal:

$$\mathbf{B} = \begin{bmatrix} \frac{1}{\sqrt{3}} & \frac{1}{\sqrt{6}} & -\frac{1}{\sqrt{2}} \\ \frac{1}{\sqrt{3}} & \frac{-2}{\sqrt{6}} & 0 \\ \frac{1}{\sqrt{3}} & \frac{1}{\sqrt{6}} & \frac{1}{\sqrt{2}} \end{bmatrix}$$

$$\mathbf{C} = \begin{bmatrix} \cos(\pi/3) & \sin(\pi/3) \\ -\sin(\pi/3) & \cos(\pi/3) \end{bmatrix}$$

2.22. Write MATLAB scripts to implement both the Gauss–Seidel and the Jacobi method and use them to solve, with an accuracy of 0.000005, the equation system $\mathbf{Ax} = \mathbf{b}$ where the elements of **A** are

$$\begin{aligned} &a_{ii} = -4 \\ &a_{ij} = 2 \text{ if } |i-j| = 1 \\ &a_{ij} = 0 \text{ if } |i-j| \geq 2 \quad \text{where} \quad i,j = 1,2,\ldots,10 \end{aligned}$$

and

$$\mathbf{b}^\top = [2\ 3\ 4\ \ldots\ 11]$$

Use initial values of $x_i = 0,\ i = 1,2,\ldots,10$. (You might also like to experiment with other initial values.) Check your results by solving the system using the MATLAB \ operator.

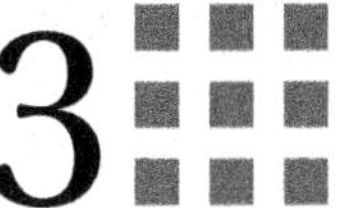

3 Solution of Nonlinear Equations

The problem of solving nonlinear equations arises frequently and naturally from the study of a wide range of practical problems. The problem may involve a system of nonlinear equations in many variables or one equation in one unknown. We shall initially confine ourselves to considering the solution of one equation in one unknown. The general form of the problem may be simply stated as finding a value of the variable x such that

$$f(x) = 0$$

where f is any nonlinear function of x. The value of x is then called a solution or root of this equation and may be one of many values satisfying the equation.

3.1 Introduction

To illustrate our discussion and provide a practical insight into the solution of nonlinear equations we shall consider an equation described by Armstrong and Kulesza (1981). These authors report a problem that arises from the study of resistive mixer circuits. Given an applied current and voltage, it is necessary to find the current flowing in part of the circuit. This leads to a simple nonlinear equation, which after some manipulation may be expressed in the form

$$x - \exp(-x/c) = 0 \text{ or equivalently } x = \exp(-x/c) \tag{3.1}$$

Here c is a given constant and x is the variable we wish to determine. The solution of such equations is not obvious, but Armstrong and Kulesza provide an approximate solution based on a series expansion that gives a reasonably accurate solution of this equation for a large range of values of c. This approximation is given in terms of c by

$$x = cu[1 - \log_e\{(1 + c)u\}/(1 + u)] \tag{3.2}$$

where $u = \log_e(1 + 1/c)$. This is an interesting and useful result since it is reasonably accurate for values of c in the five-decade range $[10^{-3}, 100]$ and gives a relatively easy way of finding the solutions of a whole family of equations generated by varying c. Although this result is useful for this particular equation, when we attempt to use this type of *ad hoc*

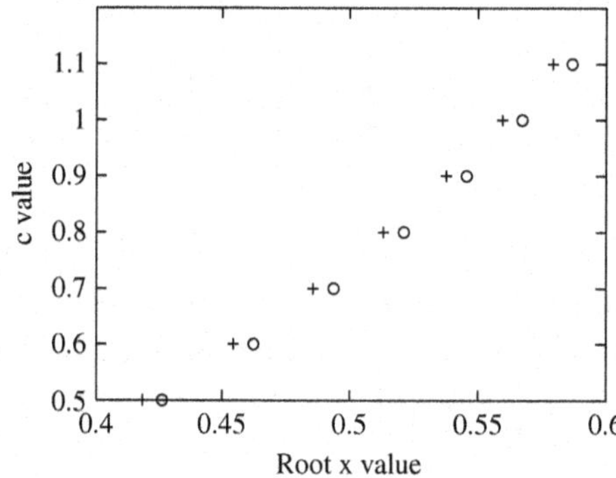

FIGURE 3.1 Solution of $x = \exp(-x/c)$. Results from the function `fzero` are indicated by $\circ$ and those from the Armstrong and Kulesza formula by $+$.

approach for the general solution of nonlinear equations, there are significant drawbacks. These are

1. *Ad hoc* approaches to the solutions of equations are rarely as successful as this example in finding a formula for the solution of a given equation; usually it is impossible to obtain such formulae.
2. Even when they exist, such formulae require considerable time and ingenuity to develop.
3. We may require greater accuracy than any ad hoc formula can provide.

To illustrate point 3 consider Figure 3.1, which is generated by the MATLAB script that follows. This figure shows the results obtained using the formula (3.2) together with the results using the MATLAB function `fzero` to solve the nonlinear equation (3.1).

```
% e3s301.m
ro = [ ]; ve = [ ]; x = [ ];
c = 0.5:0.1:1.1; u = log(1+1./c);
x = c.*u.*(1-log((1+c).*u)./(1+u));
% solve equation using MATLAB function fzero
i = 0;
for c1 = 0.5:0.1:1.1
    i = i+1;
    ro(i) = fzero(@(x) x-exp(-x/c1),1,0.00005);
end
plot(x,c,'+')
axis([0.4 0.6 0.5 1.2])
hold on
plot(ro,c,'o')
xlabel('Root x value'), ylabel('c value')
hold off
```

The function `fzero` is discussed in detail in Section 3.10. Note that the call `fzero` takes the form `fzero(@(x) x-exp(-x/c1),1,0.00005)`. This gives an accuracy of 0.00005 for the roots and uses an initial approximation of 1. The function `fzero` provides the root with up to 16-digit accuracy, if required, whereas the formula (3.2) of Armstrong and Kulesza, although faster, gives the result to one or two decimal places only. In fact, the method of Armstrong and Kulesza becomes more accurate for large values of c.

From the preceding discussion we conclude that, although occasionally ingenious alternatives may be available, in the vast majority of cases we must use algorithms which provide, with reasonable computational effort, the solutions of general problems to any specified accuracy. Before describing the nature of these algorithms in detail, we consider different types of equations and the general nature of their solutions.

3.2 The Nature of Solutions to Nonlinear Equations

We illustrate the nature of the solutions to nonlinear equations by considering two equations that we wish to solve for the variable x.

(a) $(x-1)^3(x+2)^2(x-3) = 0$ – that is,
$x^6 - 2x^5 - 8x^4 + 14x^3 + 11x^2 - 28x + 12 = 0$

(b) $\exp(-x/10)\sin(10x) = 0$

The first equation is a special type of nonlinear equation known as a polynomial equation since it involves only integer powers of the variable x and no other functions. Such polynomial equations have the important characteristic that they have n roots where n is the degree of the polynomial. In this example the highest power of x, and hence the degree of the polynomial, is six. The solutions of a polynomial may be complex or real, separate or coincident. Figure 3.2 illustrates the nature of the solutions of this equation. Although there must be six roots, three are coincident at $x = 1$ and two are coincident at $x = -2$. There is also a single root at $x = 3$. Coincident roots may present difficulties for

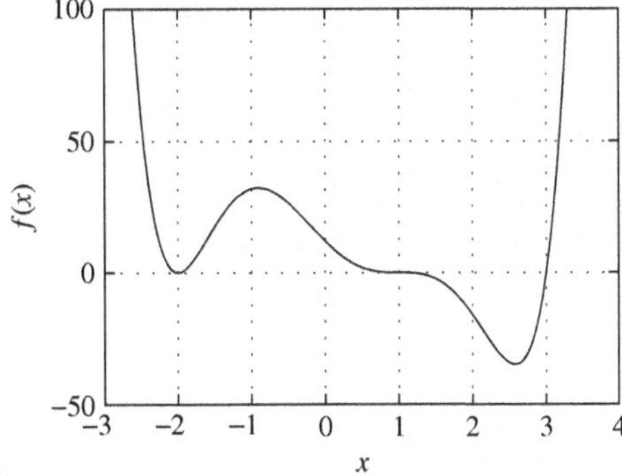

FIGURE 3.2 Plot of the function $f(x) = (x-1)^3(x+2)^2(x-3)$.

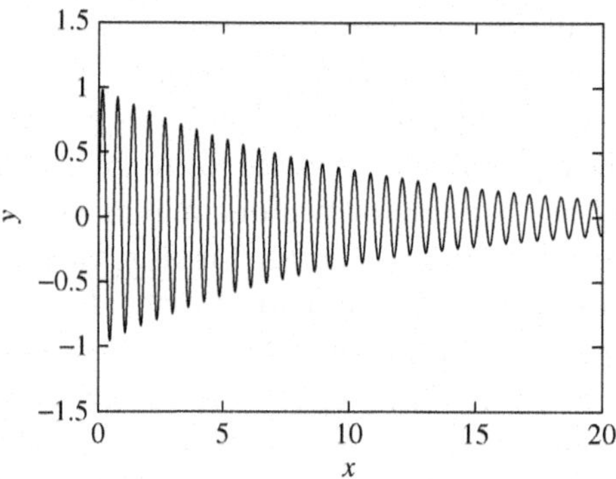

FIGURE 3.3 Plot of $f(x) = \exp(-x/10)\sin(10x)$.

some algorithms, as do roots which are very close together, so it is important to appreciate their existence. The user may require a particular root of the equation or all the roots. In the case of polynomial equations special algorithms exist to find all the roots.

The preceding second equation is a nonlinear equation involving transcendental functions. The task of finding all the roots of this class of nonlinear equation is a daunting one, since the number of roots may not be known or there may be an infinity of roots. This situation is illustrated by Figure 3.3, which shows the graph of the second equation for x in the range $[0, 20]$. If we extended the range of x, more roots would be revealed.

We now consider some simple algorithms to find a specific root of a given nonlinear equation.

3.3 The Bisection Algorithm

This simple algorithm assumes that an initial interval is known in which a root of the equation $f(x) = 0$ lies and then proceeds to reduce this interval until the required accuracy is achieved for the root. This algorithm is mentioned only briefly since it is not in practice used by itself but in conjunction with other algorithms to improve their reliability. The algorithm may be described by

<u>input</u> interval in which the root lies

<u>while</u> interval too large

1. Bisect the current interval in which the root lies.
2. Determine in which half of the interval the root lies.

<u>end</u>

<u>display</u> root

The principles on which this algorithm based is simple. Given an initial interval in which a specific root lies, the algorithm will provide an improved approximation for the root.

However, the requirement that an interval be known is sometimes difficult to achieve, and although the algorithm is reliable it is extremely slow.

Alternative algorithms have been developed that converge more rapidly; this chapter is concerned with describing some of the most important of these. All the algorithms we consider are iterative in character—that is, they proceed by repeating the same sequence of steps until the root approximation is accurate enough to satisfy the user. We now consider the general form of an iterative method, the nature of the convergence of such methods, and the problems they encounter.

3.4 Iterative or Fixed Point Methods

We are required to solve the general equation $f(x) = 0$; however, to illustrate iterative methods clearly we consider a simple example. Suppose we wish to solve the quadratic

$$x^2 - x - 1 = 0 \tag{3.3}$$

This equation can be solved by using the standard formula for solving quadratics but we take a different approach. Rearrange (3.3) as follows:

$$x = 1 + 1/x$$

Then rewrite it in iterative form using subscripts as follows:

$$x_{r+1} = 1 + 1/x_r \quad \text{for} \quad r = 0, 1, 2, \ldots \tag{3.4}$$

Assuming we have an initial approximation x_0 to the root we are seeking, we can proceed from one approximation to another using this formula. The iterates we obtain in this way may or may not converge to the solution of the original equation. This is not the only iterative procedure for attempting to solve (3.3); we can generate two others from (3.3) as follows:

$$x_{r+1} = x_r^2 - 1 \quad \text{for} \quad r = 0, 1, 2, \ldots \tag{3.5}$$

and

$$x_{r+1} = \sqrt{x_r + 1} \quad \text{for} \quad r = 0, 1, 2, \ldots \tag{3.6}$$

Starting from the same initial approximation, these iterative procedures may or may not converge to the same root. Table 3.1 shows what happens when we use the initial approximation $x_0 = 2$ with the iterative procedures (3.4), (3.5), and (3.6). It shows that iterations (3.4) and (3.6) converge but (3.5) does not.

Note that when the root is reached no further improvement is possible and the point remains fixed. Hence the roots of the equation are the *fixed points* of the iteration. To remove the unpredictability of this approach we must be able to find general conditions

Table 3.1 Difference between Exact Root and Iterate for $x^2 - x - 1 = 0$

Iteration (3.4)	Iteration (3.5)	Iteration (3.6)
−0.1180	1.3820	0.1140
0.0486	6.3820	0.0349
−0.0180	61.3820	0.0107
0.0070	3966.3820	0.0033
−0.0026	15745021.3820	0.0010

that determine when such iterative schemes converge, when they do not, and the nature of this convergence.

3.5 The Convergence of Iterative Methods

The procedure described in Section 3.4 can be applied to any equation $f(x) = 0$ and has the general form

$$x_{r+1} = g(x_r) \quad \text{for} \quad r = 0, 1, 2, \ldots \tag{3.7}$$

It is not our purpose to give the details of the derivation of convergence conditions for this form of iteration, but to point out some of the difficulties that may arise in using them even when this condition is satisfied. The detailed derivation is given in many textbooks; see, for example, Lindfield and Penny (1989). It can be shown that the approximate relation between the current error ε_{r+1} at the $(r+1)$th iteration and the previous error ε_r is given by

$$\varepsilon_{r+1} = \varepsilon_r g'(t_r)$$

where t_r is a point lying between the exact root and the current approximation to the root. Thus the error will be decreasing if the absolute value of the derivative at these points is less than 1. However, this does not guarantee convergence from all starting points and the initial approximation must be sufficiently close to the root for convergence to occur.

In the case of the specific iterative procedures (3.4) and (3.5), Table 3.2 shows how the values of the derivatives of the corresponding $g(x)$ vary with the values of the approximations to x_r. This table provides numerical evidence for the theoretical assertion in the case of iterations (3.4) and (3.5).

However, the concept of convergence is more complex than this. We need to give some answer to the crucial question: If an iterative procedure converges, how can we classify the rate of convergence? We do not derive this result but refer the reader to Lindfield and Penny (1989) and state the answer to the question. Suppose all derivatives of the function $g(x)$ of order 1 to $p-1$ are zero at the exact root a. Then the relation between the current

Table 3.2 Values of the Derivatives for Iterations Given by (3.4) and (3.5)

Iteration (3.4)	Derivative	Iteration (3.5)	Derivative
−0.1180	−0.44	1.3820	6.00
0.0486	−0.36	6.3820	16.00
−0.0180	−0.39	61.3820	126.00
0.0070	−0.38	3966.3820	7936.00

error ε_{r+1} at the $(r + 1)$th iteration and the previous error ε_r is given by

$$\varepsilon_{r+1} = (\varepsilon_r)^p g^{(p)}(t_r)/p! \tag{3.8}$$

where t_r lies between the exact root and the current approximation to the root and $g^{(p)}$ denotes the pth derivative of g. The importance of this result is that it means the current error is proportional to the pth power of the previous error and clearly, on the basis of the reasonable assumption that the errors are much smaller than 1, the higher the value of p, the faster the convergence. Such methods are said to have pth-order convergence. In general it is cumbersome to derive iterative methods of order higher than two or three and second-order methods have proved very satisfactory in practice for solving a wide range of nonlinear equations. In this case, the current error is proportional to the square of the previous error. This is often called quadratic convergence; if the error is proportional to the previous error it is called linear convergence. This provides a convenient classification for the convergence of iterative methods but avoids the difficult questions: For what range of starting values will the process converge and how sensitive is convergence to changes in the starting values?

3.6 Ranges for Convergence and Chaotic Behavior

We illustrate some of the problems of convergence by considering a specific example that highlights some of the difficulties. Short (1992) examined the behavior of the iterative process

$$x_{r+1} = -0.5(x_r^3 - 6x_r^2 + 9x_r - 6) \quad \text{for} \quad r = 0, 1, 2, \ldots$$

for solving the equation $(x - 1)(x - 2)(x - 3) = 0$. This iterative procedure clearly has the form

$$x_{r+1} = g(x_r), \quad r = 0, 1, 2, \ldots$$

and it is easy to verify that it has the following properties:

$$g'(1) = 0 \quad \text{and} \quad g''(1) \neq 0$$
$$g'(2) \neq 0$$
$$g'(3) = 0 \quad \text{and} \quad g''(3) \neq 0$$

Thus by taking $p = 2$ in result (3.8), we can expect, for appropriate starting values, quadratic convergence for the roots at $x = 1$ and $x = 3$ but at best linear convergence for the root at $x = 2$. The major problem is, however, to determine the ranges of initial approximation that will converge to the different roots. This is not an easy task but one simple way of doing this is to draw a graph of $y = x$ and $y = g(x)$. The points of intersection provide the roots. The line $y = x$ has a slope of 1, and points where the slope of $g(x)$ is less than this provide a range of initial approximations that converge to one or other of the roots.

This graphical analysis shows that points within the range 1 to 1.43 (approximately) converge to the root 1 and points in the range 2.57 (approximately) to 3 converge to the root 3. This is the obvious part of the analysis. However, Short demonstrates that there are many other ranges of convergence for this iterative procedure, many of them very narrow indeed, which lead to chaotic behavior in the iterative process. He demonstrates, for example, that taking $x_0 = 4.236067968$ will converge to the root $x = 3$ whereas taking $x_0 = 4.236067970$ converges to the root $x = 1$, a remarkable change for such a small variation in the initial approximation. This should serve as a warning to the reader that the study of convergence properties is in general not an easy task.

Figure 3.4 illustrates this point quite strikingly. It shows the graph of x and the graph of $g(x)$ where

$$g(x) = -0.5(x^3 - 6x^2 + 9x - 6)$$

The x line intersects with $g(x)$ to give the roots of the original equation. The graph also shows iterates starting from $x_0 = 4.236067968$, indicated by "$\circ$," and iterates starting from

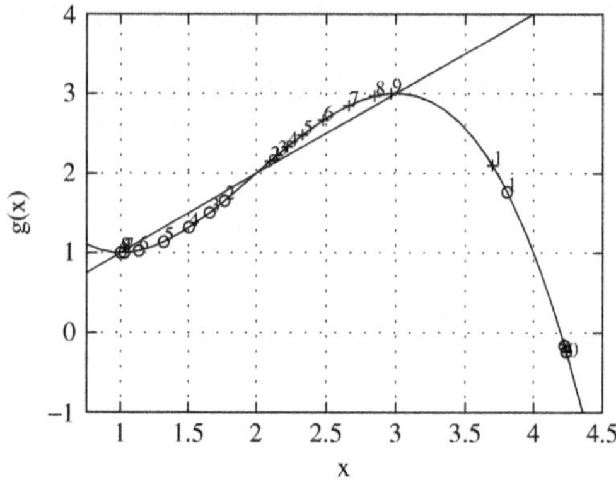

FIGURE 3.4 Iterates in the solution of $(x-1)(x-2)(x-3) = 0$ from close but different starting points.

$x = 4.236067970$, indicated by "+." The starting points are so close they are of course superimposed on the graph. However, the iterates soon take their separate paths to converge on different roots of the equation. The path indicated by "∘" converges to the root $x = 1$ and the path indicated by "+" converges to the root $x = 3$. The sequence of numbers on the graph shows the last nine iterates. The point referenced by zero is in fact all the points that are initially very close together. This is a remarkable example and users should verify these phenomena for themselves by running the following MATLAB script:

```
% e3s302.m
x = 0.75:0.1:4.5;
g = -0.5*(x.^3-6*x.^2+9*x-6);
plot(x,g)
axis([.75,4.5,-1,4])
hold on, plot(x,x)
xlabel('x'), ylabel('g(x)'), grid on
ch = ['o','+'];
num = [ '0','1','2','3','4','5','6','7','8','9'];
ty = 0;
for x1 = [4.236067970 4.236067968]
    ty = ty+1;
    for i = 1:19
        x2 = -0.5*(x1^3-6*x1^2+9*x1-6);
        % First ten points very close, so represent by '0'
        if i==10
            text(4.25,-0.2,'0')
        elseif i>10
            text(x1,x2+0.1,num(i-9))
        end
        plot(x1,x2,ch(ty))
        x1 = x2;
    end
end
hold off
```

It is interesting to note that the iterative form

$$x_{r+1} = x_r^2 + c \quad \text{for} \quad r = 0, 1, 2, \ldots$$

demonstrates strikingly chaotic behavior when the iterates are plotted in the complex plane and for complex ranges of values for c.

We now return to the more mundane task of developing algorithms that work in general for the solution of nonlinear equations. In the next section we shall consider a simple method of order 2.

3.7 Newton's Method

This method for the solution of the equation $f(x) = 0$ is based on the simple geometric properties of the tangent to the curve $f(x)$. The method requires some initial approximation to the root and that the derivative of $f(x)$ exists in the range of interest. Figure 3.5 illustrates the operation of the method. The diagram shows the tangent to the curve at the current approximation x_0. This tangent strikes the x-axis at x_1 and provides us with an improved approximation to the root. Similarly, the tangent at x_1 gives the improved approximation x_2.

The process is repeated until some convergence criterion is satisfied. It is easy to translate this geometrical procedure into a numerical method for finding the root since the tangent of the angle between the x-axis and the tangent equals

$$f(x_0)/(x_1 - x_0)$$

and the slope of this tangent itself equals $f'(x_0)$, the derivative of $f(x)$ at x_0. So we have the equation

$$f'(x_0) = f(x_0)/(x_1 - x_0)$$

Thus the improved approximation, x_1, is given by

$$x_1 = x_0 - f(x_0)/f'(x_0)$$

This may be written in iterative form as

$$x_{r+1} = x_r - f(x_r)/f'(x_r) \quad \text{where} \quad r = 0, 1, 2, \ldots \tag{3.9}$$

We note that this method is of the general iterative form

$$x_{r+1} = g(x_r) \quad \text{where} \quad r = 0, 1, 2, \ldots$$

Consequently, the discussion of Section 3.5 applies to it. On computing $g'(a)$, where a is the exact root, we find it is zero. However, $g''(a)$ is in general nonzero so the method is of order

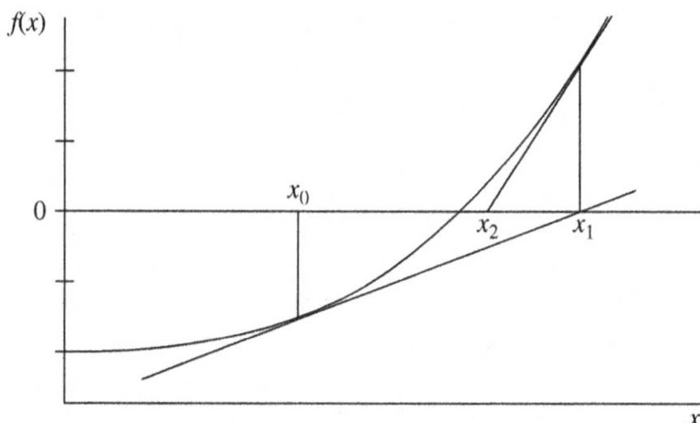

FIGURE 3.5 Geometric interpretation of Newton's method.

2 and we expect convergence to be quadratic. For a sufficiently close initial approximation, convergence to the root will be rapid.

A MATLAB function `fnewton` is supplied for Newton's method. The function that forms the left side of the equation we wish to solve *and* its derivative must be supplied by the user as functions; these become the first and second parameters of the function. The third parameter is an initial approximation to the root. The convergence criterion used is that the difference between successive approximations to the root is less than a small preset value. This value must be supplied by the user and is given as the fourth parameter of the function.

```
function [res, it] = fnewton(func,dfunc,x,tol)
% Finds a root of f(x) = 0 using Newton's method.
% Example call: [res, it] = fnewton(func,dfunc,x,tol)
% The user defined function func is the function f(x).
% The user defined function dfunc is df/dx.
% x is an initial starting value, tol is required accuracy.
it = 0; x0 = x;
d = feval(func,x0)/feval(dfunc,x0);
while abs(d) > tol
    x1 = x0-d;  it = it+1;  x0 = x1;
    d = feval(func,x0)/feval(dfunc,x0);
end
res = x0;
```

We will now find a root of the equation

$$x^3 - 10x^2 + 29x - 20 = 0$$

To use Newton's method we must define the function and its derivative as follows:

```
>> f = @(x) x.^3-10*x.^2+29*x-20;
>> df = @(x) 3*x.^2-20*x+29;
```

We may call the function `fnewton` as follows:

```
>> [x,it] = fnewton(f,df,7,0.00005)

x =

    5.0000


it =

     6
```

The progress of the iterations when solving $x^3 - 10x^2 + 29x - 20 = 0$ by Newton's method is shown in Figure 3.6.

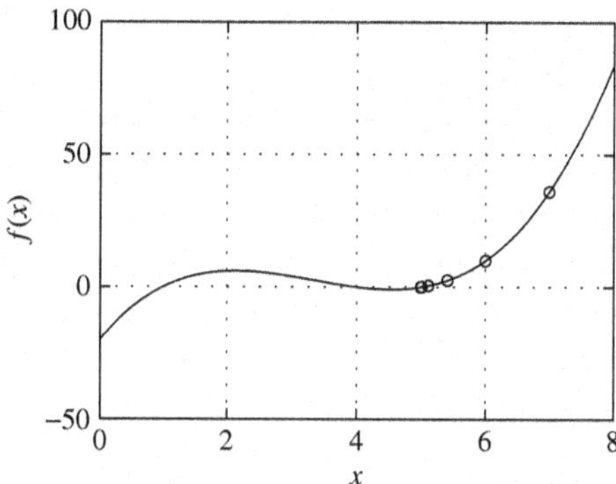

FIGURE 3.6 Plot of $x^3 - 10x^2 + 29x - 20 = 0$ with the iterates of Newton's method shown by ◦.

Table 3.3 Newton's Method to Solve $x^3 - 10x^2 + 29x - 20 = 0$ with an Initial Approximation of −2

x value	Error ε_r	$2\varepsilon_{r+1}/\varepsilon_r^2$	Approximate Second Derivative of g
−2.000000	3.000000	−0.320988	−0.395062
−0.444444	1.444444	−0.513956	−0.589028
0.463836	0.536164	−0.792621	−0.845260
0.886072	0.113928	−1.060275	−1.076987
0.993119	0.006881	−1.159637	−1.160775
0.999973	0.000027	−1.166639	−1.166643
1.000000	0.000000	−1.166639	−1.166667

Table 3.3 gives numerical results for this problem when Newton's method is used to seek a root, starting the iteration at −2. The second column of the table gives the current error ε_r by subtracting the known exact root from the current iterate. The third column contains the value of $2\varepsilon_{r+1}/\varepsilon_r^2$. This value tends to a constant as the process proceeds. From theoretical considerations, this value should approach the second derivative of the right side of the Newton iterative formula. This follows from (3.8) with $p = 2$. The final column contains the value of the second-order derivative of $g(x)$ calculated as follows. From (3.9) we have $g(x) = x - f(x)/f'(x)$. Thus from this we have

$$g'(x) = 1 - [\{f'(x)\}^2 - f''(x)f(x)]/[f'(x)]^2 = f''(x)f(x)/[f'(x)]^2$$

On differentiating again,

$$g''(x) = [\{f'(x)\}^2\{f'''(x)f(x) + f''(x)f'(x)\} - 2f'(x)\{f''(x)\}^2 f(x)]/[f'(x)]^4$$

Putting $x = a$, where a is the exact root, since $f(a) = 0$, we have

$$g''(a) = f''(a)/f'(a) \tag{3.10}$$

Thus we have a value for the second derivative of $g(x)$ when $x = a$. We note that as x approaches the root, the final column of Table 3.3, which uses this formula, gives an increasingly accurate approximation to the second derivative of $g(x)$. The table thus verifies our theoretical expectations.

We can find complex roots using Newton's method, providing our initial approximation is complex. For example, consider

$$\cos x - x = 0 \tag{3.11}$$

This equation has only one real root, which is $x = 0.7391$, but it has an infinity of complex roots. Figure 3.7 shows the distribution of the roots of (3.11) in the complex plane in the range $-30 < \text{Re}(x) < 30$. Working with complex values presents no additional difficulty in the MATLAB environment since MATLAB implements complex arithmetic and so we can use the function `fnewton` without modification to deal with these cases.

Figure 3.8 illustrates the fact that it is difficult to predict which root we will find from a given starting value. This figure shows that the starting values $15 + \jmath 10$, $15.2 + \jmath 10$, $15.4 + \jmath 10$, $15.8 + \jmath 10$, and $16 + \jmath 10$, which are close together, lead to a sequence of iterations that converge to very different roots. In one case the complete trajectory is not shown because the complex part of the intermediate iterates is well outside the range of the graph.

Newton's method requires the first derivative of $f(x)$ to be supplied by the user. To make the procedure more self-contained we can use a standard approximation to the first derivative, which takes the form

$$f'(x_r) = \{f(x_r) - f(x_{r-1})\}/(x_r - x_{r-1}) \tag{3.12}$$

FIGURE 3.7 Plot showing the complex roots of $\cos x - x = 0$.

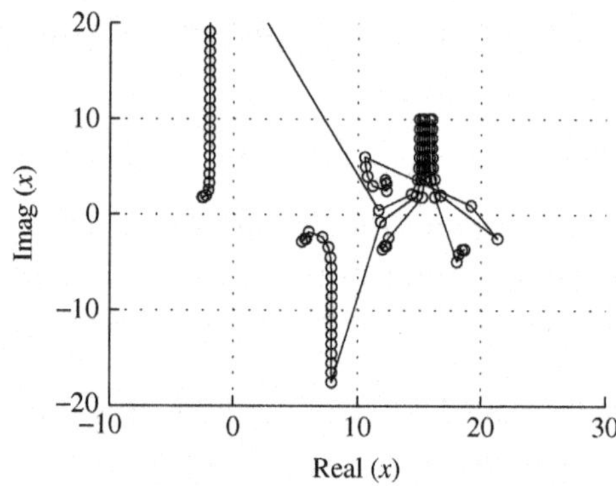

FIGURE 3.8 Plot of the iterates for five complex initial approximations for the solution of $\cos x - x = 0$ using Newton's method. Each iterate is shown by "∘."

Substituting this result in (3.9) gives the new procedure for calculating the improvements to x as

$$x_{r+1} = [x_{r-1}f(x_r) - x_r f(x_{r-1})]/[f(x_r) - f(x_{r-1})] \tag{3.13}$$

This method does not require the calculation of the first derivative of $f(x)$ but does require that we know two initial approximations to the root, x_0 and x_1. Geometrically, we have simply approximated the slope of the tangent to the curve by the slope of a secant. For this reason the method is known as the *secant method*. The convergence of this method is slower than Newton's method. Another procedure similar to the secant method is called *regula falsi*. In this method two values of x that enclose the root are chosen to start the next iteration rather than the most recent pair of x values as in the secant method.

Newton's method and the secant method work well on a wide range of problems. However, for problems where the roots of an equation are close together or equal, the convergence may be slow. We now consider a simple adjustment to Newton's method that provides good convergence even with multiple roots.

3.8 Schroder's Method

In Section 3.2 we described how coincident roots present significant problems for most algorithms. In the case of Newton's method its performance is no longer quadratic for finding a coincident root and the procedure must be modified if it is to maintain this property. The iteration for Schroder's method for finding multiple roots has a form similar to that of Newton's method given in (3.9) except for the inclusion of a multiplying factor m. Thus

$$x_{r+1} = x_r - mf(x_r)/f'(x_r) \quad \text{where} \quad r = 0, 1, 2, \ldots \tag{3.14}$$

Here m is an integer equal to the multiplicity of the root to which we are trying to converge. Since the user may not know the value of m, it may have to be found experimentally.

It can be verified by some simple but lengthy algebraic manipulation that for a function $f(x)$ with multiple roots at $x = a$, $g'(a) = 0$. Here $g(x)$ is the right side of (3.14) and a is the exact root. This modification is sufficient to preserve the quadratic convergence of Newton's method

A MATLAB function for Schroder's method, `schroder`, is provided as follows:

```
function [res, it] = schroder(func,dfunc,m,x,tol)
% Finds a multiple root of f(x) = 0 using Schroder's method.
% Example call: [res, it] = schroder(func,dfunc,m,x,tol)
% The user defined function func is the function f(x).
% The user defined function dfunc is df/dx.
% x is an initial starting value, tol is required accuracy.
% function has a root of multiplicity m.
% x is a starting value, tol is required accuracy.
it = 0; x0 = x;
d = feval(func,x0)/feval(dfunc,x0);
while abs(d)>tol
    x1 = x0-m*d; it = it+1; x0 = x1;
    d = feval(func,x0)/feval(dfunc,x0);
end
res = x0;
```

We will now use the function `schroder` to solve $(e^{-x} - x)^2 = 0$. In this case we must set the multiplying factor m to 2. We write the function `f` and its derivative `df` and call the function `schroder` as follows:

```
>> f = @(x) (exp(-x)-x).^2;
>> df = @(x) 2*(exp(-x)-x).*(-exp(-x)-1);
>> [x, it] = schroder(f,df,2,-2,0.00005)

x =

    0.5671


it =

     5
```

It is interesting to note that Newton's method took 17 iterations to solve this problem in contrast to the 5 required by Schroder's method.

When a function $f(x)$ is known to have repeated roots, an alternative to Schroder's approach is to apply Newton's method to the function $f(x)/f'(x)$ rather than to the function $f(x)$ itself. It can be easily shown by direct differentiation that if $f(x)$ has a root of any multiplicity then $f(x)/f'(x)$ will have the same root but with multiplicity 1. Thus the algorithm has the iterative form (3.9) but modified by replacing $f(x)$ with $f(x)/f'(x)$. The advantage of this approach is that the user does not have to know the multiplicity of the root that is to be found. The considerable disadvantage is that both the first- and second-order derivatives must be supplied by the user.

3.9 Numerical Problems

We now consider the following problems that arise in solving single-variable nonlinear equations.

1. Finding good initial approximations
2. Ill-conditioned functions
3. Deciding on the most suitable convergence criteria
4. Discontinuities in the equation to be solved

These problems are now examined in detail.

1. Finding an initial approximation can be difficult for some nonlinear equations and a graph can be a considerable help in supplying such a value. The advantage of working in a MATLAB environment is that the script for the graph of the function can easily be generated and input can be taken from it directly. The function `plotapp` that is defined here finds an approximation to the root of a function supplied by the user in the range given by the parameters `rangelow` and `rangeup` using a step given by the `interval`.

```
function approx = plotapp(func,rangelow,interval,rangeup)
% Plots a function and allows the user to approximate a
% particular root using the cursor.
% Example call: approx = plotapp(func,rangelow,interval,rangeup)
% Plots the user defined function func in the range rangelow to
% rangeup using a step given by interval. Returns approx to root.
approx = [ ];
x = rangelow:interval:rangeup;
plot(x,feval(func,x))
hold on, xlabel('x'), ylabel('f(x)')
title(' ** Place cursor close to root and click mouse ** ')
grid on
% Use ginput to get approximation from graph using mouse
approx = ginput(1);
fprintf('Approximate root is %8.2f\n',approx(1)), hold off
```

The script that follows shows how this function may be used with the MATLAB function `fzero` to find a root of $x - \cos x = 0$.

```
% e3s303.m
g = @(x) x-cos(x);
approx = plotapp(g,-2,0.1,2);
```

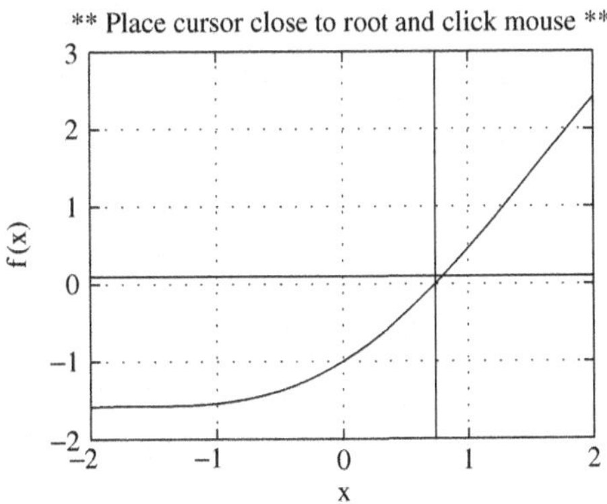

FIGURE 3.9 The cursor is shown close to the position of the root.

```
% Use this approximation and fzero to find exact root
root = fzero(g,approx(1),0.00005);
fprintf('Exact root is %8.5f\n',root)
```

Figure 3.9 gives the graph of $x - \cos x = 0$ generated by `plotapp` and shows the crosshairs cursor generated by the `ginput` function close to the root. The call `ginput(1)` means only one point is taken. The cursor can be positioned over the intersection of the curve with the $f(x) = 0$ line. This provides a useful initial approximation, the accuracy of which depends on the scale of the graph. In this example an initial approximation was found to be 0.74 and the more exact value was found using `fzero` to be 0.73909.

2. Ill-conditioning in a nonlinear equation means that small changes in the coefficients of the equation lead to unexpectedly large errors in the solutions. An interesting example of a very ill-conditioned polynomial is Wilkinson's polynomial. The MATLAB function `poly(v)` generates the coefficients of a polynomial, beginning with the coefficient of the highest power, with roots that are equal to the elements of the vector `v`. Thus `poly(1:n)` generates the coefficients of the polynomial with the roots $1, 2, \ldots, n$, which is Wilkinson's polynomial of degree $n - 1$.
3. In the design of any numerical algorithm for the solution of nonlinear equations, the termination criterion is particularly important. There are two major indicators of convergence: the difference between successive iterates and the value of the function at the current iterate. Taken separately these indicators may be misleading. For example, some nonlinear functions are such that small changes in the independent variable value may lead to large changes in the function value. In this case it may be better to monitor both indicators.
4. The function $f(x) = \sin(1/x)$ is particularly difficult to plot, and $\sin(1/x) = 0$ is very difficult to solve since it has an infinite number of roots, all clustered between 1 and -1. The function has a discontinuity at $x = 0$. Figure 3.10 attempts to illustrate

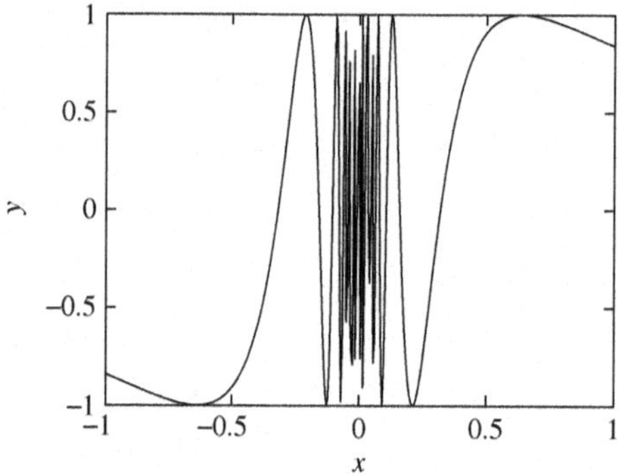

FIGURE 3.10 Plot of graph $f(x) = \sin(1/x)$. This plot is spurious in the range ± 0.2.

the behavior of this function. In fact, the graph shown does not truly represent the function and this plotting problem is discussed in more detail in Chapter 4. Near a discontinuity the function changes rapidly for small changes in the independent variable and some algorithms may have problems with this.

All the preceding points emphasize that algorithms for solving nonlinear equations need to be not only fast and efficient but robust as well. The next algorithm combines these properties and is relatively undemanding on the user.

3.10 The MATLAB Function `fzero` and Comparative Studies

Some problems may present particular difficulties for algorithms that in general work well. For example, algorithms that have fast ultimate convergence may initially diverge. One way to improve the reliability of an algorithm is to ensure that at each stage the root is confined to a known interval and the method of bisection, introduced in Section 3.3, may be used to provide an interval in which the root lies. Thus a method that combines bisection with a rapidly convergent procedure may be able to provide both rapid *and* reliable convergence.

The method of Brent combines inverse quadratic interpolation with bisection to provide a powerful method that has been found to be successful on a wide range of difficult problems. The method is easily implemented and a detailed description of the algorithm may be found in Brent (1971). Similar algorithms of comparable efficiency have been developed by Dekker (1969).

Experience with Brent's algorithm has shown it to be both reliable and efficient on a wide range of problems. A variation of this method is directly available in MATLAB and is called `fzero`. It may be used as follows:

```
x = fzero('funcname',x0,tol,trace);
```

where funcname is replaced by the name of any system function such as cos, sin, and so on, or the name of a function predefined by the user. The initial approximation is x0. The accuracy of the solution is set by tol and if trace is a value greater than 1, an output of the intermediate approximations is given. Only the first two parameters need be given and so an alternative call of this function is given by

```
x = fzero('funcname',x0);
```

To plot the function $(e^x - \cos x)^3$ and then determine some roots of $(e^x - \cos x)^3 = 0$ with tolerance 0.0005, initial approximations of 1.65 and -3, and no trace of the iterations, we use fzero as follows:

```
% e3s304.m
f = @(x) (exp(x)-cos(x)).^3;
x = -4:0.02:0.5;
plot(x,f(x)), grid on
xlabel('x'), ylabel('f(x)');
title('f(x) = (exp(x)-cos(x)).^3')
root = fzero(f,1.65,0.00005);
fprintf('A root of this equation is %6.4f\n',root)
root = fzero(f,-3,0.00005);
fprintf('A root of this equation is %6.4f\n',root)
```

The output and plot generated by this script are not given. However, the script is provided for reader experimentation.

Before we deal with the problem of finding many roots of a polynomial equation simultaneously, we present a comparative study of the MATLAB function fzero with the function fnewton. The following functions are considered:

1. $\sin(1/x) = 0$
2. $(x-1)^5 = 0$
3. $x - \tan x = 0$
4. $\cos\{(x^2+5)/(x^4+1)\} = 0$

The results of these comparative studies are given in Table 3.4. We see that fnewton is less reliable than fzero and that fzero produces accurate answers.

Table 3.4 Solution of Equations (1) through (4) with the Same Starting Point $x = -2$ and Accuracy$=0.00005$

Function	1	2	3	4
fnewton	Fail	0.999795831	Fail	−1.352673831
fzero	−0.318309886	1.000000000	−1.570796327	−1.352678708

3.11 Methods for Finding All the Roots of a Polynomial

The problem of solving polynomial equations is a special one in that these equations contain only combinations of integer powers of x and no other functions. Because of their special structure, algorithms have been developed to find all of the roots of a polynomial equation simultaneously. The function `roots` is provided in MATLAB. This function sets up the companion matrix for the polynomial and determines its eigenvalues, which can be shown to be the roots of the polynomial. For a description of the companion matrix, see Appendix A.

The following sections describe the methods of Bairstow and Laguerre but do not give a detailed theoretical justification of them. We provide a MATLAB function for Bairstow's method.

3.11.1 Bairstow's Method

Consider the polynomial

$$a_0x^n + a_1x^{n-1} + a_2x^{n-2} + \cdots + a_n = 0 \tag{3.15}$$

Since this is a polynomial equation of degree n, it has n roots. A common approach for locating the roots of a polynomial is to find all its quadratic factors. These will have the form

$$x^2 + ux + v \tag{3.16}$$

where u and v are the constants we wish to determine. Once all the quadratic factors are found it is easy to solve the quadratics to find all the roots of the equation. We now outline the major steps used in Bairstow's method for finding these quadratic factors.

If $R(x)$ is the remainder after the division of polynomial (3.15) by the quadratic factor (3.16), then there will clearly exist constants $b_0, b_1, b_2, \ldots$ such that the following equality holds:

$$(x^2 + ux + v)(b_0x^{n-2} + b_1x^{n-3} + b_2x^{n-4} + \cdots + b_{n-2}) + R(x) = x^n + a_1x^{n-1} + a_2x^{n-2} + \cdots + a_n \tag{3.17}$$

where a_0 has been taken as one and $R(x)$ will have the form $rx + s$. To ensure that $x^2 + ux + v$ is an exact factor of the polynomial (3.15), the remainder $R(x)$ must be zero. For this to be true both r and s must be zero and we must adjust u and v until this is true. Thus since both r and s depend on u and v, the problem reduces to solving the equations

$$r(u, v) = 0$$

$$s(u, v) = 0$$

To solve these equations we use an iterative method that assumes some initial approximations u_0 and v_0. Then we require improved approximations u_1 and v_1 where $u_1 = u_0 + \Delta u_0$ and $v_1 = v_0 + \Delta v_0$ such that

$$r(u_1, v_1) = 0$$
$$s(u_1, v_1) = 0$$

or r and s are as close to zero as possible.

Now we wish to find the changes Δu_0 and Δv_0 that will result in this improvement. Consequently, we must expand the two equations

$$r(u_0 + \Delta u_0, v_0 + \Delta v_0) = 0$$
$$s(u_0 + \Delta u_0, v_0 + \Delta v_0) = 0$$

using a Taylor series expansion and neglecting higher powers of Δu_0 and Δv_0. This leads to two approximating linear equations for Δu_0 and Δv_0:

$$\begin{aligned} r(u_0, v_0) + (\partial r/\partial u)_0 \Delta u_0 + (\partial r/\partial v)_0 \Delta v_0 = 0 \\ s(u_0, v_0) + (\partial s/\partial u)_0 \Delta u_0 + (\partial s/\partial v)_0 \Delta v_0 = 0 \end{aligned} \tag{3.18}$$

The subscript 0 denotes that the partial derivatives are calculated at the point u_0, v_0. Once the corrections are found, the iteration can be repeated until r and s are sufficiently close to zero. The method we have used here is a two-variable form of Newton's method, which will be described in Section 3.12.

Clearly this method requires the first-order partial derivatives of r and s with respect to u and v. The form of these is not obvious; however, they may be determined using recurrence relations derived from equating coefficients in (3.17) and then differentiating them. The details of this derivation are not given here but a clear description of the process is given by Froberg (1969). Once the quadratic factor is found, the same process is applied to the residual polynomial with the coefficients b_i to obtain the remaining quadratic factors. The details of this derivation are not provided here but a MATLAB function `bairstow` is given next.

```
function [rts,it] = bairstow(a,n,tol)
% Bairstow's method for finding the roots of a polynomial of degree n.
% Example call: [rts,it] = bairstow(a,n,tol)
% a is a row vector of REAL coefficients so that the
% polynomial is x^n+a(1)*x^(n-1)+a(2)*x^(n-2)+...+a(n).
% The accuracy to which the polynomial is satisfied is given by tol.
% The output is produced as an (n x 2) matrix rts.
% Cols 1 & 2 of rts contain the real & imag part of root respectively.
% The number of iterations taken is given by it.
it = 1;
```

```
while n>2
    %Initialise for this loop
    u = 1; v = 1; st = 1;
    while st>tol
        b(1) = a(1)-u; b(2) = a(2)-b(1)*u-v;
        for k = 3:n
            b(k) = a(k)-b(k-1)*u-b(k-2)*v;
        end
        c(1) = b(1)-u; c(2) = b(2)-c(1)*u-v;
        for k = 3:n-1
            c(k) = b(k)-c(k-1)*u-c(k-2)*v;
        end
        %calculate change in u and v
        c1 = c(n-1); b1 = b(n); cb = c(n-1)*b(n-1);
        c2 = c(n-2)*c(n-2); bc = b(n-1)*c(n-2);
        if n>3, c1 = c1*c(n-3); b1 = b1*c(n-3); end
        dn = c1-c2;
        du = (b1-bc)/dn; dv = (cb-c(n-2)*b(n))/dn;
        u = u+du; v = v+dv;
        st = norm([du dv]); it = it+1;
    end
    [r1,r2,im1,im2] = solveq(u,v,n,a);
    rts(n,1:2) = [r1 im1]; rts(n-1,1:2) = [r2 im2];
    n = n-2;
    a(1:n) = b(1:n);
end
% Solve last quadratic or linear equation
u = a(1); v = a(2);
[r1,r2,im1,im2] = solveq(u,v,n,a);
rts(n,1:2) = [r1 im1];
if n==2
    rts(n-1,1:2) = [r2 im2];
end
% ----------------------------------------------------------
function [r1,r2,im1,im2] = solveq(u,v,n,a);
% Solves x^2 + ux + v = 0 (n ~= 1) or x + a(1) = 0 (n = 1).
% Example call: [r1,r2,im1,im2] = solveq(u,v,n,a)
% r1, r2 are real parts of the roots,
% im1, im2 are the imaginary parts of the roots.
% Called by function bairstow.
if n==1
    r1 = -a(1); im1 = 0; r2 = 0; im2 = 0;
```

```
else
    d = u*u-4*v;
    if d<0
        d = -d;
        im1 = sqrt(d)/2; r1 = -u/2; r2 = r1; im2 = -im1;
    elseif d>0
        r1 = (-u+sqrt(d))/2; im1 = 0; r2 = (-u-sqrt(d))/2; im2 = 0;
    else
        r1 = -u/2; im1 = 0; r2 = -u/2; im2 = 0;
    end
end
```

Note that the MATLAB function `solveq` is nested within the function `bairstow`. The function is not stored separately and so it can only be accessed by `bairstow`. We may now use `bairstow` to solve the specific polynomial equation

$$x^5 - 3x^4 - 10x^3 + 10x^2 + 44x + 48 = 0$$

In this case, we take the coefficient vector as `c` where `c = [-3 -10 10 44 48]` and if we require accuracy of four decimal places we take `tol` as 0.00005. The script uses `bairstow` to solve the given polynomial.

```
% e3s305.m
c = [-3 -10 10 44 48];
[rts, it] = bairstow(c,5,0.00005);
for i = 1:5
    fprintf('\nroot%3.0f Real part=%7.4f',i,rts(i,1))
    fprintf(' Imag part=%7.4f',rts(i,2))
end
fprintf('\n')
```

Note how `fprintf` is used to provide a clearer output from the matrix `rts`.

```
root  1 Real part= 4.0000 Imag part= 0.0000
root  2 Real part=-1.0000 Imag part=-1.0000
root  3 Real part=-1.0000 Imag part= 1.0000
root  4 Real part=-2.0000 Imag part= 0.0000
root  5 Real part= 3.0000 Imag part= 0.0000
```

As we have indicated, MATLAB provides a function `roots` to determine the roots of a polynomial. It is interesting to compare this function with Bairstow's method. Table 3.5 gives the results of this comparison applied to specific polynomials. The problems p1 through p5 are the polynomials:

$$\text{p1}: \quad x^5 - 3x^4 - 10x^3 + 10x^2 + 44x + 48 = 0$$

$$\text{p2}: \quad x^3 - 3.001x^2 + 3.002x - 1.001 = 0$$

Table 3.5 Time Required to Obtain All Roots (in Seconds)

	`roots`	`bairstow`
p1	7	33
p2	6	19
p3	6	14
p4	10	103
p5	11	37

$$\text{p3}: \quad x^4 - 6x^3 + 11x^2 + 2x - 28 = 0$$

$$\text{p4}: \quad x^7 + 1 = 0$$

$$\text{p5}: \quad x^8 + x^7 + x^6 + x^5 + x^4 + x^3 + x^2 + x + 1 = 0$$

Both methods determine the correct roots for all problems, although the function `roots` is more efficient.

3.11.2 Laguerre's Method

Laguerre's method provides a rapidly convergent procedure for locating the roots of a polynomial. The algorithm is interesting and for this reason it is described in this section. The method is applied to a polynomial in the form

$$p(x) = x^n + a_1 x^{n-1} + a_2 x^{n-2} + \cdots + a_n$$

Starting with an initial approximation x_1, we apply the following iterative formula to the polynomial $p(x)$:

$$x_{i+1} = x_i - np(x_i)/[p'(x_i) \pm \surd\{h(x_i)\}] \quad \text{for} \quad i = 1, 2, \ldots \tag{3.19}$$

where

$$h(x_i) = (n-1)[(n-1)\{p'(x_i)\}^2 - np(x_i)p''(x_i)]$$

and n is the degree of the polynomial. The sign taken in (3.19) is determined so that it is the same as the sign of $p'(x_i)$.

It is important to give some justification for using a formula with such a complex structure. The reader will notice that if the square root term were not present in (3.19), the iterative form would be similar to that of Newton's method, (3.9), and identical to that of Schroder's method, (3.14). Thus we would have a method with quadratic convergence for the roots of the polynomial. In fact, the more complex structure of (3.19) provides third-order convergence since the error is proportional to the cube of the previous error and consequently provides faster convergence than Newton's method. Thus, given an initial approximation, the method will converge rapidly to a root of the polynomial, which we can denote by r.

To obtain the other roots of the polynomial we divide the polynomial $p(x)$ by the factor $(x-r)$, which provides another polynomial of degree $n-1$. We can then apply iteration (3.19) to this polynomial and repeat the whole procedure again. This is repeated until all roots are found to the required accuracy. The process of dividing by $(x-r)$ is known as deflation and can be performed in a simple and efficient way, described as follows.

Since we have a known factor $(x-r)$, then

$$a_0x^n + a_1x^{n-1} + a_2x^{n-2} + \cdots + a_n = (x-r)(b_0x^{n-1} + b_1x^{n-2} + b_2x^{n-3} + \cdots + b_{n-1}) \tag{3.20}$$

On equating coefficients of the powers of x on both sides we have

$$\begin{aligned} b_0 &= a_0 \\ b_i &= a_i + rb_{i-1} \quad \text{for} \quad i = 1, 2, \ldots, n-1 \end{aligned} \tag{3.21}$$

This process is known as synthetic division. Care must be taken here, particularly if the root is found to low accuracy, since ill-conditioning can magnify the effect of small errors in the coefficients of the deflated polynomial.

This completes the description of the method but a few important points should be noted. Assuming sufficient accuracy can be maintained in calculations, the method of Laguerre will converge for any value of the initial approximation. Convergence to complex roots and multiple roots can be achieved but at a slower rate because the convergence rate is linear. In the case of a complex root the value of the function $h(x_i)$ becomes negative and consequently the algorithm must be adjusted to deal with this situation. A key feature that should be considered is that the derivatives of the polynomial can be found efficiently by synthetic division.

To summarize the important features of the algorithm:

1. The algorithm is third order, thus providing rapid convergence to individual roots.
2. All roots of the polynomial can be found by using synthetic division.
3. Derivatives can be calculated efficiently using synthetic division.

3.12 Solving Systems of Nonlinear Equations

The methods considered so far have been concerned with finding one or all the roots of a nonlinear algebraic equation with one independent variable. We now consider methods for solving systems of nonlinear algebraic equations in which each equation is a function of a specified number of variables. We can write such a system in the form

$$f_i(x_1, x_2, \ldots, x_n) = 0 \quad \text{for} \quad i = 1, 2, 3, \ldots, n \tag{3.22}$$

A simple method for solving this system of nonlinear equations is based on Newton's method for the single equation. To illustrate this procedure we first consider a system of two equations in two variables:

$$\begin{aligned} f_1(x_1, x_2) &= 0 \\ f_2(x_1, x_2) &= 0 \end{aligned} \tag{3.23}$$

Given initial approximations x_1^0 and x_2^0 for x_1 and x_2, we may find new approximations x_1^1 and x_2^1 as follows:

$$\begin{aligned} x_1^1 &= x_1^0 + \Delta x_1^0 \\ x_2^1 &= x_2^0 + \Delta x_2^0 \end{aligned} \tag{3.24}$$

These approximations should be such that they drive the values of the functions closer to zero, so that

$$\begin{aligned} f_1(x_1^1, x_2^1) &\approx 0 \\ f_2(x_1^1, x_2^1) &\approx 0 \end{aligned}$$

or

$$\begin{aligned} f_1(x_1^0 + \Delta x_1^0,\ x_2^0 + \Delta x_2^0) &\approx 0 \\ f_2(x_1^0 + \Delta x_1^0,\ x_2^0 + \Delta x_2^0) &\approx 0 \end{aligned} \tag{3.25}$$

Applying a two-dimensional Taylor series expansion to (3.25) gives

$$\begin{aligned} f_1(x_1^0, x_2^0) + \{\partial f_1/\partial x_1\}^0 \Delta x_1^0 + \{\partial f_1/\partial x_2\}^0 \Delta x_2^0 + \cdots &\approx 0 \\ f_2(x_1^0, x_2^0) + \{\partial f_2/\partial x_1\}^0 \Delta x_1^0 + \{\partial f_2/\partial x_2\}^0 \Delta x_2^0 + \cdots &\approx 0 \end{aligned} \tag{3.26}$$

If we neglect terms involving powers of Δx_1^0 and Δx_2^0 higher than one, then (3.26) represents a system of two linear equations in two unknowns. The zero superscript means that the function is to be calculated at the initial approximation and Δx_1^0 and Δx_2^0 are the unknowns we wish to find. Having solved (3.26) we can obtain our new improved approximations and then repeat the process until we have obtained the accuracy we require. A common convergence criterion is to continue iterations until

$$\sqrt{(\Delta x_1^r)^2 + (\Delta x_2^r)^2} < \varepsilon$$

where r denotes the iteration number and ε is a small positive quantity preset by the user.

It is a simple step to generalize this procedure for any number of variables and equations. We may write the general system of equations as

$$\mathbf{f}(\mathbf{x}) = \mathbf{0}$$

where $\mathbf{f}$ denotes the column vector of n components $(f_1, f_2, \ldots, f_n)^\top$ and $\mathbf{x}$ is a column vector of n components $(x_1, x_2, \ldots, x_n)^\top$. Let $\mathbf{x}^{r+1}$ denote the value of $\mathbf{x}$ at the $(r+1)$th iteration; then

$$\mathbf{x}^{r+1} = \mathbf{x}^r + \Delta\mathbf{x}^r \quad \text{for} \quad r = 0, 1, 2, \ldots$$

If $\mathbf{x}^{r+1}$ is an improved approximation to $\mathbf{x}$, then

$$\mathbf{f}(\mathbf{x}^{r+1}) \approx \mathbf{0}$$

or

$$\mathbf{f}(\mathbf{x}^r + \Delta\mathbf{x}^r) \approx \mathbf{0} \tag{3.27}$$

Expanding (3.27) by using an n-dimensional Taylor series expansion gives

$$\mathbf{f}(\mathbf{x}^r + \Delta\mathbf{x}^r) = \mathbf{f}(\mathbf{x}^r) + \nabla\mathbf{f}(\mathbf{x}^r)\Delta\mathbf{x}^r + \cdots \tag{3.28}$$

where ∇ is a vector operator of partial derivatives with respect to each of the n components of $\mathbf{x}$. If we neglect higher-order terms in $(\Delta\mathbf{x}^r)^2$, this gives, by virtue of (3.27),

$$\mathbf{f}(\mathbf{x}^r) + \mathbf{J}_r\Delta\mathbf{x}^r \approx \mathbf{0} \tag{3.29}$$

where $\mathbf{J}_r = \nabla\mathbf{f}(\mathbf{x}^r)$. $\mathbf{J}_r$ is called the Jacobian matrix. The subscript r denotes that the matrix is evaluated at the point $\mathbf{x}^r$ and it can be written in component form as

$$\mathbf{J}_r = [\partial f_i(\mathbf{x}^r)/\partial x_j] \quad \text{for} \quad i = 1, 2, \ldots, n \quad \text{and} \quad j = 1, 2, \ldots, n$$

On solving (3.29) we have the improved approximation

$$\mathbf{x}^{r+1} = \mathbf{x}^r - \mathbf{J}_r^{-1}\mathbf{f}(\mathbf{x}^r) \quad \text{for} \quad r = 1, 2, \ldots$$

The matrix $\mathbf{J}_r$ may be singular and in this situation the inverse, $\mathbf{J}_r^{-1}$, cannot be calculated.

This is the general form of Newton's method. However, there are two major disadvantages with this method:

1. The method may not converge unless the initial approximation is a good one.
2. The method requires the user to provide the derivatives of each function with respect to each variable. The user must therefore provide n^2 derivatives and any computer implementation must evaluate the n functions and the n^2 derivatives at each iteration.

The MATLAB function `newtonmv` given here implements this method.

```
function [xv,it] = newtonmv(x,f,jf,n,tol)
% Newton's method for solving a system of n nonlinear equations
% in n variables.
% Example call: [xv,it] = newtonmv(x,f,jf,n,tol)
% Requires an initial approximation column vector x. tol is
% required accuracy. User must define functions f (system equations)
% and jf (partial derivatives). xv is the solution vector, the it
% parameter is number of iterations taken.
% WARNING. The method may fail, for example if initial estimates are poor.
it = 0; xv = x;
fr = feval(f,xv);
while norm(fr) > tol
    Jr = feval(jf,xv);  xv = xv-Jr\fr;
    fr = feval(f,xv);  it = it+1;
end
```

Figure 3.11 illustrates the following system of two equations in two variables:

$$\begin{aligned} x^2+y^2 &= 4 \\ xy &= 1 \end{aligned} \tag{3.30}$$

To solve the system (3.30) we define the MATLAB function by `f` and its Jacobian by `Jf` and then call `newtonmv` using initial approximations for the roots $x=3$ and $y=-1.5$ and a tolerance of 0.00005 as follows:

```
>> f = @(v) [v(1)^2+v(2)^2-4; v(1)*v(2)-1];
>> Jf = @(v) [2*v(1) 2*v(2); v(2) v(1)];
>> [rootvals,iter] = newtonmv([3 -1.5]',f,Jf,2,0.00005)
```

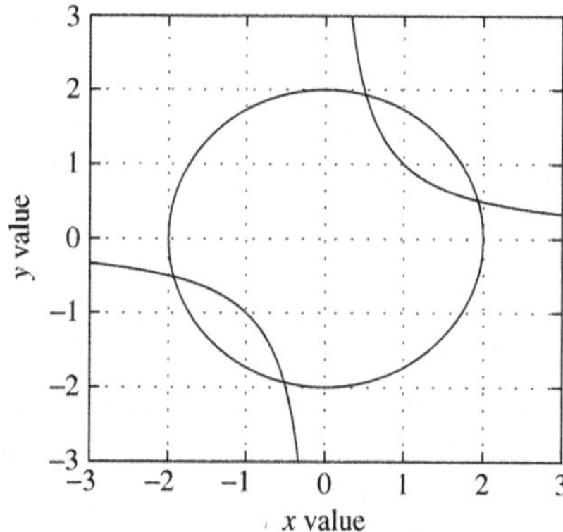

FIGURE 3.11 Plot of system (3.30). Intersections show roots.

This results in the MATLAB output

```
rootvals =
    1.9319
    0.5176

iter =
     5
```

The solution is $x = 1.9319$ and $y = 0.5176$. Clearly the user must supply a large amount of information for this function. The next section attempts to deal with this problem.

3.13 Broyden's Method for Solving Nonlinear Equations

The method of Newton described in Section 3.12 does not provide a practical procedure for solving any but the smallest systems of nonlinear equations. As we have seen, the method requires the user to provide not only the function definitions but also the definitions of the n^2 partial derivatives of the functions. Thus, for a system of 10 equations in 10 unknowns, the user must provide 110 function definitions!

To deal with this problem a number of techniques have been proposed but the group of methods that appears most successful is the class known as the quasi-Newton methods. The quasi-Newton methods avoid the calculation of the partial derivatives by obtaining approximations to them involving only the function values. The set of derivatives of the functions evaluated at any point $\mathbf{x}^r$ may be written in the form of the Jacobian matrix

$$\mathbf{J}_r = [\partial f_i(\mathbf{x}^r)/\partial x_j] \text{ for } i = 1,2,\ldots,n \quad \text{and} \quad j = 1,2,\ldots,n \tag{3.31}$$

The quasi-Newton methods provide an updating formula, which gives successive approximations to the Jacobian for each iteration. Broyden and others have shown that under specified circumstances these updating formulae provide satisfactory approximations to the inverse Jacobian. The structure of the algorithm suggested by Broyden is

1. Input an initial approximation to the solution. Set the counter r to zero.
2. Calculate or assume an initial approximation to the inverse Jacobian $\mathbf{B}^r$.
3. Calculate $\mathbf{p}^r = -\mathbf{B}^r\mathbf{f}^r$ where $\mathbf{f}^r = \mathbf{f}(\mathbf{x}^r)$.
4. Determine the scalar parameter t such that $||\mathbf{f}(\mathbf{x}^r + t_r\mathbf{p}^r)|| < ||\mathbf{f}^r||$ where the symbols $||\quad||$ denote that the norm of the vector is to be taken.
5. Calculate $\mathbf{x}^{r+1} = \mathbf{x}^r + t_r\mathbf{p}^r$.

6. Calculate $\mathbf{f}^{r+1} = \mathbf{f}(\mathbf{x}^{r+1})$. If $||\mathbf{f}^{r+1}|| < \varepsilon$ (where ε is a small preset positive quantity), then exit. If not continue with step 7.
7. Use the updating formula to obtain the required approximation to the Jacobian

   $\mathbf{B}^{r+1} = \mathbf{B}^r - (\mathbf{B}^r\mathbf{y}^r - \mathbf{p}^r)(\mathbf{p}^r)^\top\mathbf{B}^r/\{(\mathbf{p}^r)^\top\mathbf{B}^r\mathbf{y}^r\}$ where $\mathbf{y}^r = \mathbf{f}^{r+1} - \mathbf{f}^r$.
8. Set $i = i + 1$ and return to step 3.

The initial approximation to the inverse Jacobian **B** is usually taken as a scalar multiple of the unit matrix. The success of this algorithm depends on the nature of the functions to be solved and on the closeness of the initial approximation to the solution. In particular, step 4 may present major problems. It may be very expensive in computer time and to avoid this t_r is sometimes set as a constant, usually 1 or smaller. This may reduce the stability of the algorithm but speeds it up.

It should be noted that other updating formulae have been suggested and it is fairly easy to replace the Broyden formula by others in the preceding algorithm. In general, the problem of solving a system of nonlinear equations is a very difficult one. There is no algorithm that is guaranteed to work for all systems of equations. For large systems of equations the available algorithms tend to require large amounts of computer time to obtain accurate solutions.

The MATLAB function `broyden` implements Broyden's method. It should be noted that this avoids the difficulty of implementing step 4 by taking $t_r = 1$.

```
function [xv,it] = broyden(x,f,n,tol)
% Broyden's method for solving a system of n nonlinear equations
% in n variables.
% Example call: [xv,it] = broyden(x,f,n,tol)
% Requires an initial approximation column vector x. tol is required
% accuracy. User must define function f.
% xv is the solution vector, parameter it is number of iterations
% taken. WARNING. Method may fail, for example, if initial estimates
% are poor.
fr = zeros(n,1); it = 0; xv = x;
Br = eye(n); %Set initial Br
fr = feval(f, xv);
while norm(fr)>tol
    it = it+1; pr = -Br*fr; tau = 1;
    xv = xv+tau*pr;
    oldfr = fr; fr = feval(f,xv);
    % Update approximation to Jacobian using Broyden's formula
    y = fr - oldfr; oldBr = Br;
    oyp = oldBr*y-pr; pB = pr'*oldBr;
    for i = 1:n
        for j = 1:n
            M(i,j) = oyp(i)*pB(j);
```

```
            end
        end
        Br = oldBr-M./(pr'*oldBr*y);
end
```

To solve the system (3.30) using Broyden's method we call `broyden` as follows:

```
>> f = @(v) [v(1)^2+v(2)^2-4; v(1)*v(2)-1];
>> [x, iter] = broyden([3 -1.5]',f,2,0.00005)
```

This results in

```
x =
    0.5176
    1.9319

iter =
    36
```

This is a correct root of system (3.30) but it is not the same root as that found by Newton's method, even though the starting values for the iteration are the same.

As a second example we consider the following system of equations, which are taken from the MATLAB *User's Guide* (1989):

$$
\begin{aligned}
\sin x + y^2 + \log_e z &= 7 \\
3x + 2y - z^3 &= -1 \\
x + y + z &= 5
\end{aligned}
\tag{3.32}
$$

The function g, which implements (3.32), is given here

```
>> g = @(p) [sin(p(1))+p(2)^2+log(p(3))-7; 3*p(1)+2^p(2)-p(3)^3+1;
                p(1)+p(2)+p(3)-5];
```

The result of solving (3.32) is given next. The starting values used are $x = 0$, $y = 2$, and $z = 2$.

```
>> x = broyden([0 2 2]',g,3,0.00005)

x =
    0.5991
    2.3959
    2.0050
```

This shows that the method is successful for two problems and does not require the evaluation of the partial derivatives. The reader may be interested in applying the function `newtonmv` to this problem. Nine first-order partial derivatives will be required.

3.14 Comparing the Newton and Broyden Methods

We end our discussion of the solution of nonlinear systems of equations by comparing the performance of the functions `broyden` and `newtonmv`, developed in Sections 3.12 and 3.13, when solving the system (3.30). The following script calls both functions and provides the number of iterations required for convergence.

```
>> f = @(v) [v(1)^2+v(2)^2-4; v(1)*v(2)-1];
>> [x,it] = broyden([3 -1.5]',f,2,0.00005)

x =

    0.5176
    1.9319

it =

    36

>> J = @(v) [2*v(1) 2*v(2);v(2) v(1)];
>> [x,it] = newtonmv([3,-1.5]',f,J,2,0.00005)

x =

    1.9319
    0.5176

it =

     5
```

Note that although a correct solution is found in each case, it is a different root.

The first-order partial derivatives are required for the Newton method and this requires a considerable effort on the part of the user. Solving the previous problem demonstrates that the relatively simple form of the function `broyden` is attractive since it relieves the user of this effort.

In Sections 3.12 and 3.13 two relatively simple algorithms were provided for the solution of a very difficult problem. They cannot always be guaranteed to work and for large problems will converge only slowly.

3.15 Summary

The user wishing to solve nonlinear equations will find that this is an area that can present particular difficulties. It is always possible to devise or meet problems that particular algorithms either cannot solve or take a long time to solve. For example, it is just not possible for many algorithms to find the roots of the apparently trivial problem $x^{20} = 0$ very accurately. However, the algorithms described, if used with care, provide ways of solving a wide range of problems. MATLAB is well suited for this study because it allows interactive experimentation and graphical insights into the behavior of methods and functions. The reader is referred to Section 9.6 for applications of the symbolic toolbox for solving nonlinear equations. The algorithms `solve`, `fnewtsym`, and `newtmvsym` are described and applied in that section.

Problems

3.1. Omar Khayyam (who lived in the twelfth century) solved, by geometric means, a cubic equation with the form

$$x^3 - cx^2 + b^2x + a^3 = 0$$

The positive roots of this equation are the x coordinates of points of intersection in the first quadrant of the circle and parabola given in the following:

$$x^2 + y^2 - (c - a^3/b^2)x + 2by + b^2 - ca^3/b^2 = 0$$
$$xy = a^3/b$$

For $a = 1$, $b = 2$, and $c = 3$ use MATLAB to plot these two functions and note the x coordinates of the points of intersection. Using the MATLAB function `fzero`, solve the cubic equation and hence verify Omar Khayyam's method. *Hint*: You may find it helpful to use the MATLAB function `ginput`.

3.2. Use the MATLAB function `fnewton` to find a root of

$$x^{1.4} - \sqrt{x} + 1/x - 100 = 0$$

given an initial approximation 50. Use an accuracy of 10^{-4}.

3.3. Find the two real roots of $|x^3| + x - 6 = 0$ using the MATLAB function `fnewton`. Use initial approximations -1 and 1 and an accuracy of 10^{-4}. Plot the function using MATLAB to verify that the equation has only two real roots. *Hint:* Take care in finding the derivative of the function.

3.4. Explain why it is relatively difficult to find the root of $\tan x - c = 0$ when c is large. Use the MATLAB function `fnewton`, with initial approximations 1.3 and 1.4 and accuracy 10^{-4}, to find a root of this equation when $c = 5$ and $c = 10$. Compare the number of iterations required in both cases. *Hint*: A MATLAB `plot` will be useful.

3.5. Find a root of the polynomial $x^5 - 5x^4 + 10x^3 - 10x^2 + 5x - 1 = 0$ correct to four decimal places by using the MATLAB function `schroder` with $n = 5$ and a starting value $x_0 = 2$. Use MATLAB function `fnewton` to solve the same problem. Compare the result and the number of iterations using both methods. Use an accuracy of 5×10^{-7}.

3.6. Use the simple iterative method to solve the equation $x^{10} = e^x$. Express the equation in the form $x = f(x)$ in different ways and start the iterations with the initial approximation $x = 1$. Compare the efficiency of the formulae you have devised and check your answer(s) using the MATLAB function `fnewton`.

3.7. The historic Kepler's equation has the form $E - e\sin E = M$. Solve this equation for $e = 0.96727464$, the eccentricity of Halley's comet, and $M = 4.527594 \times 10^{-3}$. Use the MATLAB function `fnewton`, with an accuracy of 0.00005 and a starting value of 1.

3.8. Examine the performance of the function `fzero` for solving $x^{11} = 0$ with an initial value of -1.5 and also 1. Use an accuracy of 1×10^{-5}.

3.9. The smallest positive root of the equation

$$1 - x + x^2/(2!)^2 - x^3/(3!)^2 + x^4/(4!)^2 - \cdots = 0$$

is 1.4458. By considering in turn only the first four, five, and six terms in the series, show that a root of the truncated series approaches this result. Use the MATLAB function `fzero` to derive these results, with an initial value of 1 and an accuracy of 10^{-4}.

3.10. Reduce the following system of equations to one equation in terms of x and solve the resulting equation using the MATLAB function `fnewton`.

$$e^{x/10} - y = 0$$

$$2\log_e y - \cos x = 2$$

Use the MATLAB function `newtonmv` to solve these equations directly and compare your results. Use an initial approximation $x = 1$ for `fnewton` and approximations $x = 1, y = 1$ for `newtonmv` and accuracy 10^{-4} in both cases.

3.11. Solve the pair of equations that follow using the MATLAB function `broyden`, with the starting point $x = 10, y = -10$ and accuracy 10^{-4}.

$$2x = \sin\{(x + y)/2\}$$

$$2y = \cos\{(x - y)/2\}$$

3.12. Solve the two equations that follow using the MATLAB functions `newtonmv` and `broyden` with the starting point $x = 1, y = 2$ and accuracy 10^{-4}.

$$x^3 - 3xy^2 = 1/2$$

$$3x^2y - y^3 = \surd 3/2$$

3.13. The polynomial equation

$$x^4 - (13 + \varepsilon)x^3 + (57 + 8\varepsilon)x^2 - (95 + 17\varepsilon)x + 50 + 10\varepsilon = 0$$

has roots 1, 2, 5, $5+\varepsilon$. Use the functions `bairstow` and `roots` to find all the roots of this polynomial for $\varepsilon = 0.1, 0.01$, and 0.001. What happens as ε becomes smaller? Use an accuracy of 10^{-5}.

3.14. Employ the MATLAB function `bairstow` to find all the roots of the following polynomial using an accuracy requirement of 10^{-4}.

$$x^5 - x^4 - x^3 + x^2 - 2x + 2 = 0$$

3.15. Use the MATLAB function `roots` to find all the roots of the equation

$$t^3 - 0.5 - \sqrt{(3/2)}\imath = 0 \quad \text{where} \quad \imath = \sqrt{-1}$$

Compare with the exact solution

$$\cos\{(\pi/3 + 2\pi k)/3\} + \imath \sin\{(\pi/3 + 2\pi k)/3\} \quad \text{for} \quad k = 0, 1, 2$$

Use an accuracy of 10^{-4}.

3.16. An outline algorithm for the Illinois method for finding a root of $f(x) = 0$ (Dowell and Jarrett, 1971) is as follows:

For $k = 0, 1, 2, \ldots$
$x_{k+1} = x_k - f_k/f[x_{k-1}, x_k]$
if $f_k f_{k+1} > 0$ set $x_k = x_{k-1}$ and $f_k = gf_{k-1}$
where $f_k = f(x_k)$, $f[x_{k-1}, x_k] = (f_k - f_{k-1})/(x_k - x_{k-1})$
and $g = 0.5$.

Write a MATLAB function to implement this method. Note that the *regula falsi* method is similar but differs in that g is taken as 1.

3.17. The following iterative formulae can be used to solve the equation $x^2 - a = 0$:

$$x_{k+1} = (x_{k+1} + a/x_k)/2, \quad k = 0, 1, 2, \ldots$$

and

$$x_{k+1} = (x_{k+1} + a/x_k)/2 - (x_k - a/x_k)^2/(8x_k), \quad k = 0, 1, 2, \ldots$$

These iterative formulae are second- and third-order methods, respectively, for solving this equation. Write a MATLAB script to implement them and compare the number of iterations required to obtain the square root of 100.112 to five decimal places. For the purpose of illustration, use an initial approximation of 1000.

3.18. Show how MATLAB can be used to study chaotic behavior by considering the iteration

$$x_{k+1} = g(x_k) \quad \text{for} \quad k = 0, 1, 2, \ldots$$

where

$$g(x) = cx(1-x)$$

for different values of the constant c. This simple iteration arises from an attempt to solve a simple quadratic equation. However, its behavior is complex and for some values of c is chaotic. Write a MATLAB script to plot the value of the iterates against the iterate number for this function and study the behavior of the iterations for $c = 2.8, 3.25, 3.5$, and 3.8. Use an initial value of $x_0 = 0.7$.

3.19. For the functions solved in Problems 3.2, 3.3, and 3.7, use the MATLAB function `plotapp`, given in Section 3.9, to find approximate solutions for these functions.

3.20. It can be shown that the cubic polynomial equation

$$x^3 - px - q = 0$$

will have real roots if the inequality $p^3/q^2 > 27/4$ is satisfied. Select five pairs of values for p and q for which this inequality is satisfied and hence, using the MATLAB function `roots`, verify in each case that the roots of the equation are real.

3.21. In the sixteen century the mathematician Ioannes Colla suggested the following problem: Divide 10 into three parts such that they shall be in continued proportion to each other and the product of the first two shall be 6. Taking x, y, and z as three parts, this problem can be stated as

$$x + y + z = 10, \; x/y = y/z, \; xy = 6$$

Now by simple manipulation these equations can be expressed in terms of the specific variable y as

$$y^4 + 6y^2 - 60y + 36 = 0$$

Clearly if we can solve this equation for y then we can easily find the other variables x and z from the original equations. Use the MATLAB function `roots` to find values for y and hence solve Colla's problem.

3.22. The natural frequencies of a simply supported beam are given by the roots of the equation

$$c_1^2 - x^4 c_3^2 = 0$$

where

$$c_1 = (\sinh(x) + \sin(x))/(2x)$$

and

$$c_3 = (\sinh(x) - \sin(x))/(2x^3)$$

Substituting for c_1 and c_3 gives

$$((\sinh(x)+\sin(x))/(2x))^2 - x^4((\sinh(x)-\sin(x))/(2x^3))^2 = 0$$

When searching for the roots of this equation no difficulty is found in determining the root for trial values of x providing x is small (say $x < 10$). For values of $x > 25$ the process becomes erratic. The roots of this equation are actually $x = k\pi$ where k is a positive integer. Use the MATLAB function `fzero` with initial approximations $x = 5$ and $x = 30$ to obtain a solution close to these initial approximations for this equation. For the purpose of this exercise, do not simplify this equation.

Why are the results so poor? If you simplify the preceding equation, which equation do you obtain and what is its solution?

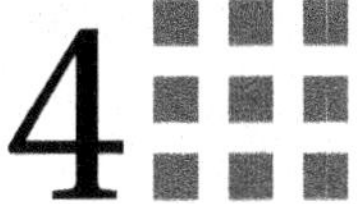

4 Differentiation and Integration

Differentiation and integration are the fundamental operations of differential calculus and occur in almost every field of mathematics, science, and engineering. Determining the derivative of a function analytically may be tedious but is relatively straightforward. The inverse of this process, that of determining the integral of a function, can often be difficult analytically or even impossible.

The difficulty of determining the analytical integral for certain functions has encouraged the development of many numerical procedures for determining approximately the value of definite integrals. In many situations the procedures work well because integration is a smoothing process and errors in the approximation tend to cancel each other. However, for certain types of functions, difficulties may arise and these will be examined as part of our discussion of specific numerical methods for the approximate evaluation of definite integrals.

4.1 Introduction

In the next section of this chapter we show how the derivative of a function may be estimated for a particular value of the independent variable. The numerical approximations for derivatives require only function values. These approximations can be used to great advantage when derivatives are required in a program. Their application saves the program user the task of determining the analytical expressions for these derivatives. In Section 4.3 and beyond we introduce the reader to a range of numerical integration methods, including methods suitable for infinite ranges of integration. Generally numerical integration works well, but there are pathological integrals that will defeat the best numerical algorithms.

4.2 Numerical Differentiation

In this section we present a range of approximations for first- and higher-order derivatives. Before we derive these approximations in detail we give a simple example that illustrates the dangers of the careless or naive use of such derivative approximations. The simplest approximation for the first-order derivative of a given function $f(x)$ arises from the formal definition of the derivative:

$$\frac{df}{dx} = \lim_{h \to 0} \left(\frac{f(x+h) - f(x)}{h} \right) \tag{4.1}$$

One interpretation of (4.1) is that the derivative of a function $f(x)$ is the slope of the tangent to the function at the point x. For small h we obtain the approximation to the derivative:

$$\frac{df}{dx} \approx \left(\frac{f(x+h) - f(x)}{h}\right) \tag{4.2}$$

This would appear to imply that the smaller the value of h, the better the value of our approximation in (4.2). The following MATLAB script plots Figure 4.1, which shows the error for various values of h.

```
% e3s401.m
g = @(x) x.^9;
x = 1; h(1) = 0.5;
hvals = [ ]; dfbydx = [ ];
for i = 1:17
    h = h/10;
    b = g(x); a = g(x+h);
    hvals = [hvals h];
    dfbydx(i) = (a-b)/h;
end;
exact = 9;
loglog(hvals,abs(dfbydx-exact),'*')
axis([1e-18 1 1e-8 1e4])
xlabel('h value'), ylabel('Error in approximation')
```

Figure 4.1 shows that for large values of h the error is large but falls rapidly as h is decreased. However, when h becomes less than about 10^{-9}, rounding errors dominate and the approximation becomes much worse. Clearly care must be taken in the choice of h. With this warning in mind we develop methods of differing accuracies for any order derivative.

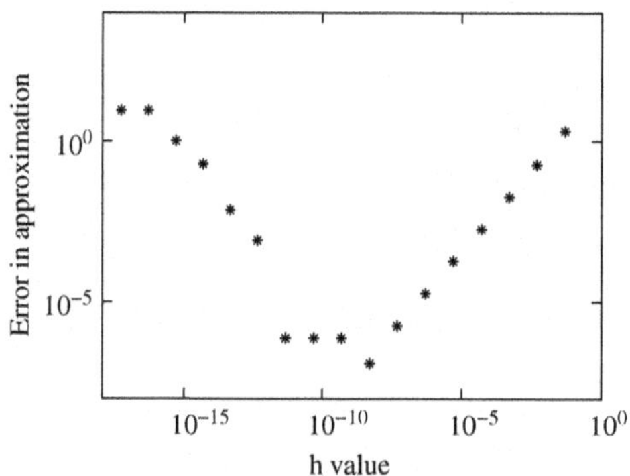

FIGURE 4.1 A log-log plot showing the error in a simple derivative approximation.

We have seen how a simple approximate formula for the first derivative can be easily obtained from the formal definition of the derivative. However, it is difficult to approximate higher derivatives and deduce more accurate formulae in this way; instead we will use the Taylor series expansion of the function $y = f(x)$. To determine the central difference approximation for the derivative of this function at x_i we expand $f(x_i + h)$:

$$f(x_i + h) = f(x_i) + hf'(x_i) + (h^2/2!)f''(x_i) + (h^3/3!)f'''(x_i) + (h^4/4!)f^{(iv)}(x_i) + \cdots \quad (4.3)$$

We sample $f(x)$ at points a distance h apart and write $x_i + h$ as x_{i+1}, and so on. We will also write $f(x_i)$ as f_i and $f(x_{i+1})$ as f_{i+1}. Thus

$$f_{i+1} = f_i + hf'(x_i) + (h^2/2!)f''(x_i) + (h^3/3!)f'''(x_i) + (h^4/4!)f^{(iv)}(x_i) + \cdots \quad (4.4)$$

Similarly

$$f_{i-1} = f_i - hf'(x_i) + (h^2/2!)f''(x_i) - (h^3/3!)f'''(x_i) + (h^4/4!)f^{(iv)}(x_i) - \cdots \quad (4.5)$$

We can find an approximation to the first derivative as follows. Subtracting (4.5) from (4.4) gives

$$f_{i+1} - f_{i-1} = 2hf'(x_i) + 2\left(h^3/3!\right)f'''(x_i) + \cdots$$

Thus, neglecting terms in h^3 and higher, we have

$$f'(x_i) = (f_{i+1} - f_{i-1})/2h \quad \text{with errors of} \quad O\left(h^2\right) \quad (4.6)$$

This is the central difference approximation and differs from (4.2), which is a forward difference approximation. Equation (4.6) is more accurate than (4.2) but in the limit as h approaches zero, the two are identical.

To determine an approximation for the second derivative we add (4.4) and (4.5) to obtain

$$f_{i+1} + f_{i-1} = 2f_i + 2\left(h^2/2!\right)f'(x_i) + 2\left(h^4/4!\right)f^{(iv)}(x_i) + \cdots$$

Thus, neglecting terms in h^4 and higher, we have

$$f''(x_i) = (f_{i+1} - 2f_i + f_{i-1})/h^2 \quad \text{with errors of} \quad O\left(h^2\right) \quad (4.7)$$

By taking more terms in the Taylor series, together with the Taylor series for $f(x + 2h)$ and $f(x - 2h)$, and so on, and performing similar manipulations, we can obtain higher derivatives and more accurate approximations if required. Table 4.1 gives examples of these formulae.

Table 4.1 Derivative Approximations

	Multipliers for $f_{i-3} \cdots f_{i+3}$							
	f_{i-3}	f_{i-2}	f_{i-1}	f_i	f_{i+1}	f_{i+2}	f_{i+3}	**Order of Error**
$2hf'(x_i)$	0	0	−1	0	1	0	0	h^2
$h^2f''(x_i)$	0	0	1	−2	1	0	0	h^2
$2h^3f'''(x_i)$	0	−1	2	0	−2	1	0	h^2
$h^4f^{(iv)}(x_i)$	0	1	−4	6	−4	1	0	h^2
$12hf'(x_i)$	0	1	−8	0	8	−1	0	h^4
$12h^2f''(x_i)$	0	−1	16	−30	16	−1	0	h^4
$8h^3f'''(x_i)$	1	−8	13	0	−13	8	−1	h^4
$6h^4f^{(iv)}(x_i)$	−1	12	−39	56	−39	12	−1	h^4

The MATLAB function `diffgen` defined in the following computes the first, second, third, and fourth derivative of a given function with errors of $O(h^4)$ for a specified value of x using data from the table.

```
function q = diffgen(func,n,x,h)
% Numerical differentiation.
% Example call: q = diffgen(func,n,x,h)
% Provides nth order derivatives, where n = 1 or 2 or 3 or 4
% of the user defined function func at the value x, using a step h.
if (n==1)|(n==2)|(n==3)|(n==4)
    c = zeros(4,7);
    c(1,:) = [ 0 1 -8 0 8 -1 0];
    c(2,:) = [ 0 -1 16 -30 16 -1 0];
    c(3,:) = [1.5 -12 19.5 0 -19.5 12 -1.5];
    c(4,:) = [ -2 24 -78 112 -78 24 -2];
    y = feval(func,x+[-3:3]*h);
    q = c(n,:)*y.';  q = q/(12*h^n);
else
    disp('n must be 1, 2, 3 or 4'), return
end
```

For example,

```
result = diffgen('cos',2,1.2,0.01)
```

determines the second derivative of $\cos(x)$ for $x = 1.2$ with $h = 0.01$ and gives `-0.3624` for the result. The following script calls the function `diffgen` four times to determine the first four derivatives of $y = x^7$ when $x = 1$:

```
% e3s402.m
g = @(x) x.^7;
h = 0.5; i = 1;
```

```
disp('     h        1st deriv  2nd deriv   3rd deriv     4th deriv');
while h>=1e-5
    t1 = h;
    t2 = diffgen(g, 1, 1, h);
    t3 = diffgen(g, 2, 1, h);
    t4 = diffgen(g, 3, 1, h);
    t5 = diffgen(g, 4, 1, h);
    fprintf('%10.5f %10.5f %10.5f %11.5f %12.5f\n',t1,t2,t3,t4,t5);
    h = h/10; i = i+1;
end
```

The output from the preceding script is

```
    h        1st deriv  2nd deriv   3rd deriv     4th deriv
 0.50000      1.43750    38.50000   191.62500     840.00000
 0.05000      6.99947    41.99965   209.99816     840.00000
 0.00500      7.00000    42.00000   210.00000     840.00001
 0.00050      7.00000    42.00000   210.00000     839.97579
 0.00005      7.00000    42.00000   209.98521    -290.13828
```

Note that as h is decreased the estimates for the first and second derivatives steadily improve, but when $h = 5 \times 10^{-4}$ the estimate for the fourth derivative begins to deteriorate. When $h = 5 \times 10^{-5}$ the estimate for the third derivative also begins to deteriorate and the fourth derivative is very inaccurate. In general we cannot predict when this deterioration will begin. It should be noted that different platforms may give different results for this value.

4.3 Numerical Integration

We will begin by examining the definite integral

$$I = \int_a^b f(x)\,dx \tag{4.8}$$

The evaluation of such integrals is often called quadrature and we will develop methods for both finite and infinite values of a and b.

The definite integral (4.8) is a summation process but it may also be interpreted as the area under the curve $y = f(x)$ from a to b. Any areas above the x-axis are counted as positive; any areas below the x-axis are counted as negative. Many numerical methods for integration are based on using this interpretation to derive approximations to the integral. Typically the interval $[a, b]$ is divided into a number of smaller subintervals, and by making simple approximations to the curve $y = f(x)$ in the subinterval, the area of the subinterval may be obtained. The areas of all the subintervals are then summed to give an approximation to the integral in the interval $[a, b]$. Variations of this technique are developed by taking groups of subintervals and fitting different degree polynomials to approximate $y = f(x)$ in each of these groups. The simplest of these methods is the trapezoidal rule.

The trapezoidal rule is based on the idea of approximating the function $y = f(x)$ in each subinterval by a straight line so that the shape of the area in the subinterval is trapezoidal. Clearly, as the number of subintervals used increases, the straight lines will approximate the function more closely. Dividing the interval from a to b into n subintervals of width h (where $h = (b-a)/n$) we can calculate the area of each subinterval since the area of a trapezium is its base times the mean of its heights. These heights are f_i and f_{i+1} where $f_i = f(x_i)$. Thus the area of the trapezium is

$$h(f_i + f_{i+1})/2 \quad \text{for} \quad i = 0, 1, 2, \ldots, n-1$$

Summing all the trapezia gives the composite trapezoidal rule for approximating (4.8):

$$I \approx h\{(f_0 + f_n)/2 + f_1 + f_2 + \cdots + f_{n-1}\} \tag{4.9}$$

The truncation error, which is the error due to the implicit approximation in the trapezoidal rule, is

$$E_n \leq (b-a)h^2 M/12 \tag{4.10}$$

where M is the upper bound for $|f''(t)|$ and t must be in the range a to b. The MATLAB function `trapz` implements this procedure and we use it in Section 4.4 to compare the performance of the trapezoidal rule with the more accurate Simpson's rule.

The level of accuracy obtained from a numerical integration procedure is dependent on three factors. The first two are the nature of the approximating function and the number of intervals used. These are controlled by the user and give rise to the truncation error, that is, the error inherent in the approximation. The third factor influencing accuracy is the rounding error, the error caused by the fact that practical computation has limited precision. For a particular approximating function the truncation error will decrease as the number of subintervals increases. Integration is a smoothing process and rounding errors do not present a major problem. However, when many intervals are used, the time to solve the problem becomes more significant because of the increased amount of computation. This problem may be reduced by writing the script efficiently.

4.4 Simpson's Rule

Simpson's rule is based on using a quadratic polynomial approximation to the function $f(x)$ over a pair of subintervals; it is illustrated in Figure 4.2. If we integrate the quadratic polynomial passing through the points (x_0, f_0), (x_1, f_1), (x_2, f_2), where $f_1 = f(x_1)$, and so on,

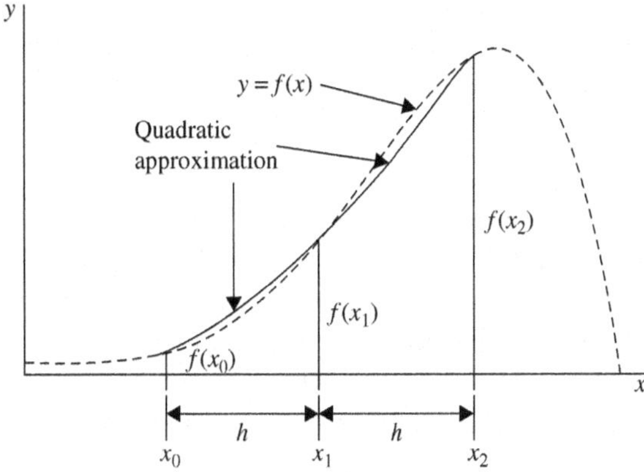

FIGURE 4.2 Simpson's rule, using a quadratic approximation over two intervals.

the following formula is obtained:

$$\int_{x_0}^{x_2} f(x)\,dx = \frac{h}{3}\left(f_0 + 4f_1 + f_2\right) \tag{4.11}$$

This is Simpson's rule for one pair of intervals. Applying the rule to all pairs of intervals in the range a to b and adding the results produces the following expression, known as the composite Simpson's rule:

$$\int_a^b f(x)\,dx = \frac{h}{3}\left\{f_0 + 4\left(f_1 + f_3 + f_5 + \cdots + f_{2n-1}\right) + 2\left(f_2 + f_4 + \cdots + f_{2n-2}\right) + f_{2n}\right\} \tag{4.12}$$

Here n indicates the number of pairs of intervals and $h = (b-a)/(2n)$. The composite rule may be also be written as a vector product:

$$\int_a^b f(x)\,dx = \frac{h}{3}\left(\mathbf{c}^\top \mathbf{f}\right) \tag{4.13}$$

where $\mathbf{c} = [1\ 4\ 2\ 4\ 2 \ldots 2\ 4\ 1]^\top$ and $\mathbf{f} = [f_1\ f_2\ f_3\ \ldots\ f_{2n}]^\top$.

The error arising from the approximation, called the truncation error, is approximated by

$$E_n = (b-a)h^4 f^{(iv)}(t)/180$$

where t lies between a and b. An upper bound for the error is given by

$$E_n \le (b-a)h^4 M/180 \tag{4.14}$$

where M is an upper bound for $|f^{(iv)}(t)|$. The upper bound for the error in the simpler trapezoidal rule, (4.10), is proportional to h^2 rather than h^4. This makes Simpson's rule superior to the trapezoidal rule in terms of accuracy at the expense of more function evaluations.

To illustrate different ways of implementing Simpson's rule we provide two alternatives, `simp1` and `simp2`. The function `simp1` creates a vector of coefficients `v` and a vector of function values `y` and multiplies the two vectors together. Function `simp2` provides a more conventional implementation of Simpson's rule. In each case the user must provide the definition of the function to be integrated, the lower and upper limits of integration, and the number of subintervals to be used. The number of subintervals must be an even number since the rule fits a function to a *pair* of subintervals.

```
function q = simp1(func,a,b,m)
% Implements Simpson's rule using vectors.
% Example call: q = simp1(func,a,b,m)
% Integrates user defined function func from a to b, using m divisions
if (m/2)~=floor(m/2)
    disp('m must be even'); return
end
h = (b-a)/m; x = a:h:b;
y = feval(func,x);
v = 2*ones(m+1,1);  v2 = 2*ones(m/2,1);
v(2:2:m) = v(2:2:m)+v2;
v(1) = 1;  v(m+1) = 1;
q = (h/3)*y*v;
```

The second nonvectorized form of this function is

```
function q = simp2(func,a,b,m)
% Implements Simpson's rule using for loop.
% Example call: q = simp2(func,a,b,m)
% Integrates user defined function
% func from a to b, using m divisions
if (m/2) ~= floor(m/2)
    disp('m must be even'); return
end
h = (b-a)/m;
s = 0; yl = feval(func,a);
for j = 2:2:m
    x = a+(j-1)*h;  ym = feval(func,x);
    x = a+j*h;   yh = feval(func,x);
    s = s+yl+4*ym+yh;  yl = yh;
end
q = s*h/3;
```

The following script calls either `simp1` or `simp2`. These functions can be used to demonstrate the effect on accuracy of the number of pairs of intervals used. The script evaluates the integral of x^7 in the range 0 to 1.

```
% e3s403.m
n = 4; i = 1;
tic
disp('    n integral value')
while n < 1025
    simpval = simp1(@(x) x.^7,0,1,n); % or simpval = simp2(etc.);
    fprintf('%5.0f %15.12f\n',n,simpval)
    n = 2*n; i = i+1;
end
t = toc;
fprintf('\ntime taken = %6.4f secs\n',t)
```

The output from this script using `simp1` is as follows:

```
 n integral value
    4  0.129150390625
    8  0.125278472900
   16  0.125017702579
   32  0.125001111068
   64  0.125000069514
  128  0.125000004346
  256  0.125000000272
  512  0.125000000017
 1024  0.125000000001

time taken = 0.0635 secs
```

On running this script, but using `simp2`, we obtain the same values for the integral but the following results for the time taken:

```
time taken = 0.1335 secs
```

Equation (4.14) shows that the truncation error will decrease rapidly for values of h smaller than 1. The preceding results illustrate this. The rounding error in Simpson's rule is due to evaluating the function $f(x)$ and the subsequent multiplications and additions. Note also that the vectorized version, `simp1`, is a little faster than `simp2`.

We now evaluate the same integral using the MATLAB function `trapz`. To call this function the user must provide a vector of function values `f`. The function `trapz(f)` estimates the integral of the function assuming unit spacing between the data points. Thus to determine the integral we multiply `trapz(f)` by the increment `h`.

```
% e3s404.m
n = 4; i = 1; f = @(x) x.^7;
tic
disp('    n  integral value')
while n<1025
    h = 1/n; x = 0:h:1;
    trapval = h*trapz(f(x));
    fprintf('%5.0f %15.12f\n',n,trapval)
    n = 2*n; i = i+1;
end
t = toc;
fprintf('\ntime taken = %4.2f secs\n',t)
```

Running this script gives

```
    n  integral value
    4  0.160339355469
    8  0.134043693542
   16  0.127274200320
   32  0.125569383381
   64  0.125142397981
  128  0.125035602755
  256  0.125008900892
  512  0.125002225236
 1024  0.125000556310

time taken = 0.06 secs
```

These results illustrate the fact that the trapezoidal rule is less accurate than the Simpson rule.

4.5 Newton–Cotes Formulae

Simpson's rule is an example of a Newton–Cotes formula for integration. Other examples of these formulae can be obtained by fitting higher-degree polynomials through the appropriate number of points. In general we fit a polynomial of degree n through $n+1$ points. The resulting polynomial can then be integrated to provide an integration formula. Here are some examples of Newton–Cotes formulae together with estimates of their truncation errors.

For $n = 3$ we have

$$\int_{x_0}^{x_3} f(x)\,dx = \frac{3h}{8}\left(f_0 + 3f_1 + 3f_2 + f_3\right) + \text{truncation error } \frac{3h^5}{80}f^{iv}(t) \tag{4.15}$$

where t lies in the interval x_0 to x_3.

For $n = 4$ we have

$$\int_{x_0}^{x_4} f(x)\,dx = \frac{2h}{45}(7f_0 + 32f_1 + 12f_2 + 32f_3 + 7f_4) + \text{truncation error } \frac{8h^7}{945} f^{(vi)}(t) \tag{4.16}$$

where t lies in the interval x_0 to x_4. Composite rules can be generated for both rules (4.15) and (4.16). The truncation errors indicate that some improvement in accuracy may be obtained by using these rules rather than Simpson's rule. However, the rules are more complex; consequently, greater computational effort is involved and rounding errors may become a more significant problem.

The MATLAB function quad uses an adaptive recursive Simpson's rule and the function quadl uses adaptive Lobatto quadrature. The MATLAB function quadgk uses an adaptive Gauss–Kronrod rule, which is particularly efficient for smooth and oscillatory integrals. The limits of integration may be infinite.

The performance of quad and quadl and simp1 (called with both 1024 and 4096 panels) are compared by determining the error when evaluating the integral e^x from 0 to n where $n = 2.5 : 2.5 : 25$ using the following script:

```
% e3s405.m
for n = 1:10; n1 = 2.5*n;
    ext = exp(n1)-1;
    err(n,1) = simp1('exp',0,n1,1024)-ext;
    err(n,2) = simp1('exp',0,n1,4096)-ext;
    err(n,3) = quadl('exp',0,n1)-ext;
    err(n,4) = quad('exp',0,n1)-ext;
end
err
```

Running this script gives the following:

```
err =
  2.2062e-012  8.8818e-015  7.2414e-009  3.9510e-009
  4.6552e-010  1.8190e-012  1.0203e-011  8.6445e-009
  2.8889e-008  1.1323e-010  2.9315e-009  1.4057e-008
  1.1129e-006  4.3437e-009  4.5475e-010  1.6258e-008
  3.3101e-005  1.2928e-007            0  1.5891e-008
  8.3618e-004  3.2666e-006 -9.3132e-010  1.8626e-008
  1.8872e-002  7.3716e-005            0  7.4506e-009
  3.9221e-001  1.5321e-003 -5.9605e-008            0
  7.6535e+000  2.9899e-002  9.5367e-007  9.5367e-007
  1.4211e+002  5.5516e-001 -1.5259e-005 -1.5259e-005
```

These results show the advantage of using adaptive subinterval sizes. Simpson's rule has a fixed interval size. For the smaller ranges of integration it performs very well but as the range of integration increases, accuracy decreases. Generally the adaptive methods maintain a much higher level of accuracy.

4.6 Romberg Integration

A major problem that arises with the nonadaptive Simpson's or Newton–Cotes rule is that the number of intervals required to provide the required accuracy is initially unknown. Clearly one approach to this problem is to double successively the number of intervals used and compare the results of applying a particular rule, as illustrated by the examples in Section 4.4. Romberg's method provides an organized approach to this problem and utilizes the results obtained by applying Simpson's rule with different interval sizes to reduce the truncation error.

Romberg integration may be formulated as follows. Let I be the exact value of the integral and T_i the approximate value of the integral obtained using Simpson's rule with i intervals. Consequently, we may write an approximation for the integral I that includes contributions from the truncation error as follows (note that the error terms are expressed in powers of h^4):

$$I = T_i + c_1 h^4 + c_2 h^8 + c_3 h^{12} + \cdots \tag{4.17}$$

If we double the number of intervals, h is halved, giving

$$I = T_{2i} + c_1 (h/2)^4 + c_2 (h/2)^8 + c_3 (h/2)^{12} + \cdots \tag{4.18}$$

We can eliminate the terms in h^4 by subtracting (4.17) from 16 times (4.18), giving

$$I = (16T_{2i} - T_i)/15 + k_2 h^8 + k_3 h^{12} + \cdots \tag{4.19}$$

Notice that the dominant or most significant term in the truncation error is now of order h^8. In general this will provide a significantly improved approximation to I. For the remainder of this discussion it is advantageous to use a double subscript notation. If we generate an initial set of approximations by successively halving the interval we may represent them by $T_{0,k}$ where $k = 0, 1, 2, 3, 4, \ldots$. These results may be combined in a similar manner to that described in (4.19) by using the general formula

$$T_{r,k} = (16^r T_{r-1,k+1} - T_{r-1,k})/(16^r - 1) \quad \text{for} \quad k = 0, 1, 2, 3 \ldots \quad \text{and} \quad r = 1, 2, 3, \ldots \tag{4.20}$$

Here r represents the current set of approximations we are generating. The calculations may be tabulated as follows:

$$\begin{array}{lllll} T_{0,0} & T_{0,1} & T_{0,2} & T_{0,3} & T_{0,4} \\ T_{1,0} & T_{1,1} & T_{1,2} & T_{1,3} & \\ T_{2,0} & T_{2,1} & T_{2,2} & & \\ T_{3,0} & T_{3,1} & & & \\ T_{4,0} & & & & \end{array}$$

In this case, the interval has been halved four times to generate the first five values in the table denoted by $T_{0,k}$. The preceding formula for $T_{r,k}$ is used to calculate the remaining values in the table and at each stage the order of the truncation error is increased by four. A common alternative is to write the preceding table with the rows and columns interchanged.

At each stage the interval size is given by

$$(b-a)/2^k \quad \text{for} \quad k = 0, 1, 2, \ldots \tag{4.21}$$

Romberg integration is implemented in the following MATLAB function, `romb`:

```
function [W T] = romb(func,a,b,d)
% Implements Romberg integration.
% Example call: W = romb(func,a,b,d)
% Integrates user defined function func from a to b, using d stages.
T = zeros(d+1,d+1);
for k = 1:d+1
    n = 2^k;  T(1,k) = simp1(func,a,b,n);
end
for p = 1:d
    q = 16^p;
    for k = 0:d-p
        T(p+1,k+1) = (q*T(p,k+2)-T(p,k+1))/(q-1);
    end
end
W = T(d+1,1);
```

We now apply the function `romb` to the evaluation of $x^{0.1}$ in the range 0 to 1. The call of the function `romb` is

```
>> [integral table] = romb(@(x) x.^0.1,0,1,5)
```

Calling this function gives the following output. Note that the best estimate is the single value in the last row of the table.

```
integral =
    0.9066

table =
    0.7887    0.8529    0.8829    0.8969    0.9034    0.9064
    0.8572    0.8849    0.8978    0.9038    0.9066         0
    0.8850    0.8978    0.9038    0.9066         0         0
    0.8978    0.9038    0.9066         0         0         0
    0.9038    0.9066         0         0         0         0
    0.9066         0         0         0         0         0
```

This integral is a surprisingly difficult one and obtaining an accurate result presents a significant problem. The exact solution to four decimal places is 0.9090 so the application of the Romberg method gives only two places of accuracy. However, taking $n = 10$ does give the answer correct to four places:

```
>> integral = romb(@(x) x.^0.1,0,1,10)

integral =
    0.9090
```

Generally the Romberg method is very efficient and accurate. For example, it evaluates the integral of e^x from 0 to 10 using five divisions of the interval more accurately and slightly more quickly than the function `quad` with the default tolerance.

An interesting exercise for the reader is to convert the function `romb` to work with the MATLAB function `trapz` instead of `simp1`.

4.7 Gaussian Integration

The common feature of the methods considered so far is that the integrand is evaluated at equal intervals within the range of integration. In contrast, Gaussian integration requires the evaluation of the integrand at specified, but unequal, intervals. For this reason Gaussian integration cannot be applied to data values that are sampled at equal intervals of the independent variable. The general form of the rule is

$$\int_{-1}^{1} f(x)\,dx = \sum_{i=1}^{n} A_i f(x_i) \tag{4.22}$$

The parameters A_i and x_i are chosen so that, for a given n, the rule is exact for polynomials up to and including degree $2n-1$. It should be noticed that the range of integration is required to be from -1 to 1. This does not restrict the integrals to which Gaussian integration can be applied since if $f(x)$ is to be integrated in the range a to b, then it can be replaced by the function $g(t)$ integrated from -1 to 1 where

$$t = (2x - a - b)\,/\,(b - a)$$

Note that in the preceding formula, when $x = a$, $t = -1$ and when $x = b$, $t = 1$.

We will now determine the four parameters A_i and x_i for $n = 2$ in (4.22). Thus (4.22) now becomes

$$\int_{-1}^{1} f(x)\,dx = A_1 f(x_1) + A_2 f(x_2) \tag{4.23}$$

This integration rule will be exact for polynomials up to and including degree 3 by ensuring that the rule is exact for the polynomials 1, x, x^2, and x^3 in turn. Thus four equations are obtained as follows:

$$
\begin{aligned}
f(x) = 1 \quad &\text{gives} \quad \int_{-1}^{1} 1dx = 1 = A_1 + A_2 \\
f(x) = x \quad &\text{gives} \quad \int_{-1}^{1} xdx = 1 = A_1x_1 + A_2x_2 \\
f(x) = x^2 \quad &\text{gives} \quad \int_{-1}^{1} x^2dx = 2/3 = A_1x_1^2 + A_2x_2^2 \\
f(x) = x^3 \quad &\text{gives} \quad \int_{-1}^{1} x^3dx = 0 = A_1x_1^3 + A_2x_2^3
\end{aligned}
\tag{4.24}
$$

Solving these equations gives

$$x_1 = -1/\sqrt{3}, \quad x_2 = 1/\sqrt{3}, \quad A_1 = 1, \quad A_2 = 1$$

Thus

$$\int_{-1}^{1} f(x)\,dx = f\left(-\frac{1}{\sqrt{3}}\right) + f\left(\frac{1}{\sqrt{3}}\right) \tag{4.25}$$

Notice that this rule, like Simpson's rule, is exact for cubic equations but requires fewer function evaluations.

A general procedure for obtaining the values of A_i and x_i is based on the fact that in the range of integration it can be shown that $x_1, x_2, \ldots, x_n$ are the roots of the Legendre polynomial of degree n. The values of A_i can then be obtained from an expression involving the Legendre polynomial of degree n, evaluated at x_i. Tables have been produced for the values of x_i and A_i for various values of n; see Abramowitz and Stegun (1965) and Olver et al. (2010). Abramowitz and Stegun provide an excellent reference not only for these functions but for a very extensive range of mathematical functions. However, this classic work is now becoming outdated and a newer handbook of mathematical functions for the twenty-first century by Olver et al. has been published with many improvements, for example, clearer, color graphics. However, this new text contains far fewer tables of functions, since most can now be rapidly computed on a personal computer.

The function `fgauss` defined as follows performs Gaussian integration. It includes a substitution so that integration in the range a to b is converted to an integration in the range -1 to 1.

```
function q = fgauss(func,a,b,n)
% Implements Gaussian integration.
% Example call: q = fgauss(func,a,b,n)
% Integrates user defined function func from a to b, using n divisions
% n must be 2 or 4 or 8 or 16.
if (n==2)|(n==4)|(n==8)|(n==16)
    c = zeros(8,4);  t = zeros(8,4);
    c(1,1) = 1;
    c(1:2,2) = [.6521451548; .3478548451];
    c(1:4,3) = [.3626837833; .3137066458; .2223810344; .1012285362];
    c(:,4 )= [.1894506104; .1826034150; .1691565193; .1495959888; ...
            .1246289712; .0951585116; .0622535239; .0271524594];
    t(1,1) = .5773502691;
    t(1:2,2) = [.3399810435; .8611363115];
    t(1:4,3) = [.1834346424; .5255324099; .7966664774; .9602898564];
    t(:,4) = [.0950125098; .2816035507; .4580167776; .6178762444; ...
            .7554044084; .8656312023; .9445750230; .9894009350];
    j = 1;
    while j<=4
        if 2^j==n; break;
        else
            j = j+1;
        end
    end
    s = 0;
    for k = 1:n/2
        x1 = (t(k,j)*(b-a)+a+b)/2;
        x2 = (-t(k,j)*(b-a)+a+b)/2;
        y = feval(func,x1)+feval(func,x2);
        s = s+c(k,j)*y;
    end
    q = (b-a)*s/2;
else
    disp('n must be equal to 2, 4, 8 or 16'); return
end
```

The following script calls the function `fgauss` to integrate $x^{0.1}$ from 0 to 1.

```
% e3s406.m
disp('  n  integral value');
for j = 1:4
    n = 2^j;
    int = fgauss(@(x) x.^0.1,0,1,n);
    fprintf('%3.0f %14.9f\n',n,int)
end
```

The output of this script is

```
 n   integral value
 2     0.916290737
 4     0.911012914
 8     0.909561226
16     0.909199952
```

Gaussian integration with $n = 16$ gives a better result than that obtained by Romberg's method with five divisions of the interval.

4.8 Infinite Ranges of Integration

Other formulae of the Gauss type are available to allow us to deal with integrals having a special form and infinite ranges of integration. These are the Gauss–Laguerre and Gauss–Hermite formulae and they take the following forms.

4.8.1 Gauss–Laguerre Formula

This method is developed from the following equation:

$$\int_0^\infty e^{-x} g(x) dx = \sum_{i=1}^{n} A_i g(x_i) \tag{4.26}$$

The parameters A_i and x_i are chosen so that, for a given n, the rule is exact for polynomials up to and including degree $2n - 1$. Considering the case when $n = 2$, we have

$$\begin{aligned}
g(x) = 1 \quad &\text{gives} \quad \int_0^\infty e^{-x} dx = 1 = A_1 + A_2 \\
g(x) = x \quad &\text{gives} \quad \int_0^\infty x e^{-x} dx = 1 = A_1 x_1 + A_2 x_2 \\
g(x) = x^2 \quad &\text{gives} \quad \int_0^\infty x^2 e^{-x} dx = 2 = A_1 x_1^2 + A_2 x_2^2 \\
g(x) = x^3 \quad &\text{gives} \quad \int_0^\infty x^3 e^{-x} dx = 6 = A_1 x_1^3 + A_2 x_2^3
\end{aligned} \tag{4.27}$$

Having evaluated the integrals on the left side of equations (4.27) we may solve for the four unknowns x_1, x_2, A_1, and A_2 so that (4.26) becomes

$$\int_0^\infty e^{-x} g(x) dx = \frac{2+\sqrt{2}}{4} g(2-\sqrt{2}) + \frac{2-\sqrt{2}}{4} g(2+\sqrt{2})$$

It can be shown that the x_i are the roots of the nth-order Laguerre polynomial and the coefficients A_i can be calculated from an expression involving the derivative of an nth-order Laguerre polynomial evaluated at x_i.

In general we wish to evaluate integrals of the form

$$\int_0^\infty f(x)\, dx$$

We may write this integral as

$$\int_0^\infty e^{-x}\{e^x f(x)\}\, dx$$

Thus, using (4.26), we have

$$\int_0^\infty f(x)\, dx = \sum_{i=1}^{n} A_i \exp(x_i) f(x_i) \tag{4.28}$$

Equation (4.28) allows integrals to be evaluated over an infinite range, assuming that the value of the integral is finite.

The Gauss–Laguerre method is implemented by the MATLAB function galag:

```
function s = galag(func,n)
% Implements Gauss-Laguerre integration.
% Example call: s = galag(func,n)
% Integrates user defined function func from 0 to inf
% using n divisions. n must be 2 or 4 or 8.
if (n==2)|(n==4)|(n==8)
    c = zeros(8,3);  t = zeros(8,3);
    c(1:2,1) = [1.533326033; 4.450957335];
    c(1:4,2) = [.8327391238; 2.048102438; 3.631146305; 6.487145084];
    c(:,3) = [.4377234105; 1.033869347; 1.669709765; 2.376924702;...
             3.208540913; 4.268575510; 5.818083368; 8.906226215];
    t(1:2,1) = [.5857864376; 3.414213562];
    t(1:4,2) = [.3225476896; 1.745761101; 4.536620297; 9.395070912];
    t(:,3) = [.1702796323; .9037017768; 2.251086630; 4.266700170;...
             7.045905402; 10.75851601; 15.74067864; 22.86313174];
```

```
        j = 1;
        while j<=3
            if 2^j==n; break
            else
                j = j+1;
            end
        end
        s = 0;
        for k = 1:n
            x = t(k,j); y = feval(func,x);
            s = s+c(k,j)*y;
        end
    else
        disp('n must be 2, 4 or 8'); return
    end
```

Sample values x_i and the product $A_i \exp(x_i)$ are given in the function definition. A more complete list may be found in Abramowitz and Stegun (1965) and Olver et al. (2010).

We will now evaluate the integral $\log_e(1+e^{-x})$ from zero to infinity. The following script evaluates the integral using the function `galag`.

```
% e3s407.m
disp(' n   integral value');
for j = 1:3
    n = 2^j;
    int = galag(@(x) log(1+exp(-x)),n);
    fprintf('%3.0f%14.9f\n',n,int)
end
```

The output is as follows:

```
n   integral value
 2   0.822658694
 4   0.822358093
 8   0.822467051
```

Note that the exact result is $\pi^2/12 = 0.82246703342411$. The eight-point integration formula is accurate to six decimal places!

4.8.2 Gauss–Hermite Formula

This method is developed from the following equation:

$$\int_{-\infty}^{\infty} \exp(-x^2)g(x)dx = \sum_{i=1}^{n} A_i g(x_i) \tag{4.29}$$

Again, the parameters A_i and x_i are chosen so that, for a given n, the rule is exact for polynomials up to and including degree $2n-1$. For the case $n=2$ we have

$$\begin{aligned}
g(x) = 1 \quad &\text{gives} \quad \int_{-\infty}^{\infty} \exp(-x^2)dx = \sqrt{\pi} = A_1 + A_2 \\
g(x) = x \quad &\text{gives} \quad \int_{-\infty}^{\infty} x\exp(-x^2)dx = 0 = A_1x_1 + A_2x_2 \\
g(x) = x^2 \quad &\text{gives} \quad \int_{-\infty}^{\infty} x^2\exp(-x^2)dx = \frac{\sqrt{\pi}}{2} = A_1x_1^2 + A_2x_2^2 \\
g(x) = x^3 \quad &\text{gives} \quad \int_{-\infty}^{\infty} x^3\exp(-x^2)dx = 0 = A_1x_1^3 + A_2x_2^3
\end{aligned} \tag{4.30}$$

We have evaluated the integrals on the left side of equations (4.30) and may now solve for the four unknowns x_1, x_2, A_1, and A_2 so that (4.29) becomes

$$\int_{-\infty}^{\infty} \exp(-x^2)g(x)dx = \frac{\sqrt{\pi}}{2}g\left(-\frac{1}{\sqrt{2}}\right) + \frac{\sqrt{\pi}}{2}g\left(\frac{1}{\sqrt{2}}\right)$$

An alternative approach is to note that x_i are the roots of the nth-order Hermite polynomial $H_n(x)$. The coefficients A_i can then be determined from an expression involving the derivative of the nth-order Hermite polynomial evaluated at x_i.

In general we wish to evaluate integrals of the form

$$\int_{-\infty}^{\infty} f(x)\,dx$$

We may write this integral as

$$\int_{-\infty}^{\infty} \exp\left(-x^2\right)\left\{\exp(x^2)f(x)\right\}dx$$

and using (4.29) we have

$$\int_{-\infty}^{\infty} f(x)dx = \sum_{i=1}^{n} A_i \exp\left(x_i^2\right)f(x_i) \tag{4.31}$$

Again, care must be taken to apply (4.31) only to functions that have a finite integral in the range $-\infty$ to ∞. Extensive tables of x_i and A_i are given in Abramowitz and Stegun (1965) and Olver et al. (2010). The MATLAB function gaherm implements Gauss–Hermite integration:

```
function s = gaherm(func,n)
% Implements Gauss-Hermite integration.
% Example call: s = gaherm(func,n)
% Integrates user defined function func from -inf to +inf,
% using n divisions. n must be 2 or 4 or 8 or 16
if (n==2)|(n==4)|(n==8)|(n==16)
    c = zeros(8,4);  t = zeros(8,4);
    c(1,1) = 1.461141183;
    c(1:2,2) = [1.059964483; 1.240225818];
    c(1:4,3) = [.7645441286; .7928900483; .8667526065; 1.071930144];
    c(:,4) = [.5473752050; .5524419573; .5632178291; .5812472754; ...
              .6097369583; .6557556729; .7382456223; .9368744929];
    t(1,1) = .7071067811;
    t(1:2,2) = [.5246476233; 1.650680124];
    t(1:4,3) = [.3811869902; 1.157193712; 1.981656757; 2.930637420];
    t(:,4) = [.2734810461; .8229514491; 1.380258539; 1.951787991; ...
              2.546202158; 3.176999162; 3.869447905; 4.688738939];
    j = 1;
    while j<=4
        if 2^j==n; break;
        else
            j = j+1;
        end
    end
    s=0;
    for k = 1:n/2
        x1 = t(k,j); x2 = -x1;
        y = feval(func,x1)+feval(func,x2);
        s = s+c(k,j)*y;
    end
else
    disp('n must be equal to 2, 4, 8 or 16'); return
end
```

We will now evaluate the integral

$$\int_{-\infty}^{\infty} \frac{dx}{(1+x^2)^2}$$

by the Gauss–Hermite method. The following script uses `gaherm` to integrate this function.

```
% e3s408.m
disp(' n   integral value');
for j = 1:4
    n = 2^j;
    int = gaherm(@(x) 1./(1+x.^2).^2,n);
    fprintf('%3.0f%14.9f\n',n,int)
end
```

The results from running this script are

```
 n   integral value
  2   1.298792163
  4   1.482336098
  8   1.550273058
 16   1.565939612
```

The exact value of this integral is $\pi/2 = 1.570796\ldots$

4.9 Gauss–Chebyshev Formula

We now consider two interesting cases where the sample points x_i and weights w_i are known in a closed or analytical form. The two integrals together with their closed forms are

$$\int_{-1}^{1} \frac{f(x)}{\sqrt{1-x^2}}\,dx = \frac{\pi}{n}\sum_{k=1}^{n} f(x_k) \quad \text{where} \quad x_k = \cos\left(\frac{(2k-1)\pi}{2n}\right) \tag{4.32}$$

$$\int_{-1}^{1} \sqrt{1-x^2} f(x)dx = \frac{\pi}{n+1}\sum_{k=1}^{n} \sin^2\left(\frac{k\pi}{n+1}\right) f(x_k) \quad \text{where} \quad x_k = \cos\left(\frac{k\pi}{n+1}\right) \tag{4.33}$$

These expressions are members of the Gauss family, in this case variations of the Gauss–Chebyshev formula. Clearly it is extremely easy to use these formulae for integrands of the required form that have a specified $f(x)$. It is simply a matter of evaluating the function at the specified points, multiplying by the appropriate factor, and summing these products. A MATLAB script or function can easily be developed and is left as an exercise for the reader (see Problem 4.11).

4.10 Gauss–Lobatto Integration

Lobatto integration or quadrature (Abramowitz and Stegun, 1965) is named after Dutch mathematician Rehuel Lobatto. It is similar to Gaussian quadrature, which we discussed previously, but the integration points include the end points of the integration interval. This has an advantage when the procedure is used in a subinterval because data can be shared between consecutive subintervals. However, Lobatto quadrature is less accurate than the Gaussian formula.

Lobatto quadrature of function $f(x)$ on interval $[-1\ 1]$ is given by the formula

$$\int_{-1}^{1} f(x)dx = \frac{2}{n(n-1)}[f(1)+f(-1)] + \sum_{i=2}^{n-1} w_i f(x_i) + R_n$$

Here the points x_i are the roots of the Legendre polynomial $P_{n-1}(x) = 0$. The weights other than for $f(1)$ and $f(-1)$, which both equal $2/(n(n-1))$, are calculated from the following formula:

$$w_i = \frac{2}{n(n-1)[P_{n-1}(x_i)]^2} \quad (x_i \neq \pm 1)$$

Clearly from this description it is an easy matter to calculate the weights required if the roots of the derivative of the Legendre polynomial are found.

The coefficients of any order Legendre polynomial can be found using Bonnet's recursion formula

$$(n+1)P_{n+1}(x) = (2n+1)xP_n(x) - nP_{n-1}(x)$$

where $P_0(x) = 1$, $P_1(x) = x$, and $P_n(x)$ is the nth Legendre polynomial. Alternatively, a recurrence relation for the polynomials can be found using the differential equation definition of the Legendre function.

The following MATLAB function is based on generating the polynomial coefficients using a recurrence formula and then finding the roots of the derivative of this polynomial using the MATLAB function `roots`. The range has been converted to any range a to b.

```
function Iv = lobattof(func,a,b,n)
% Implementation of Lobatto's method
% func is the function to be integrated from the a to b
% using n points.
% Generate Legendre polynomials based on recurrence relation
% derived from the differential equation which the Legendre polynomial
% satisfies.
```

```
% Obtain derivitive of that polynomial
% The roots of this polynomial give the Lobatto nodes
% From the nodes calculate the weights using standard algorithm
lc = [ ];
for k = 0:n-1
    if n>=2*k
        fnk = factorial(2*n-2*k);
        fnp = 2^n*factorial(k)*factorial(n-k)*factorial(n-2*k);
        lc(n-2*k+1) = (-1)^k*fnk/fnp;
    end
end
% Find coefficients of derivitive of the polynomial
lcd = [ ];
for k = 0:n-1
    if n>=2*k
        lcd(n-2*k+1) = (n-2*k)*lc(n-2*k+1);
    end
end
lcd(n) = 0;
% Obtain Lobatto points
x = roots(fliplr(lcd(2:n+1)));
x1 = sort(x,'descend');
pv = zeros(size(x));
% Calculate Lobatto weights
for k = 1:n+1
    pv = pv+lc(k)*x.^(k-1);
end
n = n+1;
w = 2./(n*(n-1)*pv.^2);
w = [2/(n*(n-1)); w; 2/(n*(n-1))];
% Transform to range a to b
x1 = (x*(b-a)+(a+b))/2;
pts = [a; x1; b];
% Implement rule for integration
Iv = (b-a)*w'*feval(func,pts)/2;
```

To test the function the following MATLAB script is used to integrate $f(x) = e^{5x}\cos(2x)$ from 0 to $\pi/2$.

```
% e3s414.m
g = @(x) exp(5*x).*cos(2*x); a = 0; b = pi/2;
```

```
for n = [2 4 8 16 32 64]
    Iv = lobattof(g,a,b,n);
    fprintf('%3.0f%19.9f\n',n,real(Iv))
end
exact = -5*(exp(2.5*pi)+1)/29;
fprintf('\n Exact %15.9f\n',exact)
```

This gives the following results:

```
  2       -674.125699610
  4       -443.869707406
  8       -444.305258005
 16       -444.305258027
 32       -444.307194507
 64        -16.994770727

 Exact   -444.305258034
```

Note that as the number of points used is increased up to 16, the integration becomes more accurate. However, above this value the accuracy decreases. This is because the function `lobattof` determines the abscissae weights by finding the roots of a polynomial. This becomes less accurate as n increases.

An alternative approach to determine the value of an integral is to subdivide the range of integration into subintervals and then apply a Lobatto rule with a small number of points to each subinterval. The following function allows the user to choose the number of points in the Lobatto integration and the number of subintervals in which the Lobatto integration is applied.

```
function s = lobattomp(func,a,b,n,m)
% n is the number of points in the Labatto quadrature
% m is the number of subintervals of the range of the integration.
h = (b-a)/m; s = 0;
for panel = 0:m-1
    a0 =a+panel*h; b0 = a+(panel+1)*h;
    s = s+lobattof(func,a0,b0,n);
end
```

The following script evaluates the error in the integration of $e^{5x}\cos(2x)$ over the range 0 to $\pi/2$. The script considers a 4-, 5-, ..., 8-point Lobatto integration applied to subintervals, the number of subintervals ranging from 2, 4, 8 to 256.

```
% e3s415.m
g = @(x) exp(5*x).*cos(2*x); a = 0; b = pi/2;
format short e
m = 2; k = 0;
while m<512
    % m is number of panels, k is the index
    k = k+1;
    p = 0;
    for n = 4:8
        % n number of Labotto points, p is index
        p = p+1;
        Integral_err(k,p) = real(lobattomp(g,a,b,n,m))+5*(exp(2.5*pi)+1)/29;
    end
    m = 2*m;
end
Integral_err
```

Running this script gives the following output. Each row gives the value of the error for the specified number of subintervals, beginning with 2, 4, 8, and so on to 256, and each column gives the value of the error for the specified number of points in the Lobatto integration, from 4, 5, ..., 8.

```
Integral_err =
   1.5122e-002  2.6320e-004  1.6910e-006  1.8372e-009 -3.7573e-011
   1.0050e-004  3.5484e-007  4.4201e-010  3.4106e-013 -1.7053e-012
   4.4719e-007  3.7181e-010  3.4106e-013  5.6843e-013 -1.0800e-012
   1.8037e-009  1.1369e-013  1.1369e-013  5.1159e-013 -1.0232e-012
   7.0486e-012 -2.2737e-013  2.2737e-013  5.1159e-013 -9.0949e-013
  -5.6843e-014 -2.2737e-013  2.8422e-013  5.1159e-013 -9.0949e-013
  -1.1369e-013 -2.2737e-013  2.2737e-013  6.2528e-013 -9.0949e-013
  -1.1369e-013 -2.8422e-013  2.2737e-013  6.2528e-013 -6.8212e-013
```

It is evident that increasing the number of subintervals (m) and increasing the number of points in the Lobatto integration (n) reduces the error in the integration. However, when the number of points in the Labatto integration and the number of subintervals increase beyond certain values the accuracy of the integration begins to decrease. The values of m and n at which this happens is problem dependent.

A further disadvantage of the Gauss formula is that the location and weight of the abscissae change as their number increases. For example, suppose we have evaluated an integral using an n-point Gauss quadrature rule. To increase the accuracy we could now increase the number of points and use the Gauss rule again, but all the points would be at a new location. An alternative strategy is to keep the existing n points and add to them $n+1$ points located at the best positions. This is the Kronrod method (Kronrod, 1965).

Thus a three-point Gauss method can be extended by keeping the three points and adding four more to give a seven-point rule. The MATLAB function quadgk implements adaptive Gauss–Kronrod quadrature.

A discussion of the family of Gaussian quadrature methods is given by Thompson (2010).

4.11 Filon's Sine and Cosine Formulae

These formulae can be applied to integrals of the form

$$\int_a^b f(x)\cos kx\,dx \quad \text{and} \quad \int_a^b f(x)\sin kx\,dx \tag{4.34}$$

The formulae are generally more efficient than standard methods for this form of integral. To derive the Filon formulae we first consider an integral of the form

$$\int_0^{2\pi} f(x)\cos kx\,dx$$

By the method of undetermined coefficients we can obtain an approximation to this integrand as follows. Let

$$\int_0^{2\pi} f(x)\ \cos x\,dx = A_1 f(0) + A_2 f(\pi) + A_3 f(2\pi) \tag{4.35}$$

Requiring that this should be exact for $f(x) = 1$, x, and x^2, we have

$$\begin{aligned} 0 &= A_1 + A_2 + A_3 \\ 0 &= A_2\pi + A_3 2\pi \\ 4\pi &= A_2\pi^2 + A_3 4\pi^2 \end{aligned}$$

Thus $A_1 = 2/\pi$, $A_2 = -4/\pi$, and $A_3 = 2/\pi$. Thus,

$$\int_0^{2\pi} f(x)\ \cos x\,dx = \frac{1}{\pi}[2f(0) - 4f(\pi) + 2f(2\pi)] \tag{4.36}$$

More general results can be developed as follows:

$$\int_0^{2\pi} f(x)\cos kx\,dx = h[A\{f(x_n)\sin kx_n - f(x_0)\sin kx_0\} + BC_e + DC_o]$$

$$\int_0^{2\pi} f(x)\ \sin kx\,dx = h[A\{f(x_0)\cos\ kx_0 - f(x_n)\cos kx_n\} + BS_e + DS_o]$$

where $h = (b-a)/n$, $q = kh$ and

$$A = \left(q^2 + q\sin 2q/2 - 2\sin^2 q\right)/q^3 \tag{4.37}$$

$$B = 2\left\{q\left(1+\cos^2 q\right) - \sin 2q\right\}/q^3 \tag{4.38}$$

$$D = 4\left(\sin q - q\cos q\right)/q^3$$

$$C_o = \sum_{i=1,\,3,\,5\ldots}^{n-1} f(x_i)\cos kx_i \tag{4.39}$$

$$C_e = \frac{1}{2}\{f(x_0)\cos kx_0 + f(x_n)\cos kx_n\} + \sum_{i=2,4,6\ldots}^{n-2} f(x_i)\cos kx_i$$

C_o and C_e are odd and even sums of cosine terms. S_o and S_e are similarly defined with respect to sine terms.

It is important to note that Filon's method, when applied to functions of the form given in (4.34), usually gives better results than Simpson's method for the same number of intervals. Approximations may be used for the expressions for A, B, and D given in (4.37), (4.38), and (4.11) by expanding them in series of ascending powers of q. This leads to the following results:

$$A = 2q^2\left(q/45 - q^3/315 + q^5/4725 - \cdots\right)$$

$$B = 2\left(1/3 + q^2/15 - 2q^4/105 + q^6/567 - \cdots\right)$$

$$D = 4/3 - 2q^2/15 + q^4/210 - q^6/11340 + \cdots$$

When the number of intervals becomes very large, h and hence q become small. As q tends to zero, A tends to zero, B tends to 2/3, and D tends to 4/3. Substituting these values into the formula for Filon's method, it can be shown that it becomes equivalent to Simpson's rule. However, in these circumstances the accuracy of Filon's rule may be worse than Simpson's rule owing to the additional complexity of the calculations.

The MATLAB function `filon` implements Filon's method for the evaluation of appropriate integrals. In the parameter list, function `func` defines $f(x)$ of (4.34) and this is multiplied by cos kx when `cas = 1` or sin kx when `cas ~= 1`. The parameters `l` and `u` specify the lower and upper limit of the integral and `n` specifies the number of divisions required. The script incorporates a modification to the standard Filon method such that the series approximation is used if q is less than 0.1 rather than (4.37) to (4.11). The justification for this is that as q becomes small, the accuracy of series approximation is sufficient and easier to compute.

```
function int = filon(func,cas,k,l,u,n)
% Implements filon's integration.
% Example call: int = filon(func,cas,k,l,u,n)
% If cas = 1, integrates cos(kx)*f(x) from l to u using n divisions.
% If cas ~= 1, integrates sin(kx)*f(x) from l to u using n divisions.
% User defined function func defines f(x).
if (n/2)~=floor(n/2)
    disp('n must be even'); return
else
    h = (u-l)/n; q = k*h;
    q2 = q*q; q3 = q*q2;
    if q<0.1
        a = 2*q2*(q/45-q3/315+q2*q3/4725);
        b = 2*(1/3+q2/15+2*q2*q2/105+q3*q3/567);
        d = 4/3-2*q2/15+q2*q2/210-q3*q3/11340;
    else
        a = (q2+q*sin(2*q)/2-2*(sin(q))^2)/q3;
        b = 2*(q*(1+(cos(q))^2)-sin(2*q))/q3;
        d = 4*(sin(q)-q*cos(q))/q3;
    end
    x = l:h:u;
    y = feval(func,x);
    yodd = y(2:2:n);  yeven = y(3:2:n-1);
    if cas == 1
        c = cos(k*x);
        codd = c(2:2:n);  co = codd*yodd';
        ceven = c(3:2:n-1);
        ce = (y(1)*c(1)+y(n+1)*c(n+1))/2;
        ce = ce+ceven*yeven';
        int = h*(a*(y(n+1)*sin(k*u)-y(1)*sin(k*l))+b*ce+d*co);
    else
        s = sin(k*x);
        sodd = s(2:2:n);  so = sodd*yodd';
        seven = s(3:2:n-1);
        se = (y(1)*s(1)+y(n+1)*s(n+1))/2;
        se = se+seven*yeven';
        int = h*(-a*(y(n+1)*cos(k*u)-y(1)*cos(k*l))+b*se+d*so);
    end
end
```

We now test the function `filon` by integrating $\sin x/x$ in the range 1×10^{-10} to 1. The lower limit is set at 1×10^{-10} to avoid the singularity at zero.

The following script uses `filon` and `filonmod` to evaluate the integral. The function `filonmod` removes the ability to switch to the series formula in `filon`. Note that from (4.34), we define $f(x) = 1/x$ for this particular problem.

```
% e3s409.m
n = 4;
g = @(x) 1./x;
disp('  n  Filon no switch  Filon with switch');
while n<=4096
    int1 = filonmod(g,2,1,1e-10,1,n);
    int2 = filon(g,2,1,1e-10,1,n);
    fprintf('%4.0f %17.8e %17.8e\n',n,int1,int2)
    n = 2*n;
end
```

Running this script gives

```
   n  Filon no switch  Filon with switch
   4    1.72067549e+006    1.72067549e+006
   8    1.08265940e+005    1.08265940e+005
  16    6.77884667e+003    6.77884667e+003
  32    4.24742208e+002    4.24742207e+002
  64    2.74361110e+001    2.74361124e+001
 128    2.60175423e+000    2.60175321e+000
 256    1.04956252e+000    1.04956313e+000
 512    9.52549009e-001    9.52550585e-001
1024    9.46489412e-001    9.46487290e-001
2048    9.46109716e-001    9.46108334e-001
4096    9.46085291e-001    9.46084649e-001
```

The exact value of the integral is 0.9460831.

In this particular problem, the switch occurs when $n = 16$. The preceding output shows that the values of the integral obtained with the switch are marginally more accurate. However, it should be noted that experiments carried out by us have shown that for a lower accuracy of computation than that supplied in the MATLAB environment, the accuracy of Filon's method, including the switch, is significantly better. The reader may find it interesting to experiment with the value of q at which the switch occurs. This is currently set at 0.1.

Finally we choose a function that is appropriate for Filon's method and compare the results with Simpson's rule. The function is $\exp(-x/2)\cos(100x)$ integrated between 0 and 2π.

The MATLAB script that implements this comparison is

```
% e3s410.m
n = 4;
disp('    n     Simpsons value     Filons value');
g1 = @(x) exp(-x/2);
g2 = @(x) exp(-x/2).*cos(100*x);
while n<=2048
    int1 = filon(g1,1,100,0,2*pi,n);
    int2 = simp1(g2,0,2*pi,n);
    fprintf('%4.0f %17.8e %17.8e\n',n,int2,int1)
    n = 2*n;
end
```

The results of this comparison are

```
   n    Simpsons value    Filons value
   4    1.91733833e+000    4.55229440e-005
   8   -5.73192992e-001    4.72338540e-005
  16    2.42801799e-002    4.72338540e-005
  32    2.92263624e-002    4.76641931e-005
  64   -8.74419731e-003    4.77734109e-005
 128    5.55127202e-004    4.78308678e-005
 256   -1.30263888e-004    4.78404787e-005
 512    4.53408415e-005    4.78381786e-005
1024    4.77161559e-005    4.78381120e-005
2048    4.78309107e-005    4.78381084e-005
```

The exact value of the integral to 10 significant digits is $4.783810813 \times 10^{-5}$. In this particular problem the switch to the series approximations does not take place because of the high value of the coefficient k. The output shows that using 2048 intervals, Filon's method is accurate to eight significant digits. In contrast, Simpson's rule is accurate to only five significant digits and its behavior is highly erratic. However, timing the evaluation of this integral shows that Simpson's method is about 25% faster than Filon's method.

4.12 Problems in the Evaluation of Integrals

The methods outlined in the previous sections are based on the assumption that the function to be integrated is well behaved. If this is not so, then the numerical methods may give poor, or totally useless, results. Problems may occur if

1. The function is continuous in the range of integration but its derivatives are discontinuous or singular.
2. The function is discontinuous in the range of integration.

3. The function has singularities in the range of integration.
4. The range of integration is infinite.

It is vital that these conditions are identified because in most cases these problems cannot be dealt with directly by numerical techniques. Consequently, some preparation of the integrand is required before the integral can be evaluated by the appropriate numerical method. Case 1 is the least serious condition but since the derivatives of polynomials are continuous, polynomials cannot accurately represent functions with discontinuous derivatives. Ideally, the discontinuity or singularity in the derivative should be located and the integral split into a sum of two or more integrals. The procedure is the same in case 2; the position of the discontinuities must be found and the integral split into a sum of two or more integrals, the ranges of which avoid the discontinuities. Case 3 can be dealt with in various ways: using a change of variable, integration by parts, and splitting the integral. In case 4 we must use a method suitable for an infinite range of integration (see Section 4.8) or make a substitution.

The following integral, taken from Fox and Mayers (1968), is an example of case 4:

$$I = \int_1^{\infty} \frac{dx}{x^2 + \cos(x^{-1})} \tag{4.40}$$

This integral can be estimated either by using function `galag` (using the substitution $y = x - 1$ to give a lower limit of zero) or by substituting $z = 1/x$. Thus $dz = -dx/x^2$ and (4.40) may be transformed as follows:

$$I = -\int_1^0 \frac{dz}{1 + z^2\cos(z)} \quad \text{or} \quad I = \int_0^1 \frac{dz}{1 + z^2\cos(z)} \tag{4.41}$$

The integral (4.41) can easily be evaluated by any standard method.

We have discussed a number of techniques for numerical integration. It must be said, however, that even the best methods have difficulty with functions that change very rapidly for small changes in the independent variable. An example of this type of function is $\sin(1/x)$. A MATLAB plot of this function is shown in Section 3.8. However, this plot does not give a true representation of the function in the range -0.1 to 0.1 because in this range the function is changing very rapidly and the number of plotting points and the screen resolution are inadequate. Indeed, as x tends to zero the frequency of the function tends to infinity. A further difficulty is that the function has a singularity at $x = 0$. If we decrease the range of x, then a small section of the function can be plotted and displayed. For example, in the range $x = 2 \times 10^{-4}$ to 2.05×10^{-4} there are approximately 19 cycles of the function $\sin(1/x)$, as shown in Figure 4.3, and in this limited range the function can be effectively sampled and plotted. Summarizing, the value of this function can change from an extreme positive to an extreme negative value for a relatively small change in x. The consequence

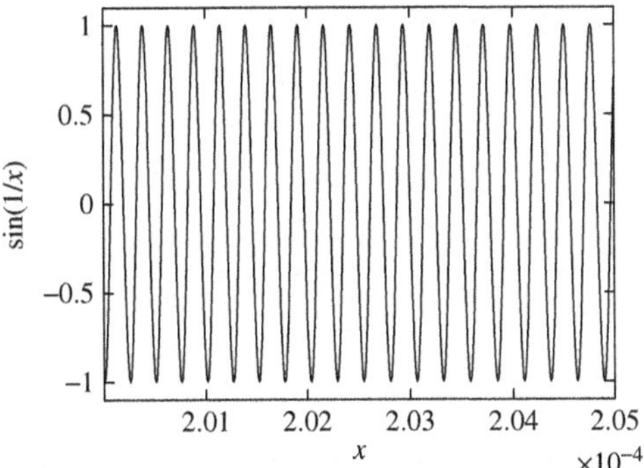

FIGURE 4.3 The function $\sin(1/x)$ in the range $x = 2 \times 10^{-4}$ to 2.05×10^{-4}. Nineteen cycles of the function are displayed.

of this is that when estimating the integral of the function, a great number of divisions of the range of integration are needed to provide the required level of accuracy, particularly for smaller values of x. For this type of problem adaptive integration methods, such as that used by the MATLAB function `quadl`, have been introduced. These methods increase the number of intervals only in those regions where the function is changing very rapidly, thus reducing the overall number of calculations required.

4.13 Test Integrals

We now compare the Gauss and Simpson methods of integration with the MATLAB function `quadl` using the following integrals:

$$\int_0^1 x^{0.001}\,dx = 1000/1001 = 0.999000999\ldots \tag{4.42}$$

$$\int_0^1 \frac{dx}{1+(230x-30)^2} = (\tan^{-1}200 + \tan^{-1}30)/230 = 0.0134924856495 \tag{4.43}$$

$$\int_0^4 x^2(x-1)^2(x-2)^2(x-3)^2(x-4)^2\,dx = 10240/693 = 14.776334776 \tag{4.44}$$

To generate the comparative results we define the function `ftable` as follows:

```
function y = ftable(fname,lowerb,upperb)
% Generates table of results.
```

```
intg = fgauss(fname,lowerb,upperb,16);
ints = simp1(fname,lowerb,upperb,2048);
intq = quadl(fname,lowerb,upperb,.00005);
fprintf('%19.8e %18.8e %18.8e \n',intg,ints,intq)
```

The following script applies this function to the three test integrals:

```
% e3s411.m
clear
disp('function       Gauss              Simpson               quadl')
fprintf('Func 1'), ftable(@(x) x.^0.001,0,1)
fprintf('Func 2'), ftable(@(x) 1./(1+(230*x-30).^2),0,1)
g = @(x) (x.^2).*((1-x).^2).*((2-x).^2).*((3-x).^2).*((4-x).^2);
fprintf('Func 3'), ftable(g,0,4)
```

The output from this script is

```
function       Gauss              Simpson              quadl
Func 1    9.99003302e-001   9.98839883e-001   9.98981017e-001
Func 2    1.46785776e-002   1.34924856e-002   1.34925421e-002
Func 3    1.47763348e+001   1.47763348e+001   1.47763348e+001
```

The integrals (4.42) and (4.43) are difficult to evaluate and Figure 4.4 shows plots of the integrands in the range of integration. Each function, at some point, changes rapidly with small changes of the independent variable, making such functions extremely difficult to integrate numerically if a high degree of accuracy is required.

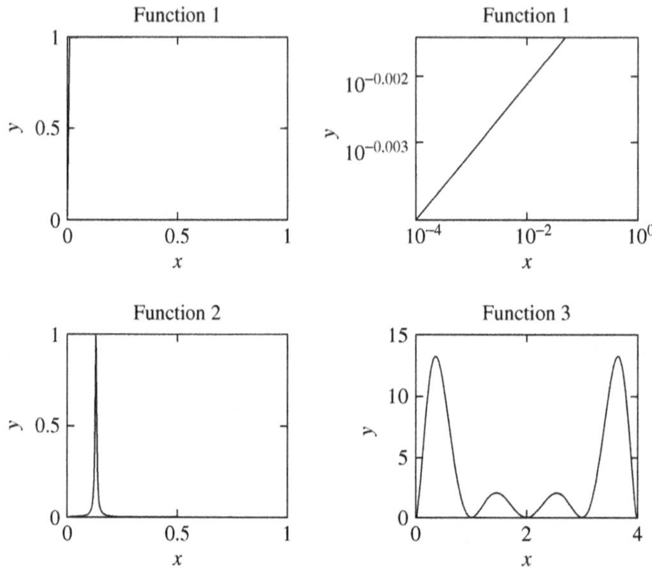

FIGURE 4.4 Plots of functions defined in script e3s411.

4.14 Repeated Integrals

In this section we confine ourselves to a discussion of repeated integrals using two variables. It is important to note that there is a significant difference between double integrals and repeated integrals. However, it can be shown that if the integrand satisfies certain requirements then double integrals and repeated integrals are equal in value. A detailed discussion of this result is given in Jeffrey (1979).

We have considered in this chapter various techniques for evaluating single integrals. The extension of these methods to repeated integrals can present considerable scripting difficulties. Furthermore, the number of computations required for the accurate evaluation of a repeated integral can be enormous. While many algorithms for the evaluation of single integrals can be extended to repeated integrals, here only extensions to the Simpson and Gauss methods with two variables are presented. These have been chosen as the best compromise between programming simplicity and efficiency.

An example of a repeated integral is

$$\int_{a_1}^{b_1} dx \int_{a_2}^{b_2} f(x,y)\,dy \tag{4.45}$$

In this notation the function is integrated with respect to x from a_1 to b_1 and with respect to y from a_2 to b_2. Here the limits of integration are constant but in some applications they may be variables.

4.14.1 Simpson's Rule for Repeated Integrals

We now apply Simpson's rule to the repeated integral (4.45) by applying it first in the y direction and then in the x direction. Consider three equispaced values of y: y_0, y_1, and y_2. On applying Simpson's rule, (4.11), to integration with respect to y in (4.45), we have

$$\int_{x_0}^{x_2} dx \int_{y_0}^{y_2} f(x,y)dy \approx \int_{x_0}^{x_2} k\left\{f(x,y_0)+4f(x,y_1)+f(x,y_2)\right\}/3\,dx \tag{4.46}$$

where $k = y_2 - y_1 = y_1 - y_0$.

Consider now three equispaced values of x: x_0, x_1, and x_2. Applying Simpson's rule again to integration with respect to x, from (4.46) we have

$$I \approx hk\left[f_{0,0}+f_{0,2}+f_{2,0}+f_{2,2}+4\left\{f_{0,1}+f_{1,0}+f_{1,2}+f_{2,1}\right\}+16f_{1,1}\right]/9 \tag{4.47}$$

where $h = x_2 - x_1 = x_1 - x_0$ and, for example, $f_{1,2} = f(x_1,y_2)$.

This is Simpson's rule in two variables. By applying this rule to each group of nine points on the surface $f(x,y)$ and summing, the composite Simpson's rule is obtained. The MATLAB function `simp2v` evaluates repeated integrals in two variables by making direct use of the composite rule.

```
function q = simp2v(func,a,b,c,d,n)
% Implements 2 variable Simpson integration.
% Example call: q = simp2v(func,a,b,c,d,n)
% Integrates user defined 2 variable function func.
% Range for first variable is a to b, and second variable, c to d
% using n divisions of each variable.
if (n/2)~=floor(n/2)
    disp('n must be even'); return
else
    hx = (b-a)/n; x = a:hx:b; nx = length(x);
    hy = (d-c)/n; y = c:hy:d; ny = length(y);
    [xx,yy] = meshgrid(x,y);
    z = feval(func,xx,yy);
    v = 2*ones(n+1,1);  v2 = 2*ones(n/2,1);
    v(2:2:n) = v(2:2:n)+v2;
    v(1) = 1;  v(n+1) = 1;
    S = v*v';  T = z.*S;
    q = sum(sum(T))*hx*hy/9;
end
```

We will now apply the function `simp2v` to evaluate the integral

$$\int_0^{10} dx \int_0^{10} y^2 \sin x \, dy$$

The graph of the function $y^2 \sin x$ is given in Figure 4.5. The following script integrates this function.

```
% e3s412.m
z = @(x,y) y.^2.*sin(x);
```

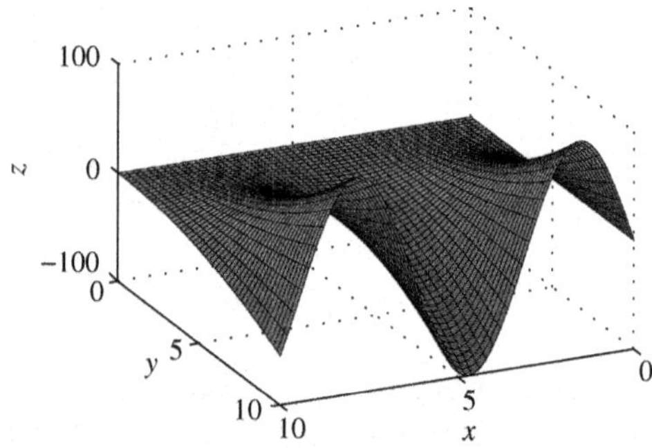

FIGURE 4.5 Graph of $z = y^2 \sin x$.

```
disp(' n      integral value');
n = 4; j = 1;
while n<=256
    int = simp2v(z,0,10,0,10,n);
    fprintf('%4.0f %17.8e\n',n,int)
    n = 2*n; j = j+1;
end
```

Running the script gives the following results:

```
   n     integral value
   4     1.02333856e+003
   8     6.23187046e+002
  16     6.13568708e+002
  32     6.13056704e+002
  64     6.13025879e+002
 128     6.13023970e+002
 256     6.13023851e+002
```

The value of this integral exact to four decimal places is 613.0238 and the number of floating-point operations tends to $7n^2$. It can be proved—see Salvadori and Baron (1961)—that when Simpson's rule is adapted to evaluate repeated integrals, the error is still of order h^4 and thus it is possible to use an extrapolation scheme similar to the Romberg method of Section 4.6.

4.14.2 Gaussian Integration for Repeated Integrals

The Gaussian method can be developed to evaluate repeated integrals with constant limits of integration. In Section 4.7 it was shown that for single integrals the integrand must be evaluated at specified points. Thus, if

$$I = \int_{-1}^{1} dx \int_{-1}^{1} f(x, y)\, dy$$

then

$$I \approx \sum_{i=1}^{n} \sum_{j=1}^{m} A_i A_j f(x_i, y_j)$$

The rules for calculating x_i, y_j, and A_i are given in Section 4.7. The MATLAB function gauss2v evaluates integrals using this technique. Because the values of x and y are chosen on the assumption that the integration takes place in the range –1 to 1, the function includes the necessary manipulations to adjust it so as to accommodate an arbitrary range of integration.

```
function q = gauss2v(func,a,b,c,d,n)
% Implements 2 variable Gaussian integration.
% Example call: q = gauss2v(func,a,b,c,d,n)
% Integrates user defined 2 variable function func,
% Range for first variable is a to b, and second variable, c to d
% using n divisions of each variable.
% n must be 2 or 4 or 8 or 16.
if (n==2)|(n==4)|(n==8)|(n==16)
    co = zeros(8,4); t = zeros(8,4);
    co(1,1) = 1;
    co(1:2,2) = [.6521451548; .3478548451];
    co(1:4,3) = [.3626837833; .3137066458; .2223810344; .1012285362];
    co(:,4) = [.1894506104; .1826034150; .1691565193; .1495959888; ...
              .1246289712;.0951585116; .0622535239; .0271524594];
    t(1,1) = .5773502691;
    t(1:2,2) = [.3399810435; .8611363115];
    t(1:4,3) = [.1834346424; .5255324099; .7966664774; .9602898564];
    t(:,4) = [.0950125098; .2816035507; .4580167776; .6178762444; ...
             .7554044084; .8656312023; .9445750230; .9894009350];
    j = 1;
    while j<=4
        if 2^j==n; break;
        else
            j = j+1;
        end
    end
    s = 0;
    for k = 1:n/2
        x1 = (t(k,j)*(b-a)+a+b)/2;  x2 = (-t(k,j)*(b-a)+a+b)/2;
        for p = 1:n/2
            y1 = (t(p,j)*(d-c)+d+c)/2;  y2 = (-t(p,j)*(d-c)+d+c)/2;
            z = feval(func,x1,y1)+feval(func,x1,y2)+feval(func,x2,y1);
            z = z+feval(func,x2,y2);
            s = s+co(k,j)*co(p,j)*z;
        end
    end
    q = (b-a)*(d-c)*s/4;
else
    disp('n must be equal to 2, 4, 8 or 16'), return
end
```

We will now consider the problem of evaluating the following integral:

$$\int_{x^2}^{x^4} dy \int_{1}^{2} x^2 y \, dx \tag{4.48}$$

Integrals of this form cannot be estimated directly by the MATLAB function gauss2v or simp2v because neither of these functions was developed to work with constant limits of integration. However, a transformation may be carried out in order to make the limits of integration constant. Let

$$y = (x^4 - x^2)z + x^2 \tag{4.49}$$

Thus when $z = 1$, $y = x^4$ and when $z = 0$, $y = x^2$ as required. Differentiating the preceding expression, we have

$$dy = (x^4 - x^2)dz$$

Substituting for y and dy in (4.48), we have

$$\int_{0}^{1} dz \int_{1}^{2} x^2 \left\{ (x^4 - x^2)z + x^2 \right\} (x^4 - x^2) \, dx \tag{4.50}$$

This integral is now in a form that can be integrated using both gauss2v and simp2v. However, we must define a MATLAB function as follows:

```
w = @(x,z) x.^2.*((x.^4-x.^2).*z+x.^2).*(x.^4-x.^2);
```

This function is used with the functions simp2v and gauss2v in the following script:

```
% e3s413.m
disp('   n  Simpson value     Gauss value')
w = @(x,z) x.^2.*((x.^4-x.^2).*z+x.^2).*(x.^4-x.^2);
n = 2; j = 1;
while n<=16
    in1 = simp2v(w,1,2,0,1,n);
    in2 = gauss2v(w,1,2,0,1,n);
    fprintf('%4.0f%17.8e%17.8e\n',n,in1,in2)
    n = 2*n; j = j+1;
end
```

Running this script gives

```
 n  Simpson value     Gauss value
 2  9.54248047e+001  7.65255915e+001
 4  8.48837042e+001  8.39728717e+001
 8  8.40342951e+001  8.39740259e+001
16  8.39778477e+001  8.39740259e+001
```

The integral is equal to 83.97402597 (=6466/77). This output shows that, in general, Gaussian integration is superior to Simpson's rule.

4.15 MATLAB Functions for Double and Triple Integration

Recent versions of MATLAB now provide the functions `dblquad` and `triplequad` for repeated integration. In this section we consider these functions and their parameters and provide examples of their use.

For double integration, which is repeated integration over two dimensions, the `dblquad` function may be used and has the general form

```
IV2 = dblquad(fname,xl,xu,yl,yu,acc)
```

where `fname` is the name of the two-variable function being integrated, which must be defined by the user; `xl` and `xu` are the lower and upper limits of the x range of integration; and similarly `yl` and `yu` are the lower and upper limits for the y range of integration. The value `acc` provides the required accuracy of the integration and is optional.

The use of `dblquad` is illustrated by the following example. Consider the integral

$$I = \int_0^1 dx \int_0^1 \frac{1}{1-xy} dy$$

We may solve this using the MATLAB function `dblquad`. It is required that the user predefine the function to be integrated; to do this we choose to use an anonymous function directly in the function parameter list. Using `dblquad` we have

```
>> I = dblquad(@(x,y) 1./(1-x.*y),0,1-1e-6,0,1-1e-6)

I =

    1.6449
```

If we try to integrate this function numerically over the exact range $x = 0$ to 1 and $y = 0$ to 1 then MATLAB gives warnings because of the singularity when $x = y = 1$ but gives the same answer.

For triple integration, which is repeated integration over three dimensions, the `triplequad` function may be used and has the general form

```
IV3 = triplequad(fname,xl,xu,yl,yu,zl,zu,acc)
```

where `fname` is the name of the three variable function being integrated, and `xl` and `xu` are the lower and upper limits of the x range of integration. Similarly `yl`, `yu` and `zl`, `zu` are the limits for the y and z range of integration. The use of `triplequad` is illustrated by the following example:

$$\int_0^1 dx \int_0^1 dy \int_0^1 64xy(1-x)^2 z\, dz$$

```
>> I3 = triplequad(@(x,y,z) 64*x.*y.*(1-x).^2.*z,0,1,0,1,0,1)

I3 =
     1.3333
```

The function `quad2d` allows the user to integrate a function of two variables (say x and y) like the function `dblquad` but additionally allows the limits in y to be functions of x. Consider the integral (4.48) repeated here:

$$\int_1^2 dx \int_{x^2}^{x^4} x^2 y\, dy$$

Using `quad2d` we have

```
>> IV = quad2d(@(x,y) x.^2.*y,1,2, @(x) x.^2,@(x) x.^4)

IV =
   83.9740
```

In the preceding example, the anonymous function `@(x,y) x.^2.*y` is the function to be integrated, 1 and 2 are the lower and upper limits of integration in the `x` variable, and `@(x) x.^2` and `@(x) x.^4` are anonymous functions defining lower and upper limits in the `y` range of integration.

4.16 Summary

In this chapter we have described simple methods for obtaining the approximate derivatives of various orders for specified functions at given values of the independent variable. The results indicate that these methods, although easy to program, are very sensitive to small changes in key parameters and should be used with considerable care. In addition, we have given a range of methods for integration. For integration, error generation is not such an unpredictable problem but we must be careful to choose the most efficient method for the integral we wish to evaluate.

The reader is referred to Sections 9.8, 9.9, and 9.10 for the application of the Symbolic Toolbox to integration and differentiation problems.

Problems

4.1. Use the function `diffgen` to find the first and second derivatives of the function $x^2 \cos x$ at $x = 1$ using $h = 0.1$ and $h = 0.01$.

4.2. Evaluate the first derivative of $\cos x^6$ for $x = 1, 2$, and 3 using the function `diffgen` and taking $h = 0.001$.

4.3. Write a MATLAB function to differentiate a given function using formulae (4.6) and (4.7). Use it to solve Problems 4.1 and 4.2.

4.4. Find the gradient of $y = \cos x^6$ at $x = 3.1, 3.01, 3.001$, and 3 using the function `diffgen` with $h = 0.001$. Compare your results with the exact result.

4.5. The approximations for partial derivatives may be defined as

$$\partial f/\partial x \approx \{f(x+h,y) - f(x-h,y)\} / (2h)$$

$$\partial f/\partial y \approx \{f(x,y+h) - f(x,y-h)\} / (2h)$$

Write a function to evaluate these derivatives. The function call should have the form

```
[pdx,pdy] = pdiff('func',x,y,h)
```

Determine the partial derivatives of $\exp(x^2 + y^3)$ at $x = 2$, $y = 1$ using this function with $h = 0.005$.

4.6. In a letter sent to Hardy, the Indian mathematician Ramanujan proposed that the number of numbers between a and b that are either squares or sums of two squares is given approximately by the integral

$$0.764 \int_a^b \frac{dx}{\sqrt{\log_e x}}$$

Test this proposition for the following pairs of values of a and b: (1,10), (1,17), and (1,30). You should use the MATLAB function `fgauss` with 16 points to evaluate the integrals required.

4.7. Verify the equality

$$\int_0^\infty \frac{dx}{(1+x^2)(1+r^2x^2)(1+r^4x^2)} = \frac{\pi(r^2+r+1)}{2(r^2+1)(r+1)^2}$$

for the values of $r = 0, 1, 2$. This result was proposed by Ramanujan. You should use the MATLAB function `galag` for your investigations, using 8 points.

4.8. Raabe established the result that

$$\int_a^{a+1} \log_e \Gamma(x) dx = a \log_e a - a + \log_e \sqrt{2\pi}$$

Verify this result for $a = 1$ and $a = 2$. Use the MATLAB function `simp1` with 32 divisions to evaluate the integrals required and the MATLAB function `gamma` to set up the integrand.

4.9. Use the MATLAB function `fgauss` with 16 points to evaluate the integral

$$\int_0^1 \frac{\log_e x dx}{1+x^2}$$

Explain why the function `fgauss` is appropriate for this problem but `simp1` is not.

4.10. Use the MATLAB function `fgauss` with 16 points to evaluate the integral

$$\int_0^1 \frac{\tan^{-1} x}{x} dx$$

Note: Integration by parts shows the integrals in Problems 4.9 and 4.10 to be the same value except for a sign.

4.11. Write a MATLAB function to implement the formulae (4.32) and (4.33) given in Section 4.9 and use your function to evaluate the following integrals using 10 points for the formula. Compare your results with the Gauss 16-point rule.

$$\textbf{(a)} \int_{-1}^1 \frac{e^x}{\sqrt{1-x^2}} dx \quad \textbf{(b)} \int_{-1}^1 e^x \sqrt{1-x^2}\, dx$$

4.12. Use the MATLAB function `simp1` to evaluate the Fresnel integrals

$$C(1) = \int_0^1 \cos\left(\frac{\pi t^2}{2}\right) dt \quad \text{and} \quad S(1) = \int_0^1 \sin\left(\frac{\pi t^2}{2}\right) dt$$

Use 32 intervals. The exact values, to seven decimal places, are $C(1) = 0.7798934$ and $S(1) = 0.4382591$.

4.13. Use the MATLAB function `filon`, with 64 intervals, to evaluate the integral

$$\int_0^{\pi} \sin x \cos kx \, dx$$

for $k = 0$, 4, and 100. Compare your results with the exact answer, $2/(1-k^2)$ if k is even and 0 if k is odd.

4.14. Solve Problem 4.13 for $k = 100$ using Simpson's rule with 1024 divisions and Romberg's methods with 9 divisions.

4.15. Evaluate the following integral using the 8-point Gauss–Laguerre method:

$$\int_0^{\infty} \frac{e^{-x}dx}{x+100}$$

Compare your answer with the exact solution 9.9019419×10^{-3} (103/10402).

4.16. Evaluate the integral

$$\int_0^{\infty} \frac{e^{-2x} - e^{-x}}{x} dx$$

using 8-point Gauss–Lagurre integration. Compare your result with the exact answer, which is $-\log_e 2 = -0.6931$.

4.17. Evaluate the following integral using the 16-point Gauss–Hermite method. Compare your answer with the exact solution $\sqrt{\pi}\exp(-1/4)$.

$$\int_{-\infty}^{\infty} \exp(-x^2)\cos x \, dx$$

4.18. Evaluate the following integrals, using Simpson's rule for repeated integrals, MATLAB function `simp2v`, with 64 divisions in each direction.

$$\textbf{(a)} \int_{-1}^{1} dy \int_{-\pi}^{\pi} x^4 y^4 dx \quad \textbf{(b)} \int_{-1}^{1} dy \int_{-\pi}^{\pi} x^{10} y^{10} dx$$

4.19. Evaluate the following integrals, using `simp2v`, with 64 divisions in each direction.

$$\textbf{(a)} \int_0^3 dx \int_1^{\sqrt{x/3}} \exp(y^3)dy \quad \textbf{(b)} \int_0^2 dx \int_0^{2-x} (1+x+y)^{-3}dy$$

4.20. Evaluate part (b) in Problems 4.17 and 4.18 using Gaussian integration, MATLAB function `gauss2v`. *Note*: To use this function the range of integration must be constant.

4.21. The definition of the sine integral Si(z) is

$$\text{Si}(z) = \int_0^z \frac{\sin t}{t} dt$$

Evaluate this integral using the 16-point Gauss method for $z = 0.5, 1$, and 2. Why does the Gaussian method work and yet the Simpson and Romberg methods fail?

4.22. Evaluate the following double integral using Gaussian integration for two variables.

$$\int_0^1 dy \int_0^1 \frac{1}{1-xy} dx$$

Compare your result with the exact answer, $\pi^2/6 = 1.6449$.

4.23. The probability, P, that a certain type of gas turbine engine will fail within a period of time of T hours is given by the equation

$$P(x < T) = \int_0^T \frac{ab^a}{(x+b)^{a+1}} dx$$

where $a = 3.5$ and $b = 8200$.

By evaluating this integral for values of $T = 500 : 100 : 2000$, draw a graph of P against T in this range. What proportion of the number of gas turbines of this type fail within 1600 hours. For more information on the probability of failure, see Percy (2011).

4.24. Consider the following integral:

$$\int_0^1 \frac{x^p - x^q}{\log_e(x)} x^r dx = \log_e \left(\frac{p+r+1}{q+r+1} \right)$$

Use the MATLAB function `quad` to verify this result for $p = 3, q = 4, r = 2$.

4.25. Consider the following three integrals:

$$A = -\int_0^1 \frac{\log_e x\, dx}{1+x^2}, \quad B = \int_0^1 \frac{\tan^{-1} x}{x} dx, \quad C = \int_0^\infty \frac{xe^{-x}}{1+e^{-2x}} dx$$

Use the MATLAB function `quad` to evaluate the two integrals A and B and hence verify that they are equal.

Use 8-point Gauss–Laguerre integration to verify that the integral C is also equal to A and B.

4.26. Use 16-point Gauss–Hermite integration to evaluate the integral

$$I = \int_{-\infty}^{\infty} \frac{\sin x}{1+x^2} dx$$

is approximately equal to zero.

4.27. Use 16-point Gauss–Hermite integration to evaluate the integral

$$I = \int_{-\infty}^{\infty} \frac{\cos x}{1+x^2} dx$$

Check your answer by comparing with the exact answer, π/e.

4.28. Use 8-point Gauss–Lagurre integration to find the value of the integral

$$I = \int_0^\infty \frac{x^{\alpha-1}}{1+x^\beta} dx$$

for values of α and $\beta = (2,3), (3,4)$. You can verify your answers using the exact value of the integral, which is $\pi/(\beta \sin(\alpha\pi/\beta))$.

4.29. An interesting relationship between the Riemann zeta function and the integral

$$S_3 = -\int_0^\infty \log_e(x)^3 e^{-x}\, dx$$

is given by

$$S_3 = \gamma^3 + \frac{1}{2}\gamma\pi^2 + 2\zeta(3)$$

where $\gamma = 0.57722$. Use the MATLAB function `quadgk` to evaluate the integral and show that it is a good estimate of S_3.

4.30. A value for the total resistance of a certain network of unit resistors has been shown to be given by $R(m, n)$, where

$$R(m,n) = \frac{1}{\pi^2} \int_0^{\pi} dx \int_0^{\pi} \frac{1 - \cos mx \cos ny}{2 - \cos x - \cos y}\, dy$$

Evaluate this integral for $R(50, 100)$ using the MATLAB functions `dblquad` and `simp2v`. Use a lower limit close to zero, say 0.0001. If zero is used the denominator in the integral is zero. For large values of m and n an approximation for this integral is given by

$$R(m,n) = \frac{1}{\pi}\left(\gamma + \frac{3}{2}\log_e 2 + \frac{1}{2}\log_e(m^2 + n^2)\right)$$

where γ is Euler's constant and can be obtained by evaluating the MATLAB expression `-psi(1)`. The function `-psi` is called the digamma function. Use this to check your result.

4.31. A value for the total resistance of a cubic network of unit resistors has been shown to be given by $R(s, m, n)$ where

$$R(s,m,n) = \frac{1}{\pi^3} \int_0^{\pi} dx \int_0^{\pi} dy \int_0^{\pi} \frac{1 - \cos sx \cos my \cos nz}{3 - \cos x - \cos y - \cos z}\, dz$$

Evaluate this integral using the MATLAB function for `triplequad` using the values $s = 2$, $m = 1$, $n = 3$. The lower limit should be set at a small nonzero value, say 0.0001.

5 Solution of Differential Equations

Many practical problems involve the study of how rates of change in two or more variables are interrelated. Often the independent variable is time. These problems give rise naturally to differential equations, which enable us to understand how the real world works and how it changes dynamically. Essentially, differential equations provide us with a model of some physical system and the solution of the differential equations enables us to predict the system's behavior. These models may be quite simple, involving one differential equation, or may involve many interrelated simultaneous differential equations.

5.1 Introduction

To illustrate how a differential equation can model a physical situation we will examine a relatively simple problem. Consider the way a hot object cools—for example, a saucepan of milk, the water in a bath, or molten iron. Each of these will cool in a different way dependent on the environment but we shall abstract only the most important features that are easy to model. To model this process by a simple differential equation we use Newton's law of cooling, which states that the *rate* at which these objects lose heat as time passes is dependent on the difference between the current temperature of the object and the temperature of its surroundings. This leads to the differential equation

$$dy/dt = K(y - s) \tag{5.1}$$

where y is the current temperature at time t, s is the temperature of the surroundings, and K is a negative constant for the cooling process. In addition we require the initial temperature, y_0, to be specified at time $t = 0$ when the observations begin. This fully specifies our model of the cooling process. We only need values for y_0, K, and s to begin our study. This type of first-order differential equation is called an *initial value problem* because we have an initial value given for the dependent variable y at time $t = 0$.

The solution of (5.1) is easily obtained analytically and will be a function of t and the constants of the problem. However, there are many differential equations that have no analytic solution or the analytic solution does not provide an explicit relation between y and t. In this situation we use numerical methods to solve the differential equation. This means that we approximate the continuous solution with an approximate discrete solution giving the values of y at specified time steps between the initial value of time

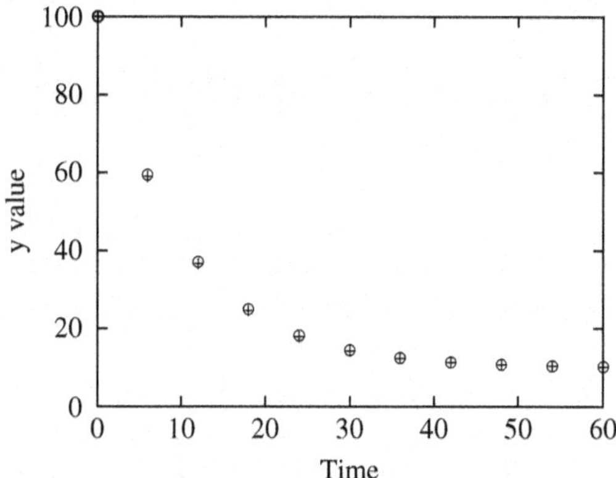

FIGURE 5.1 Exact (○) and approximate (+) solution for $dy/dt = -0.1(y - 10)$.

and some final time value. Thus we compute values of y, which we denote by y_i, for values of t denoted by t_i where $t_i = t_0 + ih$ for $i = 0, 1, \ldots, n$. Figure 5.1 illustrates the exact solution and an approximate solution of (5.1) where $K = -0.1$, $s = 10$, and $y_0 = 100$. This figure is generated using the standard MATLAB function for solving differential equations, `ode23`, from time 0 to 60 and plotting the values of y using the symbol "+." The values of the exact solution are plotted on the same graph using the symbol "○."

To use `ode23` to solve (5.1) we begin by writing a function `yprime` that defines the right side of (5.1). Then `ode23` is called in the following script and requires the initial and final values of t, 0, and 60, which must be placed in a row vector; a starting value for y of 100; and a low relative tolerance of 0.5. This tolerance is set using the `odeset` function, which allows tolerances and other parameters to be set as required.

```
% e3s501.m
yprime = @(t,y) -0.1*(y-10); %RH of diff equn.
options = odeset('RelTol',0.5);
[t y] = ode23(yprime,[0 60],100,options);
plot(t,y,'+')
xlabel('Time'), ylabel('y value'),
hold on
plot(t,90*exp(-0.1.*t)+10,'o'), % Exact solution.
hold off
```

This type of step-by-step solution is based on computing the current y_i value from a single or combination of functions of previous y values. If the value of y is calculated from a combination of more than one previous value, it is called a *multistep* method. If only one previous value is used it is called a *single-step* method. We shall now describe a simple *single-step* method known as Euler's method.

5.2 Euler's Method

The dependent variable y and the independent variable t, which we used in the preceding section, can be replaced by any variable names. For example, many textbooks use y as the dependent variable and x as the independent variable. However, for some consistency with MATLAB notation we generally use y to represent the dependent variable and t to represent the independent variable. Clearly initial value problems are not restricted to the time domain, although in most practical situations they are.

Consider the differential equation

$$dy/dt = y \tag{5.2}$$

One of the simplest approaches for obtaining the numerical solution of a differential equation is the method of Euler. This employs the Taylor series but uses only the first two terms of the expansion. Consider the following form of the Taylor series in which the third term is called the remainder term and represents the contribution of all the terms not included in the series:

$$y(t_0 + h) = y(t_0) + y'(t_0)h + y''(\theta)h^2/2 \tag{5.3}$$

where θ lies in the interval (t_0, t_1). For small values of h we may neglect the terms in h^2, and setting $t_1 = t_0 + h$ in (5.3) leads to the formula

$$y_1 = y_0 + hy'_0$$

where the prime denotes differentiation with respect to t and $y'_i = y'(t_i)$. In general,

$$y_{n+1} = y_n + hy'_n \quad \text{for} \quad n = 0, 1, 2, \ldots$$

By virtue of (5.2) this may be written

$$y_{n+1} = y_n + hf(t_n, y_n) \quad \text{for} \quad n = 0, 1, 2, \ldots \tag{5.4}$$

This is known as Euler's method and it is illustrated geometrically in Figure 5.2. This is an example of the use of a single function value to determine the next step. From (5.3) we can see that the local truncation error (i.e., the error for individual steps) is of order h^2.

The method is simple to script and is implemented in the MATLAB function `feuler` as follows:

```
function [tvals, yvals] = feuler(f,tspan, startval,step)
% Euler's method for solving
% first order differential equation dy/dt = f(t,y).
% Example call: [tvals, yvals]=feuler(f,tspan,startval,step)
% Initial and final value of t are given by tspan = [start finish].
% Initial value of y is given by startval, step size is given by step.
```

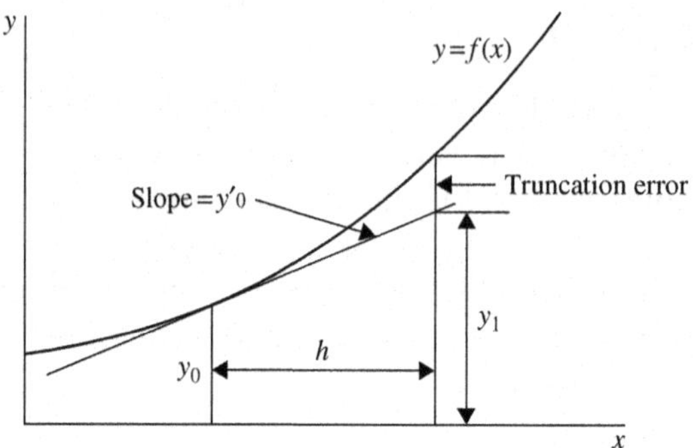

FIGURE 5.2 Geometric interpretation of Euler's method.

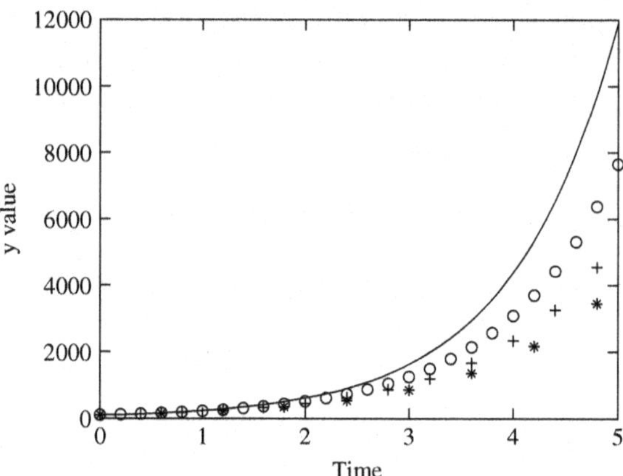

FIGURE 5.3 Points from the Euler solution of $dy/dt = y - 20$ given that $y = 100$ when $t = 0$. Approximate solutions for $h = 0.2$, 0.4, and 0.6 are plotted using o, +, and $*$, respectively. The exact solution is given by the solid line.

```
% The function f(t,y) must be defined by the user.
steps = (tspan(2)-tspan(1))/step+1;
y = startval; t = tspan(1);
yvals = startval; tvals = tspan(1);
for i = 2:steps
    y1 = y+step*feval(f,t,y); t1 = t+step;
    %collect values together for output
    tvals = [tvals, t1]; yvals = [yvals, y1];
    t = t1; y = y1;
end
```

Applying this function to the differential equation (5.1) with $K = 1$, $s = 20$, and an initial value of $y = 100$ gives Figure 5.3, which illustrates how the approximate solution varies for different values of h. The exact value computed from the analytical solution is given for comparison purposes by the solid line. Clearly, in view of the very large errors shown by Figure 5.3, the Euler method, although simple, requires a very small step h to provide reasonable levels of accuracy. If the differential equation must be solved for a wide range of values of t, the method becomes very expensive in terms of computer time because of the very large number of small steps required to span the interval of interest. In addition, the errors made at each step may accumulate in an unpredictable way. This is a crucial issue, and we discuss this in the next section.

5.3 The Problem of Stability

To ensure that errors do not accumulate we require that the method for solving the differential equation be stable. We have seen that the error at each step in Euler's method is of order h^2. This error is known as the local truncation error since it tells only how accurate the individual step is, not what the error is for a sequence of steps. The error for a sequence of steps is difficult to find since the error from one step affects the accuracy of the next in a way that is often complex. This leads us to the issue of absolute and relative stability. We now discuss these concepts and examine their effects in relation to a simple equation and explain how the results for this equation may be extended to differential equations in general.

Consider the differential equation

$$dy/dt = Ky \tag{5.5}$$

Since $f(t,y) = Ky$, Euler's method will have the form

$$y_{n+1} = y_n + hKy_n \tag{5.6}$$

Thus using this recursion repeatedly and assuming that there are no errors in the computation from stage to stage we obtain

$$y_{n+1} = (1 + hK)^{n+1} y_0 \tag{5.7}$$

For small enough h it is easily shown that this value will approach the exact value e^{Kt}.

To obtain some understanding of how errors propagate when using Euler's method let us assume that y_0 is perturbed. This perturbed value of y_0 may be denoted by y_0^a where $y_0^a = (y_0 - e_0)$ and e_0 is the error. Thus (5.7) becomes, on using this approximate value instead of y_0,

$$y_{n+1}^a = (1 + hK)^{n+1} y_0^a = (1 + hK)^{n+1}(y_0 - e_0) = y_{n+1} - (1 + hK)^{n+1} e_0$$

Consequently, the initial error will be magnified if $|1 + hK| \geq 1$. After many steps this initial error will grow and may dominate the solution. This is the characteristic of instability and in these circumstances Euler's method is said to be unstable. If, however, $|1+hK|<1$, then the error dies away and the method is said to be absolutely stable. Rewriting this inequality leads to the condition for absolute stability:

$$-2 < hK < 0 \tag{5.8}$$

This condition may be too demanding and we may be content if the error does not increase as a proportion of the y values. This is called relative stability. Notice that Euler's method is not absolutely stable for any positive value of K.

The condition for absolute stability can be generalized to an ordinary differential equation of the form of (5.2). It can be shown that the condition becomes

$$-2 < h\partial f/\partial y < 0 \tag{5.9}$$

This inequality implies that, since $h > 0$, $\partial f/\partial y$ must be negative for absolute stability. Figures 5.4 and 5.5 give a comparison of the absolute and relative error for $h = 0.1$ for the differential equation $dy/dt = y$ where $y = 1$ when $t = 0$. Figure 5.4 shows that the error is increasing rapidly and the errors are large for even relatively small step sizes. Figure 5.5 shows that the error is becoming an increasing proportion of the solution values. Thus the relative error is increasing linearly and so the method is neither relatively stable nor absolutely stable for this problem.

We have seen that Euler's method may be unstable for some values of h. For example, if $K = -100$, then Euler's method is only absolutely stable for $0 < h < 0.02$. Clearly if we are required to solve the differential equation between 0 and 10, we would require 500 steps. We now consider an improvement to this method called the trapezoidal method, which has improved stability features although it is similar in principle to Euler's method.

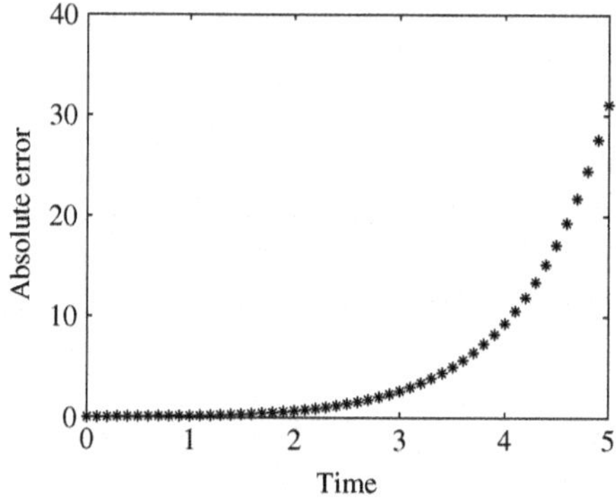

FIGURE 5.4 Absolute errors in the solution of $dy/dt = y$ where $y = 1$ when $t = 0$, using Euler's method with $h = 0.1$.

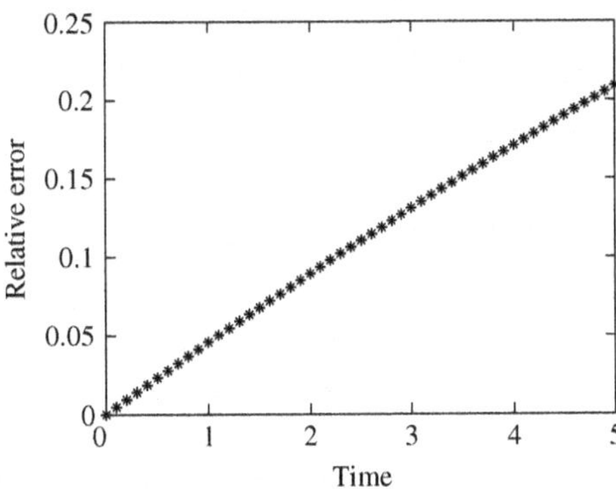

FIGURE 5.5 Relative errors in the solution of $dy/dt = y$ where $y = 1$ when $t = 0$, using Euler's method with $h = 0.1$.

5.4 The Trapezoidal Method

The trapezoidal method has the form

$$y_{n+1} = y_n + h\left\{f(t_n, y_n) + f(t_{n+1}, y_{n+1})\right\}/2 \quad \text{for} \quad n = 0, 1, 2, \ldots \tag{5.10}$$

Applying the error analysis of Section 5.3 to this problem gives us, from (5.5), that

$$y_{n+1} = y_n + h(Ky_n + Ky_{n+1})/2 \quad \text{for} \quad n = 0, 1, 2, \ldots \tag{5.11}$$

Thus expressing y_{n+1} in terms of y_n gives

$$y_{n+1} = (1 + hK/2)/(1 - hK/2)y_n \quad \text{for} \quad n = 0, 1, 2, \ldots \tag{5.12}$$

Using this result recursively for $n = 0, 1, 2, \ldots$ leads to the result

$$y_{n+1} = \left\{(1 + hK/2)/(1 - hK/2)\right\}^{n+1} y_0 \tag{5.13}$$

Now, as in Section 5.3, we can obtain some understanding of how error propagates by assuming that y_0 is perturbed by the error e_0 so that it is replaced by $y_0^a = (y_0 - e_0)$. Hence using the same procedure (5.13) becomes

$$y_{n+1} = \left\{(1 + hK/2)/(1 - hK/2)\right\}^{n+1} (y_0 - e_0)$$

This leads directly to the result

$$y_{n=1}^a = y_{n+1} - \left\{(1 + hK/2)/(1 - hK/2)\right\}^{n+1} e_0$$

Thus we conclude from this that the influence of the error term that involves e_0 will die away if its multiplier is less than unity in magnitude, that is

$$|(1+hK/2)/(1-hK/2)| < 1$$

If K is negative, then for all h the method is absolutely stable. For positive K it is not absolutely stable for any h.

This completes the error analysis of this method. However, we note that the method requires a value for y_{n+1} before we can start. An estimate for this value can be obtained by using Euler's method, that is

$$y_{n+1} = y_n + hf(t_n, y_n) \quad \text{for} \quad n = 0, 1, 2, \ldots$$

This value can now be used in the right side of (5.10) as an estimate for y_{n+1}. This combined method is often known as the Euler-trapezoidal method. The method can be written formally as

1. Start with n set at zero where n indicates the number of steps taken.
2. Calculate $y_{n+1}^{(1)} = y_n + hf(t_n, y_n)$.
3. Calculate $f(t_{n+1}, y_{n+1}^{(1)})$ where $t_{n+1} = t_n + h$.
4. For $k = 1, 2, \ldots$ calculate

$$y_{n+1}^{(k+1)} = y_n + h\left\{f(t_n, y_n) + f(t_{n+1}, y_{n+1}^{(k)})\right\}/2 \tag{5.14}$$

At step 4, when the difference between successive values of y_{n+1} is sufficiently small, increment n by 1 and repeat steps 2, 3, and 4. This method is implemented in the MATLAB function `eulertp`:

```
function [tvals, yvals] = eulertp(f,tspan,startval,step)
% Euler trapezoidal method for solving
% first order differential equation dy/dt = f(t,y).
% Example call: [tvals, yvals] = eulertp(f,tspan,startval,step)
% Initial and final value of t are given by tspan = [start finish].
% Initial value of y is given by startval, step size is given by step.
% The function f(t,y) must be defined by the user.
steps = (tspan(2)-tspan(1))/step+1;
y = startval; t = tspan(1);
yvals = startval; tvals = tspan(1);
for i = 2:steps
    y1 = y+step*feval(f,t,y);
    t1 = t+step;
    loopcount = 0; diff = 1;
    while abs(diff)>0.05
```

```
        loopcount = loopcount+1;
        y2 = y+step*(feval(f,t,y)+feval(f,t1,y1))/2;
        diff = y1-y2; y1 = y2;
    end
    %collect values together for output
    tvals = [tvals, t1]; yvals = [yvals, y1];
    t = t1; y = y1;
end
```

We use `eulertp` to study the performance of this method compared with Euler's method for solving $dy/dt = y$. The results are given in Figure 5.6, which shows graphs of the absolute errors of the two methods. The difference is clear but although the Euler-trapezoidal method gives much greater accuracy for this problem, in other cases the difference may be less marked. In addition, the Euler-trapezoidal method takes longer.

An important feature of this method is the number of iterations that are required to obtain convergence in step 4. If this is high, the method is likely to be inefficient. However, for the example we just solved, a maximum of two iterations at step 4 was required. This algorithm may be modified to use only one iteration at step 4 in (5.14). This is called Heun's method.

Finally, we examine theoretically how the error in Heun's method compares with Euler's method. By considering the Taylor series expansion of y_{n+1} we can obtain the order of the error in terms of the step size h:

$$y_{n+1} = y_n + hy'_n + h^2 y''_n/2! + h^3 y'''_n(\theta)/3! \tag{5.15}$$

where θ lies in the interval (t_n, t_{n+1}). It can be shown that y''_n may be approximated by

$$y''_n = (y'_{n+1} - y'_n)/h + O(h) \tag{5.16}$$

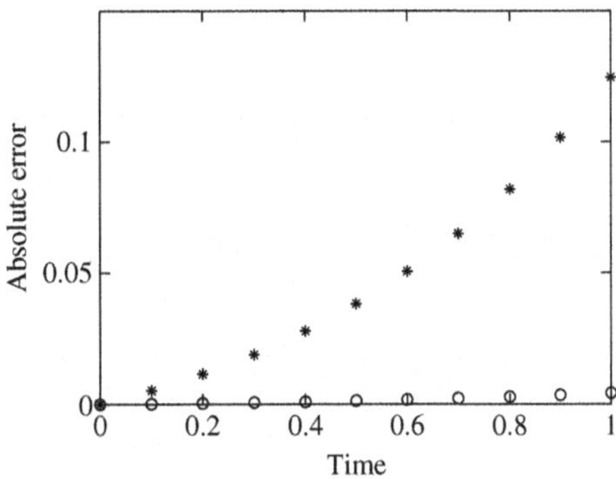

FIGURE 5.6 Absolute error in the solution of $dy/dt = y$ using Euler (*) and trapezoidal method (○). Step $h = 0.1$ and $y_0 = 1$ at $t = 0$.

Substituting this expression for y_n'' in (5.15) gives

$$\begin{aligned} y_{n+1} &= y_n + hy_n' + h(y_{n+1}' - y_n')/2! + O(h^3) \\ &= y_n + h(y_{n+1}' + y_n')/2! + O(h^3) \end{aligned}$$

This shows that the local truncation error is of order h^3 so there is a significant improvement in accuracy over the basic Euler method, which has a truncation error of order h^2.

We now describe a range of methods that will be considered under the collective title of Runge–Kutta methods.

5.5 Runge–Kutta Methods

The Runge–Kutta methods comprise a large family of methods having a common structure. Heun's method, described by (5.14) but with only one iteration of the corrector, can be recast in the form of a simple Runge–Kutta method. We set

$$k_1 = hf(t_n, y_n) \quad \text{and} \quad k_2 = hf(t_{n+1}, y_{n+1})$$

since

$$y_{n+1} = y_n + hf(t_n, y_n)$$

We have

$$k_2 = hf(t_{n+1}, y_n + hf(t_n, y_n))$$

Hence from (5.10) we have Heun's method in the form for $n = 0, 1, 2, \ldots$

$$\begin{aligned} k_1 &= hf(t_n, y_n) \\ k_2 &= hf(t_{n+1}, y_n + k_1) \end{aligned}$$

and

$$y_{n+1} = y_n + (k_1 + k_2)/2$$

This is a simple form of a Runge–Kutta method.

The most commonly used Runge–Kutta method is the classical one; it has the form for each step $n = 0, 1, 2, \ldots$

$$\begin{aligned} k_1 &= hf(t_n, y_n) \\ k_2 &= hf(t_n + h/2, y_n + k_1/2) \\ k_3 &= hf(t_n + h/2, y_n + k_2/2) \\ k_4 &= hf(t_n + h, y_n + k_3) \end{aligned} \tag{5.17}$$

and

$$y_{n+1} = y_n + \left(k_1 + 2k_2 + 2k_3 + k_4\right)/6$$

It has a global error of order h^4. The next Runge–Kutta method we consider is a variation on the formula (5.17). It is due to Gill (1951) and takes the form for each step $n = 0, 1, 2, \ldots$

$$\begin{aligned}
k_1 &= hf(t_n, y_n) \\
k_2 &= hf(t_n + h/2, y_n + k_1/2) \\
k_3 &= hf(t_n + h/2, y_n + (\sqrt{2} - 1)k_1/2 + (2 - \sqrt{2})\,k_2/2) \\
k_4 &= hf(t_n + h, y_n - \sqrt{2}k_2/2 + (1 + \sqrt{2}/2)k_3)
\end{aligned} \tag{5.18}$$

and

$$y_{n+1} = y_n + \left\{k_1 + (2 - \sqrt{2})k_2 + (2 + \sqrt{2})k_3 + k_4\right\}/6$$

Again this method is fourth order and has a local truncation error of order h^5 and a global error of order h^4.

A number of other forms of the Runge–Kutta method have been derived that have particularly advantageous properties. The equations for these methods will not be given but their important features are as follows:

1. *Merson–Runge–Kutta method* (Merson, 1957). This method has an error term of order h^5 and in addition allows an estimate of the local truncation error to be obtained at each step in terms of known values.
2. *Ralston–Runge–Kutta method* (Ralston, 1962). We have some degree of freedom in assigning the coefficients for a particular Runge–Kutta method. In this formula the values of the coefficients are chosen so as to minimize the truncation error.
3. *Butcher–Runge–Kutta method* (Butcher, 1964). This method provides higher accuracy at each step, the error being of order h^6.

Runge–Kutta methods have the general form for each step $n = 0, 1, 2, \ldots$

$$\begin{aligned}
k_1 &= hf(t_n, y_n) \\
k_i &= hf(t_n + hd_i, y_n + \sum_{j=1}^{i-1} c_{ij}k_j), \quad i = 2, 3, \ldots, p
\end{aligned} \tag{5.19}$$

$$y_{n+1} = y_n + \sum_{j=1}^{p} b_j k_j \tag{5.20}$$

The order of this general method is p.

The derivation of the various Runge–Kutta methods is based on the expansion of both sides of (5.20) as a Taylor series and equating coefficients. This is a relatively straightforward idea but involves lengthy algebraic manipulation.

We now discuss the stability of the Runge–Kutta methods. Since the instability that may arise in the Runge–Kutta methods can usually be reduced by a step size reduction, it is known as partial instability. To avoid repeated reduction of the value of h and rerunning the method, an estimate of the value of h that will provide stability for the fourth-order Runge–Kutta methods is given by the inequality

$$-2.78 < h\partial f/\partial y < 0$$

In practice $\partial f/\partial y$ may be approximated using the difference of successive values of f and y.

Finally, it is interesting to see how we can apply MATLAB to provide an elegant function for the general Runge–Kutta method given by (5.20) and (5.19). We define two vectors **d** and **b**, where **d** contains the coefficients d_i in (5.19) and **b** contains the coefficients b_j in (5.20), and a matrix **c** that contains the coefficients c_{ij} in (5.19). If the computed values of the k_j are assigned to a vector **k**, then the MATLAB statements that generate the values of the function and the new value of y are relatively simple; they will have the form

```
k(1) = step*feval(f,t,y);
for i = 2:p
    k(i)=step*feval(f,t+step*d(i),y+c(i,1:i-1)*k(1:i-1)');
end
y1 = y+b*k';
```

This is of course repeated for each step. A MATLAB function, `rkgen`, based on this follows. Since `c` and `d` are easily changed in the script, any form of the Runge–Kutta method can be implemented using this function and it is useful for experimenting with different techniques.

```
function[tvals,yvals] = rkgen(f,tspan,startval,step,method)
% Runge Kutta methods for solving
% first order differential equation dy/dt = f(t,y).
% Example call:[tvals,yvals]=rkgen(f,tspan,startval,step,method)
% The initial and final values of t are given by tspan = [start finish].
% Initial y is given by startval and step size is given by step.
% The function f(t,y) must be defined by the user.
% The parameter method (1, 2 or 3) selects
% Classical, Butcher or Merson RK respectively.
b = [ ]; c = [ ]; d = [ ];
switch method
```

```
    case 1
        order = 4;
        b = [ 1/6 1/3 1/3 1/6]; d = [0 .5 .5 1];
        c=[0 0 0 0;0.5 0 0 0;0 .5 0 0;0 0 1 0];
        disp('Classical method selected')
    case 2
        order = 6;
        b = [0.07777777778 0 0.355555556 0.13333333 ...
             0.355555556 0.0777777778];
        d = [0 .25 .25 .5 .75 1];
        c(1:4,:) = [0 0 0 0 0 0;0.25 0 0 0 0 0;0.125 0.125 0 0 0 0; ...
                    0 -0.5 1 0 0 0];
        c(5,:) = [.1875 0 0 0.5625 0 0];
        c(6,:) = [-.4285714 0.2857143 1.714286 -1.714286 1.1428571 0];
        disp('Butcher method selected')
    case 3
        order = 5;
        b = [1/6 0 0 2/3 1/6];
        d = [0 1/3 1/3 1/2 1];
        c = [0 0 0 0 0;1/3 0 0 0 0;1/6 1/6 0 0 0;1/8 0 3/8 0 0; ...
            1/2 0 -3/2 2 0];
        disp('Merson method selected')
    otherwise
        disp('Invalid selection')
end
steps = (tspan(2)-tspan(1))/step+1;
y = startval; t = tspan(1);
yvals = startval; tvals = tspan(1);
for j = 2:steps
    k(1) = step*feval(f,t,y);
    for i = 2:order
        k(i) = step*feval(f,t+step*d(i),y+c(i,1:i-1)*k(1:i-1)');
    end
    y1 = y+b*k'; t1 = t+step;
    %collect values together for output
    tvals = [tvals, t1]; yvals = [yvals, y1];
    t = t1; y = y1;
end
```

A further issue that needs to be considered is that of adaptive step size adjustment. Where a function is relatively smooth in the area of interest, a large step may be used throughout the region. If the region is such that rapid changes in y occur for small changes in t, then a small step size is required. However, for functions where both these regions

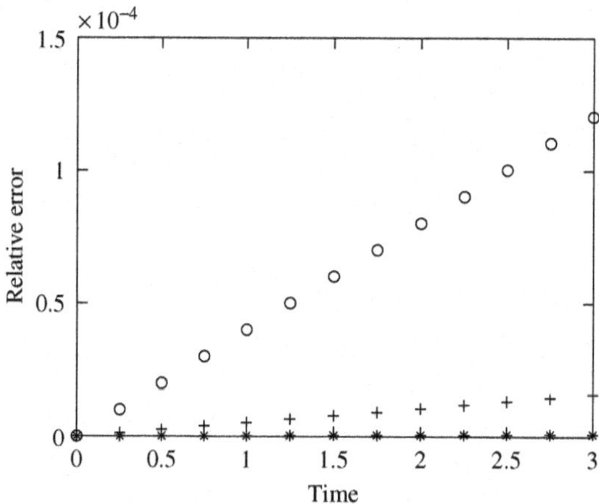

FIGURE 5.7 Relative error in the solution of $dy/dt = -y$. $*$ represents the Butcher method, + the Merson method, and ∘ the classical method.

exist, then rather than use a small step in the whole region, adaptive step size adjustment would be more efficient. The details of producing this step adjustment are not provided here; however, for an elegant discussion see Press et al. (1990). This type of procedure is implemented for Runge–Kutta methods in the MATLAB functions `ode23` and `ode45`.

Figure 5.7 plots the relative errors in the solution of the specific differential equation $dy/dt = -y$ by the classical, Merson, and Butcher methods using the following MATLAB script:

```
% e3s502.m
yprime = @(t,y) -y;
char = 'o*+';
for meth = 1:3
    [t, y] = rkgen(yprime,[0 3],1,0.25,meth);
    re = (y-exp(-t))./exp(-t);
    plot(t,re,char(meth))
    hold on
end
hold off, axis([0 3 0 1.5e-4])
xlabel('Time'), ylabel('Relative error')
```

It is clear from the graphs that the Butcher method is the best and both the Butcher and Merson methods are significantly more accurate than the classical method.

5.6 Predictor–Corrector Methods

The trapezoidal method, which has already been described in Section 5.4, is a simple example of both a Runge–Kutta method and a predictor–corrector method with a truncation error of order h^3. The predictor–corrector methods we consider now have much smaller truncation errors. As an initial example we consider the Adams–Bashforth–Moulton method. This method is based on the following equations:

$$\begin{aligned} y_{n+1} &= y_n + h(55y'_n - 59y'_{n-1} + 37y'_{n-2} - 9y'_{n-3})/24 \quad (P) \\ y'_{n+1} &= f(t_{n+1}, y_{n+1}) \quad (E) \end{aligned} \tag{5.21}$$

and

$$\begin{aligned} y_{n+1} &= y_n + h\left(9y'_{n+1} + 19y'_n - 5y'_{n-1} + y'_{n-2}\right)/24 \quad (C) \\ y'_{n+1} &= f(t_{n+1}, y_{n+1}) \quad (E) \end{aligned} \tag{5.22}$$

where $t_{n+1} = t_n + h$. In (5.21) we use the predictor equation (P), followed by a function evaluation (E). Then in (5.22) we use the corrector equation (C), followed by a function evaluation (E). The truncation error for both the predictor and corrector is $O(h^5)$. The first equation in the system (5.21) requires a number of initial values to be known before y can be calculated.

After each application of (5.21) and (5.22), that is, a complete *PECE* step, the independent variable t_n is incremented by h, n is incremented by one, and the process repeated until the differential equation has been solved in the range of interest. The method is started with $n = 3$ and consequently the values of y_3, y_2, y_1, and y_0 must be known before the method can be applied. For this reason it is called a *multipoint method.* In practice y_3, y_2, y_1, and y_0 must be obtained using a self-starting procedure such as one of the Runge–Kutta methods described in Section 5.5. The self-starting procedure chosen should have the same order truncation error as the predictor–corrector method.

The Adams–Bashforth–Moulton method is often used since its stability is relatively good. Its range of absolute stability in *PECE* mode is

$$-1.25 < h\partial f/\partial y < 0$$

Apart from the need for initial starting values, the Adams–Bashforth–Moulton method in the *PECE* mode requires less computation at each step than the fourth-order Runge–Kutta method. For a true comparison of these methods, however, it is necessary to consider how they behave over a range of problems since applying any method to some differential equations results, at each step, in a growth of error that ultimately swamps the calculation since the step is outside the range of absolute stability.

The Adams–Bashforth–Moulton method is implemented by the function `abm`. It should be noted that errors arise from the choice of starting procedure, in this case the classical Runge–Kutta method. It is, however, easy to amend this function to include the option of entering highly accurate initial values.

```
function [tvals, yvals] = abm(f,tspan,startval,step)
% Adams Bashforth Moulton method for solving
```

```
% first order differential equation dy/dt = f(t,y).
% Example call: [tvals, yvals] = abm(f,tspan,startval,step)
% The initial and final values of t are given by tspan = [start finish].
% Initial y is given by startval and step size is given by step.
% The function f(t,y) must be defined by the user.
% 3 steps of Runge--Kutta are required so that ABM method can start.
% Set up matrices for Runge--Kutta methods
b = [ ]; c = [ ]; d = [ ]; order = 4;
b = [1/6 1/3 1/3 1/6]; d = [0 .5 .5 1];
c = [0 0 0 0;0.5 0 0 0;0 .5 0 0;0 0 1 0];
steps = (tspan(2)-tspan(1))/step+1;
y = startval; t = tspan(1); fval(1) = feval(f,t,y);
ys(1) = startval; yvals = startval; tvals = tspan(1);
for j = 2:4
    k(1) = step*feval(f,t,y);
    for i = 2:order
        k(i) = step*feval(f,t+step*d(i),y+c(i,1:i-1)*k(1:i-1)');
    end
    y1 = y+b*k'; ys(j) = y1; t1 = t+step;
    fval(j) = feval(f,t1,y1);
    %collect values together for output
    tvals = [tvals,t1]; yvals = [yvals,y1];
    t = t1; y = y1;
end
%ABM now applied
for i = 5:steps
    y1 = ys(4)+step*(55*fval(4)-59*fval(3)+37*fval(2)-9*fval(1))/24;
    t1 = t+step; fval(5) = feval(f,t1,y1);
    yc = ys(4)+step*(9*fval(5)+19*fval(4)-5*fval(3)+fval(2))/24;
    fval(5) = feval(f,t1,yc);
    fval(1:4) = fval(2:5);
    ys(4) = yc;
    tvals = [tvals,t1]; yvals = [yvals,yc];
    t = t1; y = y1;
end
```

Figure 5.8 illustrates the behavior of the Adams–Bashforth–Moulton method when applied to the specific problem $dy/dt = -2y$ where $y = 1$ when $t = 0$, using a step size equal to 0.5 and 0.7 in the interval 0 to 10. It is interesting to note that for this problem, since $\partial f/\partial y = -2$, the range of steps for absolute stability is $0 \leq h \leq 0.625$. For $h = 0.5$, a value inside the range of absolute stability, the plot shows that the absolute error does die away. However, for $h = 0.7$, a value outside the range of absolute stability, the plot shows that the absolute error increases.

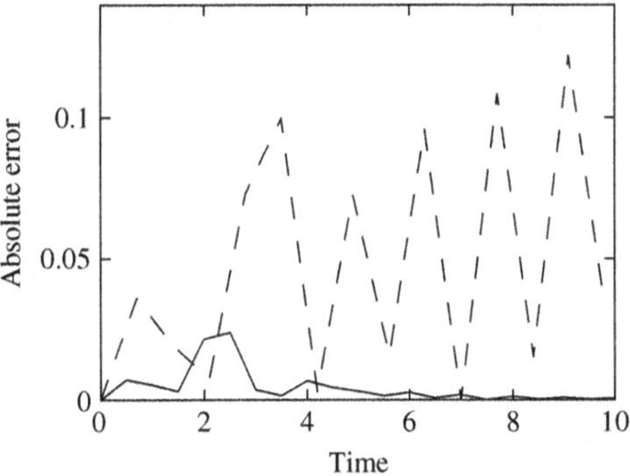

FIGURE 5.8 Absolute error in the solution of $dy/dt = -2y$ using the Adams–Bashforth–Moulton method. The *solid line* plots the errors with a step size of 0.5. The *dot-dashed line* plots the errors with step size 0.7.

5.7 Hamming's Method and the Use of Error Estimates

The method of Hamming (1959) is based on the following pair of predictor–corrector equations:

$$\begin{aligned} y_{n+1} &= y_{n-3} + 4h\left(2y'_n - y'_{n-1} + 2y'_{n-2}\right)/3 && (P) \\ y'_{n+1} &= f\left(t_{n+1}, y_{n+1}\right) && (E) \end{aligned} \qquad (5.23)$$

$$\begin{aligned} y_{n+1} &= \left\{9y_n - y_{n-2} + 3h\left(y'_{n+1} + 2y'_n - y'_{n-1}\right)\right\}/8 && (C) \\ y'_{n+1} &= f\left(t_{n+1}, y_{n+1}\right) && (E) \end{aligned}$$

where $t_{n+1} = t_n + h$.

The first equation (P) is used as the predictor and the third as the corrector (C). To obtain a further improvement in accuracy at each step in the predictor and corrector we modify these equations using expressions for the local truncation errors. Approximations for these local truncation errors can be obtained using the predicted and corrected values of the current approximation to y. This leads to the equations

$$y_{n+1} = y_{n-3} + 4h\left(y'_n - y'_{n-1} + 2y'_{n-2}\right)/3 \quad (P) \qquad (5.24)$$

$$(y^M)_{n+1} = y_{n+1} - 112(Y_P - Y_C)/121 \qquad (5.25)$$

In this equation Y_P and Y_C represent the predicted and corrected value of y at the nth step.

$$y^*_{n+1} = \left\{9y_n - y_{n-2} + 3h((y^M)'_{n+1} + 2y'_n - y'_{n-1})\right\}/8 \quad (C) \qquad (5.26)$$

In this equation $(y^M)'_{n+1}$ is the value of y'_{n+1} calculated using the modified value of y_{n+1}, which is $(y^M)_{n+1}$.

$$y_{n+1} = y^*_{n+1} + 9(y_{n+1} - y^*_{n+1}) \tag{5.27}$$

Equation (5.24) is the predictor and (5.25) modifies the predicted value by using an estimate of the truncation error. Equation (5.26) is the corrector, which is modified by (5.27) using an estimate of the truncation error. The equations in this form are each used only once before n is incremented and the steps repeated again. This method is implemented as MATLAB function `fhamming` as follows:

```
function [tvals, yvals] = fhamming(f,tspan,startval,step)
% Hamming's method for solving
% first order differential equation dy/dt = f(t,y).
% Example call: [tvals, yvals] = fhamming(f,tspan,startval,step)
% The initial and final values of t are given by tspan = [start finish].
% Initial y is given by startval and step size is given by step.
% The function f(t,y) must be defined by the user.
% 3 steps of Runge-Kutta are required so that hamming can start.
% Set up matrices for Runge-Kutta methods
b = [ ]; c =[ ]; d = [ ]; order = 4;
b = [1/6 1/3 1/3 1/6]; d = [0 0.5 0.5 1];
c = [0 0 0 0;0.5 0 0 0;0 0.5 0 0;0 0 1 0];
steps = (tspan(2)-tspan(1))/step+1;
y = startval; t = tspan(1);
fval(1) = feval(f,t,y);
ys(1) = startval;
yvals = startval; tvals = tspan(1);
for j = 2:4
    k(1) = step*feval(f,t,y);
    for i = 2:order
        k(i) = step*feval(f,t+step*d(i),y+c(i,1:i-1)*k(1:i-1)');
    end
    y1 = y+b*k'; ys(j) = y1; t1 = t+step; fval(j) = feval(f,t1,y1);
    %collect values together for output
    tvals = [tvals, t1]; yvals = [yvals, y1]; t = t1; y = y1;
end
%Hamming now applied
for i = 5:steps
    y1 = ys(1)+4*step*(2*fval(4)-fval(3)+2*fval(2))/3;
    t1 = t+step; y1m = y1;
    if i>5, y1m = y1+112*(c-p)/121; end
    fval(5) = feval(f,t1,y1m);
    yc = (9*ys(4)-ys(2)+3*step*(2*fval(4)+fval(5)-fval(3)))/8;
```

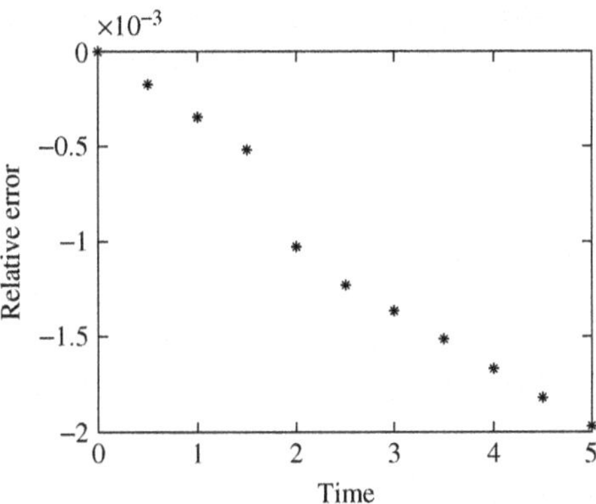

FIGURE 5.9 Relative error in the solution of $dy/dt = y$ where $y = 1$ when $t = 0$, using Hamming's method with a step size of 0.5.

```
        ycm = yc+9*(y1-yc)/121;
        p = y1; c = yc;
        fval(5) = feval(f,t1,ycm); fval(2:4) = fval(3:5);
        ys(1:3) = ys(2:4); ys(4) = ycm;
        tvals = [tvals, t1]; yvals = [yvals, ycm];
        t = t1;
    end
```

The choice of h must be made carefully so that the error does not increase without bound. Figure 5.9 shows Hamming's method used to solve the equation $dy/dt = y$. This is the problem used in Section 5.6.

5.8 Error Propagation in Differential Equations

In the preceding sections we described various techniques for solving differential equations and the order, or a specific expression, for the truncation error at each step was given. As we discussed in Section 5.3 for the Euler and trapezoidal methods, it is important to examine not only the magnitude of the error at each step but also how that error accumulates as the number of steps taken increases.

For the predictor–corrector method described in Section 5.7, it can be shown that the predictor–corrector formulae introduce additional spurious solutions. As the iterative process proceeds, for some problems the effect of these spurious solutions may be to overwhelm the true solution. In these circumstances the method is said to be unstable. Clearly we seek stable methods where the error does not develop in an unpredictable and unbounded way.

It is important to examine each numerical method to see if it is stable. In addition, if it is not stable for all differential equations we should provide tests to determine when it can be used with confidence. The theoretical study of stability for differential equations is a major undertaking and it is not intended to include a detailed analysis here. Section 5.9 summarizes the stability characteristics of specific methods and compares the performance of the major methods considered on a number of example differential equations.

5.9 The Stability of Particular Numerical Methods

A good discussion of the stability of many of the numerical methods for solving first-order differential equations is given by Ralston and Rabinowitz (1978) and Lambert (1973). Some of the more significant features, assuming all variables are real, are as follows.

1. Euler and trapezoidal methods: For a detailed discussion see Sections 5.3 and 5.4.
2. Runge–Kutta methods: Runge–Kutta methods do not introduce spurious solutions but instability may arise for some values of h. This may be removed by reducing h to a sufficiently small value. We have already described how the Runge–Kutta methods are less efficient than the predictor–corrector methods because of the greater number of function evaluations that may be required at each step by the Runge–Kutta methods. If h is reduced too far, the number of function evaluations required may make the method uneconomic. The restriction on the size of the interval required to maintain stability may be estimated from the inequality $M < h\partial f/\partial y < 0$ where M is dependent on the particular Runge–Kutta method being used and may be estimated. Clearly this emphasizes the need for careful step size adjustment during the solution process. This is efficiently implemented in the functions `ode23` and `ode45` so that this question does not present a problem when applying these MATLAB functions.
3. Adams–Bashforth–Moulton method: In *PECE* mode the range of absolute stability is given by $-1.25 < h\partial f/\partial y < 0$, implying that $\partial f/\partial y$ must be negative for absolute stability.
4. Hamming's method: In the *PECE* mode the range of absolute stability is given by $-0.5 < h\partial f\partial y < 0$, again implying that $\partial f\partial y$ must be negative for absolute stability.

Notice that the formulae given for estimating the step size can be difficult to use if f is a general function of y and t. However, in some cases the derivative of f is easily calculated, for example, when $f = Cy$ where C is a constant.

We now give some results of applying the methods discussed in previous sections to solve more general problems. The following script solves the three examples that follow by setting `example` equal to 1, 2, or 3 in the first line of the script.

```
% e3s503.m
example = 1;
switch example
    case 1
        yprime = @(t,y) 2*t*y;
        sol = @(t) 2*exp(t^2);
```

```
        disp('Solution of dy/dt = 2yt')
        t0 = 0; y0 = 2;
    case 2
        yprime = @(t,y) (cos(t)-2*y*t)/(1+t^2);
        sol = @(t) sin(t)/(1+t^2);
        disp('Solution of (1+t^2)dy/dt = cos(t)-2yt')
        t0 = 0; y0 = 0;
    case 3
        yprime = @(t,y) 3*y/t;
        disp('Solution of dy/dt = 3y/t')
        sol = @(t) t^3;
        t0 = 1; y0 = 1;
end
tf = 2; tinc = 0.25; steps = floor((tf-t0)/tinc+1);
[t,x1] = abm(yprime,[t0 tf],y0,tinc);
[t,x2] = fhamming(yprime,[t0 tf],y0,tinc);
[t,x3] = rkgen(yprime,[t0 tf],y0,tinc,1);
disp('t         abm          Hamming     Classical      Exact')
for i = 1:steps
    fprintf('%4.2f%12.7f%12.7f',t(i),x1(i),x2(i))
    fprintf('%12.7f%12.7f\n',x3(i),sol(t(i)))
end
```

Example 5.1

Solve

$$dy/dt = 2yt \quad \text{where} \quad y = 2 \quad \text{when} \quad t = 0$$

Exact solution: $y = 2\exp(t^2)$. Running the script `e3s503.m` with `example = 1` produces the following output:

```
Solution of dy/dt = 2yt
t          abm         Hamming     Classical      Exact
0.00   2.0000000   2.0000000   2.0000000   2.0000000
0.25   2.1289876   2.1289876   2.1289876   2.1289889
0.50   2.5680329   2.5680329   2.5680329   2.5680508
0.75   3.5099767   3.5099767   3.5099767   3.5101093
1.00   5.4340314   5.4294215   5.4357436   5.4365637
1.25   9.5206761   9.5152921   9.5369365   9.5414664
1.50  18.8575896  18.8690552  18.9519740  18.9754717
1.75  42.1631012  42.2832017  42.6424234  42.7618855
2.00 106.2068597 106.9045567 108.5814979 109.1963001
```

Examples 5.2 and 5.3 appear to show that there is little difference between the three methods considered and they are all fairly successful for the step size $h = 0.25$ in this range. Example 5.1 is a relatively difficult problem in which the classical Runge–Kutta method performs well.

Example 5.2

Solve

$$(1+t^2)dy/dt + 2ty = \cos t \quad \text{where} \quad y = 0 \quad \text{when} \quad t = 0$$

Exact solution: $y = (\sin t)/(1+t^2)$. Running script `e3s503.m` with `example = 2` produces the following output:

```
Solution of (1+t^2)dy/dt = cos(t)-2yt
t          abm        Hamming    Classical     Exact
0.00    0.0000000   0.0000000   0.0000000   0.0000000
0.25    0.2328491   0.2328491   0.2328491   0.2328508
0.50    0.3835216   0.3835216   0.3835216   0.3835404
0.75    0.4362151   0.4362151   0.4362151   0.4362488
1.00    0.4181300   0.4196303   0.4206992   0.4207355
1.25    0.3671577   0.3705252   0.3703035   0.3703355
1.50    0.3044513   0.3078591   0.3068955   0.3069215
1.75    0.2404465   0.2432427   0.2421911   0.2422119
2.00    0.1805739   0.1827267   0.1818429   0.1818595
```

Example 5.3

Solve

$$dy/dt = 3y/t \quad \text{where} \quad y = 1 \quad \text{when} \quad t = 1$$

Exact solution: $y = t^3$. Running script `e3s503.m` with `example = 3` produces the following output:

```
Solution of dy/dt = 3y/t
t          abm        Hamming    Classical     Exact
1.00    1.0000000   1.0000000   1.0000000   1.0000000
1.25    1.9518519   1.9518519   1.9518519   1.9531250
1.50    3.3719182   3.3719182   3.3719182   3.3750000
1.75    5.3538346   5.3538346   5.3538346   5.3593750
2.00    7.9916917   7.9919728   7.9912355   8.0000000
```

For a further comparison, we now use the MATLAB function `ode113`. This employs a predictor–corrector method based on the *PECE* approach described in Section 5.6 in relation to the Adams–Bashforth–Moulton method. However the method implemented in `ode113` is of variable order. The standard call of the function takes the form

```
[t,y] = ode113(f,tspan,y0,options);
```

where `f` is the name of the function providing the right sides of the system of differential equations; `tspan` is the range of solution for the differential equation, given as a vector `[to tfinal]`; `y0` is the vector of initial values for the differential equation at time $t = 0$; and `options` is an optional parameter providing additional settings for the differential equation such as accuracy.

To illustrate the use of this function we consider the example

$$dy/dt = 2yt \quad \text{with initial conditions} \quad y = 2 \quad \text{when} \quad t = 0$$

The call to solve this differential equation is

```
>> options = odeset('RelTol', 1e-5,'AbsTol',1e-6);
>> [t,yy] = ode113(@(t,x) 2*t*x,[0,2],[2],options); y = yy', time = t'
```

The result of executing these statements is

```
y =
 Columns 1 through 7
   2.0000    2.0000    2.0000    2.0002    2.0006    2.0026    2.0103
 Columns 8 through 14
   2.0232    2.0414    2.0650    2.0943    2.1707    2.2731    2.4048
 Columns 15 through 21
   2.5703    2.7755    3.0279    3.3373    3.7161    4.1805    4.7513
 Columns 22 through 28
   5.4557    6.3290    7.4177    8.7831   10.5069   15.5048   22.7912
 Columns 29 through 32
  34.6321   54.3997   88.3328  109.1944


time =
 Columns 1 through 7
        0    0.0022    0.0045    0.0089    0.0179    0.0358    0.0716
 Columns 8 through 14
   0.1073    0.1431    0.1789    0.2147    0.2862    0.3578    0.4293
 Columns 15 through 21
   0.5009    0.5724    0.6440    0.7155    0.7871    0.8587    0.9302
 Columns 22 through 28
   1.0018    1.0733    1.1449    1.2164    1.2880    1.4311    1.5599
 Columns 29 through 32
   1.6887    1.8175    1.9463    2.0000
```

Although a direct comparison between each step is not possible, because `ode113` uses a variable step size, we can compare the result for $t = 2$ with the results given for Example 5.1. This shows that the final y value given by `ode113` is better than those given by the other methods.

5.10 Systems of Simultaneous Differential Equations

The numerical techniques we have described for solving a single first-order differential equation can be applied, after simple modification, to solve systems of first-order differential equations. Systems of differential equations arise naturally from mathematical models of the physical world. In this section we shall introduce a system of differential equations by considering a relatively simple example. This example is based on a much simplified model of the heart introduced by Zeeman and incorporates ideas from catastrophe theory. The model is described briefly here but more detail is given in the excellent text of Beltrami (1987). The resulting system of differential equations will be solved using the MATLAB function `ode23` and the graphical facilities of MATLAB will help to clarify the interpretation of the results.

The starting point for this model of the heart is Van der Pol's equation, which may be written in the form

$$dx/dt = u - \mu(x^3/3 - x)$$
$$du/dt = -x$$

This is a system of two simultaneous equations. The choice of this differential equation reflects our wish to imitate the beat of the heart. The fluctuation in the length of the heart fiber, as the heart contracts and dilates subject to an electrical stimulus, thus pumping blood through the system, may be represented by this pair of differential equations. The fluctuation has certain subtleties that our model should allow for. Starting from the relaxed state, the contraction begins with the application of the stimulus slowly at first and then becomes faster, so giving a sufficient final impetus to the blood. When the stimulus is removed, the heart dilates slowly at first and then more rapidly until the relaxed state is again reached and the cycle can begin again.

To follow this behavior, the Van der Pol equation requires some modification so that the x variable represents the length of heart fiber and the u variable can be replaced by one that represents the stimulus applied to the heart. This is achieved by making the substitution $s = -u/\mu$, where s represents the stimulus and μ is a constant. Since ds/dt is equal to $(-du/dt)/\mu$ it follows that $du/dt = -\mu ds/dt$. Hence we obtain

$$dx/dt = \mu(-s - x^3/3 + x)$$
$$ds/dt = x/\mu$$

If these differential equations are solved for s and x for a range of time values, we find that s and x oscillate in a manner representing the fluctuations in the heart fiber length and stimulus. However, Zeeman proposed the introduction into this model of a tension factor p, where $p > 0$, in an attempt to account for the effects of increased blood pressure in terms of increased tension on the heart fiber. The model he suggested has the form

$$dx/dt = \mu(-s - x^3/3 + px)$$
$$du/dt = x/\mu$$

Although the motivation for such a modification is plausible, the effects of these changes are by no means obvious.

This problem provides an interesting opportunity to apply MATLAB to simulate the heartbeat in an experimental environment that allows us to monitor its changes under the effects of differing tension values. The following script solves the differential equations and draws various graphs.

```
% e3s504.m Solving Zeeman's Catastrophe model of the heart
clear all
p = input('enter tension value ');
simtime = input('enter runtime ');
acc = input('enter accuracy value ');
xprime = @(t,x) [0.5*(-x(2)-x(1)^3/3+p*x(1)); 2*x(1)];
options = odeset('RelTol',acc);
initx = [0 -1]';
[t x] = ode23(xprime,[0 simtime],initx,options);
% Plot results against time
plot(t,x(:,1),'--',t,x(:,2),'-')
xlabel('Time'), ylabel('x and s')
```

In the preceding function definition, $\mu = 0.5$. Figure 5.10 shows graphs of the fiber length x and the stimulus s against time for a relatively small tension factor set at 1. The graphs show that a steady periodic oscillation of fiber length for this tension value is achieved

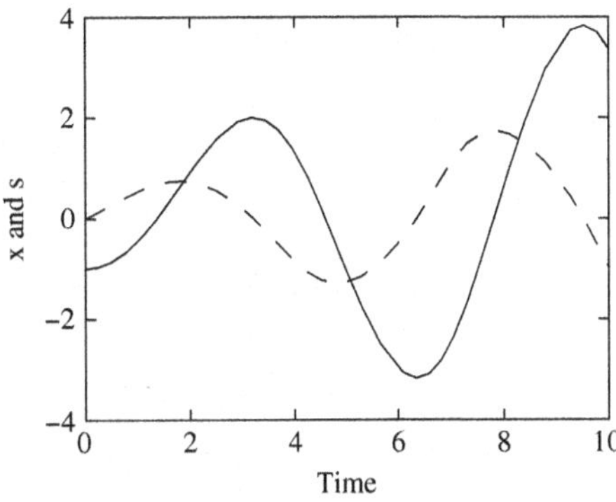

FIGURE 5.10 Solution of Zeeman's model with $p = 1$ and accuracy 0.005. The *solid line* represents s and the *dashed line* represents x.

for small stimulus values. However, Figure 5.11 plots x and s against time with the tension set at 20. This shows that the behavior of the oscillation is clearly more labored and that much larger values of stimulus are required to produce the fluctuations in fiber length for the much higher tension value. Thus, the graphs show the deterioration in the beat with increasing tension. The results parallel the expected physical effects and also give some degree of experimental support to the validity of this simple model.

A further interesting study can be made. The interrelation of the three parameters x, s, and p can be represented by a three-dimensional surface called the cusp catastrophe surface. This surface can be shown to have the form

$$-s - x^3/3 + px = 0$$

See Beltrami (1987) for a more detailed explanation. Figure 5.12 shows a series of sections of the cusp catastrophe curve for $p = 0:10:40$. The curve has a pleat that becomes increasingly pronounced in the direction of increasing p. High tension or high p value

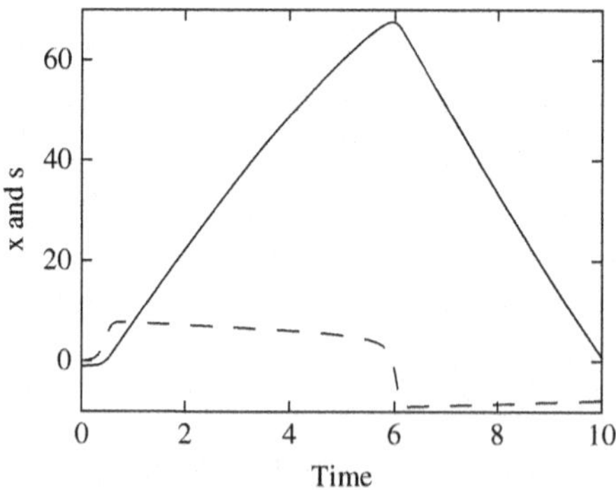

FIGURE 5.11 Solution of Zeeman's model with $p = 20$ and accuracy 0.005. The *solid line* represents s and the *dashed line* represents x.

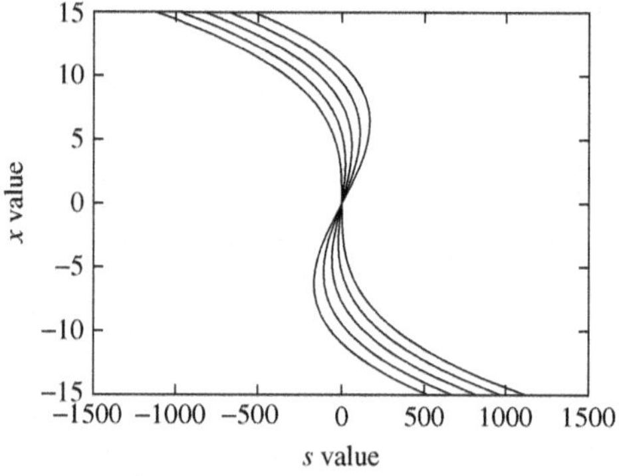

FIGURE 5.12 Sections of the cusp catastrophe curve in Zeeman's model for $p = 0:10:40$.

consequently corresponds to movement on the sharply pleated part of this surface and thus provides smaller changes in the heart fiber length relative to the stimulus.

5.11 The Lorenz Equations

As an example of a system of three simultaneous equations, we consider the Lorenz system. This system has a number of important applications including weather forecasting. The system has the form

$$dx/dt = s(y - x)$$
$$dy/dt = rx - y - xz$$
$$dz/dt = xy - bz$$

subject to appropriate initial conditions. As the parameters s, r, and b are varied through various ranges of values, the solutions of this system of differential equations vary in form. In particular, for certain values of the parameters the system exhibits chaotic behavior. To provide more accuracy in the computation process we use the MATLAB function ode45 rather than ode23. The MATLAB script for solving this problem is as follows:

```
% e3s505.m  Solution of the Lorenz equations
r = input('enter a value for the constant r ');
simtime = input('enter runtime ');
acc = input('enter accuracy value ');
xprime = @(t,x) [10*(x(2)-x(1)); r*x(1)-x(2)-x(1)*x(3); ...
                 x(1)*x(2)-8*x(3)/3];
initx = [-7.69 -15.61 90.39]';
tspan = [0 simtime];
options = odeset('RelTol',acc);
[t x] = ode45(xprime,tspan,initx,options);
% Plot results against time
figure(1), plot(t,x,'k')
xlabel('Time'), ylabel('x')
figure(2), plot(x(:,1),x(:,3),'k')
xlabel('x'), ylabel('z')
```

The results of running this script are given in Figures 5.13 and 5.14. Figure 5.13 is characteristic of the Lorenz equations and shows the complexity of the relationship between x and z. Figure 5.14 shows how x, y, and z change with time.

For $r = 126.52$ and for other large values of r the behavior of this system is chaotic. In fact for $r > 24.7$ most orbits exhibit chaotic wandering. The trajectory passes around two points of attraction, called "strange attractors," switching from one to another in an apparently unpredictable fashion. This appearance of apparently random behavior is

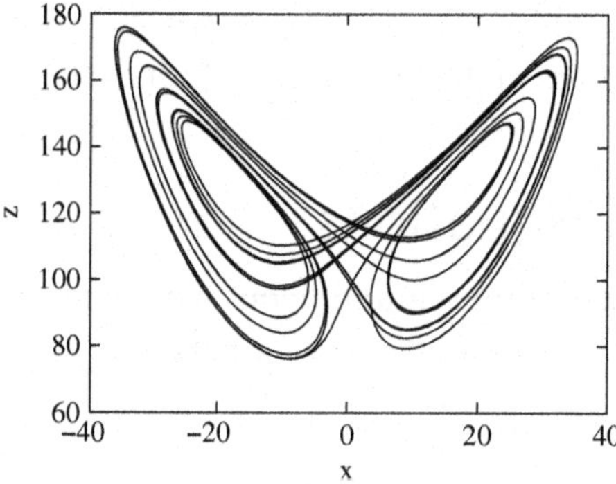

FIGURE 5.13 Solution of Lorenz equations for $r = 126.52$, using an accuracy of 0.000005 and terminating at $t = 8$.

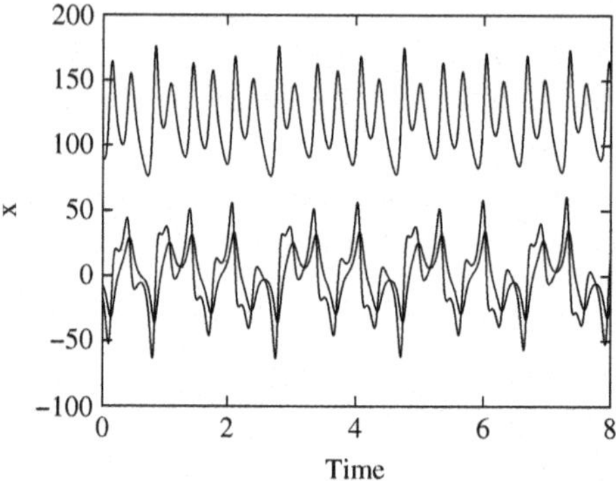

FIGURE 5.14 Solution of Lorenz equations where each variable is plotted against time. Conditions are the same as those used to generate Figure 5.13. Note the unpredictable nature of the solutions.

remarkable considering the clearly deterministic nature of the problem. However, for other values of r the behavior of the trajectories is simple and stable.

5.12 The Predator–Prey Problem

A system of differential equations that models the interaction of competing or predator–prey populations is based on the Volterra equations and may be written in the form

$$\begin{aligned} dP/dt &= K_1 P - CPQ \\ dQ/dt &= -K_2 Q + DPQ \end{aligned} \tag{5.28}$$

together with the initial conditions

$$Q = Q_0 \quad \text{and} \quad P = P_0 \quad \text{at time} \quad t = 0$$

The variables P and Q give the size of the prey and predator populations, respectively, at time t. These two populations interact and compete. K_1, K_2, C, and D are positive constants. K_1 relates to the rate of growth of the prey population P, and K_2 relates to the rate of decay of the predator population Q. It seems reasonable to assume that the number of encounters of predator and prey is proportional to P multiplied by Q and that a proportion C of these encounters will be fatal to members of the prey population. Thus the term CPQ gives a measure of the decrease in the prey population and the unrestricted growth in this population, which could occur assuming ample food, must be modified by the subtraction of this term. Similarly the decrease in the population of the predator must be modified by the addition of the term DPQ since the predator population gains food from its encounters with its prey and therefore more of the predators survive.

The solution of the differential equation depends on the specific values of the constants and will often result in nature in a stable cyclic variation of the populations. This is because as the predators continue to eat the prey, the prey population will fall and become insufficient to support the predator population, which itself then falls. However, as the predator population falls, more of the prey survive and consequently the prey population will then increase. This in turn leads to an increase in the predator population since it has more food and the cycle begins again. This cycle maintains the predator–prey populations between certain upper and lower limits. The Volterra differential equations can be solved directly but this solution does not provide a simple relation between the size of the predator and prey populations; therefore, numerical methods of solution should be applied. An interesting description of this problem is given by Simmons (1972).

We now use MATLAB to study the behavior of a system of equations of the form (5.28) applied to the interaction of the lynx and its prey, the hare. The choice of the constants K_1, K_2, C, and D is not a simple matter if we wish to obtain a stable situation where the populations of the predator and prey never die out completely but oscillate between upper and lower limits. The MATLAB script that follows uses $K_1 = 2$, $K_2 = 10$, $C = 0.001$, and $D = 0.002$, and considers the interaction of a population of lynxes and hares where it is assumed that this interaction is the crucial feature in determining the size of the two populations. With an initial population of 5000 hares and 100 lynxes, the following script uses these values to produce the graph in Figure 5.15.

```
% e3s506.m
% x(1) and x(2) are hare and lynx populations.
simtime = input('enter runtime ');
acc = input('enter accuracy value ');
fv = @(t,x) [2*x(1)-0.001*x(1)*x(2); -10*x(2)+0.002*x(1)*x(2)];
initx = [5000 100]';
options = odeset('RelTol',acc);
[t x] = ode23(fv,[0 simtime],initx,options);
plot(t,x(:,1),'k',t,x(:,2),'k--')
xlabel('Time'), ylabel('Population of hares and lynxes')
```

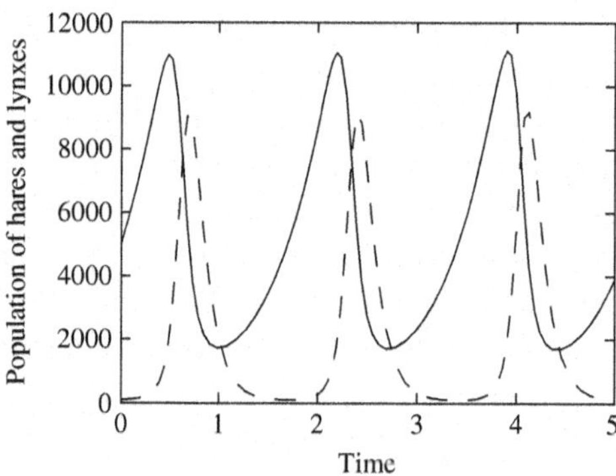

FIGURE 5.15 Variation in the population of lynxes (*dashed line*) and hares (*solid line*) against time using an accuracy of 0.005 beginning with 5000 hares and 100 lynxes.

For these parameters there is a remarkably wide variation in the populations of hares and lynxes. The lynx population, although periodically small, still recovers following a recovery of the hare population.

5.13 Differential Equations Applied to Neural Networks

Different types of neural networks have been used to solve a wide range of problems. Neural networks often consist of several layers of "neurons" that are "trained" by fixing a set of weights. These weights are found by minimizing the sum of squares of the difference between actual and required outputs. Once trained, the networks can be used to classify a range of inputs. However, here we consider a different approach that uses a neural network that may be based directly on considering a system of differential equations. This approach is described by Hopfield and Tank (1985, 1986), who demonstrated the application of neural networks to solving specific numerical problems. It is not our intention to provide the full details or proofs of this process here.

Hopfield and Tank, in their 1985 and 1986 papers, utilized a system of differential equations that take the form

$$\frac{du_i}{dt} = \frac{-u_i}{\tau} + \sum_{j=0}^{n-1} T_{ij} V_j + I_i \quad \text{for} \quad i = 0, 1, \ldots\ n-1 \tag{5.29}$$

where τ is a constant usually taken as 1. This system of differential equations represents the interaction of a system of n neurons, and each differential equation is a simple model of a single biological neuron. (This is only one of a number of possible models of a neural network.) Clearly, to establish a network of such neurons, they must be able to interact

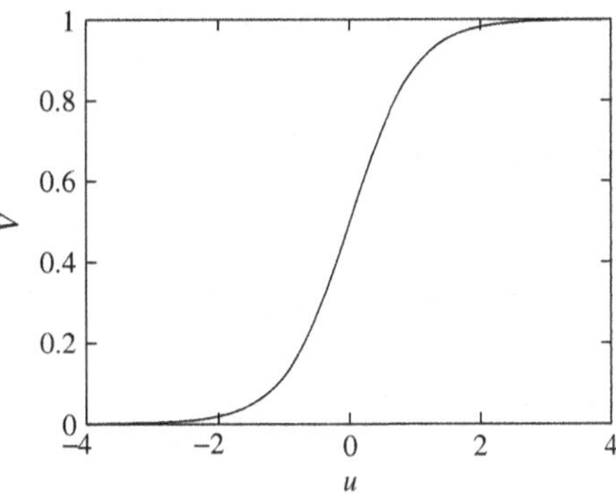

FIGURE 5.16 Plot of sigmoid function $V = (1 + \tanh u)/2$.

with each other and this interaction must be represented in the differential equations. The T_{ij} provide the strengths of the interconnections between the ith and jth neuron and the I_i provide the externally applied current to the ith neuron. These I_i may be viewed as inputs to the system. The V_j values provide the outputs from the system and are directly related to the u_j so that we may write $V_j = g(u_j)$. The function g, called a sigmoidal function, may be specified, for example, by

$$V_j = (1 + \tanh u_j)/2 \quad \text{for all} \quad j = 0, 1, \ldots\, n-1$$

A plot of this function is given in Figure 5.16.

Having provided such a model of a neural network, the question still remains: How can we show that it can be used to solve specific problems? This is the key issue and a significant problem in itself. Before we can solve a given problem using a neural network we must first reformulate our problem so that it can be solved by this approach.

To illustrate this process, Hopfield and Tank chose as an example the simple problem of binary conversion, that is, to find the binary equivalent of a given decimal number. Since there is no obvious and direct relationship between this problem and the system of differential equations (5.29) that model the neural network, a more direct link has to be established.

Hopfield and Tank have shown that the stable state solution of (5.29), in terms of the V_j, is given by the minima of the energy function

$$E = -\frac{1}{2}\sum_{i=0}^{n-1}\sum_{j=0}^{n-1} T_{ij}V_iV_j - \sum_{j=0}^{n-1} I_jV_j \tag{5.30}$$

It is an easy matter to link the solution of the binary conversion problem to the minimization of the function (5.30).

Hopfield and Tank consider the energy function

$$E = \frac{1}{2}\left\{x - \sum_{j=0}^{n-1} V_j 2^j\right\}^2 + \sum_{j=0}^{n-1} 2^{2j-1} V_j (1 - V_j) \tag{5.31}$$

Now the minimum of (5.31) will be attained when $x = \Sigma V_j 2^j$ and $V_j = 0$ or 1. Clearly the first term ensures that the required binary representation is achieved while the second term provides that the V_j take either 0 or 1 values when the value of E is minimized. On expanding this energy function (5.31) and comparing it with the general energy function (5.30) we find that if we make

$$T_{ij} = -2^{i+j} \quad \text{for} \quad i \neq j \quad \text{and} \quad T_{ij} = 0 \quad \text{when} \quad i = j$$
$$I_j = -2^{2j-1} + 2^j x$$

then the two energy functions are equivalent, apart from a constant. Thus the minimum of one gives the minimum of the other. Solving the binary conversion problem expressed in this way is thus equivalent to solving the system of differential equations (5.29) with this special choice of values for T_{ij} and I_i. In fact, by using an appropriate choice of T_{ij} and I_i, a range of problems can be represented by a neural network in the form of the system of differential equations (5.29). Hopfield and Tank have extended this process from the simple preceding example to attempting to solve the very challenging traveling salesman problem. The details of this are given in Hopfield and Tank (1985, 1986).

In MATLAB we may use `ode23` or `ode45` to solve this problem. The crucial part of this exercise is to define the function that gives the right sides of the differential equation system for the neural network. This can be done very simply using the following function `hopbin`. This function gives the right side for the differential equations that solve the binary conversion problem. In the definition of function `hopbin`, `sc` is the decimal value we wish to convert.

```
function neurf = hopbin(t,x)
global n sc
% Calculate synaptic current
I = 2.^[0:n-1]*sc-0.5*2.^(2.*[0:n-1]);
% Perform sigmoid transformation
V = (tanh(x/0.02)+1)/2;
% Compute interconnection values
p = 2.^[0:n-1].*V';
% Calculate change for each neuron
neurf = -x-2.^[0:n-1]'*sum(p)+I'+2.^(2.*[0:n-1])'.*V;
```

This function `hopbin` is called by the following script to solve the system of differential equations that define the neural network and hence simulate its operation.

```
% e3s507.m Hopfield and Tank neuron model for binary conversion problem
global n sc
n = input('enter number of neurons ');
sc = input('enter number to be converted to binary form ');
simtime = 0.2; acc = 0.005;
initx = zeros(1,n)';
options = odeset('RelTol',acc);
%Call ode45 to solve equation
[t x] = ode45('hopbin',[0 simtime],initx,options);
V = (tanh(x/0.02)+1)/2;
bin = V(end,n:-1:1);
for i = 1:n
    fprintf('%8.4f', bin(i))
end
fprintf('\n\n')
plot(t,V,'k')
xlabel('Time'), ylabel('Binary values')
```

Running this script to convert the decimal number 5 gives

```
enter number of neurons 3
enter number to be converted to binary form 5
  1.0000  0.0000  0.9993
```

together with Figure 5.17. This plot shows how the neural network model converges to the required results, that is, `V(1) = 1`, `V(2) = 0`, and `V(3) = 1` or binary number 101.

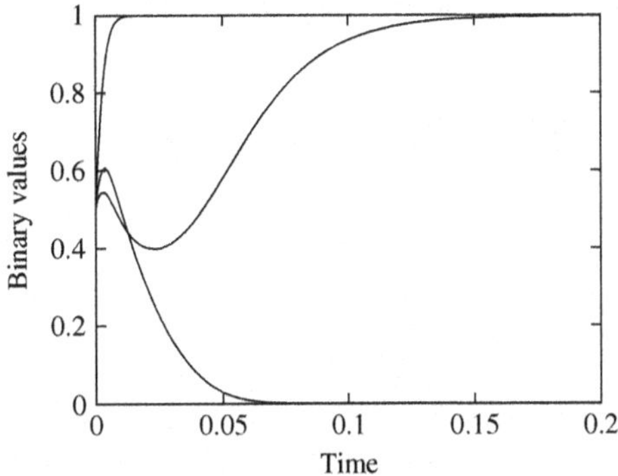

FIGURE 5.17 A neural network finds the binary equivalent of 5 using 3 neurons and an accuracy of 0.005. The three curves show the convergence to the binary digits 1, 0, and 1.

As a further example, we convert the decimal number 59 using 7 neurons as follows:

```
enter number of neurons 7
enter number to be converted to binary form 59
  0.0000  1.0000  1.0000  1.0000  0.0000  1.0000  0.9999
```

Again, a correct result.

This is an application of neural networks to a trivial problem. A real test for neural computing is the traveling salesman problem. The MATLAB neural network toolbox provides a range of functions to solve neural network problems.

5.14 Higher-Order Differential Equations

Higher-order differential equations can be solved by converting them to a system of first-order differential equations. To illustrate this consider the second-order differential equation

$$2d^2x/dt^2 + 4(dx/dt)^2 - 2x = \cos x \tag{5.32}$$

together with the initial conditions $x = 0$ and $dx/dt = 10$ when $t = 0$. If we substitute $p = dx/dt$, then (5.32) becomes

$$\begin{aligned} 2dp/dt + 4p^2 &= \cos x + 2x \\ dx/dt &= p \end{aligned} \tag{5.33}$$

with initial conditions $p = 10$ and $x = 0$ when $t = 0$.

The second-order differential equations have been replaced by a system of first-order differential equations. If we have an nth-order differential equation of the form

$$a_n d^n y/dt^n + a_{n-1} d^{n-1} y/dt^{n-1} + \cdots + a_0 y = f(t, y) \tag{5.34}$$

by making the substitutions

$$P_0 = y \quad \text{and} \quad dP_{i-1}/dt = P_i \quad \text{for} \quad i = 1, 2, \ldots, n-1 \tag{5.35}$$

(5.34) becomes

$$a_n dP_{n-1}/dt = f(t, y) - a_{n-1}P_{n-1} - a_{n-2}P_{n-2} - \cdots - a_0 P_0 \tag{5.36}$$

Now (5.35) and (5.36) together constitute a system of n first-order differential equations. Initial values will be given for (5.34) in terms of the various order derivatives P_i for $i = 1, 2, \ldots, n-1$ at some initial value t_0 and these can easily be translated into initial conditions for the system of equations (5.35) and (5.36). In general, the solutions of the original nth-order differential equation and the system of first-order differential equations, (5.35) and (5.36), are the same. In particular, the numerical solution will provide the values of y for a specified range of t. An excellent discussion of the equivalence of the solutions of the two problems is given in Simmons (1972). We can see from this description that any order differential equation of the form (5.34) with given initial values can be reduced to solving a system of first-order differential equations. This argument is easily extended to the more general nth-order differential equation by making exactly the same substitutions as in the preceding in

$$d^n y/dt^n = f(t, y, y', \dots y^{(n-1)})$$

where $y^{(n-1)}$ denotes the $(n-1)$th-order derivative of y.

5.15 Stiff Equations

When the solution of a system of differential equations contains components that change at significantly different rates for given changes in the independent variable, the equation system is said to be "stiff." When this phenomenon is present, a particularly careful choice of the step size must be made if stability is to be achieved.

We will now consider how the stiffness phenomenon arises in an apparently simple system of differential equations. Consider the following system:

$$\begin{aligned} dy_1/dt &= -by_1 - cy_2 \\ dy_2/dt &= y_1 \end{aligned} \tag{5.37}$$

This system may be written in matrix form as

$$d\mathbf{y}/dt = \mathbf{Ay} \tag{5.38}$$

The solution of this system is

$$\begin{aligned} y_1 &= A\exp(r_1 t) + B\exp(r_2 t) \\ y_2 &= C\exp(r_1 t) + D\exp(r_2 t) \end{aligned} \tag{5.39}$$

where A, B, C, and D are constants set by the initial conditions. It can easily be verified that r_1 and r_2 are the eigenvalues of the matrix $\mathbf{A}$.

If a numerical procedure is applied to solve these systems of differential equations, the success of the method will depend crucially on the eigenvalues of the matrix $\mathbf{A}$ and in particular the ratio of the smallest and largest eigenvalues. By taking various values of b and c in (5.37) we can generate many problems of the form (5.38) having solutions (5.39) where the eigenvalues r_1 and r_2 will, of course, change from problem to problem.

The purpose of the following script is to investigate how the difficulty of solving (5.37) depends on the ratio of the largest and smallest eigenvalues by comparing the time taken to solve specific problems.

```
% e3s508.m
b = [20 100 500 1000 2000]; c = [0.1 1 1 1 1]; tspan = [0 2];
options = odeset('reltol',1e-5,'abstol',1e-5);
for i = 1:length(b)
    et(i) = 0;
    eigenratio(i) = 0;
    for j = 1:100
        a = [-b(i) -c(i);1 0];
        lambda = eig(a);
        eigenratio(i) = eigenratio(i)+max(abs(lambda))/min(abs(lambda));
        v = @(t,y) a*y;
        inity = [0 1]';  time0 = clock;
        [t,y] = ode15s(v,tspan,inity,options);
        et(i) = et(i)+etime(clock,time0);
    end
end
e_ratio = eigenratio/100
time_taken = et/100
```

Running this script gives

```
e_ratio =
  1.0e+006 *
    0.0040    0.0100    0.2500    1.0000    4.0000

time_taken =
    0.0045    0.0120    0.0484    0.0962    0.1878
```

As the eigenvalue ratio increases so does the time taken to solve the problem. Problems will arise if there is a wide variation in the magnitude of the eigenvalues.

The MATLAB function `ode23s` is designed to deal specifically with stiff equations. Replacing `ode23` by `ode23s` in script `e3s508.m` and running it gives

```
e_ratio =
  1.0e+006 *
    0.0040    0.0100    0.2500    1.0000    4.0000

time_taken =
    0.0122    0.0141    0.0122    0.0111    0.0108
```

Note the interesting difference between these results and the output from the script using the function `ode23`. Using `ode23` the time increases markedly with the size of the eigenratio whereas with `ode23s` there is little difference between the time taken to solve the differential equations, no matter what the eigenratio.

Another alternative exists for solving stiff differential equations, `ode15s`. This is a variable order method and has the advantage that it can be used when the matrix **A** of (5.38) is time dependent. Replacing `ode23` by `ode15s` in script `e3s508.m` and running it we obtain the following output:

```
e_ratio =
  1.0e+006 *
    0.0040    0.0100    0.2500    1.0000    4.0000

time_taken =
    0.0136    0.0155    0.0158    0.0144    0.0141
```

Clearly there is little difference between the two stiff solvers.

As an example of a matrix with widely spaced eigenvalues we can take the 8×8 Rosser matrix; this is available in MATLAB as `rosser`. The sequence of statements

```
>> a = rosser; lambda = eig(a);
>> eigratio = max(abs(lambda))/min(abs(lambda))

eigratio =
  1.8480e+015
```

produces a matrix with eigenvalue ratios of order 10^{16}. Thus a system of ordinary first-order differential equations involving this matrix would be pathologically difficult to solve. The significance of the eigenvalue ratio in relation to the required step size can be generalized to systems of many equations. Consider the system of n equations

$$d\mathbf{y}/dt = \mathbf{A}\mathbf{y} + \mathbf{P}(t) \tag{5.40}$$

where $\mathbf{y}$ is an n component column vector, $\mathbf{P}(t)$ is an n component column vector of functions of t, and $\mathbf{A}$ is an $n \times n$ matrix of constants. It can be shown that the solution of this system takes the form

$$\mathbf{y}(t) = \sum_{i=1}^{n} v_i \mathbf{d}_i \exp(r_i t) + \mathbf{s}(t) \tag{5.41}$$

Here $r_1, r_2, \ldots$ are the eigenvalues and $\mathbf{d}_1$, $\mathbf{d}_2, \ldots$ the eigenvectors of $\mathbf{A}$. The vector function $\mathbf{s}(t)$ is the particular integral of the system, sometimes called the steady-state solution since for negative eigenvalues the exponential terms should die away with increasing t. If it is assumed that the $r_k < 0$ for $k = 1, 2, 3, \ldots$ and we require the steady-state solution of system (5.40), then any numerical method applied to solve this problem may face significant difficulties, as we have seen. We must continue the integration until the exponential components have been reduced to negligible levels and yet we must take sufficiently small steps to ensure stability, thus requiring many steps over a large interval. This is the most significant effect of stiffness.

The definition of stiffness can be extended to any system of the form (5.40). The stiffness ratio is defined as the ratio of the largest and smallest eigenvalues of **A** and gives a measure of the stiffness of the system. The methods used to solve stiff problems must be based on stable techniques. The MATLAB function `ode23s` uses continuous step size adjustment and therefore is able to deal with such problems, although the solution process may be slow. If we use a predictor–corrector method, not only must this method be stable but the corrector must also be iterated to convergence. An interesting discussion of this topic is given by Ralston and Rabinowitz (1978). Specialized methods have been developed for solving stiff problems, and Gear (1971) has provided a number of techniques that have been reported to be successful.

5.16 Special Techniques

A further set of predictor–corrector equations may be generated by making use of an interpolation formula due to Hermite. An unusual feature of these equations is that they contain second-order derivatives. It is usually the case that the calculation of second-order derivatives is not particularly difficult and consequently this feature does not add a significant amount of work to the solution of the problem. However, it should be noted that in using a computer program for this technique the user has to supply not only the function on the right side of the differential equation but its derivative as well. To the general user this may be unacceptable.

The equations for Hermite's method take the form

$$
\begin{aligned}
y_{n+1}^{(1)} &= y_n + h\left(y'_n - 3y'_{n-1}\right)/2 + h^2\left(17y''_n + 7y''_{n-1}\right)/12 \\
y_{n+1}^{*\,(1)} &= y_{n+1}^{(1)} + 31\left(y_n - y_n^{(1)}\right)/30 \\
y_{n+1}'^{(1)} &= f\left(t_{n+1}, y_{n+1}^{*\,(1)}\right)
\end{aligned}
\tag{5.42}
$$

For $k = 1, 2, 3, \ldots$

$$
y_{n+1}^{(k+1)} = y_n + h\left(y_{n+1}'^{\,(k)} + y'_n\right)/2 + h^2\left(-y_{n+1}''^{\,(k)} + y''_n\right)/12
$$

This method is stable and has a smaller truncation error at each step than Hamming's method. Thus it may be worthwhile accepting the additional effort required by the user. We note that since we have

$$
dy/dt = f(t, y)
$$

then

$$d^2y/dt^2 = df/dt$$

and thus y_n'' and so on are easily calculated as the first derivative of f. The MATLAB function `fhermite` implements this method, and the script follows. Note that in this function, the function `f` must provide both the first and second derivatives of y.

```
function [tvals, yvals] = fhermite(f,tspan,startval,step)
% Hermite's method for solving
% first order differential equation dy/dt = f(t,y).
% Example call: [tvals, yvals] = fhermite(f,tspan,startval,step)
% The initial and final values of t are given by tspan = [start finish].
% Initial value of y is given by startval, step size is given by step.
% The function f(t,y) and its derivative must be defined by the user.
% 3 steps of Runge-Kutta are required so that hermite can start.
% Set up matrices for Runge-Kutta methods
b = [ ]; c = [ ]; d = [ ];
order = 4;
b = [1/6 1/3 1/3 1/6]; d = [0 0.5 0.5 1];
c = [0 0 0 0;0.5 0 0 0;0 0.5 0 0;0 0 1 0];
steps = (tspan(2)-tspan(1))/step+1;
y = startval; t = tspan(1);
ys(1) = startval; w = feval(f,t,y); fval(1) = w(1); df(1) = w(2);
yvals = startval; tvals = tspan(1);
for j = 2:2
    k(1) = step*fval(1);
    for i = 2:order
        w = feval(f,t+step*d(i),y+c(i,1:i-1)*k(1:i-1)');
        k(i) = step*w(1);
    end
    y1 = y+b*k'; ys(j) = y1; t1 = t+step;
    w = feval(f,t1,y1); fval(j) = w(1); df(j) = w(2);
    %collect values together for output
    tvals = [tvals, t1]; yvals = [yvals, y1];
    t = t1; y = y1;
end
%hermite now applied
h2 = step*step/12; er = 1;
for i = 3:steps
    y1 = ys(2)+step*(3*fval(1)-fval(2))/2+h2*(17*df(2)+7*df(1));
    t1 = t+step; y1m = y1; y10 = y1;
    if i>3, y1m = y1+31*(ys(2)-y10)/30; end
```

```
        w = feval(f,t1,y1m); fval(3) = w(1); df(3)=w(2);
        yc = 0; er = 1;
        while abs(er)>0.0000001
            yp = ys(2)+step*(fval(2)+fval(3))/2+h2*(df(2)-df(3));
            w = feval(f,t1,yp); fval(3) = w(1); df(3) = w(2);
            er = yp-yc; yc = yp;
        end
        fval(1:2) = fval(2:3); df(1:2) = df(2:3);
        ys(2) = yp;
        tvals = [tvals, t1]; yvals = [yvals, yp];
        t = t1;
    end
```

Figure 5.18 gives the error when solving the specific equation $dy/dt = y$ using the same step size and starting point as for Hamming's method—see Figure 5.9. For this particular problem Hermite's method performs better than Hamming's method.

Finally we compare the Hermite, Hamming, and Adams–Bashforth–Moulton methods for the difficult problem

$$dy/dt = -10y \quad \text{given} \quad y = 1 \quad \text{when} \quad t = 0$$

The following script implements these comparisons:

```
% e3s509.m
vg = @(t,x) [-10*x 100*x];
v = @(t,x) -10*x;
disp('Solution of dx/dt = -10x')
t0 = 0; y0 = 1;
tf = 1; tinc = 0.1; steps = floor((tf-t0)/tinc+1);
[t,x1] = abm(v,[t0 tf],y0,tinc);
[t,x2] = fhamming(v,[t0 tf],y0,tinc);
[t,x3] = fhermite(vg,[t0 tf],y0,tinc);
disp('t         abm          Hamming     Hermite      Exact');
for i = 1:steps
    fprintf('%4.2f%12.7f%12.7f',t(i),x1(i),x2(i))
    fprintf('%12.7f%12.7f\n',x3(i),exp(-10*(t(i))))
end
```

Note that for the function `fhermite` we must supply both the first and second derivatives of y with respect to t. For the first derivative, we have directly $dy/dt = -10y$ but the second derivative d^2y/dt^2 is given by $-10dy/dt = -10(-10y) = 100y$. Consequently, the function takes the form

```
vg = @(t,x) [-10*x 100*x];
```

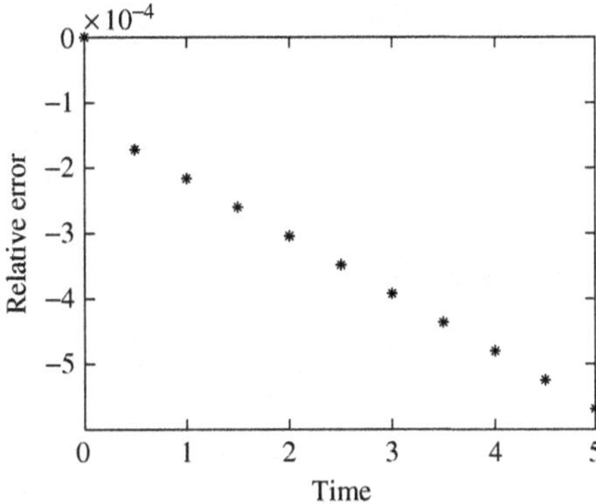

FIGURE 5.18 Relative error in the solution of $dy/dt = y$ using Hermite's method with an initial condition $y = 1$ when $t = 0$ and a step of 0.5.

The functions `abm` and `fhamming` require only the first derivative of y with respect to t, and we define the function as follows:

```
v = @(t,x) -10*x;
```

Running the preceding script provides the following results, demonstrating the superiority of the Hermite method.

```
Solution of dx/dt = -10x
t         abm         Hamming      Hermite      Exact
0.00   1.0000000   1.0000000   1.0000000   1.0000000
0.10   0.3750000   0.3750000   0.3750000   0.3678794
0.20   0.1406250   0.1406250   0.1381579   0.1353353
0.30   0.0527344   0.0527344   0.0509003   0.0497871
0.40  -0.0032654   0.0109440   0.0187528   0.0183156
0.50  -0.0171851   0.0070876   0.0069089   0.0067379
0.60  -0.0010598   0.0131483   0.0025454   0.0024788
0.70   0.0023606   0.0002607   0.0009378   0.0009119
0.80  -0.0063684   0.0006066   0.0003455   0.0003355
0.90  -0.0042478   0.0096271   0.0001273   0.0001234
1.00   0.0030171  -0.0065859   0.0000469   0.0000454
```

One feature that may be used to improve many of the methods discussed previously is step size adjustment. This means that we adjust the step size h according to the progress of the iteration. One criterion for adjusting h is to monitor the size of the truncation error. If the truncation error is smaller than the accuracy requirement, we can increase h; however, if the truncation error is too large, we can reduce h. Step size adjustment can lead to considerable additional work; for example, if a predictor–corrector method is used, new initial values must be calculated. The following method is an interesting alternative to this kind of procedure.

5.17 Extrapolation Techniques

The extrapolation method described in this section is based on a similar procedure to that used in Romberg integration, introduced in Chapter 4. The procedure begins by obtaining successive initial approximations for y_{n+1} using a modified mid-point method. The interval sizes used for obtaining these approximations are calculated from

$$h_i = h_{i-1}/2 \quad \text{for} \quad i = 1, 2, \ldots \tag{5.43}$$

with the initial value h_0 given.

Once these initial approximations have been obtained, we can use (5.44), the extrapolation formula, to obtain improved approximations.

$$T_{m,\,k} = (4^m T_{m-1,\,k+1} - T_{m-1,\,k})/(4^m - 1) \quad \text{for} \quad m = 1, 2, \ldots \quad \text{and} \quad k = 1, 2, \ldots\, s - m \tag{5.44}$$

The calculations are set out in an array in much the same way as the calculations for Romberg's method for integration described in Chapter 4. When $m = 0$, the values of $T_{0,k}$ for $k = 0, 1, 2, \ldots, s$ are taken as the successive approximations to the values of y_{n+1} using the h_i values obtained from (5.43).

The formula for calculating the approximations used for the initial values $T_{0,k}$ in the preceding array are computed using the following equations:

$$y_1 = y_0 + hy_0' \quad y_{n+1} = y_{n-1} + 2hy_n' \quad \text{for} \quad n = 1, 2 \ldots, N_k \tag{5.45}$$

Here $k = 1, 2, \ldots$ and N_k is the number of steps taken in the range of interest so that $N_k = 2^k$ as the size of the interval is halved each time. The distance $2h$ between y_{n+1} and y_{n-1} values may lead to significant variations in the magnitude of the error. Because of this, instead of using the final value of y_{n+1} given by (5.45), Gragg (1965) has suggested that at the final step these values be smoothed using the intermediate value y_n. This leads to the following values for $T_{0,k}$:

$$T_{0,\,k} = \left(y_{N-1}^k + 2y_N^k + y_{N+1}^k\right)/4$$

where the superscript k denotes the value at the kth division of the interval.

Alternatives to the method of Gragg are available for finding the initial values in the function `rombergx` and various combinations of predictor–correctors may be used. It should be noted, however, that if the corrector is iterated until convergence is achieved, this will improve the accuracy of the initial values but at considerable computational expense for smaller step sizes, that is, for larger N values. The following MATLAB function `rombergx` implements the extrapolation method.

```
function [v W] = rombergx(f,tspan,intdiv,inity)
% Solves dy/dt = f(t,y) using Romberg's method.
% Example call: [v W] = rombergx(f,tspan,intdiv,inity)
% The initial and final values of t are given by tspan = [start finish].
% Initial value of y is given by inity.
% The number of interval divisions is given by intdiv.
% The function f(t,y) must be defined by the user.
W = zeros(intdiv-1,intdiv-1);
for index = 1:intdiv
    y0 = inity; t0 = tspan(1);
    intervals = 2^index;
    step = (tspan(2)-tspan(1))/intervals;
    y1 = y0+step*feval(f,t0,y0);
    t = t0+step;
    for i = 1:intervals
        y2 = y0+2*step*feval(f,t,y1);
        t = t+step;
        ye2 = y2; ye1 = y1; ye0 = y0; y0 = y1; y1 = y2;
    end
    tableval(index) = (ye0+2*ye1+ye2)/4;
end
for i = 1:intdiv-1
    for j = 1:intdiv-i
        table(j) = (tableval(j+1)*4^i-tableval(j))/(4^i-1);
        tableval(j) = table(j);
    end
    tablep = table(1:intdiv-i);
    W(i,1:intdiv-i) = tablep;
end
v = tablep;
```

We can now call this function to solve $dx/dt = -10x$ with $x = 1$ at $t = 0$. The following MATLAB statement solves this differential equation when $t = 0.5$:

```
>> [fv P] = rombergx(@(t,x) -10*x,[0 0.5],7,1)

fv =
    0.0067
```

```
P =

   -2.5677    0.2277    0.1624    0.0245    0.0080    0.0068
    0.4141    0.1580    0.0153    0.0069    0.0067         0
    0.1539    0.0131    0.0068    0.0067         0         0
    0.0125    0.0068    0.0067         0         0         0
    0.0068    0.0067         0         0         0         0
    0.0067         0         0         0         0         0
```

The final value, 0.0067, is better than any of the results achieved for this problem by other methods presented in this chapter. It must be noted that only the final value is found; other values in a given interval can be obtained if intermediate ranges are considered.

This completes our discussion of those types of differential equations known as initial value problems. In Chapter 6 we consider a different type of differential equation known as a boundary value problem.

5.18 Summary

This chapter has defined a range of MATLAB functions for solving differential equations and systems of differential equations that supplement those provided in MATLAB. We have demonstrated how these functions may be used to solve a wide variety of problems.

Problems

5.1. A radioactive material decays at a rate that is proportional to the amount that remains. The differential equation that models this process is

$$dy/dt = -ky \quad \text{where} \quad y = y_0 \quad \text{when} \quad t = t_0$$

Here y_0 represents the mass at time t_0. Solve this equation for $t = 0$ to 10 given that $y_0 = 50$ and $k = 0.05$, using

(a) The function `feuler`, with $h = 1, 0.1, 0.01$
(b) The function `eulertp`, with $h = 1, 0.1$
(c) The function `rkgen`, set for the classical method, with $h = 1$

Compare your results with the exact solution, $y = 50\exp(-0.05t)$.

5.2. Solve $y' = 2xy$ with initial conditions $y_0 = 2$ when $x_0 = 0$ in the range $x = 0$ to 2. Use the classical, Merson, and Butcher variants of the Runge–Kutta method, all implemented in function `rkgen`, with step $h = 0.2$. Note that the exact solution is $y = 2\exp(x^2)$.

5.3. Repeat Problem 5.1 using the following predictor–corrector methods with $h = 2$ and for $t = 0$ to 50:

(a) Adams–Bashforth–Moulton's method, function `abm`

(b) Hamming's method, function `fhamming`

5.4. Express the following second-order differential equation as a pair of first-order equations:

$$xy'' - y' + 8x^3y^3 = 0$$

with the initial conditions $y = 1/2$ and $y' = -1/2$ at $x = 1$. Solve the pair of first-order equations using both `ode23` and `ode45` in the range 1 to 4. The exact solution is given by $y = 1/(1 + x^2)$.

5.5. Use function `fhermite` to solve

(a) Problem 5.1 with $h = 1$

(b) Problem 5.2 with $h = 0.2$

(c) Problem 5.2 with $h = 0.02$

5.6. Use the MATLAB function `rombergx` to solve the following problems. In each case use eight divisions.

(a) $y' = 3y/x$ with initial conditions $x = 1$, $y = 1$. Determine y when $x = 20$.

(b) $y' = 2xy$ with initial conditions $x = 0$, $y = 2$. Determine y when $x = 2$.

5.7. Consider the predator–prey problem described in Section 5.12. This problem may be extended to consider the effect of culling on the interacting populations by subtracting a term from both equations in (5.28) as follows:

$$dP/dt = K_1P - CPQ - S_1P$$

$$dQ/dt = K_2Q - DPQ - S_2Q$$

Here S_1 and S_2 are constants that provide the culling level for the populations. Use `ode45` to solve this problem with $K_1 = 2$, $K_2 = 10$, $C = 0.001$, and $D = 0.002$ and initial values of the population $P = 5000$ and $Q = 100$. Assuming that S_1 and S_2 are equal, experiment with values in the range 1 to 2. There is a wealth of experimental opportunity in this problem and the reader is encouraged to investigate different values of S_1 and S_2.

5.8. Solve the Lorenz equations given in Section 5.11 for $r = 1$, using `ode23`.

5.9. Use the Adams–Bashforth–Moulton method to solve $dy/dt = -5y$, with $y = 50$ when $t = 0$, in the range $t = 0$ to 6. Try step sizes h of 0.1, 0.2, 0.25, and 0.4. Plot the error against t for each case. What can you deduce from these results with regard to the stability of the method? The exact answer is $y = 50e^{-5t}$.

5.10. The following first-order differential equation represents the growth in a population in an environment that can support a maximum population of K:

$$dN/dt = rN(1 - N/K)$$

where $N(t)$ is the population at time t and r is a constant. Given $N = 100$ when $t = 0$, use the MATLAB function `ode23` to solve this differential equation in the range 0 to 200 and plot a graph of N against time. Take $\mathrm{K} = 10{,}000$ and $\mathrm{r} = 0.1$.

5.11. The Leslie–Gower predator–prey problem takes the form

$$dN_1/dt = N_1(r_1 - cN_1 - b_1N_2)$$
$$dN_2/dt = N_2\left(r_2 - b_2\left(N_2/N_1\right)\right)$$

where $N_1 = 15$ and $N_2 = 15$ at time $t = 0$. Use `ode45` to solve this equation given $r_1 = 1$, $r_2 = 0.3$, $c = 0.001$, $b_1 = 1.8$, and $b_2 = 0.5$. Plot N_1 and N_2 in the range $t = 0$ to 40.

5.12. By setting $u = dx/dt$, reduce the following second-order differential equation to two first-order differential equations.

$$\frac{d^2x}{dt^2} + k\left(\frac{1}{v_1} + \frac{1}{v_2}\right)\frac{dx}{dt} = 0$$

where $x = 0$ and $dx/dt = 10$ when $t = 0$. Use the MATLAB function `ode45` to solve this problem given that $v_1 = v_2 = 1$ and $k = 10$.

5.13. A model for a conflict between guerilla, g_2, and government forces, g_1, is given by the equations

$$dg_1/dt = -cg_2$$
$$dg_2/dt = -rg_2g_1$$

Given that the government forces number 2000 and the guerilla forces number 700 at time $t = 0$, use the function `ode45` to solve this system of equations, taking $c = 30$ and $r = 0.01$. You should solve the equations over the time interval 0 to 0.6. Plot a graph of the solution showing the changes in government and guerilla forces over time.

5.14. The following differential equation provides a simple model of a suspension system. The constant m gives the mass of moving parts, the constant k relates to stiffness of the suspension system, and the constant c is a measure of the damping in the system. F is a constant force applied at $t = 0$.

$$m\frac{d^2x}{dt^2} + c\frac{dx}{dt} + kx = F$$

Given that $m = 1$, $k = 4$, $F = 1$, and both $x = 0$ and $dx/dt = 0$ when $t = 0$, use `ode23` to determine the response, $x(t)$ for $t = 0$ to 8 and plot a graph of x against t in each case.

Assume the following values for c:

(**a**) $c = 0$ (**b**) $c = 0.3\surd(mk)$ (**c**) $c = \surd(mk)$

(**d**) $c = 2\surd(mk)$ (**e**) $c = 4\surd(mk)$

Comment on the nature of your solutions. The exact solutions are as follows:

For cases (a), (b), and (c)

$$x(t) = \frac{F}{k}\left[1 - \frac{1}{\sqrt{1-\zeta^2}} e^{-\zeta\omega_n t} \cos(\omega_d t - \phi)\right]$$

For case (d)

$$x(t) = \frac{F}{k}\left[1 - (1 + \omega_n t)\, e^{-\omega_n t}\right]$$

where

$$\omega_n = \sqrt{k/m}, \quad \zeta = c/(2\sqrt{mk}), \quad \omega_d = \omega_n\sqrt{1-\zeta^2}$$

$$\phi = \tan^{-1}\left(\zeta \Big/ \sqrt{1-\zeta^2}\right)$$

For case (e)

$$x(t) = \frac{F}{k}\left[1 + \frac{1}{2q}\left(s_2 e^{s_1 t} - s_1 e^{s_2 t}\right)\right]$$

where

$$q = \omega_n\sqrt{\zeta^2 - 1}, \; s_1 = -\zeta\omega_n + q, \; s_2 = -\zeta\omega_n - q$$

Plot these solutions and compare them with your numerical solutions.

5.15. Gilpin's system for modeling the behavior of three interacting species is given by the differential equations

$$\begin{aligned}
dx_1/dt &= x_1 - 0.001x_1^2 - 0.001kx_1x_2 - 0.01x_1x_3 \\
dx_2/dt &= x_2 - 0.001kx_1x_2 - 0.001x_2^2 - 0.001x_2x_3 \\
dx_3/dt &= -x_3 + 0.005x_1x_3 + 0.0005x_2x_3
\end{aligned}$$

Given $x_1 = 1000$, $x_2 = 300$, and $x_3 = 400$ at time $t = 0$, and taking $k = 0.5$, use `ode45` to solve this system of equations in the range $t = 0$ to $t = 50$ and plot the behavior of the population of the three species against time.

5.16. A problem that arises in planet formation is where a range of objects called planetesimals coagulate to form larger objects and this coagulation continues until a stable state is reached where a number of planetary size objects have been created. To simulate this situation we assume that a minimum size object exists of mass m_1 and the masses of all other objects are integral multiples of the mass of this object. Thus there are n_k objects of mass m_k where $m_k = km_1$. Then the manner in which the number of objects of specific mass changes over time t is given by the *Coagulation Equation* as follows:

$$\frac{dn_k}{dt} = \frac{1}{2}\sum_{i+j=k} A_{ij}n_in_j - n_k\sum_{i=k+1}^{\max k} A_{ki}n_i$$

The values A_{ij} are the probabilities of collisions between the objects i and j. A simple interpretation of this equation is that number of bodies of mass n_k is increased by collisions between bodies of lesser mass but decreased by collisions with larger bodies.

As an exercise write out the equations for this system for the case where there are only three different sizes of planetesimals and assign A_{ij} equal to $n_in_j/(1000(n_i+n_j))$. Note that the division by 1000 ensures that impacts are relatively rare, which seems a plausible asumption in the vast volume of space considered. The initial values for the numbers of planetesimals n_1, n_2, n_3 are taken as 200, 25, and 1, respectively.

Solve the resulting system using the MATLAB function `ode45` using a time interval of 2 units. Study the case where the values of the collision probabilities are calculated using the varying values of the number of planetesimals as time varies. Plot graphs of your results.

5.17. The following example studies the effect of life on a planetary environment. A relatively simple way of studying these effects is to consider the concept of daisy world. This envisages a world inhabited by only two life forms; white and black daisies. This situation can be modeled as a pair of differential equations where the area covered by the black daisies a_b and the area covered by the white daisies a_w changes with time t as follows:

$$da_b/dt = a_b(x\beta_b - \gamma)$$

$$da_w/dt = a_w(x\beta_w - \gamma)$$

where $x = 1 - a_b - a_w$ represents the area not covered by either daisy assuming the total area of the planet is represented by unity. The value of γ gives the death rate for the daisies and β_b and β_w give the growth rate for the black and white daisies respectively. This is related to the energy they receive from the planetary Sun or the local temperature. Consequently, an empirical formula may be given for these values as follows:

$$\beta_b = 1 - 0.003265(295.5 - T_b)^2$$

and

$$\beta_w = 1 - 0.003265(295.5 - T_w)^2$$

where the values of T_b and T_w lie in the range 278 to 313 K, where K denotes degrees Kelvin. Outside this range, growth is assumed zero. Taking $\gamma = 0.3$, $T_b = 295$ K, $T_w = 285$ K, and initial values for $a_b = 0.2$, $a_w = 0.3$, solve the system of equations for $t = [0, 10]$ using the MATLAB function `ode45`. Plot graphs of the changes in a_b and a_w with respect to time. It should be noted that the extent of the areas covered by the black and white daisies will affect the overall temperature of the planet, since white and dark areas react differently in the way they absorb energy from the Sun.

6 Boundary Value Problems

In Chapter 5, we examined methods for solving initial value problems. The solution of these equations depends on the nature of the equation and the initial conditions. In this chapter algorithms for solving certain boundary value problems and problems with both boundary and initial values are given. The solution of a boundary value problem in one independent variable must satisfy specified conditions at two points, and the solution of a boundary value problem in two independent variables must satisfy specified conditions along a curve or set of lines enclosing a specified region.

Although not considered in this chapter, a further important boundary value problem is one with three independent variables—for example, Laplace's equation in three dimensions. In this case the solution must satisfy specified conditions over a surface enclosing a specified volume. Note that in a mixed boundary and initial value problem one independent variable, usually time, will be associated with one or more initial values and the remaining independent variables will depend on boundary values.

6.1 Classification of Second-Order Partial Differential Equations

In this chapter we restrict the discussion to second-order differential equations in one or two independent variables; Figure 6.1 shows how these equations may be classified. The general form of these equations for one and two independent variables is given by (6.1) and (6.2), respectively.

$$A(x)\frac{d^2z}{dx^2}+f\left(x,z,\frac{dz}{dx}\right)=0 \tag{6.1}$$

$$A(x,y)\frac{\partial^2 z}{\partial x^2}+B(x,y)\frac{\partial^2 z}{\partial x\partial y}+C(x,y)\frac{\partial^2 z}{\partial y^2}+f\left(x,y,z,\frac{\partial z}{\partial x},\frac{\partial z}{\partial y}\right)=0 \tag{6.2}$$

These equations are linear in the second-order terms, but the terms

$$f\left(x,z,\frac{dz}{dx}\right) \quad \text{and} \quad f\left(x,y,z,\frac{\partial z}{\partial x},\frac{\partial z}{\partial y}\right)$$

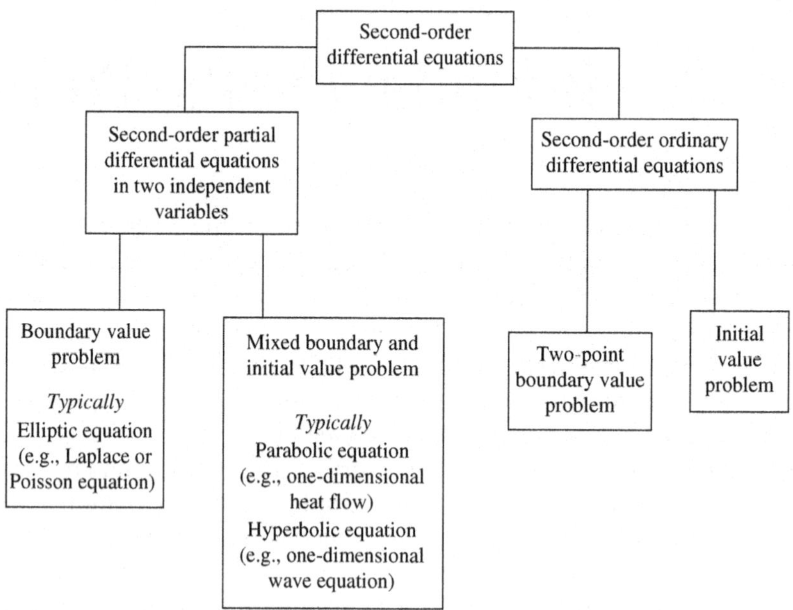

FIGURE 6.1 Second-order differential equations with one or two independent variables and their solutions.

may be linear or nonlinear. In particular, (6.2) is classified as an elliptic, parabolic, or hyperbolic partial differential equation as follows:

If $B^2 - 4AC < 0$, the equation is elliptic.

If $B^2 - 4AC = 0$, the equation is parabolic.

If $B^2 - 4AC > 0$, the equation is hyperbolic.

Since the coefficients A, B, and C are, in general, functions of the independent variables, the classification of (6.2) may vary in different regions of the domain in which the problem is defined. We will commence with a study of (6.1).

6.2 The Shooting Method

An initial value problem and a two-point boundary value problem derived from the same differential equation may have the same solution. For example, consider the differential equation

$$\frac{d^2y}{dx^2} + y = \cos 2x \tag{6.3}$$

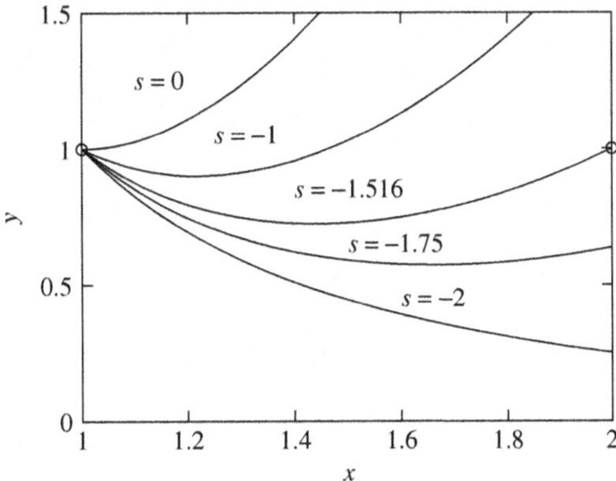

FIGURE 6.2 Solutions for $x^2(d^2y/dx^2) - 6y = 0$ with $y = 1$ and $dy/dx = s$ when $x = 1$, for trial values of s.

Given the initial conditions that when $x = 0$, then $y = 0$ and $dy/dx = 1$, the solution of (6.3) is

$$y = (\cos x - \cos 2x)/3 + \sin x$$

However, this solution also satisfies (6.3) with the two boundary conditions $x = 0$, $y = 0$ and $x = \pi/2$, $y = 4/3$.

This observation provides a useful method of solving two-point boundary value problems called "the shooting method." As an example, consider the equation

$$x^2 \frac{d^2y}{dx^2} - 6y = 0 \tag{6.4}$$

with boundary conditions $y = 1$ when $x = 1$ and $y = 1$ when $x = 2$. We will treat this problem as an initial value problem where $y = 1$ when $x = 1$ and assume trial values for dy/dx when $x = 1$, denoted by s. Figure 6.2 shows the solution for various trial values of s. When $s = -1.516$, the solution satisfies the required boundary condition that $y = 1$ when $x = 2$. The solution for (6.4) can be found by changing it into a pair of first-order differential equations and using any appropriate numerical method described in Chapter 5. Equation (6.4) is equivalent to

$$\begin{aligned} dy/dx &= z \\ dz/dx &= 6y/x^2 \end{aligned} \tag{6.5}$$

We must determine the slope dy/dx that gives the correct boundary condition. This could be achieved by trial and error, but this is tedious and in practice we can use interpolation. The following script solves (6.5) for four trial slopes using the MATLAB function `ode45`. Vector `s` contains trial values of the slope dy/dx at $x = 1$. Vector `b` contains the corresponding values of y when $x = 2$, computed by `ode45`. From these values of y we can interpolate to determine the value of s required to give $y = 1$ when $x = 2$. The interpolation is carried out using the function aitken (described in Chapter 7). Finally, this interpolated value of slope, s0, is used in ode45 to determine the correct solution to (6.5).

```
% e3s601.m
f = @(x,y) [y(2); 6*y(1)/x^2];
option = odeset('RelTol',0.0005);
s = -1.25:-0.25:-2; s0 = [ ];
ncase = length(s); b = zeros(1,ncase);
for i = 1:ncase
    [x,y] = ode45(f,[1 2],[1 s(i)],option);
    [m,n] = size(y);
    b(1,i) = y(m,1);
end
s0 = aitken(b,s,1)
[x,y] = ode45(f,[1 2],[1 s0],option);
[x y(:,1)]
```

The right sides of the differential equations (6.5) are defined in the first line of this script. Running this script gives

```
s0 =
   -1.5161

ans =
    1.0000    1.0000
    1.0111    0.9836
    1.0221    0.9679
    1.0332    0.9529
    1.0442    0.9386
    1.0692    0.9084
    1.0942    0.8812
    1.1192
```

This output is very lengthy and part of it has therefore been deleted. The final stages are as follows:

```
              0.9293
    1.9442    0.9501
    1.9582    0.9622
    1.9721    0.9745
    1.9861    0.9871
    2.0000    1.0000
```

The interpolated value of the slope is -1.5161. The first column of ans gives the values of x and the second gives the corresponding values of y.

While the shooting method is not particularly efficient, it does have the advantage of being able to solve nonlinear boundary value problems. We now examine an alternative method for solving boundary problems: the finite difference method.

6.3 The Finite Difference Method

Chapter 4 shows how derivatives can be approximated by the use of finite differences. We can use the same approach for the solution of certain types of differential equations. The method effectively replaces the differential equation by a set of approximate difference equations. The central difference approximations for the first and second derivatives of z with respect to x are given by (6.6) and (6.7), which follow. In these and subsequent equations the operator D_x represents d/dx, $\mathrm{D}_x^2 = d^2/dx^2$, and so on. The subscript x is omitted where there is no danger of confusion. Thus at z_i,

$$\mathrm{D}z_i \approx (-z_{i-1} + z_{i+1})/(2h) \tag{6.6}$$

$$\mathrm{D}^2 z_i \approx (z_{i-1} - 2z_i + z_{i+1})/h^2 \tag{6.7}$$

In (6.6) and (6.7), h is the distance between the nodal points (see Figure 6.3) and these approximating formulae have errors of order h^2. Higher-order approximations can be generated that have errors of order h^4, but we do not require them. To achieve the same degree of accuracy we can make h smaller.

We can also determine the approximations for unevenly spaced nodal points. For example, it can be shown that (6.6) and (6.7) become

$$\mathrm{D}z_i \approx \frac{1}{h\beta(\beta+1)} \left\{ -\beta^2 z_{i-1} - \left(1-\beta^2\right) z_i + z_{i+1} \right\} \tag{6.8}$$

$$\mathrm{D}^2 z_i \approx \frac{2}{h^2\beta(\beta+1)} \left\{ \beta z_{i-1} - (1+\beta) z_i + z_{i+1} \right\} \tag{6.9}$$

where $h = x_i - x_{i-1}$ and $\beta h = x_{i+1} - x_i$. Note that when $\beta = 1$, (6.8) and (6.9) simplify to (6.6) and (6.7), respectively. Approximation (6.8) has an error of order h^2, regardless of the value of β, and (6.9) has an error of order h for $\beta \neq 1$ and h^2 for $\beta = 1$.

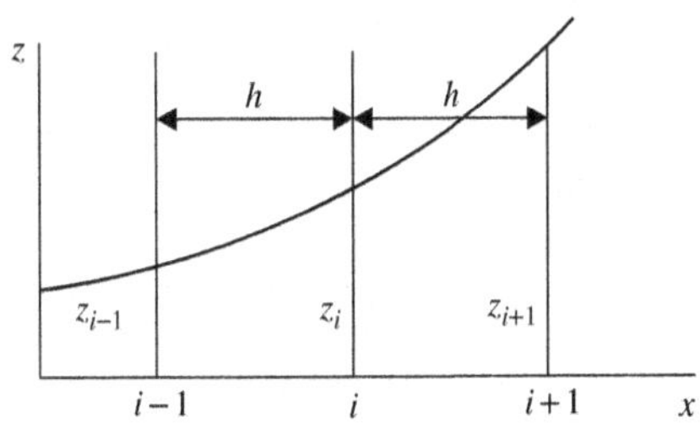

FIGURE 6.3 Equispaced nodal points.

Equations (6.6) through (6.9) are central difference approximations; that is, the approximation for a derivative uses values of the function on either side of the point at which the derivative is to be determined. These are generally the most accurate approximations, but in some situations it is necessary to use forward or backward difference approximations. For example, the forward difference approximation for $\mathrm{D}z_i$ is

$$\mathrm{D}z_i \approx (-z_i + z_{i+1})/h \text{ with an error of order } h \tag{6.10}$$

The backward difference approximation for $\mathrm{D}z_i$ is

$$\mathrm{D}z_i \approx (-z_{i-1} + z_i)/h \text{ with an error of order } h \tag{6.11}$$

To determine solutions for partial differential equations we require the finite difference approximation for various partial derivatives in two or more variables. These approximations can be derived by combining some of the preceding equations. For example, we can determine the finite difference approximation for $\partial^2 z/\partial x^2 + \partial^2 z/\partial y^2$ (i.e., $\nabla^2 z$) from the approximation (6.7) or (6.9). To avoid double subscripts we use the notation applied to the mesh shown in Figure 6.4. Thus, from (6.7),

$$\nabla^2 z_i \approx (z_l - 2z_i + z_r)/h^2 + (z_a - 2z_i + z_b)/k^2 \approx \left\{r^2 z_l + r^2 z_r + z_a + z_b - 2\left(1 + r^2\right) z_i\right\} / \left(r^2 h^2\right) \tag{6.12}$$

where $r = k/h$. If $r = 1$, then (6.12) becomes

$$\nabla^2 z_i \approx (z_l + z_r + z_a + z_b - 4z_i)/h^2 \tag{6.13}$$

These central difference approximations for $\nabla^2 z_i$ have an error of $O(h^2)$.

The finite difference approximation for the second-order mixed derivative of z with respect to x and y, $\partial^2 z/\partial x \partial y$ or D_{xy}, is determined by applying (6.6) in the x direction to

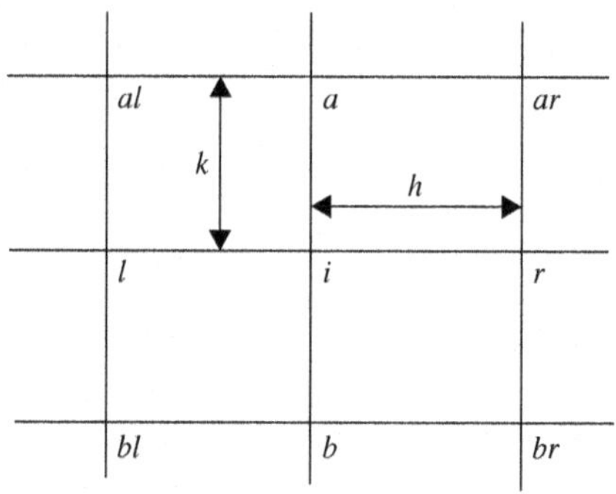

FIGURE 6.4 Grid mesh in rectangular coordinates.

each term of (6.6) in the y direction:

$$\mathrm{D}_{xy}z_i \approx \left[(z_r - z_l)_a/(2h) - (z_r - z_l)_b/(2h)\right]/(2k) \approx (z_{ar} - z_{al} - z_{br} + z_{bl})/(4hk) \tag{6.14}$$

We can develop the finite difference approximations in other coordinate systems such as skew and polar coordinates, and we can have uneven spacing of the node points in any direction; see Salvadori and Baron (1961).

6.4 Two-Point Boundary Value Problems

Before considering the application of finite difference methods to solve a differential equation, we first consider the nature of the solution. We begin by considering the following second-order inhomogeneous differential equation in one independent variable:

$$\left(1+x^2\right)\frac{d^2z}{dx^2} + x\frac{dz}{dx} - z = x^2 \tag{6.15}$$

subject to the boundary conditions $x=0$, $z=1$ and $x=2$, $z=2$. The solution of this equation is

$$z = -\frac{\sqrt{5}}{6}x + \frac{1}{3}\left(1+x^2\right)^{1/2} + \frac{1}{3}\left(2+x^2\right) \tag{6.16}$$

This is the only solution that satisfies both the equation and its boundary conditions. In contrast to this, consider the solution of the second-order homogeneous equation

$$x\frac{d^2z}{dx^2} + \frac{dz}{dx} + \lambda x^{-1}z = 0 \tag{6.17}$$

subject to the conditions that $z=0$ at $x=1$ and $dz/dx=0$ at $x=e$ (where $e=2.7183\ldots$). If λ is a given constant, this homogeneous equation has the trivial solution $z=0$. However, if λ is an unknown, then we can determine values of λ to give nontrivial solutions for z. Equation (6.17) is then a characteristic value or eigenvalue problem. Solving (6.17) gives an infinite number of solutions for λ and z as follows:

$$z_n = \sin\left\{(2n+1)\frac{\pi}{2}\log_e|x|\right\},\ \lambda_n = \{(2n+1)\pi/2\}^2 \quad \text{where} \quad n=0, 1, 2, \ldots \tag{6.18}$$

The values of λ that satisfy (6.18) are called characteristic values or eigenvalues, and the corresponding values of z are called characteristic functions or eigenfunctions. This particular type of boundary value problem is called a characteristic value or eigenvalue problem. It has arisen because both the differential equation and the specified boundary conditions are homogeneous.

The application of finite differences to the solution of boundary value problems is now illustrated by Examples 6.1 and 6.2.

Example 6.1

Determine an approximate solution for (6.15). We begin by multiplying (6.15) by $2h^2$ and writing d^2z/dx^2 as D^2z, and so on, to give

$$2(1+x^2)(h^2D^2z)+xh(2hDz)-2h^2z=2h^2x^2 \tag{6.19}$$

Using (6.6) and (6.7), we can replace (6.19) by

$$2\left(1+x_i^2\right)(z_{i-1}-2z_i+z_{i+1})+x_ih(-z_{i-1}+z_{i+1})-2h^2z_i=2h^2x_i^2 \tag{6.20}$$

Figure 6.5 shows x divided into four segments ($h=1/2$) with nodes numbered 1 to 5. Applying (6.20) to nodes 2, 3, and 4 gives

At node 2: $2(1+0.5^2)(z_1-2z_2+z_3)+0.25(-z_1+z_3)-0.5z_2=0.5(0.5^2)$

At node 3: $2(1+1.0^2)(z_2-2z_3+z_4)+0.50(-z_2+z_4)-0.5z_3=0.5(1.0^2)$

At node 4: $2(1+1.5^2)(z_3-2z_4+z_5)+0.75(-z_3+z_5)-0.5z_4=0.5(1.5^2)$

The problem boundary conditions are $x=0, z=1$ and $x=2, z=2$. Thus $z_1=1$ and $z_5=2$. Using these values, the preceding equations can be simplified and written in matrix form:

$$\begin{bmatrix} -44 & 22 & 0 \\ 28 & -68 & 36 \\ 0 & 46 & -108 \end{bmatrix} \begin{bmatrix} z_2 \\ z_3 \\ z_4 \end{bmatrix} = \begin{bmatrix} -17 \\ 4 \\ -107 \end{bmatrix}$$

This equation system can easily be solved using MATLAB as follows:

```
>> A = [-44 22 0;28 -68 36;0 46 -108];
>> b = [-17 4 -107].';
>> y = A\b

y =
    0.9357
    1.0987
    1.4587
```

Note that the rows in the preceding matrix equation can always be scaled in order to make the coefficient matrix symmetrical. This is important in a large problem.

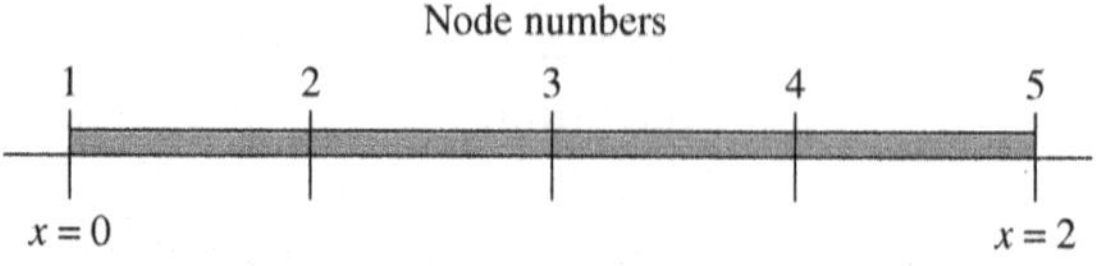

FIGURE 6.5 Node numbering used in the solution of (6.15).

To increase the accuracy of the solution we must increase the number of nodal points that consequently decrease h. However, formulating the finite difference approximation by hand for a large number of nodes is a tedious and error-prone process. The MATLAB function `twopoint` implements the process of solving the second-order boundary problem comprising differential equation (6.21) together with appropriate boundary conditions.

$$C(x)\frac{d^2z}{dx^2}+D(x)\frac{dz}{dx}+E(x)z=F(x) \tag{6.21}$$

The user must supply a vector listing the values of nodal points chosen. These do not have to be equispaced. The user must also supply vectors listing the values of $C(x)$, $D(x)$, $E(x)$, and $F(x)$ for the nodal points. Finally, the user must provide the boundary conditions, which can be in terms of either z or dz/dx.

```
function y = twopoint(x,C,D,E,F,flag1,flag2,p1,p2)
% Solves 2nd order boundary value problem
% Example call: y = twopoint(x,C,D,E,F,flag1,flag2,p1,p2)
% x is a row vector of n+1 nodal points.
% C, D, E and F are row vectors
% specifying C(x), D(x), E(x) and F(x).
% If y is specified at node 1, flag1 must equal 1.
% If y' is specified at node 1, flag1 must equal 0.
% If y is specified at node n+1, flag2 must equal 1.
% If y' is specified at node n+1, flag2 must equal 0.
% p1 & p2 are boundary values (y or y') at nodes 1 and n+1.
n = length(x)-1;
h(2:n+1) = x(2:n+1)-x(1:n);
h(1) = h(2); h(n+2) = h(n+1);
r(1:n+1) = h(2:n+2)./h(1:n+1);
s = 1+r;
if flag1==1
    y(1) = p1;
else
    slope0 = p1;
end
if flag2==1
    y(n+1) = p2;
else
    slopen = p2;
end
W = zeros(n+1,n+1);
if flag1==1
    c0 = 3;
    W(2,2) = E(2)-2*C(2)/(h(2)^2*r(2));
    W(2,3) = 2*C(2)/(h(2)^2*r(2)*s(2))+D(2)/(h(2)*s(2));
    b(2) = F(2)-y(1)*(2*C(2)/(h(2)^2*s(2))-D(2)/(h(2)*s(2)));
```

```
else
    c0=2;
    W(1,1) = E(1)-2*C(1)/(h(1)^2*r(1));
    W(1,2) = 2*C(1)*(1+1/r(1))/(h(1)^2*s(1));
    b(1) = F(1)+slope0*(2*C(1)/h(1)-D(1));
end
if flag2==1
    c1 = n-1;
    W(n,n) = E(n)-2*C(n)/(h(n)^2*r(n));
    W(n,n-1) = 2*C(n)/(h(n)^2*s(n))-D(n)/(h(n)*s(n));
    b(n) = F(n)-y(n+1)*(2*C(n)/(h(n)^2*s(n))+D(n)/(h(n)*s(n)));
else
    c1 = n;
    W(n+1,n+1) = E(n+1)-2*C(n+1)/(h(n+1)^2*r(n+1));
    W(n+1,n) = 2*C(n+1)*(1+1/r(n+1))/(h(n+1)^2*s(n+1));
    b(n+1) = F(n+1)-slopen*(2*C(n+1)/h(n+1)+D(n+1));
end
for i = c0:c1
    W(i,i) = E(i)-2*C(i)/(h(i)^2*r(i));
    W(i,i-1) = 2*C(i)/(h(i)^2*s(i))-D(i)/(h(i)*s(i));
    W(i,i+1) = 2*C(i)/(h(i)^2*r(i)*s(i))+D(i)/(h(i)*s(i));
    b(i) = F(i);
end
z = W(flag1+1:n+1-flag2,flag1+1:n+1-flag2)\b(flag1+1:n+1-flag2)';
if flag1==1 & flag2==1, y = [y(1); z; y(n+1)]; end
if flag1==1 & flag2==0, y = [y(1); z]; end
if flag1==0 & flag2==1, y = [z; y(n+1)]; end
if flag1==0 & flag2==0, y = z; end
```

We can use this function to solve (6.15) for nine nodes using the following script:

```
% e3s602.m
x = 0:.2:2;
C = 1+x.^2; D = x; E = -ones(1,11); F = x.^2;
flag1 = 1; p1 = 1; flag2 = 1; p2 = 2;
z = twopoint(x,C,D,E,F,flag1,flag2,p1,p2);
B = 1/3; A = -sqrt(5)*B/2;
xx = 0:.01:2;
zz = A*xx+B*sqrt(1+xx.^2)+B*(2+xx.^2);
plot(x,z,'o',xx,zz)
xlabel('x'); ylabel('z')
```

This script outputs the graph of Figure 6.6.

The results from the finite difference analysis are very accurate. This is because the solution of the boundary problem, given by (6.16), is well approximated by a low-order polynomial.

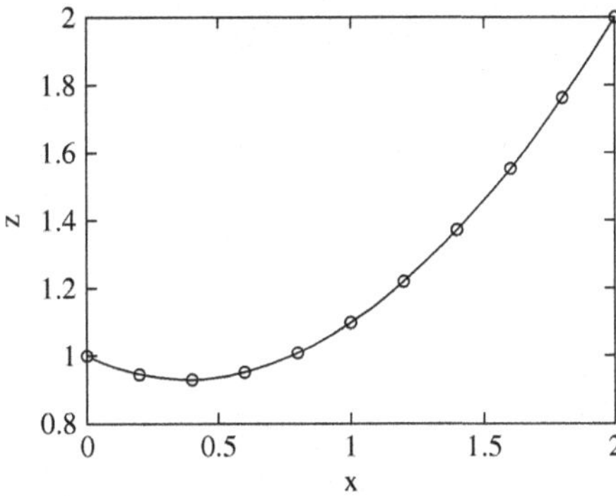

FIGURE 6.6 Finite difference solution of $(1+x^2)(d^2z/dx^2)+xdz/dx-z=x^2$. The $\circ$ indicates the finite difference estimate; the *continuous line* is the exact solution.

Example 6.2

Determine the approximate solution of (6.17) subject to the boundary conditions that $z=0$ at $x=1$ and $dz/dx=0$ at $x=e$. The exact eigensolutions are given by $\lambda_n=\{(2n+1)\pi/2\}^2$ and $z_n(x)=\sin\{(2n+1)(\pi/2)\log_e|x|\}$, where $n=0,1,\ldots,\infty$. We will use the node-numbering scheme shown in Figure 6.7. To apply the boundary condition at $x=e$ we must consider the finite difference approximation for Dz at node 5 (i.e., at $x=e$) and make $\mathrm{D}z_5=0$. Applying (6.6), we have

$$2h\mathrm{D}z_5=-z_4+z_6=0 \tag{6.22}$$

Note that we have been forced to introduce a fictitious node, node 6. However, from (6.22), $z_6=z_4$.

Multiplying (6.17) by $2h^2$ gives

$$2x(h^2\mathrm{D}^2z)+h(2h\mathrm{D}z)=-\lambda 2x^{-1}h^2z$$

Thus,

$$2x_i(z_{i-1}-2z_i+z_{i+1})+h(-z_{i-1}+z_{i+1})=-\lambda 2x_i^{-1}h^2z_i$$

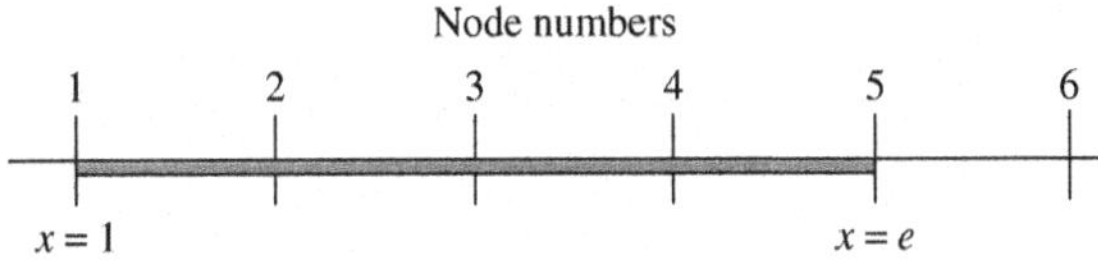

FIGURE 6.7 Node numbering used in the solution of (6.17).

Now $L = e - 1 = 1.7183$; thus $h = L/4 = 0.4296$. Applying (6.19) to nodes 2 through 5, we have

At node 2: $2(1.4296)(z_1 - 2z_2 + z_3) + 0.4296(-z_1 + z_3) = -2\lambda(1.4296)^{-1}(0.4296)^2 z_2$

At node 3: $2(1.8591)(z_2 - 2z_3 + z_4) + 0.4296(-z_2 + z_4) = -2\lambda(1.8591)^{-1}(0.4296)^2 z_3$

At node 4: $2(2.2887)(z_3 - 2z_4 + z_5) + 0.4296(-z_3 + z_5) = -2\lambda(2.2887)^{-1}(0.4296)^2 z_4$

At node 5: $2(2.7183)(z_4 - 2z_5 + z_6) + 0.4296(-z_4 + z_6) = -2\lambda(2.7183)^{-1}(0.4296)^2 z_5$

Letting $z_1 = 0$ and $z_6 = z_4$ leads to

$$\begin{bmatrix} -5.7184 & 3.2887 & 0 & 0 \\ 3.2887 & -7.4364 & 4.1478 & 0 \\ 0 & 4.1478 & -9.1548 & 5.0070 \\ 0 & 0 & 10.8731 & -10.8731 \end{bmatrix} \begin{bmatrix} z_2 \\ z_3 \\ z_4 \\ z_5 \end{bmatrix}$$

$$= \lambda \begin{bmatrix} -0.2582 & 0 & 0 & 0 \\ 0 & -0.1985 & 0 & 0 \\ 0 & 0 & -0.1613 & 0 \\ 0 & 0 & 0 & -0.1358 \end{bmatrix} \begin{bmatrix} z_2 \\ z_3 \\ z_4 \\ z_5 \end{bmatrix}$$

We can solve these equations using MATLAB as follows:

```
>> A = [-5.7184 3.2887 0 0;3.2887 -7.4364 4.1478 0;
   0 4.1478 -9.1548 5.0070; 0 0 10.8731 -10.8731];
>> B = diag([-0.2582 -0.1985 -0.1613 -0.1358]);
>> [u lambda] = eig(A,B)

u =

   -0.5424    1.0000   -0.4365    0.0169
   -0.8362    0.1389    1.0000   -0.1331
   -0.9686   -0.6793   -0.3173    0.5265
   -1.0000   -0.9112   -0.8839   -1.0000


lambda =

    2.5110         0         0         0
         0   20.3774         0         0
         0         0   51.3254         0
         0         0         0  122.2197
```

The exact values for the lowest four eigenvalues are 2.4674, 22.2066, 61.6850, and 120.9027. The graph of Figure 6.8 shows the first two eigenfunctions $z_0(x)$ and $z_1(x)$ and the estimates derived from the first and second columns of the preceding array u. Note that the values of u have been scaled to make those corresponding to the node z_5 either 1 or -1. The following script evaluates and plots the exact eigenfunctions $z_0(x)$ and $z_1(x)$ and plots the scaled sample points that estimate these functions.

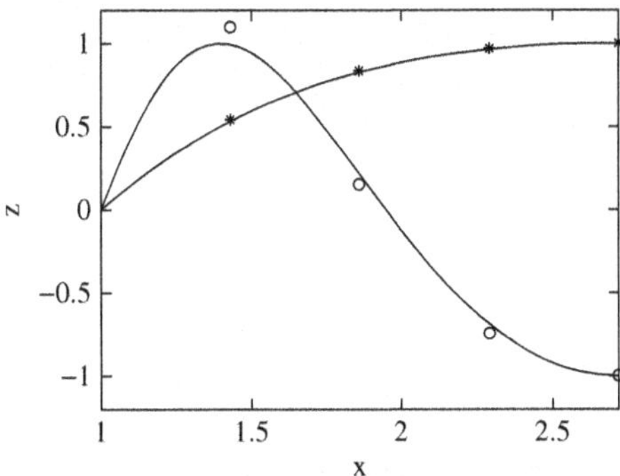

FIGURE 6.8 The finite difference estimates for the first (∗) and second (○) eigenfunctions of $x(d^2z/dx^2)+dz/dx+\lambda z/x=0$. *Solid lines* show the exact eigenfunctions $z_0(x)$ and $z_1(x)$.

```
% e3s603.m
x = 1:.01:exp(1);
% compute eigenfunction values scaled to 1 or -1.
z0 = sin((1*pi/2)*log(abs(x)));
z1 = sin((3*pi/2)*log(abs(x)));
% plot eigenfuctions
plot(x,z0,x,z1),  hold on
% Discrete approximations to eigenfunctions
% Scaled to 1 or -1.
u0 = [0.5424 0.8362 0.9686 1];
u1 = (1/0.9112)*[1 0.1389 -0.6793 -0.9112];
% determine x values for plotting
r = (exp(1)-1)/4;
xx = [1+r 1+2*r 1+3*r 1+4*r];
plot(xx,u0,'*',xx,u1,'o'),  hold off
axis([1 exp(1) -1.2 1.2])
xlabel('x'), ylabel('z')
```

6.5 Parabolic Partial Differential Equations

The general second-order partial differential equation in terms of the independent variables x and y is given by (6.2). The equation is repeated here, except that y has been replaced by t.

$$A(x,t)\frac{\partial^2 z}{\partial x^2}+B(x,t)\frac{\partial^2 z}{\partial x\partial t}+C(x,t)\frac{\partial^2 z}{\partial t^2}+f\left(x,t,z,\frac{\partial z}{\partial x},\frac{\partial z}{\partial t}\right)=0 \tag{6.23}$$

This equation will be a parabolic equation if $B^2 - 4AC = 0$. Parabolic equations are not defined in a closed domain, but propagate in an open domain. For example, the one-dimensional heat-flow equation, which describes heat flow assuming no energy generation, is

$$K\frac{\partial^2 u}{\partial x^2} = \frac{\partial u}{\partial t}, \quad 0 < x < L \quad \text{and} \quad t > 0 \tag{6.24}$$

where K is the thermal diffusivity and u is the temperature of the material. Comparing (6.24) with (6.23), we see that A, B, and C of (6.23) are K, 0, and 0, respectively, so that the term $B^2 - 4AC$ is zero and the equation is parabolic.

To solve this equation, boundary conditions must be specified at $x = 0$ and $x = L$ and initial conditions when $t = 0$ must be given. To develop a finite difference solution we divide the spatial domain into n sections, each of length h, so that $h = L/n$, and consider as many time steps as required, each time step of duration k. A finite difference approximation for (6.24) at node (i,j) is obtained by replacing $\partial^2 u/\partial x^2$ by the central difference approximation (6.7) and $\partial u/\partial t$ by the forward difference approximation (6.10) to give

$$K\left(\frac{u_{i-1,j} - 2u_{i,j} + u_{i+1,j}}{h^2}\right) = \left(\frac{-u_{i,j} + u_{i,j+1}}{k}\right) \tag{6.25}$$

or

$$u_{i,j+1} = u_{i,j} + \alpha(u_{i-1,j} - 2u_{i,j} + u_{i+1,j}), \quad i = 0, 1, \ldots, n; \quad j = 0, 1, \ldots \tag{6.26}$$

In (6.26), $\alpha = Kk/h^2$. Node (i,j) is the point $x = ih$ at time jk. Equation (6.26) allows us to determine $u_{i,j+1}$, that is, u at time $j+1$ from values of u at time j. Values of $u_{i,0}$ are provided by the initial conditions; values of $u_{0,j}$ and $u_{n,j}$ are obtained from the boundary conditions. This method of solution is called the explicit method.

In the numeric solution of parabolic partial differential equations, solution stability and convergence are important. It can be proved that when using the explicit method we must make $\alpha \leq 0.5$ to ensure a steady decay of the entire solution. This requirement means that the grid separation in time must sometimes be very small, necessitating a very large number of time steps.

An alternative finite difference approximation for (6.24) is obtained by considering node $(i,j+1)$. We again approximate $\partial^2 u/\partial x^2$ by the central difference approximation (6.7), but we approximate $\partial u/\partial t$ by the backward difference approximation (6.11) to give

$$K\left(\frac{u_{i-1,j+1} - 2u_{i,j+1} + u_{i+1,j+1}}{h^2}\right) = \left(\frac{-u_{i,j} + u_{i,j+1}}{k}\right) \tag{6.27}$$

This equation is identical to (6.25) except that approximation is made at the $(j+1)$th time step instead of at the jth time step. Rearranging (6.27) with $\alpha = Kk/h^2$ gives

$$(1 + 2\alpha)u_{i,j+1} - \alpha(u_{i+1,j+1} + u_{i-1,j+1}) = u_{i,j} \tag{6.28}$$

where $i = 0, 1, \ldots, n$; $j = 0, 1, \ldots$. The three variables on the left side of this equation are unknown. However, if we have a grid of $n+1$ spatial points, then at time $j+1$ there are $n-1$ unknown nodal values and two known boundary values. We can assemble the set of $n-1$ equations of the form of (6.28) as follows:

$$\begin{bmatrix} \gamma & -\alpha & 0 & \cdots & 0 \\ -\alpha & \gamma & -\alpha & \cdots & 0 \\ 0 & -\alpha & \gamma & \cdots & 0 \\ \vdots & \vdots & \vdots & & \vdots \\ 0 & 0 & 0 & \cdots & -\alpha \\ 0 & 0 & 0 & \cdots & \gamma \end{bmatrix} \begin{bmatrix} u_{1,j+1} \\ u_{2,j+1} \\ u_{3,j+1} \\ \vdots \\ u_{n-2,j+1} \\ u_{n-1,j+1} \end{bmatrix} = \begin{bmatrix} u_{1,j} + \alpha u_0 \\ u_{2,j} \\ u_{3,j} \\ \vdots \\ u_{n-2,j} \\ u_{n-1,j} + \alpha u_n \end{bmatrix}$$

where $\gamma = 1 + 2\alpha$. Note that u_0 and u_n are the known boundary conditions, assumed to be independent of time. By solving the preceding equation system, we determine $u_1, u_2, \ldots, u_{n-1}$ at time step $j+1$ from $u_1, u_2, \ldots, u_{n-1}$ at time step j. This approach is called the implicit method. Compared with the explicit method, each time step requires more computation; however, the method has the significant advantage that it is unconditionally stable. However, although stability does not place any restriction on α, h and k must be chosen to keep the discretization error small to maintain accuracy.

The following function `heat` implements an implicit finite difference solution for the parabolic differential equation (6.24).

```
function [u alpha] = heat(nx,hx,nt,ht,init,lowb,hib,K)
% Solves parabolic equ'n.
% e.g. heat flow equation.
% Example call: [u alpha] = heat(nx,hx,nt,ht,init,lowb,hib,K)
% nx, hx are number and size of x panels
% nt, ht are number and size of t panels
% init is a row vector of nx+1 initial values of the function.
% lowb & hib are boundaries at low and hi values of x.
% Note that lowb and hib are scalar values.
% K is a constant in the parabolic equation.
alpha = K*ht/hx^2;
A = zeros(nx-1,nx-1); u = zeros(nt+1,nx+1);
u(:,1) = lowb*ones(nt+1,1);
u(:,nx+1) = hib*ones(nt+1,1);
u(1,:) = init;
A(1,1) = 1+2*alpha; A(1,2) = -alpha;
for i = 2:nx-2
    A(i,i) = 1+2*alpha;
    A(i,i-1) = -alpha; A(i,i+1) = -alpha;
end
```

```
A(nx-1,nx-2) = -alpha; A(nx-1,nx-1) = 1+2*alpha;
b(1,1) = init(2)+init(1)*alpha;
for i = 2:nx-2, b(i,1) = init(i+1); end
b(nx-1,1) = init(nx)+init(nx+1)*alpha;
[L,U] = lu(A);
for j = 2:nt+1
    y = L\b; x = U\y;
    u(j,2:nx) = x'; b = x;
    b(1,1) = b(1,1)+lowb*alpha;
    b(nx-1,1) = b(nx-1,1)+hib*alpha;
end
```

We now use the function heat to study how the temperature distribution in a brick wall varies with time. The wall is 0.3 m thick and is initially at a uniform temperature of 100°C. For the brickwork, $K = 5 \times 10^{-7}\,\mathrm{m/s^2}$. If the temperature of both surfaces is suddenly lowered to 20°C and kept at this temperature, we wish to plot the subsequent variation of temperature through the wall at 440 s (7.33 min) intervals for 22,000 s (366.67 min).

To study this problem we will use a mesh with 15 subdivisions of x and 50 subdivisions of t.

```
% e3s604.m
K = 5e-7; thick = 0.3; tfinal = 22000;
nx = 15; hx = thick/nx;
nt = 50; ht = tfinal/nt;
init = 100*ones(1,nx+1); lowb = 20; hib = 20;
[u al] = heat(nx,hx,nt,ht,init,lowb,hib,K);
alpha = al, surfl(u)
axis([0 nx+1 0 nt+1 0 120])
view([-217 30]), xlabel('x - node nos.')
ylabel('Time - node nos.'), zlabel('Temperature')
```

Running this script gives

```
alpha =
    0.5500
```

together with the plot shown in Figure 6.9.

The plot shows how the temperature across the wall decreases with time. Figure 6.10 shows the variation of temperature with time at the center of the wall, calculated by both the implicit method (using the MATLAB function heat) and the explicit method using the same mesh size. In the latter case a MATLAB function is not provided. The solution determined using the explicit method becomes unstable with increasing time. We expect this because the mesh size has been chosen to make $\alpha = 0.55$.

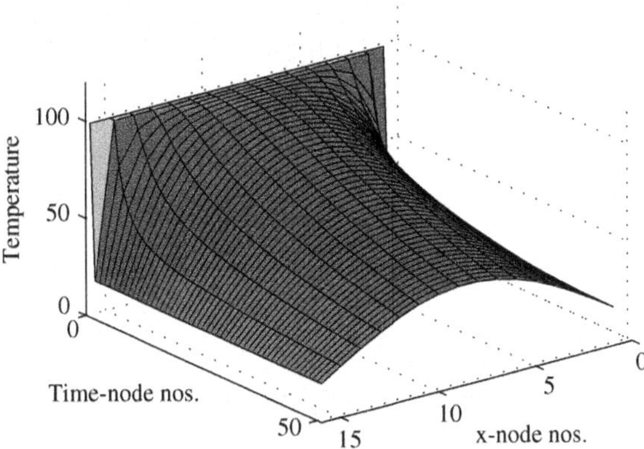

FIGURE 6.9 Plot showing how the distribution of temperature through a wall varies with time.

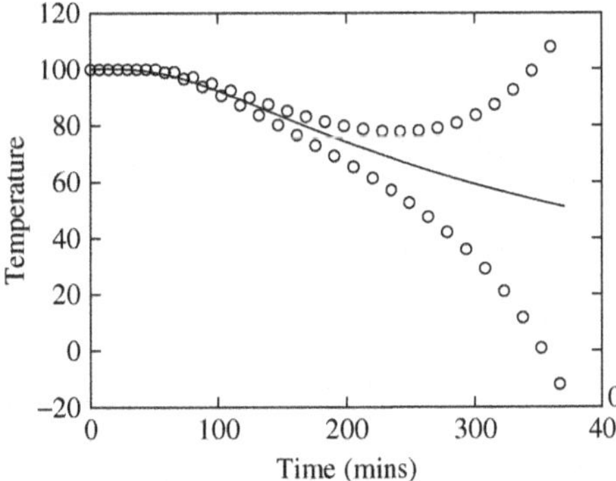

FIGURE 6.10 Variation in temperature in the center of a wall. The steadily decaying solution was generated using the implicit method of solution; the oscillating solution was generated using the explicit method of solution.

6.6 Hyperbolic Partial Differential Equations

Consider the following equation:

$$c^2\frac{\partial^2 u}{\partial x^2} = \frac{\partial^2 u}{\partial t^2}, \quad 0 < x < L \quad \text{and} \quad t > 0 \tag{6.29}$$

This is the one-dimensional wave equation, and like the heat-flow problem of Section 6.5, its solution usually propagates in an open domain. Equation (6.29) describes the wave in a taut string where c is the velocity of propagation of the waves in the string. Comparing (6.29) with (6.23), we see that $B^2 - 4AC = -4c^2(-1)$. Since c^2 is positive, $B^2 - 4AC > 0$ and the equation is hyperbolic. Equation (6.29) is subject to boundary conditions at $x = 0$ and $x = L$ and also subject to initial conditions when $t = 0$.

We now develop equivalent finite difference approximations for these equations. Divide L into n sections so that $h = L/n$ and consider time steps of duration k. Approximating (6.29) by central finite difference approximations based on (6.7) at node (i,j), we have

$$c^2\left(\frac{u_{i-1,j} - 2u_{i,j} + u_{i+1,j}}{h^2}\right) = \left(\frac{u_{i,j-1} - 2u_{i,j} + u_{i,j+1}}{k^2}\right)$$

or

$$(u_{i-1,j} - 2u_{i,j} + u_{i+1,j}) - (1/\alpha^2)\,(u_{i,j-1} - 2u_{i,j} + u_{i,j+1}) = 0$$

where $\alpha^2 = c^2k^2/h^2$, $i = 0, 1, \ldots, n$, and $j = 0, 1, \ldots$. Node (i,j) is the point $x = ih$ at time $t = jk$. Rearranging the preceding equation gives

$$u_{i,j+1} = \alpha^2\,(u_{i-1,j} + u_{i+1,j}) + 2(1 - \alpha^2)u_{i,j} - u_{i,j-1} \tag{6.30}$$

When $j = 0$, equation (6.30) becomes

$$u_{i,\,1} = \alpha^2\,(u_{i-1,\,0} + u_{i+1,\,0}) + 2(1 - \alpha^2)u_{i,\,0} - u_{i,\,-1} \tag{6.31}$$

To solve a hyperbolic partial differential equation, initial values of $u(x)$ and $\partial u/\partial t$ must be specified. Let these values be U_i and V_i, respectively, where $i = 0, 1, \ldots, n$. We can replace $\partial u/\partial t$ by its central finite difference approximation based on (6.6) as follows:

$$V_i = (-u_{i,-1} + u_{i,1})/(2k)$$

Thus,

$$-u_{i,-1} = 2kV_i - u_{i,1} \tag{6.32}$$

In (6.31) we replace $u_{i,0}$ by U_i and $u_{i,-1}$ by using (6.32) to give

$$u_{i,1} = \alpha^2\,(U_{i-1} + U_{i+1}) + 2(1 - \alpha^2)U_i + 2kV_i - u_{i,1}$$

so that

$$u_{i,1} = \alpha^2\,(U_{i-1} + U_{i+1})\,/2 + (1 - \alpha^2)U_i + kV_i \tag{6.33}$$

Equation (6.33) is the starting equation and allows us to determine the values of u at time step $j = 1$. Once we obtain these values, we can use (6.30) to provide an explicit method of solution. In order to ensure stability, the parameter α should be equal to or less than one. However, if α is less than one, the solution becomes less accurate.

The following function, fwave, implements an explicit finite difference solution for (6.29).

```
function [u alpha] = fwave(nx,hx,nt,ht,init,initslope,lowb,hib,c)
% Solves hyperbolic equ'n, e.g. wave equation.
% Example: [u alpha] = fwave(nx,hx,nt,ht,init,initslope,lowb,hib,c)
% nx, hx are number and size of x panels
% nt, ht are number and size of t panels
% init is a row vector of nx+1 initial values of the function.
% initslope is a row vector of nx+1 initial derivatives of
% the function.
% lowb is a column vector of nt+1 boundary values at the
% low value of x.
% hib is a column vector of nt+1 boundary values at hi value of x.
% c is a constant in the hyperbolic equation.
alpha = c*ht/hx;
u = zeros(nt+1,nx+1);
u(:,1) = lowb; u(:,nx+1) = hib; u(1,:) = init;
for i = 2:nx
    u(2,i) = alpha^2*(init(i+1)+init(i-1))/2+(1-alpha^2)*init(i) ...
    +ht*initslope(i);
end
for j = 2:nt
    for i = 2:nx
        u(j+1,i)=alpha^2*(u(j,i+1)+u(j,i-1))+(2-2*alpha^2)*u(j,i) ...
        -u(j-1,i);
    end
end
```

We now use the fwave function to examine the effect of displacing the boundary at one end of a taut string by 10 units in a positive direction for the time period $t = 0.1$ to $t = 4$ units.

```
% e3s605.m
T = 4; L = 1.6;
nx = 16; nt = 40; hx = L/nx; ht = T/nt;
c = 1; t = 0:nt;
hib = zeros(nt+1,1); lowb = zeros(nt+1,1);
lowb(2:5,1) = 10;
init = zeros(1,nx+1); initslope = zeros(1,nx+1);
[u al] = fwave(nx,hx,nt,ht,init,initslope,lowb,hib,c);
alpha = al, surfl(u)
axis([0 16 0 40 -10 10])
xlabel('Position along string')
ylabel('Time'), zlabel('Vertical displacement')
```

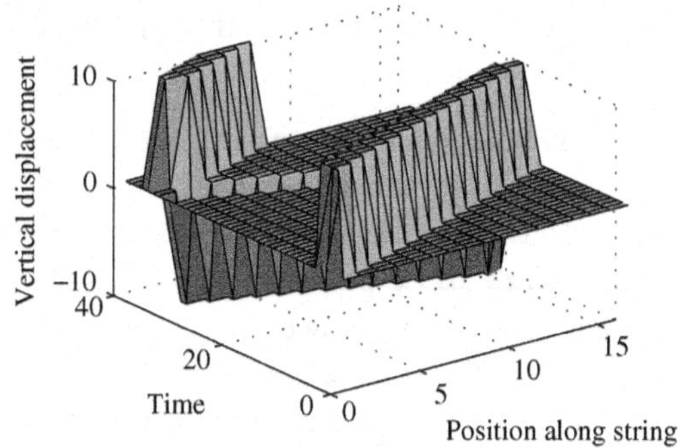

FIGURE 6.11 Solution of (6.29) subject to specific boundary and initial conditions.

Running this script produces the following output, together with Figure 6.11.

```
alpha =
     1
```

The figure shows that the disturbance at the boundary travels along the string. At the other boundary, it is reflected and becomes a negative disturbance. This process of reflection and reversal continues at each boundary. The disturbance travels at a velocity c and its shape does not change. Similarly, pressure fluctuations do not change as they travel along a speaking tube; if pressure fluctuations representing the sound "HELLO" enter the tube, the sound "HELLO" is detected at the other end. In practice, energy loss, which is not included in this model, would cause the amplitude of the disturbance to decay to zero over a period of time.

6.7 Elliptic Partial Differential Equations

The solution of a second-order elliptic partial differential equation is determined over a closed region, and the shape of the boundary and its condition at every point must be specified. Some important second-order elliptic partial differential equations, which arise naturally in the description of physical systems, are

$$\text{Laplace's equation: } \nabla^2 z = 0 \tag{6.34}$$

$$\text{Poisson's equation: } \nabla^2 z = F(x,y) \tag{6.35}$$

$$\text{Helmholtz's equation: } \nabla^2 z + G(x,y)z = F(x,y) \tag{6.36}$$

where $\nabla^2 z = \partial^2 z/\partial x^2 + \partial^2 z/\partial y^2$ and $z(x,y)$ is an unknown function. Note that the Laplace and Poisson equations are special cases of Helmholtz's equation. In general, these equations must satisfy boundary conditions that are specified in terms of either the function value or the derivative of the function normal to the boundary. Furthermore, a problem can have mixed boundary conditions. If we compare (6.34), (6.35), and (6.36) to the standard second-order partial differential equation in two variables, that is,

$$A(x,y)\frac{\partial^2 z}{\partial x^2}+B(x,y)\frac{\partial^2 z}{\partial x\partial y}+C(x,y)\frac{\partial^2 z}{\partial y^2}+f\left(x,y,z,\frac{\partial z}{\partial x},\frac{\partial z}{\partial y}\right)=0$$

we see that in each case $A=C=1$ and $B=0$, so that $B^2-4AC<0$, confirming that the equations are elliptic.

The Laplace equation is homogeneous, and if a problem has boundary conditions that are also homogeneous then the solution, $z=0$, will be trivial. Similarly in (6.35), if $F(x,y)=0$ and the problem boundary conditions are homogeneous, then $z=0$. However, in (6.36) we can scale $G(x,y)$ by a factor λ, so that (6.36) becomes

$$\nabla^2 z+\lambda G(x,y)z=0 \tag{6.37}$$

This is a characteristic or eigenvalue problem, and we can determine values of λ and corresponding nontrivial values of $z(x,y)$.

The elliptic equations (6.34) through (6.37) can only be solved in a closed form for a limited number of situations. For most problems, it is necessary to use a numerical approximation. Finite difference methods are relatively simple to apply, particularly for rectangular regions. We will now use the finite difference approximation for $\nabla^2 z$, given by (6.12) or (6.13), to solve some elliptic partial differential equations over a rectangular domain.

Example 6.3

Laplace's equation. Determine the distribution of temperature in a rectangular plane section, subject to a temperature distribution around its edges as follows:

$$x=0, T=100y; \quad x=3, T=250y; \quad y=0, T=0; \quad \text{and} \quad y=2, T=200+(100/3)x^2$$

The section shape, the boundary temperature distribution section and two chosen nodes are shown in Figure 6.12.

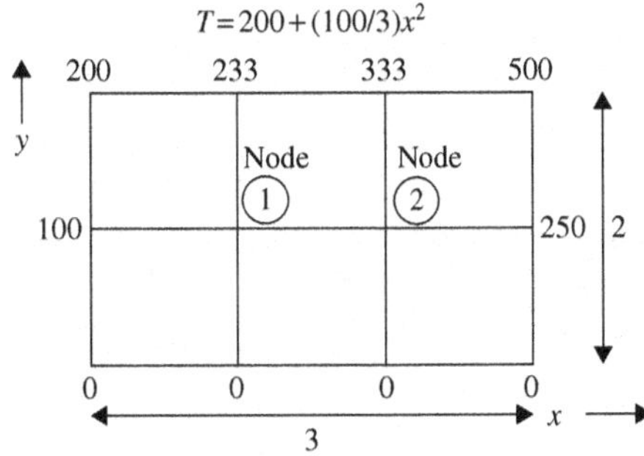

FIGURE 6.12 Temperature distribution around a plane section. Locations of nodes 1 and 2 are shown.

The temperature distribution is described by Laplace's equation. Solving this equation by the finite difference method, we apply (6.13) to nodes 1 and 2 of the mesh shown in Figure 6.12. This gives

$$(233.33 + T_2 + 0 + 100 - 4T_1)/h^2 = 0$$
$$(333.33 + 250 + 0 + T_1 - 4T_2)/h^2 = 0$$

where T_1 and T_2 are the unknown temperatures at nodes 1 and 2, respectively, and $h = 1$. Rearranging these equations gives

$$\begin{bmatrix} -4 & 1 \\ 1 & -4 \end{bmatrix} \begin{bmatrix} T_1 \\ T_2 \end{bmatrix} = \begin{bmatrix} -333.33 \\ -583.33 \end{bmatrix}$$

Solving this equation, we have $T_1 = 127.78$ and $T_2 = 177.78$.

If we require a more accurate solution of Laplace's equation, then we must use more nodes and the computation burden increases rapidly. The following MATLAB function `ellipgen` uses the finite difference approximation (6.12) to solve the general elliptic partial differential equations (6.34) through (6.37) for a rectangular domain only. The function is also limited to problems in which the boundary value is specified by values of the function $z(x, y)$, not its derivative. If the user calls the function with 10 arguments, the function solves (6.34) through (6.36); see Examples 6.4 and 6.5. Calling it with six arguments causes it to solve (6.37); see Example 6.6.

```
function [a,om] = ellipgen(nx,hx,ny,hy,G,F,bx0,bxn,by0,byn)
% Function either solves:
% nabla^2(z)+G(x,y)*z = F(x,y) over a rectangular region.
% Function call: [a,om]=ellipgen(nx,hx,ny,hy,G,F,bx0,bxn,by0,byn)
% hx, hy are panel sizes in x and y directions,
% nx, ny are number of panels in x and y directions.
% F and G are (nx+1,ny+1) arrays representing F(x,y), G(x,y).
% bx0 and bxn are row vectors of boundary conditions at x0 and xn
% each beginning at y0. Each is (ny+1) elements.
% by0 and byn are row vectors of boundary conditions at y0 and yn
% each beginning at x0. Each is (nx+1) elements.
% a is an (nx+1,ny+1) array of sol'ns, inc the boundary values.
% om has no interpretation in this case.
% or the function solves
% (nabla^2)z+lambda*G(x,y)*z = 0 over a rectangular region.
% Function call: [a,om]=ellipgen(nx,hx,ny,hy,G,F)
% hx, hy are panel sizes in x and y directions,
% nx, ny are number of panels in x and y directions.
% G are (ny+1,nx+1) arrays representing G(x,y).
```

```
% In this case F is a scalar and specifies the
% eigenvector to be returned in array a.
% Array a is an (ny+1,nx+1) array giving an eigenvector,
% including the boundary values.
% The vector om lists all the eigenvalues lambda.
nmax = (nx-1)*(ny-1); r = hy/hx;
a = zeros(ny+1,nx+1); p = zeros(ny+1,nx+1);
if nargin==6
    ncase = 0; mode = F;
end
if nargin==10
    test = 0;
    if F==zeros(nx+1,ny+1), test = 1; end
    if bx0==zeros(1,ny+1), test = test+1; end
    if bxn==zeros(1,ny+1), test = test+1; end
    if by0==zeros(1,nx+1), test = test+1; end
    if byn==zeros(1,nx+1), test = test+1; end
    if test==5
        disp('WARNING - problem has trivial solution, z = 0.')
        disp('To obtain eigensolution use 6 parameters only.')
        return
    end
    bx0 = bx0(1,ny+1:-1:1); bxn = bxn(1,ny+1:-1:1);
    a(1,:) = byn; a(ny+1,:) = by0;
    a(:,1) = bx0'; a(:,nx+1) = bxn'; ncase = 1;
end
for i = 2:ny
    for j = 2:nx
        nn = (i-2)*(nx-1)+(j-1);
        q(nn,1) = i; q(nn,2) = j; p(i,j) = nn;
    end
end
C = zeros(nmax,nmax); e = zeros(nmax,1); om = zeros(nmax,1);
if ncase==1, g = zeros(nmax,1); end
for i = 2:ny
    for j = 2:nx
        nn = p(i,j); C(nn,nn) = -(2+2*r^2); e(nn) = hy^2*G(j,i);
        if ncase==1, g(nn) = g(nn)+hy^2*F(j,i); end
        if p(i+1,j)~=0
            np = p(i+1,j); C(nn,np) = 1;
        else
            if ncase==1, g(nn) = g(nn)-by0(j); end
        end
```

```
            if p(i-1,j)~=0
                np = p(i-1,j); C(nn,np) = 1;
            else
                if ncase==1, g(nn) = g(nn)-byn(j); end
            end
                if p(i,j+1)~=0
                    np = p(i,j+1); C(nn,np) = r^2;
                else
                if ncase==1, g(nn) = g(nn)-r^2*bxn(i); end
            end
            if p(i,j-1)~=0
                np = p(i,j-1); C(nn,np) = r^2;
            else
                if ncase==1, g(nn) = g(nn)-r^2*bx0(i); end
            end
        end
end
if ncase==1
    C = C+diag(e); z = C\g;
    for nn = 1:nmax
        i = q(nn,1); j = q(nn,2); a(i,j) = z(nn);
    end
else
    [u,lam] = eig(C,-diag(e));
    [om,k] = sort(diag(lam)); u = u(:,k);
    for nn = 1:nmax
        i = q(nn,1); j = q(nn,2);
        a(i,j) = u(nn,mode);
    end
end
```

We now give examples of the application of the `ellipgen` function.

Example 6.4

Use the function `ellipgen` to solve Laplace's equation over a rectangular region subject to the boundary conditions shown in Figure 6.12. The following script calls the function to solve this problem using a 12×12 mesh. The example is the same as Example 6.3, but a finer mesh is used in the solution.

```
% e3s606.m
Lx = 3; Ly = 2;
nx = 12; ny = 12; hx = Lx/nx; hy = Ly/ny;
```

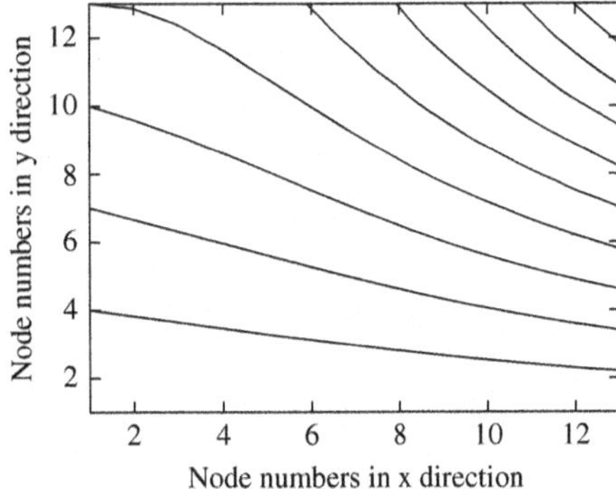

FIGURE 6.13 Finite difference estimate for the temperature distribution for the problem defined in Figure 6.12.

```
by0 = 0*[0:hx:Lx];
byn = 200+(100/3)*[0:hx:Lx].^2;
bx0 = 100*[0:hy:Ly];
bxn = 250*[0:hy:Ly];
F = zeros(nx+1,ny+1); G = F;
a = ellipgen(nx,hx,ny,hy,G,F,bx0,bxn,by0,byn);
aa = flipud(a); contour(aa,'k')
xlabel('Node numbers in x direction');
ylabel('Node numbers in y direction');
```

The output from this script is the contour plot shown in Figure 6.13. The temperature is not shown on the contour plot; if required, it can be obtained from aa.

Example 6.5

Poisson's equation. Determine the deflection of a uniform square membrane, held at its edges and subject to a distributed load, which can be approximated to a unit load at each node. This problem is described by Poisson's equation, (6.35), where $F(x,y)$ specifies the load on the membrane. We use the following script to determine the deflection of this membrane using the MATLAB function `ellipgen`.

```
% e3s607.m
Lx = 1; Ly = 1;
nx = 18; ny = 18; hx = Lx/nx; hy = Ly/ny;
by0 = zeros(1,nx+1); byn = zeros(1,nx+1);
bx0 = zeros(1,ny+1); bxn = zeros(1,ny+1);
F = -ones(nx+1,ny+1); G = zeros(nx+1,ny+1);
a = ellipgen(nx,hx,ny,hy,G,F,bx0,bxn,by0,byn);
surfl(a)
axis([1 nx+1 1 ny+1 0 0.1])
```

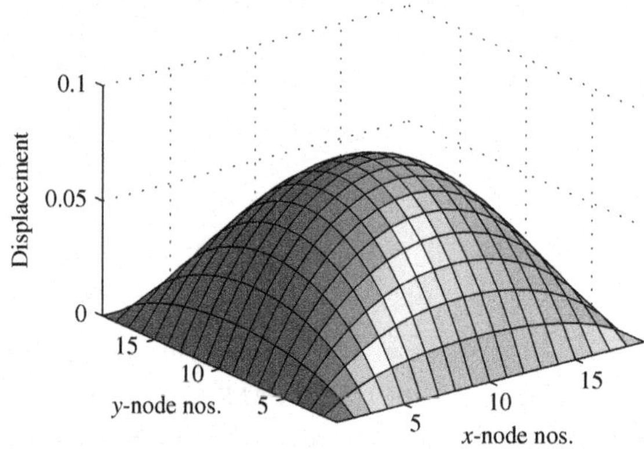

FIGURE 6.14 Deflection of a square membrane subject to a distributed load.

```
xlabel('x-node nos.'), ylabel('y-node nos.')
zlabel('Displacement')
max_disp = max(max(a))
```

Running this script gives the output shown in Figure 6.14 together with

```
max_disp =
    0.0735
```

This compares with the exact value of 0.0737.

Example 6.6

Characteristic value problem. Determine the natural frequencies and mode shapes of a freely vibrating square membrane held at its edges. This problem is described by the eigenvalue problem (6.37). The natural frequencies are related to the eigenvalues, and the mode shapes are the eigenvectors. The following MATLAB script determines the eigenvalues and vectors. It calls the function `ellipgen` and outputs a list of eigenvalues and provides Figure 6.15, showing the second mode shape of the membrane.

```
% e3s608.m
Lx = 1; Ly = 1.5;
nx = 20; ny = 30; hx = Lx/nx; hy = Ly/ny;
G = ones(nx+1,ny+1); mode = 2;
[a,om] = ellipgen(nx,hx,ny,hy,G,mode);
eigenvalues = om(1:5), surf(a)
view(140,30)
axis([1 nx+1 1 ny+1 -1.2 1.2])
xlabel('x - node nos.'), ylabel('y - node nos.')
zlabel('Relative displacement')
```

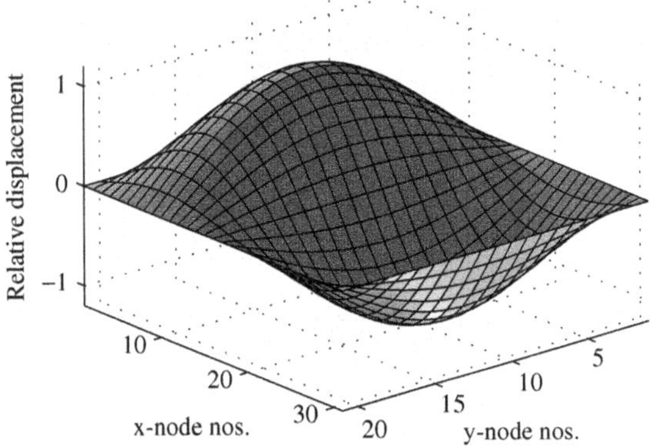

FIGURE 6.15 Finite difference approximation of the second mode of vibration of a uniform rectangular membrane.

Table 6.1 Finite Difference Approximations versus Exact Eigenvalues for Uniform Rectangular Membrane

FD Approximation	Exact	Error (%)
14.2318	14.2561	0.17
27.3312	27.4156	0.31
43.5373	43.8649	0.75
49.0041	49.3480	0.70
56.6367	57.0244	0.70

Running this script gives

```
eigenvalues =
   14.2318
   27.3312
   43.5373
   49.0041
   56.6367
```

These eigenvalues compare with the exact values given in Table 6.1.

6.8 Summary

In this chapter, we examined the application of finite difference methods to a broad range of second-order ordinary and partial differential equations. A major problem in the development of scripts is the difficulty of accounting for the wide variety of boundary conditions and boundary shapes that can occur. Software packages have been developed to solve partial differential equations that arise in computational fluid dynamics and continuum mechanics, using either finite difference or finite element methods; however, they are both complex and expensive because they allow the user total freedom to define boundary shapes and conditions.

Problems

6.1. Classify the following second-order partial differential equations:

$$\frac{\partial^2 y}{\partial t^2} + a\frac{\partial^2 y}{\partial x \partial t} + \frac{1}{4}(a^2 - 4)\frac{\partial^2 y}{\partial x^2} = 0$$

$$\frac{\partial u}{\partial t} - \frac{\partial}{\partial x}\left(A(x,t)\frac{\partial u}{\partial x}\right) = 0$$

$$\frac{\partial^2 \varphi}{\partial x^2} = k\frac{\partial^2 (\varphi^2)}{\partial y^2} \quad \text{where} \quad k > 0$$

6.2. Use the shooting method to solve $y'' + y' - 6y = 0$, where the prime denotes differentiation with respect to x, given the boundary conditions $y(0) = 1$ and $y(1) = 2$. Note that an illustrative script for the shooting method is given in Section 6.2. Use trial slopes in the range $-3:0.5:2$. Compare your results with those you obtain using the finite difference method with 10 divisions. The finite difference method is implemented by the function `twopoint`. Note that the exact solution is

$$y = 0.2657\exp(2x) + 0.7343\exp(-3x)$$

6.3. **(a)** Use the shooting method to solve $y'' - 62y' + 120y = 0$, where the prime denotes differentiation with respect to x, given the boundary conditions $y(0) = 0$ and $y(1) = 2$. Solve this equation by applying the shooting method, using trial slopes in the range $-0.5:0.1:0.5$. Note that the exact solution is

$$y = 1.751302152539304 \times 10^{-26}\{\exp(60x) - \exp(2x)\}$$

(b) By substituting $x = 1 - p$ in the original differential equation, show that $y'' + 62y' + 120y = 0$, where the prime denotes differentiation with respect to p. Note that the boundary conditions of this problem are $y(0) = 2$ and $y(1) = 0$. Solve this equation by applying the shooting method, using trial slopes in the range 0 to -150 in steps of -30 at $p = 0$. Note that a very good approximation to the solution is $y = 2\exp(-60p)$.

Compare the two answers you obtain for (a) and (b). Note that an illustrative script for the shooting method is given in Section 6.2. Also solve (a) and (b) using the finite difference method, implemented in `twopoint`. Use 10 divisions and repeat with 50 divisions. You should plot your answers and compare with a plot of the exact solution.

6.4. Solve the boundary value problem $xy'' + 2y' - xy = e^x$ given that $y(0) = 0.5$ and $y(2) = 3.694528$, using the finite difference method implemented by the function `twopoint`. Use 10 divisions in the finite difference solution and plot the results, together with the exact solution, $y = \exp(x)/2$.

6.5. Determine the finite difference equivalence of the characteristic value problem defined by $y'' + \lambda y = 0$, where $y(0) = 0$ and $y(2) = 0$. Use 20 divisions in the finite difference method. Then solve the finite difference equations using the MATLAB function `eig` to determine the dominant value of λ, that is, the dominant eigenvalue.

6.6. Solve the parabolic equation (6.24) with $K = 1$, subject to the following boundary conditions: $u(0, t) = 0$, $u(1, t) = 10$, and $u(x, 0) = 0$ for all x except $x = 1$. When $x = 1$, $u(1, 0) = 10$. Use the function `heat` to determine the solution for $t = 0$ to 0.5 in steps of 0.01 with 20 divisions of x. You should plot the solution for ease of visualization.

6.7. Solve the wave equation (6.29) with $c = 1$, subject to the following boundary and initial conditions: $u(t, 0) = u(t, 1) = 0$, $u(0, x) = \sin(\pi x) + 2\sin(2\pi x)$, and $u_t(0, x) = 0$, where the subscript t denotes partial differentiation with respect to t. Use the function `fwave` to determine the solution for $t = 0$ to 4.5 in steps of 0.05, and use 20 divisions of x. Plot your results and compare with a plot of the exact solution, which is given by

$$u = \sin(\pi x)\cos(\pi t) + 2\sin(2\pi x)\cos(2\pi t)$$

6.8. Solve the equation

$$\nabla^2 V + 4\pi^2(x^2 + y^2)V = 4\pi \cos\{\pi(x^2 + y^2)\}$$

over the square region $0 \le x \le 0.5$ and $0 \le y \le 0.5$. The boundary conditions are

$$V(x, 0) = \sin(\pi x^2), \quad V(x, 0.5) = \sin\{\pi(x^2 + 0.25)\}$$
$$V(0, y) = \sin(\pi y^2), \quad V(0.5, y) = \sin\{\pi(y^2 + 0.25)\}$$

Use the function `ellipgen` to solve this equation with 15 divisions of x and y. Plot your results and compare with a plot of the exact solution, which is given by $V = \sin\{\pi(x^2 + y^2)\}$.

6.9. Solve the eigenvalue problem $\nabla^2 z + \lambda G(x, y)z = 0$ over a rectangular region bounded by $0 \le x \le 1$ and $0 \le y \le 1.5$, $z = 0$ at all boundaries. Use the function `ellipgen` with six divisions in both directions. The function $G(x, y)$ over this grid is given by the MATLAB statements `G = ones(10,7); G(4:7,3:5) = 3*ones(4,3);`.

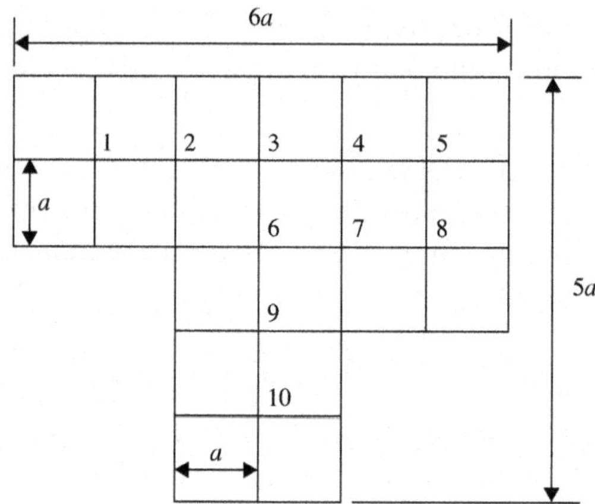

FIGURE 6.16 Region for Problem 6.10.

This represents a membrane with a central area thicker than its periphery. The eigenvalues are related to the natural frequencies of this membrane.

6.10. Solve Poisson's equation $\nabla^2\phi + 2 = 0$ over the region in Figure 6.16 with $a = 1$ at the boundary $\phi = 0$. You will have to assemble the finite difference equation by hand, applying (6.13) to the 10 nodes, and then use MATLAB to solve the resulting linear equation system.

7 Fitting Functions to Data

In this chapter we consider a variety of methods for fitting functions to data, describe some of the MATLAB functions that are available for this purpose, and develop some additional ones. The application of these functions is illustrated by appropriate examples.

7.1 Introduction

We fit functions to two general classes of data: data that is exact and data that we know contains errors. When we fit a function to exact data, we fit to the points exactly. When we fit a function to data that is known to contain errors, we try to obtain the best fit to the trend of the data, using some criterion. The user must exercise skill in making a sensible choice of the function to fit.

We begin by examining polynomial interpolation, which is an example of fitting to exact data.

7.2 Interpolation Using Polynomials

Suppose y is some unknown function of x. Given a table of values of x and y, we may wish to obtain a value of y corresponding to a value of x that is not tabulated. Interpolation implies that the untabulated value of x is within the range of the tabulated data. If the untabulated x is outside this range, the process is called *extrapolation* and is often less accurate.

The simplest form of interpolation is linear interpolation. In this method only the pair of data points enclosing the required value are used. Thus if (x_0,y_0) and (x_1,y_1) are two adjacent data points in a tabulation, to obtain the value of y corresponding to an x where $x_0 < x < x_1$, we fit the straight line $y = ax + b$ to these points and evaluate y as follows:

$$y = [y_0(x_1 - x) + y_1(x - x_0)]/(x_1 - x_0) \tag{7.1}$$

We may use the MATLAB function `interp1` for this purpose. For example, consider the function $y = x^{1.9}$, tabulated at $x = 1,2,\ldots,5$. If we require estimates of y for $x = 2.5$ and 3.8, we may use `interp1` setting the third parameter as `'linear'` to obtain linear interpolation as follows:

```
>> x = 1:5;
>> y = x.^1.9;
>> interp1(x,y,[2.5,3.8],'linear')

ans =
        5.8979    12.7558
```

The exact answers are $y = 5.7028$ and $y = 12.6354$ corresponding to $x = 2.5$ and 3.8, respectively. For some applications this may be sufficiently accurate.

Interpolation becomes more accurate when more of the tabulated data values are used because we can use a higher-degree polynomial. A polynomial of degree n can be adjusted to pass through $n+1$ data points. We do not need to know the coefficients of the polynomial explicitly, but they are used implicitly in the procedure to estimate y for a given value of x. For example, MATLAB allows cubic interpolation by calling `interp1` with the third parameter set as `'cubic'`. The following example implements cubic interpolation using the same data as the previous example.

```
>> interp1(x,y,[2.5 3.8],'cubic')

ans =
        5.6938    12.6430
```

The cubic interpolation gives a much more accurate result.

An algorithm that provides an efficient method for fitting any degree polynomial to data is Aitken's algorithm. In this procedure a sequence of polynomial functions are fitted to the data. As the degree of the polynomial is increased, more of the data points are used and the accuracy of the interpolation improves.

Aitken's algorithm proceeds as follows. Suppose we have five pairs of data values labeled 1, 2, …, 5 and we wish to determine y^*, the value of y corresponding to a given x^*. Initially the algorithm determines straight lines (i.e., first-degree polynomials) that pass through data points 1 and 2, 1 and 3, 1 and 4, and 1 and 5, as shown in Figure 7.1(a). These four straight lines allow the procedure to determine four, probably poor, estimates for y^*.

Using $x_2, x_3, \ldots, x_5$ from the tabulated data and the four estimates of y^* determined from the first-degree polynomial, the algorithm repeats the preceding procedure using these new points, but this now provides second-degree polynomials through the sets of data points $\{1,2,3\}$, $\{1,2,4\}$, and $\{1,2,5\}$, as shown in Figure 7.1(b). From these second-degree polynomials, the procedure determines three improved estimates for y^*.

Using x_3, x_4, x_5 from the tabulated data and the three new estimates for y^* obtained from the second-degree polynomials, the algorithm computes the third-degree polynomials passing through the sets of data points $\{1, 2, 3, 4\}$ and $\{1, 2, 3, 5\}$, as shown in

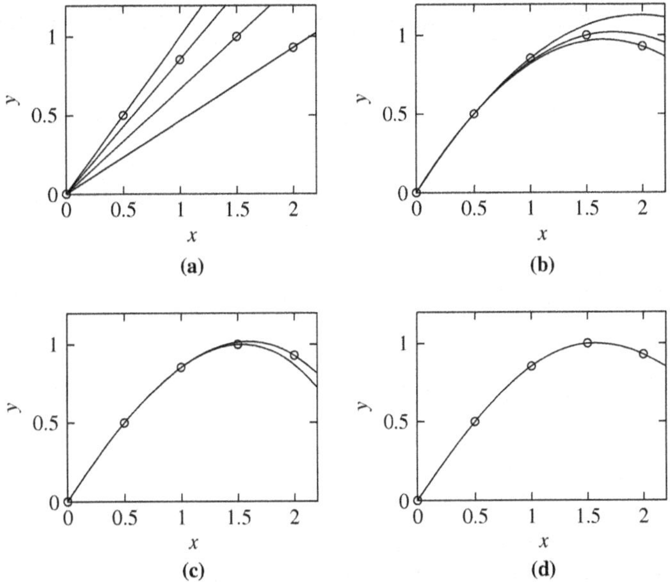

FIGURE 7.1 Increasing the degree of the polynomial fit: (a) 1st degree, (b) 2nd degree, (c) 3rd degree, and (d) 4th degree.

Figure 7.1(c) to allow the procedure to determine two further improved estimates for y^*. Finally a fourth-degree polynomial is computed that fits all the data. This fourth-degree polynomial provides the best estimate for y^*, as shown in Figure 7.1(d).

Aitken's algorithm has two advantages. First of all it is very efficient. Each new estimate for y^* requires only two multiplications and one division so that for $n+1$ data points, the estimate using all the data requires $n(n+1)$ multiplications and $n(n+1)/2$ divisions. It is interesting to note that if we attempted to determine the coefficients of a polynomial passing through $n+1$ data points by assembling a set of $n+1$ linear equations, then in addition to the computation required to assemble the equations, we would require $(n+1)^3/2$ multiplications and divisions to solve them. The second advantage of Aitken's algorithm is that the process of fitting higher- and higher-degree polynomials to more and more of the data can be terminated when we note that the estimate of y^* is no longer changing significantly.

The following MATLAB function `aitken` implements Aitken's algorithm. The user must provide a set of data for the vectors `x` and `y`. The function then determines a value of `y` corresponding to `xval`. The function provides the best value obtained and, if required, a table showing all the intermediate values obtained.

```
function [Q R] = aitken(x,y,xval)
% Aitken's method for interpolation.
% Example call: [Q R] = aitken(x,y,xval)
% x and y give the table of values. Parameter xval is
% the value of x at which interpolation is required.
```

```
% Q is interpolated value, R gives table of intermediate results.
n = length(x); P = zeros(n);
P(1,:) = y;
for j = 1:n-1
    for i = j+1:n
        P(j+1,i) = (P(j,i)*(xval-x(j))-P(j,j)*(xval-x(i)))/(x(i)-x(j));
    end
end
Q = P(n,n); R = [x.' P.'];
```

We now use this function to determine the reciprocal of 1.03 from a table of 10 equispaced values of x in the range 1 to 2 and $y = 1/x$. The following script calls the function `aitken` to solve this example:

```
% e3s701.m
x = 1:.2:2; y = 1./x;
[interpval table] = aitken(x,y,1.03);
fprintf('Interpolated value= %10.8f\n\n',interpval)
disp('Table = ')
disp(table)
```

Running this script gives the following output:

```
interpolated value= 0.97095439

Table =
   1.0000   1.0000        0        0        0        0        0
   1.2000   0.8333   0.9750        0        0        0        0
   1.4000   0.7143   0.9786   0.9720        0        0        0
   1.6000   0.6250   0.9813   0.9723   0.9713        0        0
   1.8000   0.5556   0.9833   0.9726   0.9713   0.9710        0
   2.0000   0.5000   0.9850   0.9729   0.9714   0.9711   0.9710
```

Notice that the first column in this table contains the tabulated x values, the second column contains the tabulated y values, and the remaining columns give successively higher-degree polynomial interpolants generated by Aitken's method. The zeros in this table are padding: The number of estimates in each column decreases as the estimates use more of the data. The exact value is $y = 0.970873786$; thus the Aitken interpolated value of $y = 0.97095439$ is correct to four decimal places. Linear interpolation gives 0.9750, a much poorer result with an error of approximately 0.2%.

Aitken's method provides an interpolated value of y for a given value of x by fitting a polynomial to the data, but the coefficients of the polynomial are not determined explicitly. Conversely, we can fit a polynomial explicitly to the data, determine its coefficients, and then determine the required interpolated value by evaluating the polynomial. This approach may be less computationally efficient. The MATLAB function `polyfit(x,y,n)` fits a polynomial of degree n through the data given by x and y and returns the coefficients of descending powers of x. For an exact fit, n must equal $m-1$ where m is the number of data points. The polynomial represented by `p` can then be evaluated using the `polyval` function. For example, to determine the reciprocal of 1.03 from a table of six equispaced values of x in the range 1 to 2 and $y = 1/x$, we have

```
% e3s702.m
x = 1:.2:2; y = 1./x;
p = polyfit(x,y,5)
interpval = polyval(p,1.03);
fprintf('interpolated value = %10.8f\n',interpval)
```

Running this script gives

```
p =
   -0.1033    0.9301   -3.4516    6.7584   -7.3618    4.2282

interpolated value = 0.97095439
```

Thus

$$y = -0.1033x^5 + 0.9301x^4 - 3.4516x^3 + 6.7584x^2 - 7.3618x + 4.2282$$

The interpolated value is identical to that given by Aitken's method, as indeed it must be (except for possible rounding errors in the computation) because there is only one polynomial that passes through all six data points and both methods have used it. We use the MATLAB function `polyfit` again in Section 7.8.

7.3 Interpolation Using Splines

The spline is used to connect data points to each other using a curve that appears to the eye to be smooth, either for the purpose of visualization in design drawings or for interpolation. It has certain advantages over the use of a high-degree polynomial, which has a tendency to oscillate between data values.

We begin with an historical example of ship design. Ships' hulls have always curved in a complex manner in two dimensions. Figure 7.2 shows hull sections for a 74-gun British warship, circa 1813. The data points are taken from the original plans, and splines have

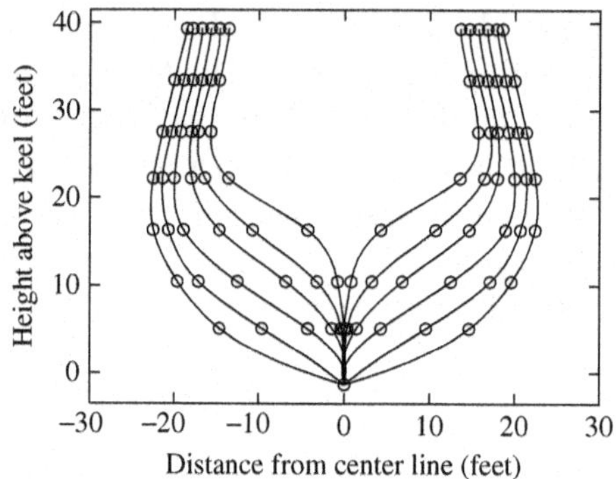

FIGURE 7.2 Use of splines to define cross-sections of a ship's hull.

been used to join the data points together smoothly. Each line shows a section of the ship; the innermost line is close to the stern, and the outermost is near amidships. The graph gives a clear impression of the way the ship builder chose to reduce the ship's cross-section toward the stern.

Polynomials of varying degrees are used for splines, but here we only consider the cubic spline. The cubic spline is a series of cubic polynomials joining data points or "knots." Suppose we have n data points joined by $n-1$ polynomials. Each cubic polynomial has four unknown coefficients so that there are $4(n-1)$ coefficients to be determined. Obviously each polynomial must pass through the two data points it joins. This provides $2(n-1)$ equations that must be satisfied. In order that the polynomials join together smoothly, we require both continuity of slope (y') and curvature (y'') between adjacent polynomials at the $n-2$ internal data points. This gives $2(n-2)$ extra equations, making a total of $4n-6$ equations. With these equations we can determine an identical number of coefficients uniquely, and so two further equations are required in order to determine *all* the unknown coefficients. The two remaining conditions can be chosen arbitrarily, but usually one of the following is used:

1. If the slope of the required curve is known at the outer ends, we can impose these two constraints. More often than not, these slopes are not known.
2. We can make the curvature at the outer ends zero, that is, $y_1'' = y_n'' = 0$. (These are called natural splines but have no particular advantage.)
3. We can make the curvature at x_1 and x_n equal to x_2 and x_{n-1}, respectively.
4. We can make the curvature at x_1 a linear extrapolation of the curvature at x_2 and x_3. Similarly, we make the curvature at x_n a linear extrapolation of the curvature at x_{n-1} and x_{n-2}.
5. We can make y''' continuous at x_2 and x_{n-1}. Since at any internal point, y, y', y'', and y''' are always made continuous, adding this condition is equivalent to using the same polynomial in the two outer panels. This is called the "not a knot" condition and is used in the MATLAB implementation of the function `spline`.

Table 7.1 Data for Spline Fit

x	0	1	2	3	4
y	3	1	0	2	4

We now illustrate the two uses of the MATLAB function `spline` applied to the small set of data given in Table 7.1. Running the script

```
% e3s703.m
x = 0:4; y = [3 1 0 2 4];
xval = 1.5; yval = spline(x,y,xval)
p = spline(x,y)
```

gives

```
yval =
    0.1719

p =
      form: 'pp'
    breaks: [0 1 2 3 4]
     coefs: [4x4 double]
    pieces: 4
     order: 4
       dim: 1
```

where `yval` is the interpolated value. There may be some occasions when the user wishes to know the values of the coefficient of the polynomials. In this case the p-p form is required, where the abbreviation p-p means piecewise polynomial. The variable `p` is a structure array that provides this information. In particular,

```
>> c = p.coefs

c =
    0.5417   -1.1250   -1.4167    3.0000
    0.5417    0.5000   -2.0417    1.0000
   -0.7083    2.1250    0.5833         0
   -0.7083   -0.0000    2.7083    2.0000
```

In this example, the coefficients of the polynomials are combined with the powers of x as follows:

$$
\begin{aligned}
y &= c_{11}x^3 + c_{12}x^2 + c_{13}x + c_{14}, && 0 \le x \le 1 \\
y &= c_{21}(x-1)^3 + c_{22}(x-1)^2 + c_{23}(x-1) + c_{24}, && 1 \le x \le 2 \\
y &= c_{31}(x-2)^3 + c_{32}(x-2)^2 + c_{33}(x-2) + c_{34}, && 2 \le x \le 3 \\
y &= c_{41}(x-3)^3 + c_{42}(x-3)^2 + c_{43}(x-3) + c_{44}, && 3 \le x \le 4
\end{aligned}
$$

It is not necessary for the MATLAB user to know the details of how the p-p values are interpreted. MATLAB provides a function `ppval` that evaluates a composite polynomial provided its p-p values are known. If `x` and `y` are vectors of data, then `y1 = spline(x,y,x1)` is equivalent to the statements `p = spline(x,y); y2 = ppval(p,x1)`.

The following script gives a plot of the spline fit to the data of Table 7.1.

```
% e3s704.m
x = 0:4; y = [3 1 0 2 4];
xx = 0:.1:4; yy = spline(x,y,xx);
plot(x,y,'o',xx,yy)
axis([0 4 -1 4])
xlabel('x'), ylabel('y')
```

Running this script generates Figure 7.3.

Section 7.2 described how polynomials are used in interpolation. However, their use is not always appropriate. When the data points are widely spaced, and when there are sudden changes in the y values, then polynomials can give very poor results. For example, the nine data points in Figure 7.4 are taken from the function

$$y = 2\{1 + \tanh(2x)\} - x/10$$

This function changes abruptly, and if an eighth-degree polynomial is fitted to the data it oscillates and the path between data points bears no relationship to the true path. In contrast the spline fit is reasonably smooth and close to the true function.

The reader should note that the MATLAB function `interp1` can also be used to fit splines to data. The call `interp1(x,y,xi,'spline')` is identical to `spline(x,y,xi)`.

A special type of spline is the Bézier curve. This is a cubic function defined by four points. The two end points are used, together with two "control" points. The slope of the curve at one end is a tangent to the line between that end point and one of the control points. Similarly, the slope at the other end point is a tangent to the line between that end

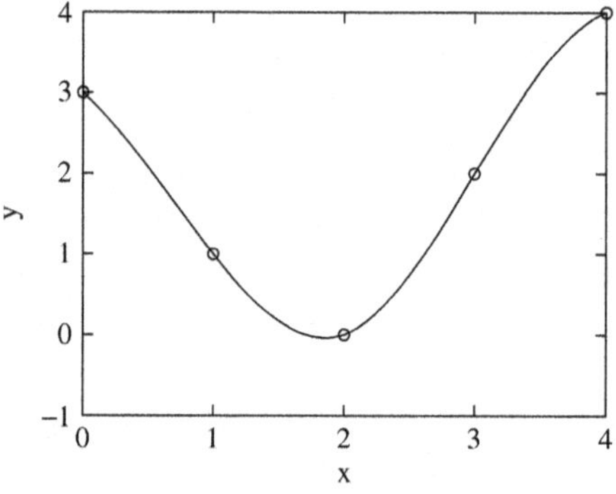

FIGURE 7.3 Spline fit to the data of Table 7.1 (denoted by ○).

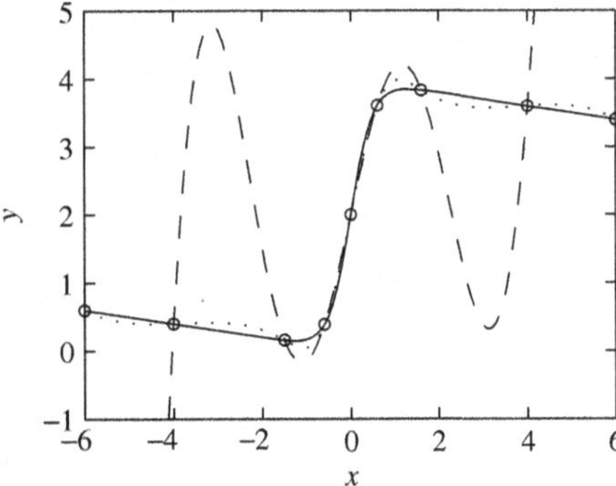

FIGURE 7.4 The *solid curve* shows the function $y = 2\{1 + \tanh(2x)\} - x/10$. The *dashed line* shows an eighth-degree polynomial fit; the *dotted line* shows a spline fit.

point and the other control point. In interactive computer graphics, the positions of the control points can be moved on the screen in order to adjust the slope of the curve at the end points.

7.4 Fourier Analysis of Discrete Data

Fourier analysis in its various forms is an important tool for the scientist or engineer engaged in the interpretation of data where a knowledge of the frequencies present in the data or function may give some insight into the mechanism that has generated it. For continuous periodic functions, the frequency content is determined from the coefficients of the terms in the well-known Fourier series; for nonperiodic functions, it is determined from the Fourier integral transform. In an analogous manner, the frequency content of a sequence of data can be determined by Fourier analysis, in this case from the *discrete* Fourier transform (DFT). The harmonic functions fit the data exactly, but here the purpose is most likely to determine the harmonic content of the data rather than to interpolate new data values.

The data can come from many sources. For example, the radial forces acting at discrete points around a cylinder constitute a sequence of data that must be periodic. The most frequently occurring form of data is a time series in which the value of some quantity is given at equal intervals of time—for example, data sampled from a signal from a transducer—and for this reason the analysis that follows is developed in terms of the independent variable t, which represents time. It must be stressed, however, that the DFT can be applied to any data regardless of the domain from which it originates. Determining the DFT for a sequence of data points is straightforward, although the computation is tedious.

We begin by defining a periodic function. A function $y(t)$ is periodic if it has the property that for any value of time t, $y(t) = y(t + T)$ where T is the time period, typically measured in seconds. The reciprocal of the period is equal to the frequency, denoted by f and measured in cycles/second. In the SI system of units, 1 Hertz (Hz) is defined as 1 cycle/second.

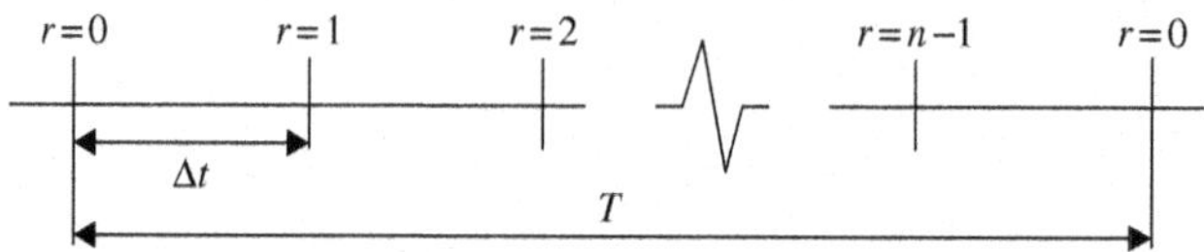

FIGURE 7.5 Numbering scheme for data points.

If we are concerned with a periodic function $z(x)$, where x is a spatial variable, then for any value of x, $z(x) = z(x+X)$ where X is the spatial period or wavelength, typically measured in meters. The frequency $f = 1/X$ is then measured in cycles/meter.

We now examine how to fit a finite set of trigonometric functions to n data points (t_r, y_r) where $r = 0, 1, 2, \ldots, n-1$. We assume the data points are equispaced and the number of data points, n, is even. Data values may be complex but in most practical situations they are real. The data points are numbered as shown in Figure 7.5. The point following the $(n-1)$th is assumed to equal the value of the zero point. Thus the DFT assumes the data is periodic with a period T equal to the range of the data.

Let the relationship between y_r and t_r be given by a finite set of sine and cosine functions as follows:

$$y_r = \frac{1}{n}\left[A_0 + \sum_{k=1}^{m-1}\{A_k \cos(2\pi k t_r/T) + B_k \sin(2\pi k t_r/T)\} + A_m \cos(2\pi m t_r/T)\right] \tag{7.2}$$

where $r = 0, 1, 2, \ldots, n-1$, $m = n/2$ and T is the range of the data as shown in Figure 7.5. The n coefficients A_0, A_m, A_k, and B_k (where $k = 1, 2, \ldots, m-1$) must be determined. Since we have n data values and n unknown coefficients, (7.2) can be made to fit the data exactly. The factor $1/n$ in (7.2) is omitted by some authors, and omitting it has the effect of reducing the size of the coefficients A_0, A_m, A_k, and B_k by the factor n. The reason for choosing $m+1$ coefficients multiplied by a cosine function (including $\cos 0$, which equals one and in fact multiplies A_0) and $m-1$ coefficients multiplied by a sine function in (7.2) will become apparent.

Each sine or cosine term of (7.2) represents k complete cycles in the range of the data T. Thus the period of each sine term is T/k, where $k = 1, 2, \ldots, (m-1)$, and the period of each cosine term is T/k, where $k = 1, 2, \ldots, m$. The corresponding frequencies are given by k/T. Thus the frequencies present in (7.2) are $1/T$, $2/T, \ldots,$ m/T. Letting Δf be the frequency increment between components and f_{max} be the maximum frequency, then

$$\Delta f = 1/T \tag{7.3}$$

and

$$f_{max} = m\Delta f = (n/2)\Delta f = n/(2T) \tag{7.4}$$

The data values t_r are equally spaced in the range T and may be expressed as

$$t_r = rT/n, \quad r = 0,1,2,\ldots,n-1 \tag{7.5}$$

Letting Δt be the sampling interval (see Figure 7.5), then

$$\Delta t = T/n \tag{7.6}$$

Let T_0 be the period corresponding to f_{max}, the maximum frequency in (7.2). Then, from (7.4),

$$f_{max} = 1/T_0 = n/(2T)$$

Thus $T = T_0 n/2$. Substituting this relationship in (7.6), we have $\Delta t = T_0/2$. This tells us that the maximum frequency component in the DFT contains two samples of data per cycle. The maximum frequency, f_{max}, is called the Nyquist frequency, and the corresponding sampling rate is called the Nyquist sampling rate.

A harmonic with a frequency that is exactly equal to the Nyquist frequency cannot be properly detected because at this frequency the DFT has a cosine term but no corresponding sine term. This result has an important implication when data is sampled from a continuously varying function or signal. It implies that there must be *more than two* data samples per cycle at the highest frequency *present in the function or signal.* If there are frequencies in the signal higher than the Nyquist frequency, then, because of the periodic nature of the DFT itself, they appear as frequency components in the DFT at a lower frequency. This phenomenon is called "aliasing." For example, if data is sampled at 0.005 second intervals, that is, 200 samples per second, then the Nyquist frequency, f_{max}, is 100 Hz. A frequency of 125 Hz in this signal would appear as a frequency component at 75 Hz. A frequency of 225 Hz would appear as 25 Hz. The relationship between frequencies in a signal and the frequency components in the DFT is shown in Figure 7.6. Frequency aliasing should be avoided because it makes it difficult or impossible to relate the frequency components in the DFT to their physical causes.

We now return to the task of determining the n coefficients A_0, A_m, A_k, and B_k in (7.2). Replacing $t_r = rT/n$ in (7.2), we obtain

$$y_r = \frac{1}{n}\left[A_0 + \sum_{k=1}^{m-1}\{A_k \cos(2\pi kr/n) + B_k \sin(2\pi kr/n)\} + A_m \cos(\pi r)\right] \tag{7.7}$$

where $r = 0,1,2,\ldots,n-1$. It was previously noted that the coefficients B_0 and B_m are absent from (7.2). It is now clear that had we introduced these coefficients they would be multiplied by $\sin(0)$ and $\sin(\pi r)$, both of which are zero.

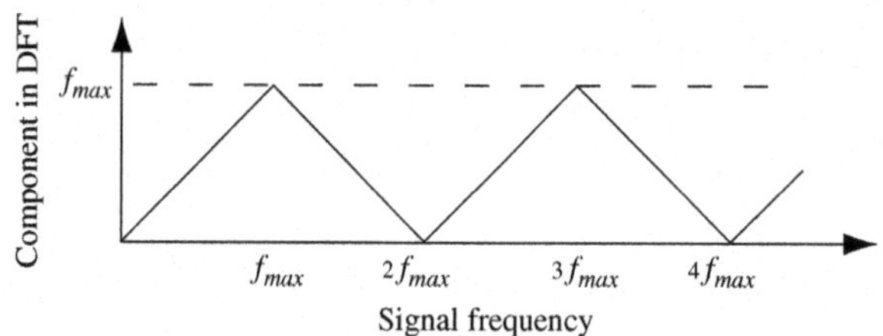

FIGURE 7.6 Relationship between a signal frequency and its component in the DFT derived by sampling using a Nyquist frequency f_{max}.

In (7.7) the n unknown coefficients are real. However, (7.7) can be expressed more concisely in terms of complex exponentials with complex coefficients. Using the fact that

$$\cos(2\pi kr/n) = \{\exp(\imath 2\pi kr/n) + \exp(-\imath 2\pi kr/n)\}/2$$
$$\sin(2\pi kr/n) = \{\exp(\imath 2\pi kr/n) - \exp(-\imath 2\pi kr/n)\}/2\imath$$

and

$$\exp\{\imath 2\pi(n-k)r/n\} = \exp(-\imath 2\pi kr/n)$$

where $k = 1,2,\ldots,m-1$ and $\imath = \sqrt{-1}$, then it can be shown that (7.7) reduces to

$$y_r = \frac{1}{n}\sum_{k=0}^{n-1} Y_k \exp(\imath 2\pi kr/n), \quad r = 0,\ 1\ 2,\ldots,\ n-1 \tag{7.8}$$

In (7.8),

$$Y_0 = A_0 \quad \text{and} \quad Y_m = A_m, \quad \text{where} \quad m = n/2$$
$$Y_k = (A_k - \imath B_k)/2 \quad \text{and} \quad Y_{n-k} = (A_k + \imath B_k)/2, \quad \text{for} \quad k = 1,2,\ldots,m-1$$

Note that if y_r is real, then A_k and B_k are also real, so that Y_{n-k} is the complex conjugate of Y_k, for $k = 1,2,\ldots,(n/2-1)$. To find the values of the unknown complex coefficients of (7.8) we make use of the following orthogonal property of exponential functions sampled at n equispaced points:

$$\sum_{r=0}^{n-1} \exp(\imath 2\pi rj/n)\exp(\imath 2\pi rj/n) = \begin{cases} 0 & \text{if } |j-k| \neq 0,\ n,\ 2n \\ n & \text{if } |j-k| = 0,\ n,\ 2n \end{cases} \tag{7.9}$$

Multiplying (7.8) by $\exp(-\imath 2\pi rj/n)$, summing over the n values of r, and then using (7.9), an expression for the unknown coefficients can be found:

$$Y_k = \sum_{r=0}^{n-1} y_r \exp(\imath 2\pi kr/n) \quad k = 0,\ 1,\ 2,\ldots,\ n-1 \tag{7.10}$$

If we let $W_n = \exp(-\iota 2\pi/n)$, where W_n is a complex constant, then (7.10) becomes

$$Y_k = \sum_{r=0}^{n-1} y_r W_n^{kr} \quad k = 0,\ 1,\ 2, \ldots,\ n-1 \tag{7.11}$$

Note that in (7.11) W_n is raised to the power kr. Alternatively, we can write (7.11) in matrix notation, giving

$$\mathbf{Y} = \mathbf{W}\mathbf{y} \tag{7.12}$$

where W_n^{kr} is the element of the $(k+1)$th row, $(r+1)$th column of **W** since k and r both start at zero. Note that **W** is an $n \times n$ array of complex coefficients. **Y** is a *vector* of the complex Fourier coefficients and in this instance we are departing from our usual convention that emboldened uppercase letters represent arrays.

We can obtain the coefficients Y_k from the equispaced data (t_r, y_r) by using (7.10), (7.11), or (7.12). These equations are alternative statements of the DFT. Furthermore, by replacing k in (7.10) by $k + np$, where p is any integer, it can be shown that $Y_{k+np} = Y_k$. Thus the DFT is periodic over the range n. The inverse of the DFT is called the inverse discrete Fourier transform (IDFT) and is implemented by (7.8). By replacing r in (7.8) by $r + np$, where p is an integer, it can be shown that the IDFT is also periodic over the range n. Both y_r and Y_k may be complex, although, as previously stated, the samples y_r are usually real. These transforms constitute a pair: If the data values are transformed by the DFT to determine the coefficients Y_k, then they can be recovered in their entirety by means of the IDFT.

To evaluate the coefficients of the DFT it would appear convenient to use (7.12). Although using these equations is satisfactory for small sequences of data, calculating the DFT for n real data points requires $2n^2$ multiplications. Thus, to transform a sequence of 4096 data points would require approximately 33 million multiplications. In 1965 this situation was dramatically changed with the publication of the fast Fourier transform (FFT) algorithm (Cooley and Tukey, 1965). The FFT algorithm is extremely efficient, and approximately $2n \log_2 n$ multiplications are required to compute the FFT for real data. With this development, allied to the developments in computing hardware that have occurred in the past forty years, it is now possible to compute the FFT for a relatively large number of data points on a personal computer.

Many refinements have been made to the basic FFT algorithm since it was first formulated, and several variants have been developed. Here one of the simplest forms of the algorithm is outlined.

To develop the basic FFT algorithm one further restriction must be placed on the data. In addition to the data being equispaced, the number of data points must be an integer power of 2. This allows a sequence of data to be successively subdivided. For example, 16 data points can be divided into two sequences of 8, four sequences of 4, and finally eight sequences of only 2 data points. A crucial relationship on which the FFT algorithm is based is now developed from (7.10) as follows. Let y_r be the sequence of n data points for which we require the DFT. We can subdivide y_r into two sequences of $n/2$ data points u_r and v_r as follows:

$$\left.\begin{aligned} u_r &= y_{2r} \\ v_r &= y_{2r+1} \end{aligned}\right\} \tag{7.13}$$

Note that alternate points in the original data sequence are placed in different subsets. We now determine the DFTs of the data sets u_r and v_r from (7.10), with n replaced by $n/2$:

$$\left.\begin{aligned} U_k &= \sum_{r=0}^{n/2-1} u_r \exp\{-\imath 2\pi kr/(n/2)\} \\ V_k &= \sum_{r=0}^{n/2-1} v_r \exp\{-\imath 2\pi kr/(n/2)\} \end{aligned}\right\} k = 0,\ 1,\ 2, \ldots,\ n/2-1 \tag{7.14}$$

The DFT, Y_k, for the original data sequence y_r is given by using (7.10) as follows:

$$\begin{aligned} Y_k &= \sum_{r=0}^{n-1} y_r \exp(-\imath 2\pi kr/n) \\ &= \sum_{r=0}^{n/2-1} y_{2r} \exp\{-\imath 2\pi k 2r/n\} + \sum_{r=0}^{n/2-1} y_{2r+1} \exp\{-\imath 2\pi k(2r+1)/n\} \end{aligned}$$

where $k = 0, 1, 2, \ldots, n$. Substituting for y_{2r} and y_{2r+1} from (7.13), we have

$$Y_k = \sum_{r=0}^{n/2-1} u_r \exp\{-\imath 2\pi kr/(n/2)\} + \exp(-\imath 2\pi k/n) \sum_{r=0}^{n/2-1} v_r \exp\{-\imath 2\pi kr/(n/2)\}$$

Comparing this equation with (7.14), we see that

$$Y_k = U_k + \exp(-\imath 2\pi k/n) V_k = U_k + (W_n^k) V_k \tag{7.15}$$

where $W_n^k = \exp(-\imath 2\pi k/n)$ and $k = 0, 1, 2, \ldots, n/2-1$.

Equation (7.15) provides only half of the required DFT. However, using the fact that U_k and V_k are periodic in k, it can be proved that

$$Y_{k+n/2} = U_k - \exp(-\imath 2\pi k/n) V_k = U_k - (W_n^k) V_k \tag{7.16}$$

We can use (7.15) and (7.16) to determine efficiently the DFT of the original data from the DFTs of subsets composed of alternate points of the original data. Of course, we can determine the DFTs of each subset of data by further subdividing these subsets until the final division leaves subsets consisting of a single data point. For a sequence of data comprising a single data point, we see from (7.10) with $n = 1$ that the DFT is equal to the value of the single data point. This is essentially how the FFT algorithm works.

In the preceding discussion we started from a sequence of data and continuously subdivided it (with alternate points in different subsets) until the subdivisions produced single

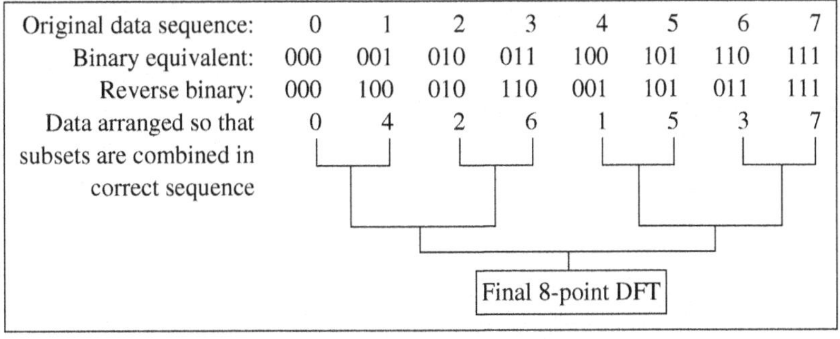

FIGURE 7.7 Stages in the FFT algorithm.

data points. What we require is a method of starting with single data points and ordering them in such a way that successively combining the DFTs of the subsets ultimately forms the required DFT of the original data. This can be achieved by the "bit reversed algorithm," and we illustrate it and the subsequent stages of the FFT by assuming a sequence of eight data points, y_0 to y_7. To determine the correct order for combining the data we express the subscript denoting the original position of each data point as a binary number and reverse the order of the digits (or bits). This reversed-order binary number determines the position of each data point in the reordered sequence and is shown for eight data points in Figure 7.7. The diagram also shows the stages of the FFT algorithm, which repeatedly uses (7.15) and (7.16) as follows:

Stage 1 Determine $\mathbf{Y}_{04}$ from Y_0 and Y_4, determine $\mathbf{Y}_{26}$ from Y_2 and Y_6, determine $\mathbf{Y}_{15}$ from Y_1 and Y_5, determine $\mathbf{Y}_{37}$ from Y_3 and Y_7.

Stage 2 Determine $\mathbf{Y}_{0246}$ from $\mathbf{Y}_{04}$ and $\mathbf{Y}_{26}$, determine $\mathbf{Y}_{1357}$ from $\mathbf{Y}_{15}$ and $\mathbf{Y}_{37}$.

Stage 3 Determine $\mathbf{Y}_{01234567}$ from $\mathbf{Y}_{0246}$ and $\mathbf{Y}_{1357}$.

Note that there are three stages in this process. For an n point DFT, the number of stages equals $\log_2 n$. In this small example, $n = 8$ and hence $\log_2 8 = \log_2 2^3 = 3$. Thus the process requires three stages.

MATLAB provides both the function `fft` to determine the DFT of a sequence of data values using the FFT algorithm, and the function `ifft` to determine the IDFT using a slight modification of the FFT algorithm. Thus to determine the DFT of the data in `y` we use the `fft` function as the following script illustrates:

```
% e3s705.m
v = 0:15;
y = [2.8 -0.77 -2.2 -3.1 -4.9 -3.2 4.83 -2.5 3.2 ...
    -3.6 -1.1 1.2 -3.2 3.3 -3.4 4.9];
s = sum(y), Y = fft(y);
[v' Y.']
```

Running this script gives the following results:

```
s =

    -7.7400

ans =

          0              -7.7400
     1.0000               3.2959 + 8.3851i
     2.0000              13.9798 +10.9313i
     3.0000               8.0796 - 6.6525i
     4.0000              -0.2300 + 4.7700i
     5.0000               4.3150 + 6.8308i
     6.0000              14.2202 + 1.4713i
     7.0000             -17.2905 +15.0684i
     8.0000              -0.2000
     9.0000             -17.2905 -15.0684i
    10.0000              14.2202 - 1.4713i
    11.0000               4.3150 - 6.8308i
    12.0000              -0.2300 - 4.7700i
    13.0000               8.0796 + 6.6525i
    14.0000              13.9798 -10.9313i
    15.0000               3.2959 - 8.3851i
```

We have already noted that for real data Y_{n-k} is the complex conjugate of Y_k, for $k = 1, 2, \ldots, (n/2 - 1)$. The preceding results illustrate this relationship and in this case $Y_{15}, Y_{14}, \ldots, Y_9$ are the complex conjugates of $Y_1, Y_2, \ldots, Y_7$, respectively, and provide no extra information. Note also that Y_0 is equal to the sum of the original data values y_r.

We now give examples of the use of the `fft` function to examine the frequency content of data sequences sampled from continuous functions.

Example 7.1

Determine the DFT of a sequence of 64 equispaced data points, sampled at intervals of 0.05 s from the function $y = 0.5 + 2\sin(2\pi f_1 t) + \cos(2\pi f_2 t)$, where $f_1 = 3.125$ Hz and $f_2 = 6.25$ Hz. The following script calls the `fft` function and displays the resulting DFT in various ways:

```
% e3s706.m
clf
nt = 64; dt = 0.05; T = dt*nt
df = 1/T, fmax = (nt/2)*df
t = 0:dt:(nt-1)*dt;
y = 0.5+2*sin(2*pi*3.125*t)+cos(2*pi*6.25*t);
f = 0:df:(nt-1)*df;  Y = fft(y);
figure(1)
```

```
subplot(121), bar(real(Y),'r')
axis([0 63 -100 100])
xlabel('Index k'), ylabel('real(DFT)')
subplot(122), bar(imag(Y),'r')
axis([0 63 -100 100])
xlabel('Index k'), ylabel('imag(DFT)')
fss = 0:df:(nt/2-1)*df;
Yss = zeros(1,nt/2); Yss(1:nt/2) = (2/nt)*Y(1:nt/2);
figure(2)
subplot(221), bar(fss,real(Yss),'r')
axis([0 10 -3 3])
xlabel('Frequency (Hz)'), ylabel('real(DFT)')
subplot(222), bar(fss,imag(Yss),'r')
axis([0 10 -3 3])
xlabel('Frequency (Hz)'), ylabel('imag(DFT)')
subplot(223), bar(fss,abs(Yss),'r')
axis([0 10 -3 3])
xlabel('Frequency (Hz)'), ylabel('abs(DFT)')
```

Running the preceding script gives

```
T =
    3.2000

df =
    0.3125

fmax =
    10
```

together with Figures 7.8 and 7.9. Note that in the script we have used the `bar` rather than the `plot` statement to emphasize the discrete nature of the DFT. Figure 7.8 shows the amplitudes of the 64 real and imaginary components of the DFT plotted against the index number k. Note that components 63 to 33 are the complex conjugates of components 1 to 31. While these plots display the DFT, the amplitude and frequency of the harmonic components in the original signal cannot easily be recognized. To achieve this the DFT must be scaled and displayed as shown in Figure 7.8. In the real part of the DFT there are components at $k = 0$, 20, and 44, each with an amplitude of 32, and in the imaginary part of the DFT there are components at $k = 10$ and 54, with amplitudes of 64 and −64, respectively. Since they contain no extra information, we ignore the components above $k = 32$ (i.e., $k = 44$ and 54) and consider only the components in the range $k = 0, 1, \ldots, 31$; in this case specifically $k = 0$, 10, and 20. We can convert the DFT index number to frequency by multiplying by Δf (= 0.3125 Hz) to give components at 0 Hz, 3.125 Hz, and 6.25 Hz, respectively. We now scale the DFT in the range $k = 1, 2, \ldots, 31$ by dividing it by $(n/2)$, in this case by 32.

The plots of the 31 scaled DFT components (most of which are zero) corresponding to frequencies in the range 0 to 9.6875 Hz are shown in Figure 7.9. We now see that the real component at 6.25 Hz has an amplitude of 1 and the imaginary component at 3.125 Hz has

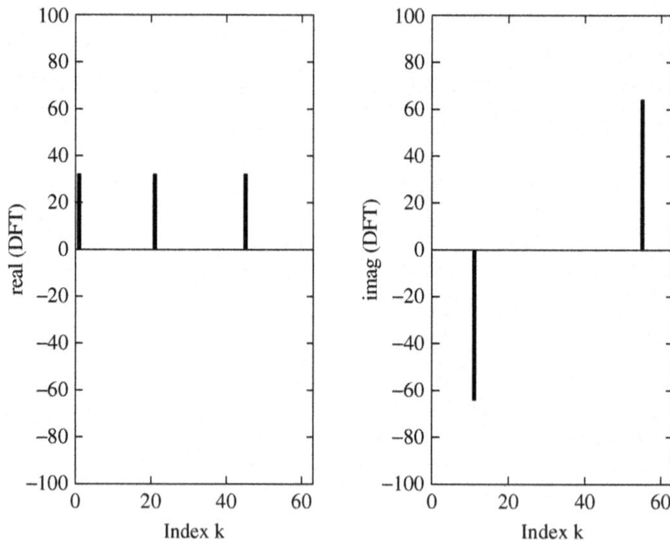

FIGURE 7.8 Plots of the real and imaginary part of the DFT.

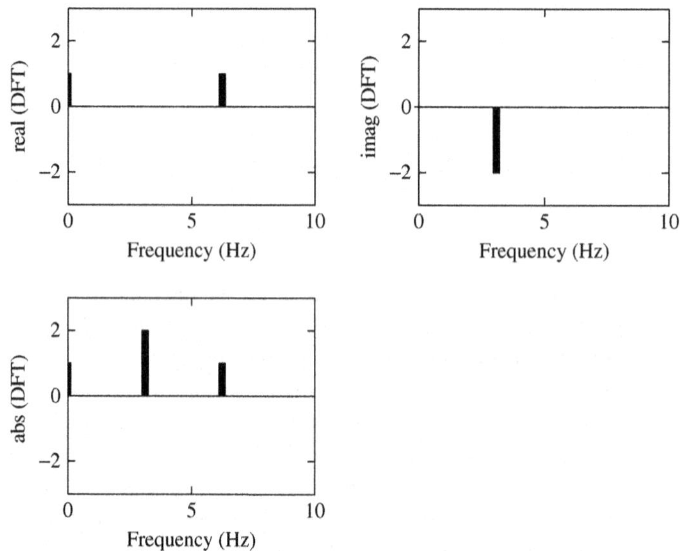

FIGURE 7.9 Frequency spectra.

an amplitude of −2. These components correspond, respectively, to the cosine component and the negative of the sine component in the original signal from which the data was sampled. If we only wish to know the amplitude of the frequency components, then we can display the absolute values of the scaled DFT. The component at $f = 0$ Hz is equal to *twice* the mean value of the data; in this case we have $2 \times 0.5 = 1$. These plots are called frequency spectra or periodograms. If sampling is over an integer number of cycles of the harmonics present in the signal, the amplitude of the components in the scaled DFT equal the amplitude of the corresponding harmonics, as shown in this example. If the sampling is not over an integer number of cycles of any harmonic present in the signal, then the component in the DFT closest to the frequency of the harmonic is reduced in amplitude and spread into other frequencies. This phenomenon is called "smearing" or "leakage" and is further discussed in Problem 7.15.

Example 7.2

We now determine the spectrum of a sequence of 512 data points sampled over a period of 2 seconds from the function

$$y = 0.2\cos(2\pi f_1 t) + 0.35\sin(2\pi f_2 t) + 0.3\sin(2\pi f_3 t) + \text{random noise}$$

where $f_1 = 20$ Hz, $f_2 = 50$ Hz, and $f_3 = 70$ Hz. The random noise is normally distributed with a standard deviation of 0.5 and a mean of zero. The following script plots the time series and the DFT scaled by the factor $n/2$.

```
% e3s707.m
clf
f1 = 20; f2 = 50; f3 = 70;
nt = 512; T = 2; dt = T/nt
t_final = (nt-1)*dt; df = 1/T
fmax = (nt/2)*df;
t = 0:dt:t_final;
dt_plt = dt/25;
t_plt = 0:dt_plt:t_final;
y_plt = 0.2*cos(2*pi*f1*t_plt)+0.35*sin(2*pi*f2*t_plt) ...
                                         +0.3*sin(2*pi*f3*t_plt);
y_plt = y_plt+0.5*randn(size(y_plt));
y = y_plt(1:25:(nt-1)*25+1); f = 0:df:(nt/2-1)*df;
figure(1);
subplot(211), plot(t_plt,y_plt)
axis([0 0.04 -3 3])
xlabel('Time (sec)'), ylabel('y')
yf = fft(y);
yp(1:nt/2) = (2/nt)*yf(1:nt/2);
subplot(212), plot(f,abs(yp))
axis([0 fmax 0 0.5])
xlabel('Frequency (Hz)'), ylabel('abs(DFT)');
```

Running the preceding script gives

```
dt =

    0.0039


df =

    0.5000
```

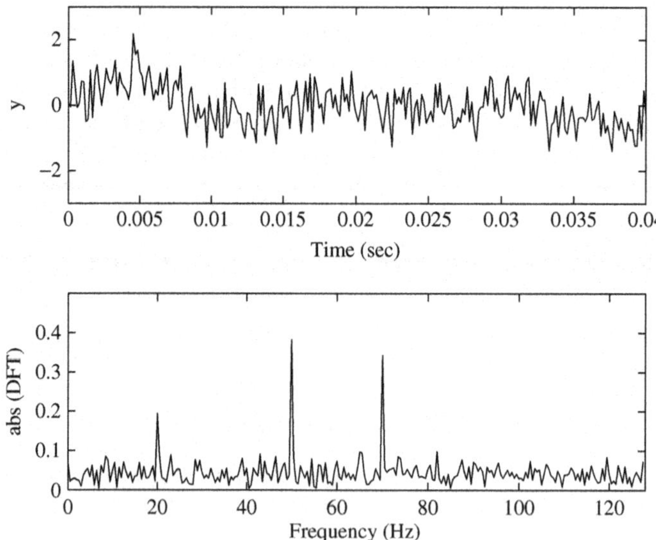

FIGURE 7.10 Signal and frequency spectrum showing frequency components at 20, 50, and 70 Hz.

together with the graphical output shown in Figure 7.10. The lower plot of Figure 7.10 shows that random noise in the signal does not prevent the frequency components 20, 50, and 70 Hz from revealing themselves in the spectrum. These components are not obviously visible in the original time series data shown in the upper plot of Figure 7.10.

Example 7.3

Determine the spectrum of a triangular wave of amplitude ± 1 and period 1 second, sampled at 1/32 second intervals over one cycle. The following script outputs the DFT scaled by the factor $n/2$.

```
% e3s708.m
nt = 32; T = 1, dt = T/nt
t = 0:dt:(nt-1)*dt;
df = 1/T, fmax = nt/(2*T)
f = 0:df:df*(nt/2-1);
y = 0.125*[8 7 6 5 4 3 2 1 0 -1 -2 -3 -4 -5 -6 -7 -8 ...
    -7 -6 -5 -4 -3 -2 -1 0 1 2 3 4 5 6 7];
Yss = zeros(1,nt/2); Y = fft(y);
Yss(1:nt/2) = (2/nt)*Y(1:nt/2);
[f' abs(Yss)']
```

Running this script gives

```
T =
       1


dt =
     0.0313


df =
       1


fmax =
     16


ans =
          0         0
     1.0000    0.8132
     2.0000         0
     3.0000    0.0927
     4.0000         0
     5.0000    0.0352
     6.0000         0
     7.0000    0.0194
     8.0000         0
     9.0000    0.0131
    10.0000         0
    11.0000    0.0100
    12.0000         0
    13.0000    0.0085
    14.0000         0
    15.0000    0.0079
```

The Fourier series for the triangular wave of this example is

$$f(t) = \frac{8}{\pi^2}\left(\cos(2\pi t) + \frac{1}{3^2}\cos(6\pi t) + \frac{1}{5^2}\cos(10\pi t) + \frac{1}{7^2}\cos(14\pi t) + \cdots\right)$$

The first eight frequency components in the scaled DFT at frequencies 1, 3, 5 Hz, and so on, are not equal to $8/\pi^2$, $8/(3\pi)^2$, $8/(5\pi)^2$ (i.e., 0.8106, 0.0901, 0.0324), and so on, because of the effect of aliasing. A triangular wave contains an infinite number of harmonics and because of aliasing these appear as components in the DFT as shown in Table 7.2. Thus the size of the 3 Hz component in the DFT is $(8/\pi^2)(1/3^2 + 1/29^2 + 1/35^2 + 1/61^2 + \cdots)$. By summing a large number of terms down the columns of Table 7.2, the terms in the DFT are obtained. The DFT as shown in the preceding is correct, and the inverse DFT recovers the original data. However,

Table 7.2 Coefficients of Aliased Harmonics

f	$3f$	$5f$	$7f$	$9f$	$11f$	$13f$	$15f$
$8/\pi^2$	$8/(3\pi)^2$	$8/(5\pi)^2$	$8/(7\pi)^2$	$8/(9\pi)^2$	$8/(11\pi)^2$	$8/(13\pi)^2$	$8/(15\pi)^2$
$8/(31\pi)^2$	$8/(29\pi)^2$	$8/(27\pi)^2$	$8/(25\pi)^2$	$8/(23\pi)^2$	$8/(21\pi)^2$	$8/(19\pi)^2$	$8/(17\pi)^2$
$8/(33\pi)^2$	$8/(35\pi)^2$	$8/(37\pi)^2$	$8/(39\pi)^2$	$8/(41\pi)^2$	$8/(43\pi)^2$	$8/(45\pi)^2$	$8/(47\pi)^2$
$8/(63\pi)^2$	$8/(61\pi)^2$	$8/(59\pi)^2$	$8/(57\pi)^2$	$8/(55\pi)^2$	$8/(53\pi)^2$	$8/(51\pi)^2$	$8/(49\pi)^2$
$8/(65\pi)^2$	$8/(67\pi)^2$	etc.					

when using it to provide information about the contribution of frequency components in the original data, the DFT must be interpreted with care.

Example 7.4

Determine the DFT of a sequence of 128 data points sampled from a signal at intervals of 0.0625 seconds. The signal has a constant amplitude of 1 unit which, after 10 samples, is switched to zero.

The following script determines the DFT for the data.

```
% e3s709.m
clf
nt = 128; nb = 10;
y = [ones(1,nb) zeros(1,nt-nb)];
dt = 0.0625; T = dt*nt
df = 1/T, fmax = (nt/2)*df;
f = 0:df:(nt/2-1)*df;
yf = fft(y);
yp = (2/nt)*yf(1:nt/2);
figure(1), bar(f,abs(yp),'w')
axis([0 fmax 0 0.2])
xlabel('Frequency (Hz)'), ylabel('abs(DFT)')
```

Running this script gives

```
T =
     8

df =
     0.1250
```

together with the graphical output shown in Figure 7.11. The plot shows that the frequency spectrum is continuous and the largest components are clustered near the zero frequency. This is in contrast to the spectra of Examples 7.1 and 7.2, which show sharp peaks due to the presence

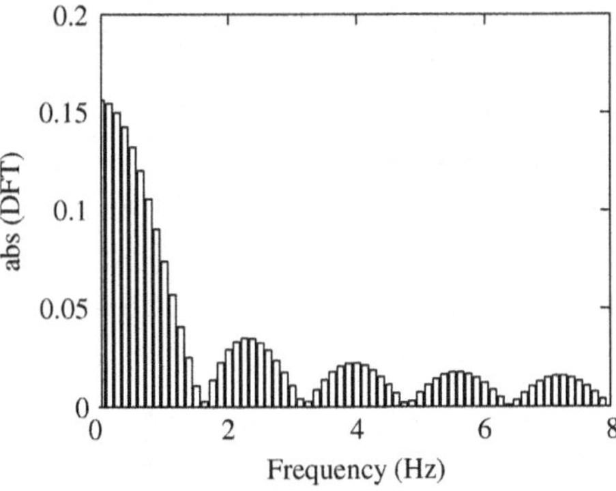

FIGURE 7.11 Spectrum of a sequence of data.

of periodic components in the original data. Note that because the original signal is a step and not periodic, the amplitude of its DFT is dependent on the sampling period.

In this section we have provided examples of how the DFT (computed using the FFT) can be used to study the distribution of frequency components in data. There are other applications of the DFT. It is sometimes used for interpolation, as is any procedure that fits mathematical functions to a sequence of data. MATLAB provides the function `interpft` to allow interpolation using the DFT.

The advances in computer hardware have extended the range of problems to which the DFT can usefully be applied, and this in turn has encouraged the development of new and powerful variants of the FFT algorithm. A detailed description of the FFT algorithm is given by Brigham (1974), and a straightforward introduction that emphasizes the practical problems in using this type of analysis is given by Ramirez (1985).

7.5 Multiple Regression: Least Squares Criterion

We now consider the problem of fitting a function to a relatively large amount of data that contains errors. It would not be sensible, nor computationally possible, to fit a very high-degree polynomial to a large amount of experimental data as may be done for interpolation. What is required is a function that smooths out fluctuations in the data due to errors and reveals any underlying trend. We therefore adjust the coefficients of a chosen function to provide a "best fit" according to some criterion. For example, the criterion may be to minimize the maximum error, the sum of the modulus of the errors, or the sum of the squares of the errors between the chosen function and the actual data points. The least squares method is the most widely used of these, criteria and we now examine how this process is carried out.

Suppose we have an n component vector of observations $\mathbf{y}$ and p separate vectors of explanatory variables, $\mathbf{x}_1, \mathbf{x}_2, \ldots, \mathbf{x}_p$. The variables $\mathbf{x}_1, \mathbf{x}_2, \ldots, \mathbf{x}_p$ are generally independent variables that are measured with negligible error or even controlled in an experiment and $\mathbf{y}$ is a single dependent variable, uncontrolled and containing random measurement errors. In some cases $\mathbf{x}_1$, $\mathbf{x}_2$, and so on, may be different functions of a single explanatory variable—in which case we call them *predictors* (see Sections 7.8 and 7.9). Our basic model consists of a *regression equation* and is said to be a regression of $\mathbf{y}$ upon the explanatory variables $\mathbf{x}_1, \mathbf{x}_2, \ldots, \mathbf{x}_p$. We assume a simple linear regression model, and so for the ith observation we have

$$y_i = \beta_0 + \beta_1 x_{1i} + \beta_2 x_{2i} + \cdots + \beta_p x_{pi} + \varepsilon_i, \quad i = 1, 2, \ldots, n \tag{7.17}$$

where β_j ($j = 0, 1, 2, \ldots, p$) are the unknown coefficients and ε_i are random errors. Initially we simply assume that these random errors are identically distributed with a zero mean and a common unknown variance, σ^2, and that they are independent of each other. This implies that

$$\beta_0 + \beta_1 x_{1i} + \beta_2 x_{2i} + \cdots + \beta_p x_{pi}$$

represents the mean value of y_i.

Our main task is to find an estimate b_j for each unknown β_j so that we can estimate the mean of y_i by fitting a function of the form

$$\hat{y}_j = b_0 + b_1 x_{1i} + b_2 x_{2i} + \cdots + b_p x_{pi}, \quad i = 1, 2, \ldots, n \tag{7.18}$$

The difference between the observed and the fitted value for the ith observation

$$e_i = y_i - \hat{y}_i \tag{7.19}$$

is called the *residual* and is an estimate of the corresponding ε_i. Our criterion for choosing the b_j estimates is that they should minimize the sum of the squares of the residuals, which is often called the *sum of squares of the errors* and is denoted by *SSE*. Thus

$$SSE = \sum_{i=1}^{n} e_i^2 = \sum_{i=1}^{n} (y_i - \hat{y}_i)^2 \tag{7.20}$$

To perform the necessary calculations efficiently, we rewrite the model in matrix form as follows. We define $\mathbf{b}$ as a $(p+1) \times 1$ vector such that

$$\mathbf{b} = [b_0 \quad b_1 \ \ldots \ b_p]^\top$$

Similarly we define **e**, **y**, **u**, and $\mathbf{x}_j$ to be $n \times 1$ vectors as follows:

$$\begin{aligned}
\mathbf{e} &= [e_1 \quad e_2 \ldots e_n]^\top \\
\mathbf{y} &= [y_1 \quad y_2 \ldots y_n]^\top \\
\mathbf{u} &= [1 \quad 1 \ldots 1]^\top \\
\mathbf{x}_j &= [x_{j1} \quad x_{j2} \ldots x_{jn}]^\top, \quad j = 1, 2, \ldots, p
\end{aligned}$$

From these vectors the $n \times (p+1)$ matrix, **X**, can be defined as follows:

$$\mathbf{X} = [\mathbf{u} \quad \mathbf{x}_1 \quad \mathbf{x}_2 \ldots \mathbf{x}_p]$$

This matrix corresponds to the coefficients of the equation system (7.18). From (7.19) the residuals may then be written as

$$\mathbf{e} = \mathbf{y} - \mathbf{X}\mathbf{b}$$

and the *SSE*, from (7.20), is then

$$SSE = \mathbf{e}^\top \mathbf{e} = (\mathbf{y} - \mathbf{X}\mathbf{b})^\top (\mathbf{y} - \mathbf{X}\mathbf{b}) \tag{7.21}$$

Differentiating the *SSE* with respect to **b**, we get a vector of partial derivatives, as follows:

$$\frac{\partial}{\partial \mathbf{b}}(SSE) = -2\mathbf{X}^\top(\mathbf{y} - \mathbf{X}\mathbf{b})$$

Matrix differentiation is described in Appendix A. Equating this derivative to zero we have $\mathbf{X}\mathbf{b} = \mathbf{y}$ and this overdetermined system of equations could be solved directly to obtain the coefficients **b** using the MATLAB operator \. However, in this instance it is more convenient to proceed as follows. Premultiplying $\mathbf{X}\mathbf{b} = \mathbf{y}$ by $\mathbf{X}^\top$, we obtain what are called the *normal equations*:

$$\mathbf{X}^\top \mathbf{X}\mathbf{b} = \mathbf{X}^\top \mathbf{y}$$

The formal solution of these equations is then

$$\mathbf{b} = (\mathbf{X}^\top \mathbf{X})^{-1} \mathbf{X}^\top \mathbf{y} = \mathbf{C}\mathbf{X}^\top \mathbf{y} \tag{7.22}$$

where

$$\mathbf{C} = (\mathbf{X}^\top \mathbf{X})^{-1} \tag{7.23}$$

Note that **C** is a $(p+1) \times (p+1)$ square matrix.

Using the expression for **b** in (7.22), the vector of fitted values corresponding to **y** is

$$\hat{\mathbf{y}} = \mathbf{X}\mathbf{b} = \mathbf{X}\mathbf{C}\mathbf{X}^\top \mathbf{y}$$

So, defining $\mathbf{H} = \mathbf{X}\mathbf{C}\mathbf{X}^{\top}$, we can write

$$\hat{\mathbf{y}} = \mathbf{H}\mathbf{y} \tag{7.24}$$

The matrix $\mathbf{H}$, which converts $\mathbf{y}$ to $\hat{\mathbf{y}}$ ("y-hat") is called the *hat matrix* and plays an important role in the interpretation of the regression model. Among its important properties is that it is idempotent (see Appendix A).

From (7.22) it can be shown that the minimum value of *SSE* is given by

$$SSE = \mathbf{y}^{\top}(\mathbf{I} - \mathbf{H})\mathbf{y} \tag{7.25}$$

where $\mathbf{I}$ is an $n \times n$ identity matrix. Our original data consisted of n sets of observations and we have introduced $p+1$ constraints into the system to estimate the parameters $\beta_0, \beta_1, \ldots, \beta_p$, so there are now $(n-p-1)$ degrees of freedom. Statistical theory shows that by dividing the minimum *SSE* from (7.25) by the number of degrees of freedom, we obtain an unbiased estimate of the unknown error variance, σ^2, which is

$$s^2 = \frac{SSE}{n-p-1} = \frac{\mathbf{y}^{\top}(\mathbf{I} - \mathbf{H})\mathbf{y}}{n-p-1} \tag{7.26}$$

On its own, the value of s obtained by fitting a single model is not very informative. However, we can also use $\mathbf{H}$ to evaluate the overall goodness of fit on an absolute scale and to examine how good each b_j is as an estimate of the corresponding β_j.

The most widely used measure of the overall goodness of fit is the *coefficient of determination*, R^2, which is defined by

$$R^2 = \sum_{i=1}^{n} (\hat{y}_i - \bar{y})^2 \Big/ \sum_{i=1}^{n} (y_i - \bar{y})^2$$

where

$$\bar{y} = \frac{1}{n}\sum_{i=1}^{n} y_i$$

Thus $\bar{y}$ is the mean of the observed y values. Using matrix notation, we can evaluate R^2 from the equivalent definition

$$R^2 = \frac{\mathbf{y}^{\top}(\mathbf{H} - \mathbf{u}\mathbf{u}^{\top}/n)\mathbf{y}}{\mathbf{y}^{\top}(\mathbf{I} - \mathbf{u}\mathbf{u}^{\top}/n)\mathbf{y}} \tag{7.27}$$

The value of R^2 will lie between 0 and 1 and represents the proportion of the total observed variance of $\mathbf{y}$ that is accounted for by the explanatory variables. Thus a value close to 1 indicates that nearly all the observed variance is accounted for and we have a good fit. However, on its own this does not necessarily indicate that the model is satisfactory because the value of R^2 can always be increased by introducing more explanatory variables, even though their introduction can have other effects that are very undesirable. In the next section we briefly consider some methods for deciding which explanatory variables should be included in the model.

7.6 Diagnostics for Model Improvement

To see which variables might be removed to improve our original model, we now examine the b_j estimates in more detail to check whether they indicate that the corresponding β_j coefficients are nonzero, which would confirm that the corresponding x_j really do contribute to explaining y, or in the case of b_0, whether it is appropriate to include the constant term β_0 in the model.

If we make the assumption that the random errors, ε_i, are normally distributed, it can be shown that each of the b_j estimates behave as if they were observed values of normal random variables whose means are the β_j and whose covariance matrix is $s^2\mathbf{C}$ where $\mathbf{C}$ is defined by (7.23). Thus the covariance matrix is

$$s^2\mathbf{C} = s^2 \begin{bmatrix} c_{00} & c_{01} & c_{02} & \cdots & c_{0p} \\ & c_{11} & c_{12} & \cdots & c_{1p} \\ & & c_{22} & \cdots & c_{2p} \\ & & & & \vdots \\ & & & & c_{pp} \end{bmatrix}$$

Note that we numbered the rows and columns of $\mathbf{C}$ from zero to p. The matrix $\mathbf{C}$ is symmetric and so the subdiagonal elements are not shown.

The variance of the distribution for b_j is s^2 times the corresponding diagonal element of $\mathbf{C}$, and the *standard error* (*SE*) of b_j is the square root of this variance:

$$SE(b_j) = s\sqrt{c_{jj}}$$

If β_j is really zero, the statistic

$$t = b_j/SE(b_j)$$

has a Student's t-distribution with $n - p - 1$ degrees of freedom. Thus a formal hypothesis test may be carried out to check whether it is reasonable to assume that β_j is zero and hence the predictor x_j does not make a significant contribution to explaining y. However, as an initial guide to whether x_j should be included in the regression model, it is usually sufficient to check whether the magnitude of the corresponding t-statistic is numerically greater than about 2. If $|t| > 2$, then x_j should be left in the model; otherwise, consideration should be given to removing it.

When there is more than one explanatory variable or predictor in the original regression, there is a possibility that two or more of these may be highly correlated with each other. This situation is called *multicollinearity*; when it occurs the columns in $\mathbf{X}$ corresponding to the correlated variables are almost linearly related, and this causes $\mathbf{X}^\top\mathbf{X}$ and its inverse $\mathbf{C}$ to be ill-conditioned. Although we shall be able to solve the normal equations for $\mathbf{b}$ as long as $\mathbf{X}^\top\mathbf{X}$ does not actually become singular, the solution is very sensitive to small changes in the data and there will be large off-diagonal elements in $\mathbf{C}$, indicating highly correlated b_j estimates. It is therefore worth calculating the *correlation matrix*, which shows the correlation between y and each of the explanatory variables and the correlations between all pairs of explanatory variables. As in the case of the covariance matrix, we number the rows and columns of the correlation matrix from 0 to p. Thus the correlation matrix is

$$\begin{array}{cc} & \begin{array}{cccc} y & x_1 & x_2 & \end{array} \\ \begin{array}{c} y \\ x_1 \\ x_2 \\ \\ \end{array} & \begin{pmatrix} r_{00} & r_{01} & r_{02} & \cdots \\ r_{10} & r_{11} & r_{12} & \cdots \\ r_{20} & r_{21} & r_{22} & \cdots \\ \cdots & \cdots & \cdots & \cdots \end{pmatrix} \end{array}$$

where we define the typical element of the correlation matrix r_{ij} to be the correlation between x_i and x_j and r_{0j} to be the correlation between y and x_j.

In situations such as *polynomial regression*, which is considered in Section 7.8, there will always be high correlations between the predictors, but examination of the t-statistics may indicate that some of the predictors do not contribute significantly to explaining y. Those with the smallest $|t|$ and the smallest $|r_{0j}|$ are the most obvious candidates for discarding.

Another set of statistics that are useful in this context are the *variance inflation factors* (*VIFs*). To find the *VIF* for x_j, we regress x_j upon the other $p-1$ explanatory variables and calculate the coefficient of determination, R_j^2. The corresponding *VIF* is

$$VIF_j = \frac{1}{1-R_j^2}$$

If x_j is almost entirely explained by the other variables, R_j^2 will be close to 1 and VIF_j will be large. A good working rule is to regard any x_j with $VIF_j > 10$ as a candidate for removal.

Note that when there are only two explanatory variables, they will always have equal variance inflation factors; if these are greater than 10, it is best to discard the variable that has the smallest correlation with y. When the model contains predictors that are different functions of some common explanatory variable, the corresponding *VIF* values may be very large, as in the case of polynomial regression described in Section 7.8. Such models would normally be considered for physical data when predictors of this type could be shown to have a causal relationship with y.

The following MATLAB function `mregg2` implements multiple regression; the diagnostics necessary for model improvement are also computed.

```
function [s_sqd R_sqd b SE t VIF Corr_mtrx residual] = mregg2(Xd,con)
% Multiple linear regression, using least squares.
% Example call:
%   [s_sqd R_sqd bt SEt tt VIFt Corr_mtrx residual] = mregg2(Xd,con)
% Fits data to y = b0 + b1*x1 + b2*x2 + ... bp*xp
% Xd is a data array. Each row of X is a set of data.
% Xd(1,:) = x1(:), Xd(2,:) = x2(:), ... Xd(p+1,:) = y(:).
% Xd has n columns corresponding to n data points and p+1 rows.
% If con = 0, no constant is used, if con ~= 0, constant term is used.
% Output arguments:
% s_sqd = Error variance, R_Sqd = R^2.
% b is the row of coefficients b0 (if con~=0), b1, b2, ... bp.
% SE is the row of standard error for the coeff b0 (if con~=0),
%  b1, b2, ... bp.
% t is the row of the t statistic for the coeff b0 (if con~=0),
%  b1, b2, ... bp.
% VIF is the row of the VIF for the coeff b0 (if con~=0), b1, b2, ... bp.
% Corr_mtrx is the correlation matrix.
% residual is an arrray of 4 columns and n rows.
% For each row i, the residual
% array contains the value of y(i), the residual(i), the standardized
% residual(i) and the Cook distance(i) where i is the ith data value.
if con==0
    cst = 0;
else
    cst = 1;
end
[p1,n] = size(Xd);
p = p1-1;  pc = p+cst;
y = Xd(p1,:)';
if cst==1
    w = ones(n,1);
    X = [w Xd(1:p,:)'];
else
    X = Xd(1:p,:)';
end
C = inv(X'*X);  b = C*X'*y; b = b.';
H = X*C*X';  SSE = y'*(eye(n)-H)*y;
s_sqd = SSE/(n-pc); Cov = s_sqd*C;
Z = (1/n)*ones(n);
num = y'*(H-Z)*y; denom = y'*(eye(n)-Z)*y;
R_sqd = num/denom;
```

```
SE = sqrt(diag(Cov)); SE = SE.';
t = b./SE;
% Compute correlation matrix
V(:,1) = (eye(n)-Z)*y;
for j = 1:p
    V(:,j+1) = (eye(n)-Z)*X(:,j+cst);
end
SS = V'*V; D = zeros(p+1,p+1);
for j=1:p+1
    D(j,j) = 1/sqrt(SS(j,j));
end
Corr_mtrx = D*SS*D;
% Compute VIF
for j = 1+cst:pc
    ym = X(:,j);
    if cst==1
        Xm = X(:,[1 2:j-1,j+1:p+1]);
    else
        Xm = X(:,[1:j-1,j+1:p]);
    end
    Cm = inv(Xm'*Xm); Hm = Xm*Cm*Xm';
    num = ym'*(Hm-Z)*ym; denom = ym'*(eye(n)-Z)*ym;
    R_sqr(j-cst) = num/denom;
end
VIF = 1./(1-R_sqr); VIF = [0 VIF];
% Analysis of residuals
ee = zeros(length(y),1);  sr = zeros(length(y),1);
cd = zeros(length(y),1);
if nargout>7
    ee = (eye(n)-H)*y;
    s = sqrt(s_sqd);
    sr = ee./(s*sqrt(1-diag(H)));
    cd = (1/pc)*(1/s^2)*ee.^2.*(diag(H)./((1-diag(H)).^2));
    residual = [y ee sr cd];
end
```

To use `mregg2` it is necessary to provide the $(p+1) \times n$ data array, `Xd`, where p is the number of explanatory variables and n is the number of data sets. Rows 1 to p contain the values of the explanatory variables, x_1 to x_p, and row $p+1$ contains the corresponding value of y. If the parameter `con` is set to zero, then the constant term is removed from the regression model; otherwise, it is included. Examples of the use of the function `mregg2` are given in Sections 7.7 and 7.8.

The multiple regression model has wider applications. For example:

1. *Polynomial regression.* Here y is a polynomial function of a single variable x, so that x_j in the general model is replaced by x^j. Thus we have

$$y_i = \beta_0 + \beta_1 x_i + \beta_2 x_i^2 + \cdots + \beta_p x_i^p + \varepsilon_i, \quad i = 1, 2, \ldots, n \tag{7.28}$$

Although this is no longer linear in the explanatory variable x, it is still linear in the β_j coefficients and so the theory for the linear regression model still applies. We can use `mregg2` to carry out polynomial regression where the rows of data are $x, x^2, x^3, \ldots$ and the last row of data is y. (See Examples 7.7, 7.8, and 7.9).

2. *Multiple polynomial regression.* Suppose that we wished to fit data to the following regression model:

$$y_i = \beta_0 + \beta_1 x_1 + \beta_2 x_2 + \beta_3 x_1^2 + \beta_4 x_2^2 + \beta_5 x_1 x_2 + \varepsilon_i, \quad i = 1, 2, \ldots, n$$

In this case the five predictors are x_1, x_2, x_1^2, x_2^2, and x_1x_2. The predictors are still linear in the β_j coefficients. To use the function `mregg2` the six rows of the data array must contain the values of x_1, x_2, x_1^2, x_2^2, x_1x_2, and y, respectively.

7.7 Analysis of Residuals

Besides considering the contributions made by each of the explanatory variables or predictors, it is important to consider how well the model fits at each data point and whether our assumptions about the error distribution are valid.

We recall from (7.24) that

$$\hat{\mathbf{y}} = \mathbf{Hy}$$

so we may write the residual vector as

$$\mathbf{e} = \mathbf{y} - \mathbf{Hy} = (\mathbf{I} - \mathbf{H})\,\mathbf{y}$$

It can be shown that the diagonal elements of $s^2(\mathbf{I} - \mathbf{H})$ represent the variances of the individual residuals, so the standard deviation of e_i is $s\sqrt{1 - h_{ii}}$. Since the standard deviation varies from one data point to another, it is difficult to make a direct comparison between residuals at different points. However, if we *standardize* the residuals by dividing each by its standard deviation, we obtain statistics that are similar to the t-ratios that we used for analyzing $\mathbf{b}$. Thus the standardized residual r_i is

$$r_i = \frac{e_i}{s\sqrt{1 - h_{ii}}} \quad i = 1, 2, \ldots, n$$

To distinguish this from other kinds of standardized residuals, it is sometimes called the *Studentized residual.* If the assumptions behind our model are correct, the average value of the standardized residual should be close to zero but if any $|r_i|$ is larger than about 2 this may indicate one or more of the following:

1. The point where this occurs is an *outlier.*
2. The assumption about equal error variance at all points is incorrect.
3. There has been a fault in specifying the model.

In this context, an outlier is an observation that was not obtained under the same conditions as the others. It is not always easy to distinguish points that result from mistakes in observation from those that correspond to values of the explanatory variables that lie far away from those used for the other observations. Such points tend to have a large influence on the fitting process.

A useful statistic for measuring the influence of a particular point is the *Cook's distance,* which combines the size of the squared residual with the distance of a particular point from the mean value, called the *leverage.* This is the corresponding diagonal element of **H** determined by the values of the explanatory variables.

$$\text{Cook's distance, } d_i = \left(\frac{1}{p+1}\right)\frac{e_i^2}{s^2}\left[\frac{h_{ii}}{(1-h_{ii})^2}\right] \quad i = 1, 2, \ldots, n$$

Any point for which $d_i > 1$ will have a considerable effect on the regression and should be checked in detail. The point may have been correctly observed and provide information that is very useful in model building, but the modeler should be aware of its influence.

Example 7.5

Fit a regression model to the following data. To save space the data is given in the form required by the function `mregg2`. The first, second, and third rows of the matrices are the values of the explanatory variables x_1, x_2, and x_3, respectively, and the fourth row contains the corresponding values of y. The following script implements this.

```
% e3s710.m
X0 = [1.00 1.00 1.00 1.00 1.00 2.00 2.00 2.00;
      2.00 2.00 4.00 4.00 6.00 2.00 2.00 4.00;
        0 1.00 0 1.00 2.00 0 1.00 0;
     -2.52 -2.71 -8.34 -8.40 -14.60 -0.62 -0.47 -6.49];
X1 = [2.00 3.00 3.00 3.00 3.00 3.00 3.00 3.00;
      6.00 2.00 2.00 2.00 4.00 6.00 6.00 6.00;
        0 0 1.00 2.00 1.00 0 1.00 2.00;
    -12.46 1.36 1.40 1.60 -4.64 -10.34 -10.43 -10.30];
Xd = [X0 X1];
[s_sqd R_sqd b SE t VIF Corr_mtrx res] = mregg2(Xd,1);
```

```
fprintf('Error variance = %7.4f      R_squared = %7.4f \n\n',s_sqd,R_sqd)
fprintf('                         Coeff      SE      t_ratio     VIF \n')
fprintf('Constant    :          %7.4f %7.4f %8.2f \n',b(1),SE(1),t(1))
fprintf('Coeff x1    :          %7.4f %7.4f %8.2f %8.2f\n',b(2),SE(2),t(2),VIF(2))
fprintf('Coeff x2    :          %7.4f %7.4f %8.2f %8.2f\n',b(3),SE(3),t(3),VIF(3))
fprintf('Coeff x3    :          %7.4f %7.4f %8.2f %8.2f\n\n',b(4),SE(4),t(4),VIF(4))
fprintf('Correlation matrix \n')
disp(Corr_mtrx)
fprintf('\n         y          Residual    St Residual    Cook dist\n')
for i = 1:length(Xd)
    fprintf('%12.4f %12.4f %12.4f %12.4f\n',res(i,1), ...
                                 res(i,2), res(i,3), res(i,4))
end
```

Running this script gives

```
Error variance =  0.0147      R_squared =  0.9996

                     Coeff      SE      t_ratio      VIF
Constant    :       1.3484  0.1006     13.40
Coeff x1    :       2.0109  0.0358     56.10      1.03
Coeff x2    :      -2.9650  0.0179   -165.43      1.03
Coeff x3    :      -0.0001  0.0412     -0.00      1.04

Correlation matrix
    1.0000    0.2278   -0.9437   -0.0944
    0.2278    1.0000    0.1064    0.1459
   -0.9437    0.1064    1.0000    0.1459
   -0.0944    0.1459    0.1459    1.0000

         y          Residual    St Residual    Cook dist
     -2.5200        0.0508        0.4847        0.0196
     -2.7100       -0.1390       -1.3223        0.1426
     -8.3400        0.1609        1.5062        0.1617
     -8.4000        0.1010        0.9274        0.0507
    -14.6000       -0.1688       -1.9402        0.8832
     -0.6200       -0.0601       -0.5458        0.0157
     -0.4700        0.0901        0.8035        0.0270
     -6.4900        0.0000        0.0002        0.0000
    -12.4600       -0.0399       -0.3835        0.0130
      1.3600       -0.0909       -0.8808        0.0730
      1.4000       -0.0508       -0.4736        0.0154
      1.6000        0.1493        1.5774        0.3958
     -4.6400       -0.1607       -1.4227        0.0755
    -10.3400        0.0692        0.6985        0.0602
    -10.4300       -0.0206       -0.1930        0.0026
    -10.3000        0.1095        1.1123        0.1589
```

The coefficient of x_3 is small, and, more important, the corresponding absolute value of the t-ratio is very small (it is in fact not zero but -0.0032). This suggests that x_3 does not make a significant contribution to the model and can be removed.

If we change the last value of y to -8.3 (and showing the analysis of the residuals only), we have

```
        y          Residual   St Residual   Cook dist
   -2.5200          0.3499        0.7758       0.0503
   -2.7100         -0.0756       -0.1670       0.0023
   -8.3400          0.3095        0.6735       0.0323
   -8.4000          0.0140        0.0300       0.0001
  -14.6000         -0.6418       -1.7153       0.6903
   -0.6200          0.1361        0.2876       0.0044
   -0.4700          0.0507        0.1052       0.0005
   -6.4900          0.0457        0.0941       0.0003
  -12.4600         -0.1447       -0.3232       0.0093
    1.3600          0.0024        0.0054       0.0000
    1.4000         -0.1930       -0.4184       0.0120
    1.6000         -0.2284       -0.5610       0.0501
   -4.6400         -0.4534       -0.9331       0.0325
  -10.3400         -0.1384       -0.3247       0.0130
  -10.4300         -0.4638       -1.0077       0.0713
   -8.3000          1.4308        3.3791       1.4664
```

For the observation $y = -8.3$ we see that the residual, the standard residual, and Cook's distance are all large, compared with the values for the rest of the data. Either we have a recording error in this particular observation or the data is correct and thus the model we are using fits this particular data point poorly.

Example 7.6

Using the same data as Example 7.5, fit a regression model using the explanatory variables x_1 and x_2 only. The data array `Xd` is not shown in the following script.

```
% e3s711.m
X0 ....
X1 ....
Xd ....
Xd = [X0 X1];
[s_sqd R_sqd b SE t VIF Corr_mtrx] = mregg2(Xd([1 2 4],:),1);
fprintf('Error variance = %7.4f     R_squared = %7.4f \n\n',s_sqd,R_sqd)
fprintf('                    Coeff      SE      t_ratio      VIF \n')
fprintf('Constant    :     %7.4f %7.4f   %8.2f \n',b(1),SE(1),t(1))
fprintf('Coeff x1    :     %7.4f %7.4f   %8.2f %8.2f\n',b(2),SE(2),t(2),VIF(2))
```

```
fprintf('Coeff x2    :      %7.4f %7.4f  %8.2f %8.2f\n\n',b(3),SE(3),t(3),VIF(3))
fprintf('Correlation matrix \n')
disp(Corr_mtrx)
```

Running this script gives

```
Error variance =  0.0135       R_squared =  0.9996

                      Coeff       SE      t_ratio      VIF
Constant    :        1.3483  0.0960       14.04
Coeff x1    :        2.0109  0.0341       58.91       1.01
Coeff x2    :       -2.9650  0.0171     -173.71       1.01

Correlation matrix
    1.0000    0.2278   -0.9437
    0.2278    1.0000    0.1064
   -0.9437    0.1064    1.0000
```

This is a better model than that derived in Example 7.5 because the absolute values of the t-ratios are now all greater than 2. In fact, the original data was generated from the model

$$y = 1.5 + 2x_1 - 3x_2 + \text{random errors}$$

Thus x_3 was not linked to the model used to generate the data and variations in x_3 only appeared to influence y because of the random errors in the measurement of y.

■ ■ ■

Note that the standard errors (*SEs*) can be used to construct confidence intervals for b_j. In this case the true values of each β_j lie within the 95% confidence interval: that is, the interval within which we expect that β_j would lie with a probability of 0.95 if we did not know its true value. The precise width of the 95% confidence interval depends on the number of degrees of freedom (see Section 7.5), but to a reasonable approximation it lies between $b_j - 2SE(b_j)$ and $b_j + 2SE(b_j)$.

More comprehensive presentations of aspects of multiple regression, model improvement, and regression analysis can be found in Draper and Smith (1998), Walpole and Myers (1993), and Anderson, Sweeney and Williams (1993).

7.8 Polynomial Regression

The polynomial regression model is given by (7.28) and is repeated here:

$$y_i = \beta_0 + \beta_1 x_i + \beta_2 x_i^2 + \cdots + \beta_p x_i^p + \varepsilon_i, \quad i = 1, 2, \ldots, n$$

Although this is no longer linear in the explanatory variable x, it is still linear in the β_j coefficients and so the theory for the linear regression model still applies.

The processes of fitting, checking the b_j estimates, and deciding whether some predictors can be discarded may all be done in the same manner as in the general case described in Section 7.5, and diagnostics for model improvement and residual analysis also follow the general case described in Sections 7.6 and 7.7. It will inevitably be found that there are quite high correlations between the predictors, which are now all powers of the same explanatory variable. As discussed in Section 7.6, these high correlations between predictors can result in an ill-conditioned coefficient matrix $\mathbf{X}^\top\mathbf{X}$. For a large number of data points this matrix tends to the Hilbert matrix.

In Chapter 2 it was shown that the Hilbert matrix is very ill-conditioned. To illustrate the influence of this on the accuracy of computations, we note that the number of decimal places lost when working with an ill-conditioned matrix **A** is given approximately by the MATLAB expression `log10(cond(A))`. Thus if we were fitting a fifth-degree polynomial, the number of decimal places that could be lost may be estimated by `log10(cond(hilb(5)))`. This equals 5.6782; that is, five or six of the 16 significant digits that MATLAB uses are lost. One way to avoid this difficulty is to formulate the problem so that no system of linear equations has to be solved. An ingenious way of doing this is to use orthogonal polynomials. We will not describe this method here but refer the reader to Lindfield and Penny (1989). However, the worst effects of ill-conditioning can be avoided provided that p (which now represents the degree of the fitted polynomial) is kept reasonably small.

If the diagnostics that we have developed in Sections 7.5, 7.6, and 7.7 are required, then we can use the function `mregg2` given in Section 7.6 for polynomial regression; in this case the data must be prepared as follows. The first row of the array `Xd` contains the values of x, the second row contains the values of x^2, and so on. The last row contains the corresponding values of y. In interpreting the output from `mregg2`, note that all the *VIF* values are inevitably high because the powers of x are correlated with each other. The usual rule about discarding predictors with $VIF > 10$ should be ignored since we have good reason to suppose that a polynomial model is appropriate.

If the diagnostics are not required, the calculation of the b_j estimates can be done using the MATLAB function `polyfit`. This uses the method of least squares to fit a polynomial of specified degree to given data. The following examples illustrate some of these issues.

Example 7.7

Fit a cubic polynomial to the following data, which has been generated from $y = 2 + 6x^2 - x^3$ with added random errors. The random errors have a normal distribution with a zero mean value and a standard deviation of 1. The following script calls the MATLAB function `polyfit` to determine the coefficients of the cubic polynomial followed by `polyval` to evaluate it for plotting. It then calls the function `mregg2` to compute a regression model using x, x^2, and x^3 as the explanatory variables.

```
% e3s712.m
x = 0:.25:6;
```

```
y = [1.7660 2.4778 3.6898 6.3966 6.6490 10.0451 12.9240 15.9565 ...
     17.0079 21.1964 24.1129 25.5704 28.2580 32.1292 32.4935 34.0305 ...
     34.0880 32.9739 31.8154 30.6468 26.0501 23.4531 17.6940 9.4439 ...
     1.7344];
xx = 0:.02:6;
p = polyfit(x,y,3), yy = polyval(p,xx);
plot(x,y,'o',xx,yy)
axis([0 6 0 40]), xlabel('x'), ylabel('y')
Xd = [x; x.^2; x.^3; y];
[s_sqd R_sqd b SE t VIF Corr_mtrx] = mregg2(Xd,1);
fprintf('Error variance = %7.4f      R_squared = %7.4f \n\n',s_sqd,R_sqd)
fprintf('                          Coeff      SE      t_ratio      VIF \n')
fprintf('Constant    :          %7.4f %7.4f %8.2f \n',b(1),SE(1),t(1))
fprintf('Coeff x     :          %7.4f %7.4f %8.2f %8.2f\n',b(2),SE(2),t(2),VIF(2))
fprintf('Coeff x^2   :          %7.4f %7.4f %8.2f %8.2f\n',b(3),SE(3),t(3),VIF(3))
fprintf('Coeff x^3   :          %7.4f %7.4f %8.2f %8.2f\n\n',b(4),SE(4),t(4),VIF(4))
fprintf('Correlation matrix \n')
disp(Corr_mtrx)
```

Running this script gives the following output, together with the graph shown in Figure 7.12. The polynomial coefficients from `polyfit` are in the order of descending powers of x. The diagnostic output from `mregg2` is self-explanatory.

```
p =
   -0.9855    5.8747    0.1828    2.2241

Error variance =  0.5191       R_squared =  0.9966

                  Coeff       SE      t_ratio      VIF
Constant    :     2.2241  0.4997      4.45
Coeff x     :     0.1828  0.7363      0.25      84.85
Coeff x^2   :     5.8747  0.2886     20.36     502.98
Coeff x^3   :    -0.9855  0.0316    -31.20     202.10
```

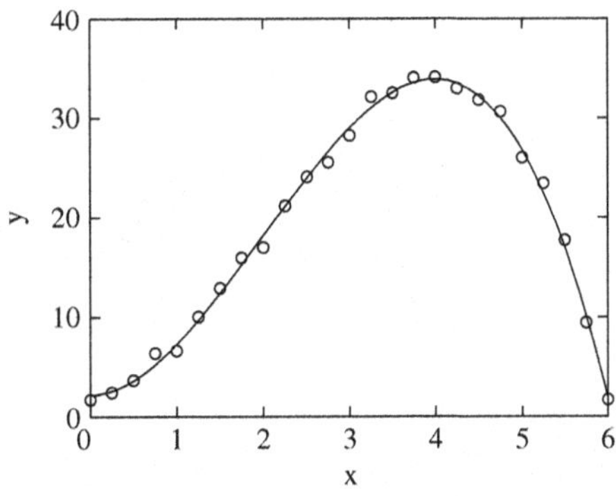

FIGURE 7.12 Fitting a cubic polynomial to data. Data points are denoted by ○.

```
Correlation matrix
    1.0000    0.4917    0.2752    0.1103
    0.4917    1.0000    0.9659    0.9128
    0.2752    0.9659    1.0000    0.9858
    0.1103    0.9128    0.9858    1.0000
```

Note that the absolute value of the t-ratio for the explanatory variable x is less than 2, indicating that x should be removed (see Example 7.8).

The cubic polynomial that fits the data is

$$\hat{y} = 2.2241 + 0.1828x + 5.8747x^2 - 0.9855x^3$$

Example 7.8

Fit a cubic polynomial to the data of Example 7.7 but use the explanatory variables x^2 and x^3 only. The following script solves this problem:

```
% e3s713.m
x = 0:.25:6; y = 2+6*x.^2-x.^3;
y = y+randn(size(x)); Xd = [x.^2; x.^3; y];
[s_sqd R_sqd b SE t VIF Corr_mtrx] = mregg2(Xd,1);
fprintf('Error variance = %7.4f      R_squared = %7.4f \n\n',s_sqd,R_sqd)
fprintf('                        Coeff      SE      t_ratio      VIF \n')
fprintf('Constant    :      %7.4f %7.4f %8.2f \n',b(1),SE(1),t(1))
fprintf('Coeff x^2   :      %7.4f %7.4f %8.2f %8.2f\n',b(2),SE(2),t(2),VIF(2))
fprintf('Coeff x^3   :      %7.4f %7.4f %8.2f %8.2f\n\n',b(3),SE(3),t(3),VIF(3))
fprintf('Correlation matrix \n')
disp(Corr_mtrx)
```

Running this script gives

```
Error variance =  0.4970      R_squared =  0.9965

                  Coeff      SE      t_ratio      VIF
Constant    :    2.3269  0.2741       8.49
Coeff x^2   :    5.9438  0.0750      79.21     35.52
Coeff x^3   :   -0.9926  0.0130     -76.61     35.52

Correlation matrix
    1.0000    0.2752    0.1103
    0.2752    1.0000    0.9858
    0.1103    0.9858    1.0000
```

Thus our improved model (compared with Example 7.7) is

$$\hat{y} = 2.1793 + 6.0210x^2 - 1.0084x^3$$

The true β_j are well within the 95% confidence limits given by this improved model. The error variance of 0.4970 is somewhat less than the unit variance of the random errors initially added to the data.

Example 7.9

Fit a third- and a fifth-degree polynomial to data generated from the function

$$y = \sin\{1/(x + 0.2)\} + 0.2x$$

contaminated with random noise, normally distributed with a standard deviation of 0.06 to simulate measurement errors as follows:

```
>> xs = [0:0.05:0.25 0.25:0.2:4.85];
>> us = sin(1./(xs+1))+0.2*xs+0.06*randn(size(xs));
>> save testdata1 xs us
```

The 30 data values are stored in the file `testdata1` so that it can be used in an example in Section 7.9. The following script loads the data, and fits and plots the least squares polynomial.

```
% e3s714.m
load testdata
xx = 0:.05:5;
t1 = 'Error variance = %7.4f      R_squared = %7.4f \n\n';
t2 = '                          Coeff       SE      t_ratio        VIF \n';
t3 = 'Constant    :        %7.4f %7.4f %8.2f \n';
t4 = 'Coeff x     :        %7.4f %7.4f %8.2f %12.2f\n';
t4a = 'Coeff x     :        %7.4f %7.4f %8.2f \n';
t5 = 'Coeff x^2   :        %7.4f %7.4f %8.2f %12.2f\n';
t6 = 'Coeff x^3   :        %7.4f %7.4f %8.2f %12.2f\n';
t7 = 'Coeff x^4   :        %7.4f %7.4f %8.2f %12.2f\n';
t8 = 'Coeff x^5   :        %7.4f %7.4f %8.2f %12.2f\n';
t9 = 'Correlation matrix \n';
p = polyfit(xs,us,3), yy = polyval(p,xx);
Xd = [xs; xs.^2; xs.^3; us];
[s_sqd R_sqd b SE t VIF Corr_mtrx] = mregg2(Xd,1);
fprintf(t1,s_sqd,R_sqd), fprintf(t2)
fprintf(t3,b(1),SE(1),t(1))
fprintf(t4,b(2),SE(2),t(2),VIF(2))
fprintf(t5,b(3),SE(3),t(3),VIF(3))
fprintf([t6 '\n'],b(4),SE(4),t(4),VIF(4))
fprintf(t9), disp(Corr_mtrx)
```

```
[s_sqd R_sqd b SE t VIF Corr_mtrx] = mregg2(Xd,0);
fprintf(t1,s_sqd,R_sqd), fprintf(t2)
fprintf(t4a,b(1),SE(1),t(1))
fprintf(t5,b(2),SE(2),t(2),VIF(2))
fprintf([t6 '\n'],b(3),SE(3),t(3),VIF(3))
fprintf(t9), disp(Corr_mtrx)
plot(xs,us,'ko',xx,yy,'k'), hold on
axis([0 5 -2 2])
p = polyfit(xs,us,5), yy = polyval(p,xx);
Xd = [xs; xs.^2; xs.^3; xs.^4; xs.^5; us];
[s_sqd R_sqd b SE t VIF Corr_mtrx] = mregg2(Xd,1);
fprintf(t1,s_sqd,R_sqd), fprintf(t2)
fprintf(t3,b(1),SE(1),t(1))
fprintf(t4,b(2),SE(2),t(2),VIF(2))
fprintf(t5,b(3),SE(3),t(3),VIF(3))
fprintf(t6,b(4),SE(4),t(4),VIF(4))
fprintf(t7,b(5),SE(5),t(5),VIF(5))
fprintf([t8 '\n'],b(6),SE(6),t(6),VIF(6))
fprintf(t9)
disp(Corr_mtrx)
plot(xx,yy,'k--'), xlabel('x'), ylabel('y'), hold off
```

Figure 7.13 shows the result of fitting a third- and a fifth-degree polynomial to the data and clearly displays the inadequacies of these polynomial approximations. The polynomials oscillate about the points and do not fit the data satisfactorily. In Section 7.9 we see that we can improve the fit by using different functions. The output from the script is as follows:

```
p =
     0.0842    -0.6619     1.5324    -0.0448

Error variance =  0.0980     R_squared =  0.6215
```


FIGURE 7.13 Fitting third- and fifth-degree polynomials (that is, a full line and a dashed line, respectively) to a sequence of data. Data points are denoted by ∘.

```
                       Coeff       SE       t_ratio          VIF
Constant      :      -0.0448   0.1402       -0.32
Coeff x       :       1.5324   0.3248        4.72          79.98
Coeff x^2     :      -0.6619   0.1708       -3.87         478.23
Coeff x^3     :       0.0842   0.0239        3.52         193.93

Correlation matrix
    1.0000     0.5966     0.4950     0.4476
    0.5966     1.0000     0.9626     0.9049
    0.4950     0.9626     1.0000     0.9847
    0.4476     0.9049     0.9847     1.0000
```

The absolute value of the t-ratio for the constant terms is low, implying that it should not be included in the cubic model. The script also fits a cubic equation without a constant term to the data and gives the following output:

```
Error variance =  0.0947       R_squared =  0.6200

                       Coeff       SE       t_ratio          VIF
Coeff x       :       1.4546   0.2116        6.87
Coeff x^2     :      -0.6285   0.1329       -4.73          35.13
Coeff x^3     :       0.0801   0.0199        4.02         299.50

Correlation matrix
    1.0000     0.5966     0.4950     0.4476
    0.5966     1.0000     0.9626     0.9049
    0.4950     0.9626     1.0000     0.9847
    0.4476     0.9049     0.9847     1.0000
```

This is a more robust model. Finally, the script fits a fifth-degree polynomial to the data and gives the following, final, output:

```
p =
    0.0434    -0.5856     2.8998    -6.3340     5.7099    -0.5789

Error variance =  0.0341       R_squared =  0.8783

                       Coeff       SE       t_ratio          VIF
Constant      :      -0.5789   0.1122       -5.16
Coeff x       :       5.7099   0.6443        8.86         904.01
Coeff x^2     :      -6.3340   0.9052       -7.00       38560.71
Coeff x^3     :       2.8998   0.4918        5.90      234903.50
Coeff x^4     :      -0.5856   0.1137       -5.15      262672.06
Coeff x^5     :       0.0434   0.0094        4.62       38084.24
```

```
Correlation matrix
    1.0000    0.5966    0.4950    0.4476    0.4172    0.3942
    0.5966    1.0000    0.9626    0.9049    0.8511    0.8041
    0.4950    0.9626    1.0000    0.9847    0.9555    0.9232
    0.4476    0.9049    0.9847    1.0000    0.9918    0.9742
    0.4172    0.8511    0.9555    0.9918    1.0000    0.9949
    0.3942    0.8041    0.9232    0.9742    0.9949    1.0000
```

Note the large values of *VIF*, caused by the regression being based on different functions of a single explanatory variable.

We now illustrate a difficulty that can sometime arise when trying to fit a polynomial function to data. To illustrate the problem we will begin by simulating experimental data based on the following relationship:

$$y = \frac{1}{\sqrt{0.02 + (4 - x^2)^2}} \tag{7.29}$$

Data values are generated by sampling this function from $x = 1$ to $x = 3$ in increments of 0.05, and small random errors are added to simulate measurement errors. The results of attempting to fit polynomials to these data values are shown in Figure 7.14. The plot shows that as the degree of the polynomial is increased from 4 to 8, and finally to 12, the polynomial fits the data better in the sense that the total least squares error decreases, but the higher-degree polynomials tend to oscillate between the data points. Thus even a twelfth-degree polynomial does not accurately represent the data, nor does it give us any insight into the underlying mathematical relationship between x and y. We return to this problem in Section 7.11.

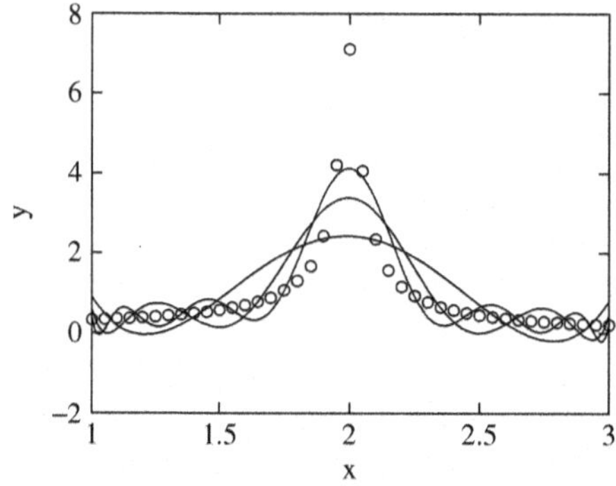

FIGURE 7.14 Polynomials of degree 4, 8, and 12 attempting to fit a sequence of data indicated by ◦ in the graph.

7.9 Fitting General Functions to Data

We now consider a regression based on (7.17) with the separate explanatory variables x_j replaced by predictors that are various functions ϕ_j of a single explanatory variable x:

$$y_i = \beta_0 + \beta_1\varphi_1(x_i) + \beta_2\varphi_2(x_i) + \cdots + \beta_p\varphi_p(x_i) + \varepsilon_i$$

The analysis presented in Section 7.5 extends directly to this regression model. Hence we can use the MATLAB function mregg2 to fit a set of any prescribed functions to data.

Consider again Example 7.9 in Section 7.8. We will fit the following function (or model) to the data:

$$\hat{y} = b_1 \sin\{1/(x+0.2)\} + b_2 x$$

This function has been chosen because the data was originally generated from it with $b_1 = 1$ and $b_2 = 0.2$ with normally distributed random noise added. Note that there is no constant term in our model. The following script calls the function mregg2. Note how the first row of the data matrix for mregg2 contains the values of $\sin(1/(x+0.2))$ and the second row contains the values of x.

```
% e3s715.m
load testdata
Xd = [sin(1./(xs+0.2)); xs; us];
[s_sqd R_sqd b SE t VIF Corr_mtrx] = mregg2(Xd,0);
fprintf('Error variance = %7.4f\n\n',s_sqd)
fprintf('                        Coeff      SE     t_ratio\n')
fprintf('sin(1/(x+0.2)):       %7.4f %7.4f %8.2f \n',b(1),SE(1),t(1))
fprintf('Coeff x          :       %7.4f %7.4f %8.2f \n\n',b(2),SE(2),t(2))
fprintf('Correlation matrix \n')
disp(Corr_mtrx)
xx = 0:.05:5; yy = b(1)*sin(1./(xx+0.2))+b(2)*xx;
plot(xs,us,'o',xx,yy,'k')
axis([0 5 -1.5 1.5]), xlabel('x'), ylabel('y')
```

Running this script gives

```
Error variance =  0.0044

                    Coeff     SE     t_ratio
sin(1/(x+0.2)):     0.9354  0.0257    36.46
Coeff x        :     0.2060  0.0053    38.55

Correlation matrix
    1.0000    0.7461    0.5966
    0.7461    1.0000   -0.0734
    0.5966   -0.0734    1.0000
```

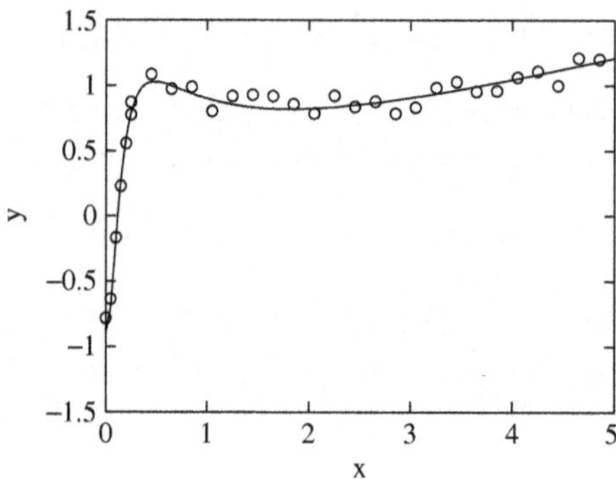

FIGURE 7.15 Data sampled from the function $y = \sin[1/(x+0.2)] + 0.2x$. Data points denoted by "○".

and the graph of Figure 7.15. The function that fits the data in a least squares sense is given by $\hat{y} = 0.9354\sin\{1/(x+0.2)\} + 0.2060x$. This is very close to the original function. Note also that the error variance of 0.0044 compares quite well with 0.0036, the variance of the noise that was added to simulate measurement errors. If a constant term is included in the model, it has a very low absolute t-ratio, indicating that it should be removed.

7.10 Nonlinear Least Squares Regression

We now consider the problem of fitting data to a function where the relationship between the unknown coefficients is nonlinear. We still use the least squares criterion, and there are several methods for fitting data to this type of model. Here we present a very simple iterative method based on a Taylor series.

Let $y = f(x, \mathbf{a})$ where f is a nonlinear function of the unknown coefficients $\mathbf{a}$. To determine these coefficients, let a trial set of coefficents be $\mathbf{a}^{(0)}$. Thus

$$y = f\left(x, \mathbf{a}^{(0)}\right)$$

This trial solution will not satisfy the requirement that the sum of the squares of the errors is a minimum. However, we can adjust the coefficients $\mathbf{a}^{(0)}$ in order to minimize the sum of the squares of the errors. Thus, letting the improved coefficients be $\mathbf{a}^{(1)}$, where

$$\mathbf{a}^{(1)} = \mathbf{a}^{(0)} + \Delta\mathbf{a}$$

we have

$$y = f\left(x, \mathbf{a}^{(1)}\right) = f\left(x, \mathbf{a}^{(0)} + \Delta\mathbf{a}\right)$$

Expanding the function as a Taylor series, and retaining only the first derivative terms in the Taylor series we have

$$y \approx f\left(x, \mathbf{a}^{(0)}\right) + \sum_{k=0}^{m} \Delta a_k \left[\partial f / \partial a_k\right]^{(0)}$$

Let $f_i^{(0)} = f\left(x_i, \mathbf{a}^{(0)}\right)$. The error between the function and y_i is given by

$$\varepsilon_i = y_i - f_i^{(0)} - \sum_{k=0}^{m} \Delta a_k \left[\partial f_i / \partial a_k\right]^{(0)}, \quad i = 1, 2, \ldots, n$$

Thus, the sum of the squares of the errors is

$$S = \sum_{i=0}^{n} \left\{ y_i - f_i^{(0)} - \sum_{k=0}^{m} \Delta a_k \left[\frac{\partial f_i}{\partial a_k}\right]^{(0)} \right\}^2$$

To determine the minimum of the sum of the squares of the errors, we have

$$\frac{\partial S}{\partial(\Delta a_p)} = -2 \sum_{i=0}^{n} \left\{ y - f_i^{(0)} - \sum_{k=0}^{m} \Delta a_k \left[\frac{\partial f_i}{\partial a_k}\right]^{(0)} \right\} \left[\frac{\partial f_i}{\partial a_p}\right]^{(0)} = 0, \quad p = 0, 1, \ldots, m$$

Rearranging we have

$$\sum_{i=0}^{n} \left(y - f_i^{(0)}\right) \left[\frac{\partial f_i}{\partial a_p}\right]^{(0)} = \sum_{k=0}^{m} \Delta a_k \left\{ \sum_{i=0}^{n} \left[\frac{\partial f_i}{\partial a_k}\right]^{(0)} \left[\frac{\partial f_i}{\partial a_p}\right]^{(0)} \right\}, \quad p = 0, 1, \ldots, m$$

These equations can be expressed in matrix notation as follows:

$$\mathbf{K}(\Delta \mathbf{a}) = \mathbf{b}$$

where $\Delta \mathbf{a}$ has elements $\Delta a_p, p = 0, 1, \ldots, m$ and

$$K_{pk} = \sum_{i=0}^{n} \left[\frac{\partial f_i}{\partial a_k}\right]^{(0)} \left[\frac{\partial f_i}{\partial a_p}\right]^{(0)}$$

$$b_p = \sum_{i=0}^{n} \left(y - f_i^{(0)}\right) \left[\frac{\partial f_i}{\partial a_p}\right]^{(0)}, \quad p, k = 0, 1, \ldots, m$$

Solving for $\Delta \mathbf{a}$ we can then determine new values for the coefficients from

$$\mathbf{a}^{(1)} = \mathbf{a}^{(0)} + \Delta \mathbf{a}$$

Because we have discarded higher-order terms in the Taylor series, $\mathbf{a}^{(1)}$ is not the exact solution, but it is a better solution than $\mathbf{a}^{(0)}$. We therefore iterate until the norm of $\Delta\mathbf{a}$ is less than a prescribed tolerance.

The following function `nlls` implements the preceding method for fitting a given nonlinear function to a set of data.

```
function [a iter] = nlls(f,df,x,y,a0,err)
% Data given by vectors x and y are to be fitted to the function f(a)
% with an error of err. Function f(a) has n variables, a(1) ... a(n).
% a0 is a vector of n trial values for the unknown paramenters a.
% Function df is a column vector [df/da(1); df/da(2); .... df/da(n)].
iter = 0;  n = length(a0); a = a0;
v = 10*err*ones(1,n);
while norm(v,2) > err
    p = feval(df,x,a);  q = y-feval(f,x,a);
    A = p*p'; b = q*p'; v = A\b';
    a = a + v';  iter = iter+1;
end
```

The next script fits the function $y = a_1 e^{a_2 x} + a_3 e^{a_4 x}$ to 16 data points using the function `nlls`.

```
% e3s718
p = @(x,a) a(1)*exp(a(2)*x)+a(3)*exp(a(4)*x);
dp = @(x,a) [exp(a(2)*x); a(1)*x.*exp(a(2)*x);
              exp(a(4)*x); a(3)*x.*exp(a(4)*x)];
x = [-10:2:0  1:1:10]; xn = length(x);
xp = -10:0.05:10;
y = [26.56 21.60 18.14 17.00 14.46 17.38 15.07 16.76 ...
    16.90 17.32 18.61 20.79 21.65 25.22 26.16 27.84];
a = [7 -0.3 7 0.3];
[a iter] = nlls(p,dp,x,y,a,1e-5)
plot(x,y,'o',xp,p(xp,a))
xlabel('x'), ylabel('y')
```

Running this script gives the following results:

```
a =
    5.4824   -0.1424   10.0343    0.0991


iter =
     7
```

together with Figure 7.16.

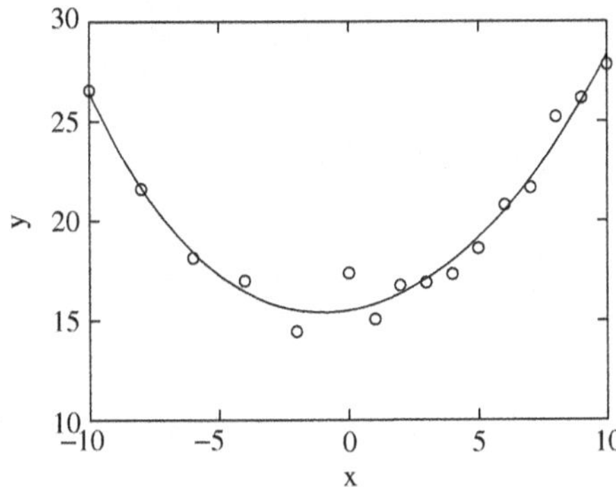

FIGURE 7.16 Fitting $y = a_1 e^{a_2 x} + a_3 e^{a_4 x}$ to data values indicated by "∘".

7.11 Transforming Data

We now consider an alternative approach to the problem of fitting data to functions where the relationship between the unknown coefficients is nonlinear. This is to transform both the data and the function so that an equivalent function provides a linear relationship between y and the unknown coefficients. The only difficulty with this method is that no general rule can be given to provide a suitable transform; indeed, such a transform may not even be possible. Consider the problem of fitting data to the following function:

$$\hat{y} = \frac{1}{\sqrt{a_0 + (a_1 - a_2 x^2)^2}} \tag{7.30}$$

Letting $\hat{Y} = 1/\hat{y}^2$ and $X = x^2$, we have

$$\begin{aligned} \hat{Y} &= 1/\hat{y}^2 = a_0 + (a_1 - a_2 x^2)^2 = (a_0 + a_1^2) - 2a_1 a_2 x^2 + a_2^2 x^4 \\ \hat{Y} &= a_0 + a_1^2 - 2a_1 a_2 X + a_2^2 X^2 \\ \hat{Y} &= b_0 + b_1 X + b_2 X^2 \end{aligned} \tag{7.31}$$

Thus $\hat{Y}$ is a quadratic in X. If the data values are transformed by letting $Y_i = 1/y_i^2$ and $X_i = x_i^2$, then the process of fitting $Y = f(X)$ to these transformed data values will be a standard least squares polynomial fit giving b_0, b_1, and b_2. Hence estimates of a_0, a_1, and a_2 can be easily determined. Note, however, that the residual of the errors, $e_i = Y_i - \hat{Y}_i$, will not provide a good estimate of the measurement errors, $y_i - \hat{y}_i$, because of the transformations.

We illustrate the preceding process by considering a sequence of data values related by (7.29) to which we have added normally distributed random errors having a zero mean value and a standard deviation of 1%. We can transform these data points using (7.31), and the following script generates the required data, transforms the data points, and fits a polynomial to them.

```
% e3s716.m
x = 1:.05:3; xx = 1:.005:3;
```

```
y = [0.3319   0.3454   0.3614   0.3710   0.3857   0.4030   0.4372 ...
     0.4605   0.4971   0.5232   0.5753   0.6363   0.6953   0.7782 ...
     0.8793   1.0678   1.3024   1.6688   2.4233   4.2046   7.0961 ...
     4.0581   2.3354   1.5663   1.1583   0.9278   0.7764   0.6480 ...
     0.5741   0.4994   0.4441   0.4005   0.3616   0.3286   0.3051 ...
     0.2841   0.2645   0.2407   0.2285   0.2104   0.2025];
Y = 1./y.^2; X = x.^2; XX = xx.^2;
p = polyfit(X,Y,2)
YY = polyval(p,XX);
for i = 1:length(xx)
    if YY(i)<0
        disp('Transformation fails with this data set');
        return
    end
end
figure(1), plot(X,Y,'o',XX,YY)
axis([1 9 0 25]), xlabel('X'), ylabel('Y')
yy = 1./sqrt(YY);
figure(2), plot(x,y,'o',xx,yy)
axis([1 3 -2 8]), xlabel('x'), ylabel('y')
```

Running this script gives the following results:

```
p =
    0.9944   -7.9638   15.9688
```

together with the plots shown in Figures 7.17 and 7.18. From the output of the script we see that the relationship between X and $\hat{Y}$ is

$$\hat{Y} = 0.9944X^2 - 7.9638X + 15.9688$$

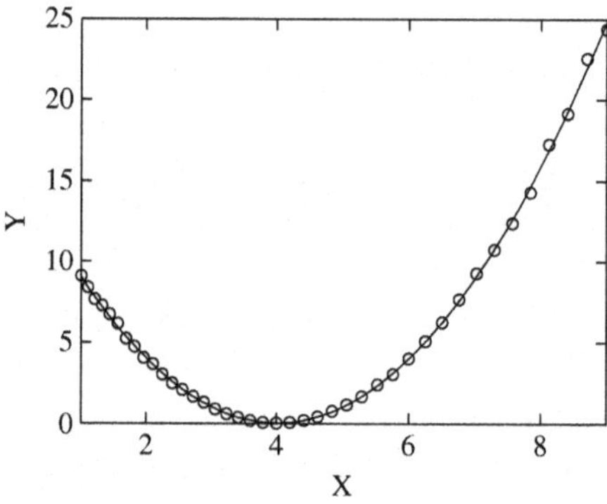

FIGURE 7.17 Fitting transformed data denoted by "∘" to a quadratic function.

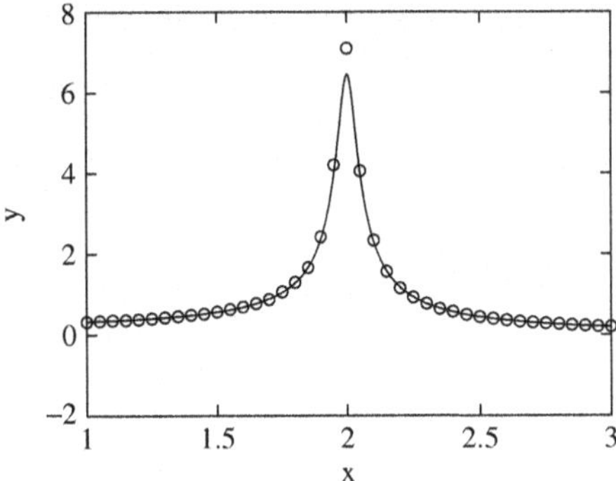

FIGURE 7.18 Fitting (7.30) to the given data denoted by ∘.

We can deduce the values of the unknown coefficients by comparing the preceding equation with (7.31) to give

$$a_2^2 = 0.9944, \quad \text{hence} \quad a_2 = \pm 0.9972 \quad \text{(take positive value as in (7.29))}$$

$$-2a_1 a_2 = 7.9748, \quad \text{hence } a_1 = 3.9931$$

$$a_1^2 + a_0 = 15.9688, \quad \text{hence } a_0 = 0.0149$$

We use these values in the original function (7.30) and fit it to the given data. This is shown in Figure 7.18. This function provides a much better fit than the polynomials shown earlier in Figure 7.14. However, even this fit does not pass through the peak value. This is caused by the sensitivity of the process to small random errors in the data. If the random errors are removed, the fit is exact. If the script is rerun with the random errors, the fit may be worse, and if the size of the random errors is increased, the process may fail. This is because in the region of $x = 2$, the value of y is essentially only dependent on a_0. This has a small value that may vary in sign.

Table 7.3 lists some functions with a nonlinear relationship between $\hat{y}$ and the coefficients of the function and the corresponding transformations that linearize these relationships so they have the form $\hat{Y} = BX + C$.

The following script implements the first two relationships shown in Table 7.3. It also determines the sum of the squares of the errors for the two original relationships and plots a graph of the data and the two fitted functions.

```
% e3s717.m
x = 0.2:0.2:4;
y = 2*exp(0.5*x).*(1+0.2*rand(size(x)));
X = log(x);  Y = log(y);
```

Table 7.3 Functions with Nonlinear Relationships

Original equation	Substitution	Transformed equation
$y = ax^b$	$Y = \log_e(y), X = \log_e(x)$	$Y = A + bX$ so that $a = e^A$
$y = ae^{bx}$	$Y = \log_e(y)$	$Y = A + bx$ so that $a = e^A$
$y = axe^{bx}$	$Y = \log_e(y/x)$	$Y = A + bx$ so that $a = e^A$
$y = a + \log_e(bx)$	$Y = e^y$	$Y = A + bx$ so that $a = \log_e(A)$
$y = 1/(a + bx)$	$Y = 1/y$	$Y = a + bx$
$y = 1/(a + bx)^2$	$Y = 1/\sqrt(y), X = 1/x$	$Y = a + bx$
$y = x/(b + ax)$	$Y = 1/y, X = 1/x$	$Y = a + bX$
$y = ax/(b + x)$	$Y = 1/y, X = 1/x$	$Y = A + BX$ so that $a = 1/A, b = B/A$

```
% Case 1: Fit y = a*x^b
v = polyfit(X,Y,1);
A1 = v(2); b1 = v(1); a1 = exp(A1);
e1 = y-a1*x.^b1; s1 = e1*e1';
fprintf('\n y = %8.4f*x^(%8.4f): SSE = %8.4f',a1,b1,s1)
% case 2: Fit y = a*exp(b*x)
v = polyfit(x,Y,1);
A2 = v(2); b2 = v(1); a2 = exp(A2);
e2 = y-a2*exp(b2*x); s2 = e2*e2';
fprintf('\n y = %8.4f*exp(%8.4f*x): SSE = %8.4f \n',a2,b2,s2)
% Plotting
n = length(x);
r = x(n)-x(1); inc = r/100;
xp = [x(1):inc:x(n)];
yp1 = a1*xp.^b1;  yp2 = a2*exp(b2*xp);
plot(x,y,'ko',xp,yp1,'k:',xp,yp2,'k')
xlabel('x'), ylabel('f(x)')
```

Running this script gives

```
 y =   4.5129*x^(  0.6736): SSE =  78.3290
 y =   2.2129*exp(  0.5021*x): SSE =    2.0649
```

Figure 7.19 and the preceding output confirms that the best fit is, as one would expect, the exponential function. Clearly this is a warning to the user to use discrimination in selecting the function to fit. This MATLAB function could be adapted to fit data to a wide range of mathematical functions.

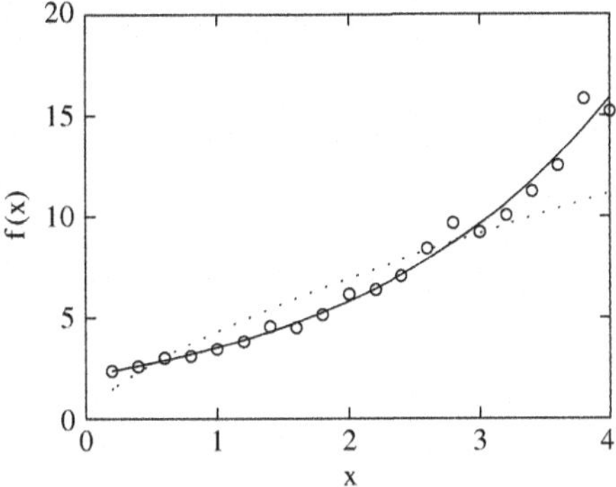

FIGURE 7.19 This graph shows the original data and the fits obtained from $y = be^{(ax)}$ (*full line*) and $y = ax^b$ (*dotted line*).

7.12 Summary

Methods have been described for fitting functions to data for the purposes of interpolation. These have included Aitken's method and spline fits. For periodic data, we have examined the fast Fourier transform. Finally, we have discussed least squares approximations to experimental data using polynomial and more general functions.

The reader who wishes to study the application of splines may find the Mathworks Spline toolbox useful.

Problems

7.1. The following tabulation gives values of the complete elliptic integral

$$E(\alpha) = \int_0^{\pi/2} \sqrt{(1 - \sin^2\alpha \sin^2\theta)}\, d\theta$$

α	0°	5°	10°	15°	20°	25°	30°
$E(\alpha)$	1.57079	1.56780	1.55888	1.54415	1.52379	1.49811	1.46746

Determine $E(\alpha)$ for $\alpha = 2°$, $13°$, and $27°$ using the MATLAB function `aitken`.

7.2. Generate a table of values of $f(x) = x^{1.4} - \sqrt{x} + 1/x - 100$ for $x = 20:2:30$. Find the value of x corresponding to $f(x) = 0$ using the MATLAB function `aitken`. This is an example of inverse interpolation since we are finding the value of x corresponding to a given value of $f(x)$. In particular, this gives an approximation to the root of the equation $f(x) = 0$. Compare your solution with that of Problem 3.2 in Chapter 3.

7.3. Given $x = -1 : 0.2 : 1$, calculate values of y from $y = \sin^2(\pi x/2)$. Using the data you have, calculate:

(a) Generate quadratic and quartic polynomials to fit this data using the least squares MATLAB function `polyfit`. Display the data and the curve fitted. *Hint*: Example 7.6 in Section 7.7 gives some guidance.

(b) Fit a cubic spline to the data using the MATLAB function `spline`. Display the data and the fitted `spline`. Compare the quality of this spline fit with the two graphs from (a).

7.4. For the data of Problem 7.3, determine the values of y for $x = 0.85$ using the MATLAB function `interp1` for a linear, spline, and cubic interpolating function. Also use the MATLAB function `aitken`.

7.5. Fit a cubic spline and a fifth-degree polynomial to the following data.

x	−2	0	2	3	4	5
y	4	0	−4	−30	−40	−50

Plot the data points, the spline, and the polynomial on the same graph. Which curve appears to give the more realistic representation of any underlying function from which the data might have been taken?

7.6. For the data given by the vectors $x = 0 : 0.25 : 3$ and

$$y = [6.3806\ 7.1338\ 9.1662\ 11.5545\ 15.6414\ 22.7371\ 32.0696 \ldots \\ 47.0756\ 73.1596\ 111.4684\ 175.9895\ 278.5550\ 446.4441]$$

fit the following functions:

(a) $f(x) = a + be^x + ce^{2x}$ using the MATLAB function `mregg2`

(b) $f(x) = a + b/(1+x) + c/(1+x)^2$ using the MATLAB function `mregg2`

(c) $f(x) = a + bx + cx^2 + dx^3$ using the MATLAB functions `polyfit` or `mregg2`

You should plot the three trial functions and the data. How well do these functions fit the data? The data values were in fact generated from $f(x) = 3 + 2e^x + e^{2x}$ with a small amount of random noise added.

7.7. The following values of x and corresponding values of y_u and y_l define an airfoil section:

$$x = [0\ 0.005\ 0.0075\ 0.0125\ 0.025\ 0.05\ 0.1\ 0.2\ 0.3\ 0.4 \ldots 0.5\ 0.6\ 0.7\ 0.8\ 0.9\ 1]$$

$$y_u = [0\ \ 0.0102\ \ 0.0134\ \ 0.0170\ \ 0.0250\ \ 0.0376\ \ 0.0563\ \ 0.0812 \ldots \\ 0.0962\ 0.1035\ 0.1033\ 0.0950\ 0.0802\ 0.0597\ 0.0340\ 0]$$

$$y_l = [0\ \ -0.0052\ \ -0.0064\ \ -0.0063\ \ -0.0064\ \ -0.0060\ \ -0.0045 \ldots \\ -0.0016\ 0.0010\ 0.0036\ 0.0070\ 0.0121\ 0.0170\ 0.0199\ 0.0178\ 0]$$

The (x, y_u) coordinates define the upper surface and the (x, y_l) coordinates define the lower surface. Use the MATLAB function `spline` to fit separate splines to the upper and lower surfaces and plot the results as a single figure.

7.8. Consider the approximation

$$\prod_{p<P}\left(1+\frac{1}{p}\right)\approx C_1+C_2\log_e P$$

where the product is taken of all the prime numbers p less than a prime number P. Write a script to generate these products from the list of prime numbers provided and fit the function $C_1 + C_2 \log_e P$ to the points given by the primes P and the corresponding values of the products using the MATLAB function `polyfit`. Generate a list of prime numbers using the MATLAB function `primes(103)`.

7.9. The gamma function may be approximated by a fifth-degree polynomial as

$$\Gamma(x+1) = a_0 + a_1x + a_2x^2 + a_3x^3 + a_4x^4 + a_5x^5$$

Use the MATLAB function `gamma` to generate values of $\Gamma(x+1)$ for $x = 0:0.1:1$. Then, using the MATLAB function `polyfit`, fit a fifth-degree polynomial to this data. Compare your answers with the approximation for the gamma function given by Abramowitz and Stegun (1965), which gives $a_0 = 1$, $a_1 = -0.5748666$, $a_2 = 0.9512363$, $a_3 = -0.6998588$, $a_4 = 0.4245549$, and $a_5 = -0.1010678$. These coefficients give an accuracy for the gamma function in the range $0 \le x \le 1$ of less than or equal to 5×10^{-5}.

7.10. Generate a table of values of z from the function

$$z(x,y) = 0.5\left(x^4 - 16x^2 + 5x\right) + 0.5\left(y^4 - 16y^2 + 5y\right)$$

in the range $x = -4:0.2:4$ and $y = -4:0.2:4$. Use this data and the MATLAB function `interp2` to interpolate a value for z at $x = y = -2.9035$. Use both linear and cubic interpolation and check your answer by direct substitution in the function. This point gives the global minimum of this function.

7.11. The difference between the mean Sun and the real Sun is called the equation of time. Thus the value of the equation of time

$$E = (\text{mean Sun time} - \text{real Sun time})$$

The following values represent E in minutes at 20 equispaced intervals during the year, beginning January 1.

$$\begin{aligned}E = [&-3.5\quad -10.5\quad -14.0\quad -14.25\quad -9.0\quad -4.0\quad 1.0\quad 3.5\quad 3.0\ldots\\ &-0.25\quad -3.5\quad -6.25\quad -5.5\quad -1.75\quad 4.0\quad 10.5\quad 15.0\quad 16.25\quad 12.75\quad 6.5]\end{aligned}$$

Plot a graph of the data values E against time of year. Then use the function `interpft` to interpolate 300 points and plot E over a period of one year. (Use the MATLAB `help` function to obtain information on `interpft`). Finally, use the command `[x,y]=ginput(4)` to read from the graph the values of the two minimum and two maximum values of E. At what times do these maxima and minima occur?

7.12. Determine the real and imaginary parts of the DFT, using the MATLAB function `fft`, for the following periodic data where the 32 data points are sampled at intervals of 0.1 second. Examine the amplitude and frequency of its components. What conclusions can you draw from these results?

$$\begin{aligned} y = [&2 \ -0.404 \ 0.2346 \ 2.6687 \ -1.4142 \ -1.0973 \ 0.8478 \ -2.37 \ 0 \ldots \\ &2.37 \ -0.8478 \ 1.0973 \ 1.4142 \ -2.6687 \ -0.2346 \ 0.404 \ -2 \ldots \\ &1.8182 \ 1.7654 \ -1.2545 \ 1.4142 \ -0.3169 \ -2.8478 \ 0.9558 \ldots \\ &0 \ -0.9558 \ 2.8478 \ 0.3169 \ -1.4142 \ 1.2545 \ -1.7654 \ -1.8182] \end{aligned}$$

7.13. Determine the DFT of $y = 32\sin^5(2\pi ft)$ where $f = 30$ Hz. Use 512 points sampled over 1 second. From the imaginary part of the DFT, estimate the coefficients a_0, a_1, a_2 in the relationship

$$32\sin^5(2\pi ft) = a_0 \sin[2\pi ft] + a_1 \sin[2\pi(3f)t] + a_2 \sin[2\pi(5f)t]$$

Repeat the process for $y = 32\sin^6(2\pi ft)$ where $f = 30$ Hz. Use 512 points sampled over 1 second. From the real part of the DFT, estimate the coefficients b_0, b_1, b_2, b_3 in the relationship

$$32\sin^6(2\pi ft) = b_0 + b_1\cos[2\pi(2f)t] + b_2\cos[2\pi(4f)t] + b_3\cos[2\pi(6f)t]$$

7.14. Determine the DFT of a set of 512 data points sampled over a 1 second period from

$$y = \sin(2\pi f_1 t) + 2\sin(2\pi f_2 t)$$

where $f_1 = 30$ Hz and $f_2 = 400$ Hz. Explain why there is a large component in the spectrum at 112 Hz.

7.15. Determine the DFT of a set of 256 data points sampled over 1 second from $y(t) = \sin(2\pi ft)$ for $f = 25$, 30.27, and 35.49 Hz. Plot the absolute value of the DFT against frequency for all three values of f in the same figure. It will be noted that even though the amplitude of the sine function from which the samples are taken is the same in each case, the frequency components corresponding to f have different amplitudes. This is because, in the case of the 30.27 Hz and 35.49 Hz waves the sampling is not over an integer number of periods of y. This phenomenon is known as "leakage" or "smearing," and part of the pure sine wave seems to have smeared into adjacent frequencies. Its effect may be reduced by applying a "window" to the data. The Hanning window is $w(t) = 0.5\{1 - \cos(2\pi t/T)\}$ where T is the sampling period. Multiply $y(t)$ by $w(t)$ and determine the DFT of the resulting data. Plot the absolute value of this DFT against frequency for all three values of f in the same figure. Note that the amplitude variation of the frequency components corresponding to f and the smearing into other frequencies has been reduced significantly.

7.16. The following 32 data points are sampled over a period of 0.0625 second.

$$y = [0\ 0.9094\ 0.4251\ -0.6030\ -0.6567\ 0.2247\ 0.6840\ 0.1217 \ldots$$
$$-0.5462\ -0.3626\ 0.3120\ 0.4655\ -0.0575\ -0.4373\ -0.1537 \ldots$$
$$0.3137\ 0.2822\ -0.1446\ -0.3164\ -0.0204\ 0.2694\ 0.1439\ -0.1702 \ldots$$
$$-0.2065\ 0.0536\ 0.2071\ 0.0496\ -0.1594\ -0.1182\ 0.0853 \ldots$$
$$0.1441 - 0.0078]$$

(a) Determine the DFT and estimate the frequency of the most significant component present in the data. What is the frequency increment in the DFT?

(b) To the end of the existing data, add an additional 480 zero values, thus increasing the number of data points to 512. This process is called "zero padding" and is used to improve the frequency resolution in the DFT. Determine the DFT of the new data set and estimate the frequency of the most significant component. What is the frequency increment in the DFT?

7.17. The cost of producing an electronic component varied over a four-year period as follows:

Year	0	1	2	3
Cost	$30.2	$25.8	$22.2	$20.2

Assuming the equation relating production cost and time is (a) a cubic and (b) a quadratic polynomial, estimate the cost of production in year 6. A small error was discovered in the data. The cost of production in year 2 should have been $22.5 and in year 3, $20.5. Recompute the estimated production cost in year 6 using a cubic and a quadratic equation as before. What conclusions can you draw from the results?

7.18. From the following table of values of the gamma function, use inverse interpolation to find the value of x in the range $x = 2$ to $x = 3$ that makes $\Gamma(x) = 1.3$. Use the MATLAB function `interp1` with the cubic option selected; also use the function `aiken`.

x	2	2.2	2.4	2.6	2.8	3
$\Gamma(x)$	1.0000	1.1018	1.2422	1.4296	1.6765	2.0000

7.19. The following table gives the value of the integral

$$I = \int_0^{\pi/2} \frac{d\varphi}{\sqrt{1 - \sin^2\alpha \sin^2\varphi}}$$

for various values of α. (This integral is the complete elliptical integral of the first kind.)

α	0°	5°	10°	15°	20°	25°
I	1.57080	1.57379	1.58284	1.59814	1.62003	1.64900

Using polynomial interpolation, find I when $\alpha = 2°$. Then use inverse interpolation to find the value of α such that $I = 1.58$. In both cases use the MATLAB function `interp1` with the cubic option selected; also use the function `aiken`.

7.20. It is required to find a formula to calculate the number of nodes in one corner of a cube. If n is the number of equally spaced nodes on an edge of the cube and f_n is the number of nodes on three half-faces, including nodes on the face diagonals, then the following table shows values of f_n for given values of n.

n	1	2	3	4
f_n	1	4	10	20

By fitting a cubic function to this data (using `polyfit`), find a general formula for the relationship between f_n and n and verify that when $n = 5, f_n = 35$.

7.21. It is required to fit a regression model of the form $z = f(x, y)$ to the following data:

x	0.5	1.0	1.0	2.0	2.5	2.0	3.0	3.5	4.0
y	2.0	4.0	5.0	2.0	4.0	5.0	2.0	4.0	5.0
z	−0.19	−0.32	−1.00	3.71	4.49	2.48	6.31	7.71	8.51

(a) Use the function `mregg2` to generate a model of the form $z = a + bx + cy$ and also $z = a + bx + cy + dxy$. Which model do you consider the best fit to the data? It is important to consider the differences in the error variance.

(b) By analysis of the residuals (particularly the Cook distance), decide whether any data point could be considered to be an *outlier*.

7.22. One of the data values given in Problem 7.21 has been found to be in error. The value of z corresponding to $x = 4, y = 5$ should have been recorded as 9.51, not 8.51, a common human error. Use the function `mregg2` to generate models of the form $z = a + bx + cy$ and also $z = a + bx + cy + dxy$. Again, assess the quality of the models.

7.23. Using the following table, obtain the pressure across a shock wave when the upstream Mach number is 4.4 by

(a) Linear interpolation
(b) Aitken's method
(c) Spline interpolation

Mach no.	1.00	2.00	3.00	4.00	5.00
p_2/p_1	1.00	4.50	10.33	18.50	29.00

7.24. MATLAB has available test data sets, including data collected for sunspot activity. This may be obtained using the MATLAB statement

```
load sunspot.dat
```

The set of data `sunspot(:,1)` gives the year of the observed sunspot activity and `sunspot(:,2)` the Wolfer number, which indicates the level of sunspot activity for that year. Let `wolfer = sunspot(:,2)`. Generate a simple plot of the the Wolfer number against the year. To further analyze this data take the fast Fourier transform of the variable `wolfer`. Scaling this data will help in its interpretation. Do this by using the transformations `Power = abs(Y(1:N/2)).^2` and `freq = (1:N/2)/N;` where `N` is the length of the vector `Y`. Plot `freq` against `power`.

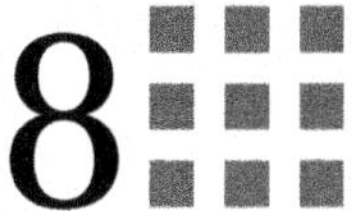

8 Optimization Methods

The purpose of this chapter is to bring together a selection of algorithms for optimizing linear and nonlinear functions that have applications in science and engineering. We deal with constrained linear optimization problems and both constrained and unconstrained nonlinear optimization problems.

8.1 Introduction

The major techniques of optimization considered in this chapter are:

1. The solution of linear programming problems by interior point methods
2. The optimization of single-variable nonlinear functions
3. The solution of nonlinear optimization problems and systems of linear equations using conjugate gradient methods
4. The solution of constrained nonlinear optimization problems using the sequential unconstrained minimization technique (SUMT)
5. The solution of nonlinear optimization problems using the genetic algorithm and the method of simulated annealing

It is not our intention to describe fully the theoretical basis for these methods but to give some indication of the ideas that lie behind them. We begin with a discussion of linear programming problems.

8.2 Linear Programming Problems

Linear programming is normally considered to be an operational research (OR) method but has a very wide range of applications. A detailed description of the problem and associated theory is beyond the scope of this text but this information can be obtained from Dantzig (1963) and Sultan (1993). The problem may be expressed in standard form as

$$\begin{aligned} \text{Minimize}\, f &= \mathbf{c}^\top \mathbf{x} \\ \text{subject to}\, \mathbf{A}\mathbf{x} &= \mathbf{b} \\ \text{and}\ \mathbf{x} &\geq \mathbf{0} \end{aligned} \tag{8.1}$$

where **x** is the column vector of n components that we wish to determine. Note that each element of **x** is constrained to be greater than zero. This is a common requirement in this type of optimization because most practical optimization problems will require nonnegative values for **x**. For example, if each element of **x** is the number of workers of a particular skill set employed by an organization, the number of workers in any group cannot be negative. The given constants of the system are provided by an m component column vector **b**, an $m \times n$ matrix **A**, and an n component column vector **c**. Clearly all the equations and the function we wish to minimize are linear in form. The problem is an optimization problem and in general it represents the requirement to minimize a linear function, $\mathbf{c}^\top \mathbf{x}$, called the objective function, subject to satisfying a system of linear equalities.

The importance of this type of problem lies in the fact that it corresponds to the general aim of optimizing the use of scarce resources to meet a specific objective. Although we have given the standard form, many other forms of this problem arise that are easily converted to this standard form. For example, the constraints may initially be inequalities and these can be converted to equalities by adding or subtracting additional variables introduced into the problem. The objective may be to maximize the function rather than minimize it. Again this is easily converted by changing the sign of the **c** vector.

Some practical examples where linear programming has been applied are

1. The hospital diet problem, requiring food costs to be minimized while dietary constraints are satisfied
2. The problem of minimizing cutting pattern loss
3. The problem of optimizing profit subject to constraints on the availability of specified materials
4. The problem of optimizing the routing of telephone calls

An important numerical algorithm for solving this problem is called the simplex method; see Dantzig (1963). This was applied to wartime problems of troop and material distribution. However, here we consider more recent developments that have provided new algorithms that are theoretically better. These are based on the work of Karmarkar (1984), who produced an algorithm that differed greatly in principle from that of Dantzig. While the theoretical complexity of Dantzig's method is exponential in the number of variables of the problem, some versions of Karmarkar's algorithm have a complexity that is of the order of the cube of the number of variables. It has been reported that for some problems this leads to substantial saving of computational effort. Here we describe an algorithm due to Barnes (1986) that provides an elegant modification of Karmarkar's algorithm but preserves its fundamental principles.

We do not describe the theoretical details of these complex algorithms but it is useful to compare, in broad terms, the nature of the Karmarkar and Dantzig algorithms. The simplex method of Dantzig is best illustrated by considering a simple linear programming problem as follows. In a factory producing electronic components, let x_1 be the number of batches of resistors and x_2 the number of batches of capacitors produced. Each batch of resistors manufactured gains 7 units of profit and each batch of capacitors gains 13 units of profit. Each is manufactured in a two-stage process. Stage 1 is limited to 18 units of time per week and stage 2 is limited to 54 units of time per week. A batch of resistors requires 1 unit of time in stage 1 and 5 units of time in stage 2. A batch of capacitors requires 3 units of time in the first stage and 6 in the second. The aim of the manufacturer is to maximize profitability while meeting the time constraints; this leads to the following linear programming problem.

$$\text{Maximize } z = 7x_1 + 13x_2 \text{ (where } z \text{ is the profit)}$$

subject to

$$\begin{aligned} x_1 + 3x_2 &\leq 18 \qquad \text{(stage 1 process)} \\ 5x_1 + 6x_2 &\leq 54 \qquad \text{(stage 2 process)} \\ \text{and} \quad x_1, x_2 &\geq 0 \end{aligned}$$

To see how the simplex algorithm works we give a geometric interpretation of this problem in Figure 8.1. In this figure the region lying under the shaded lines and confined by the x_1- and x_2-axes represents the feasible region. This is the region in which all possible solutions to the problem lie. Clearly there is an infinity of such points. Fortunately, it can be shown that the only true candidates for the optimum solution are the points that lie at the vertices of the feasible region. In fact, we can find this optimum using simple geometric principles. The objective function is represented in Figure 8.1 by the dashed line of constant slope and variable intercept proportional to the value of the objective function. If we move this line parallel to itself until it just leaves the feasible region, it leaves at the vertex that gives the maximum value of the objective function. Clearly, beyond this point the values of x_1 and x_2 no longer satisfy the constraints. For this problem the optimum solution is given by $x_1 = 6$, $x_2 = 4$ so that the profit $z = 94$.

Although this provides a solution for this simple two-variable problem, linear programming problems often involve thousands or hundreds of thousands of variables. For practical problems a well-specified numerical algorithm is required. This is provided by

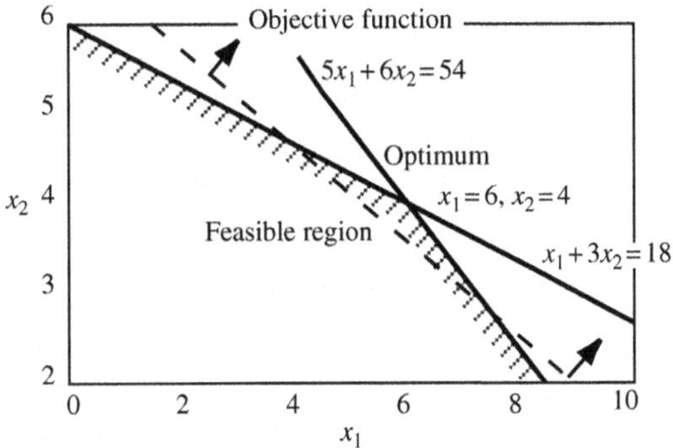

FIGURE 8.1 Graphical representation of an optimization problem. The *dashed line* represents the objective function and the *solid lines* represent the constraints.

Dantzig's simplex algorithm. We do not describe this in detail here but the general principle of its operation is to generate a sequence of points that correspond mathematically to the vertices of the multidimensional feasible region. The algorithm proceeds from one vertex to another, each time improving the value of the objective function, until the optimum is found. These points are all on the surface of the feasible region and for larger problems there may be a huge number of them.

The algorithm proposed by Karmarkar deals with the linear programming problem in a different way. The algorithm was developed at AT&T to solve very large linear programming problems concerned with routing telephone calls in the Pacific Basin. This algorithm transforms the problem to a more convenient form and then searches through the interior of the feasible region using a good direction of search toward its surface. Because this type of algorithm uses interior points, it is often described as an *interior point* method. Since its discovery, many improvements and modifications have been made to this algorithm and here we describe a form which, although conceptually complex, leads to a remarkably simple and elegant linear programming algorithm. This formulation was given by Barnes (1986).

The Barnes algorithm may be applied to any linear programming problem once it is converted to the form of (8.1). However, one important initial modification is required to ensure the algorithm starts at an interior point $\mathbf{x}^0 > \mathbf{0}$. This modification is achieved by introducing an additional column, that is, a new last column, to the **A** matrix, the elements of which are the **b** vector minus the sum of the columns of the **A** matrix. We associate an additional variable with this additional column and, in order that we do not have a superfluous variable in the solution, we introduce an extra element in the vector **c**. We make the value of this element very large to ensure that the new variable is driven to zero when the optimum is reached. Now we find that $\mathbf{x}^0 = [1\ 1\ 1\ \dots\ 1]^\top$ satisfies this set of constraints and clearly $\mathbf{x}^0 \geq \mathbf{0}$. We now describe the Barnes algorithm:

- *Step* 0: Assuming n variables in the original problem,

$$\text{set } a(i,n+1) = b(i) - \sum_j a(i,j) \quad \text{and} \quad c(n+1) = 10000$$

$$\mathbf{x}^0 = [1\ 1\ 1\ \dots\ 1], \qquad k = 0$$

- *Step* 1: Set $\mathbf{D}^k = \text{diag}(\mathbf{x}^k)$ and compute an improved point using the equation

$$\mathbf{x}^{k+1} = \mathbf{x}^k - \frac{s(\mathbf{D}^k)^2(\mathbf{c} - \mathbf{A}^\top \lambda^k)}{\text{norm}(\mathbf{D}^k(\mathbf{c} - \mathbf{A}^\top \lambda^k))}$$

  where the vector λ^k is given by

$$\lambda^k = (\mathbf{A}(\mathbf{D}^k)^2\mathbf{A}^\top)^{-1}\mathbf{A}(\mathbf{D}^k)^2\mathbf{c}$$

The step s is chosen such that

$$s = \min\left\{\frac{\text{norm}\left(\left(\mathbf{D}^k\right)\left(\mathbf{c}-\mathbf{A}^\top\lambda^k\right)\right)}{x_j^k\left(c_j-\mathbf{A}_j^\top\lambda^k\right)}\right\} - \alpha$$

where $\mathbf{A}_j$ is the jth column of the matrix $\mathbf{A}$ and α is a small preset constant value. Here the minimum is taken for the values

$$\left(c_j-\mathbf{A}_j^\top\lambda^k\right) > 0 \text{ only}$$

Note also that λ^k provides an approximation for the solution of the dual problem (see for example, Problems 8.1 and 8.2).

- *Step* 2: Stop if the *primal* and *dual* values of the objective functions are approximately equal. Else set $k = k+1$ and repeat from step 1.

Note that in Ludwig step 2 we use an important result in linear programming. This is that every primal problem (i.e., the original problem) has a corresponding dual problem and if a solution exists, the optimal values of their objective functions are equal. There are several other termination criteria that could be used and Barnes suggested a more complex but more reliable one.

The algorithm provides an iterative improvement starting from the initial point $\mathbf{x}^0$ by taking the maximum step that ensures that $\mathbf{x}^k > \mathbf{0}$ in the normalized direction given by $(\mathbf{D}^k)^2(\mathbf{c}-\mathbf{A}^\top\lambda^k)$. It is this direction that is the crucial element of the algorithm. This direction is a projection of the objective function coefficients into the constraint space. For a proof that this direction reduces the objective function, while ensuring the constraints are satisfied, the reader is referred to Barnes (1986).

The reader should be warned that this algorithm is deceptively simple. In fact, the computation of the direction is very difficult for large problems. This is because the algorithm requires the solution of an extremely ill-conditioned equation system. Many alternatives have been suggested for finding the direction of search, including the use of a conjugate gradient method that is discussed in Section 8.6. The MATLAB function `barnes` provided here solves the ill-conditioned equation system in a direct manner using the MATLAB \ operator. The function `barnes` is easily modified to use the conjugate gradient solver given in Section 8.6.

```
function [xsol,basic,objective] = barnes(A,b,c,tol)
% Barnes' method for solving a linear programming problem
% to minimize c'x subject to Ax = b. Assumes problem is non-degenerate.
% Example call: [xsol,basic]=barnes(A,b,c,tol)
% A is the matrix of coefficients of the constraints.
% b is the right-hand side column vector and c is the row vector of
% cost coefficients. xsol is the solution vector, basic is the
```

```
% list of basic variables.
x2 = [ ];  x = [ ];
[m n] = size(A);
% Set up initial problem
aplus1 = b-sum(A(1:m,:)')';
cplus1 = 1000000;
A = [A aplus1]; c = [c cplus1]; B = [ ];
n = n+1;
x0 = ones(1,n)'; x = x0;
alpha = .0001; lambda = zeros(1,m)';
iter = 0;
% Main step
while abs(c*x-lambda'*b)>tol
    x2 = x.*x;
    D = diag(x); D2 = diag(x2); AD2 = A*D2;
    lambda = (AD2*A')\(AD2*c');
    dualres = c'-A'*lambda;
    normres = norm(D*dualres);
    for i = 1:n
        if dualres(i)>0
            ratio(i) = normres/(x(i)*(c(i)-A(:,i)'*lambda));
        else
            ratio(i)=inf;
        end
    end
    R = min(ratio)-alpha;
    x1 = x-R*D2*dualres/normres;
    x = x1;
    basiscount = 0;
    B = [ ]; basic = [ ];
    cb = [ ];
    for k = 1:n
        if x(k)>tol
            basiscount = basiscount+1;
            basic = [basic k];
        end
    end
    % Only used if problem non-degenerate
    if basiscount==m
        for k = basic
            B = [B A(:,k)];   cb = [cb c(k)];
        end
```

```
        primalsol = b'/B';
        xsol = primalsol;
        break
    end
    iter = iter+1;
end
objective = c*x;
```

We now solve the linear programming problem

$$\text{Maximize } z = 2x_1 + x_2 + 4x_3$$

subject to

$$\begin{aligned} x_1 + x_2 + x_3 &\le 7 \\ x_1 + 2x_2 + 3x_3 &\le 12 \\ x_1, x_2, x_3 &\ge 0 \end{aligned}$$

The requirements that $x_1, x_2, x_3 \ge 0$ are called nonnegativity constraints. This linear programming problem can be easily transformed to the standard form by adding new positive-valued variables, called slack variables, to the left sides of the inequalities and changing the signs of the coefficients in the objective function so that it is converted to a minimization problem subject to equality constraints as follows:

$$\text{Minimize } -z = -(2x_1 + x_2 + 4x_3)$$

subject to

$$\begin{aligned} x_1 + x_2 + x_3 + x_4 &= 7 \\ x_1 + 2x_2 + 3x_3 + x_5 &= 12 \\ x_1, x_2, x_3, x_4, x_5 &\ge 0 \end{aligned}$$

The variables x_4 and x_5 are called the slack variables and they represent the difference between the available resources and the resources used. Note that if the constraints were of the form greater than or equal to zero, we would subtract slack variables to produce equality. These subtracted variables are sometimes called surplus variables. Thus we have

$$\mathbf{c} = \begin{bmatrix} -2 & -1 & -4 & 0 & 0 \end{bmatrix}$$

We use the following script to solve this problem.

```
% e3s801.m
c = [-2 -1 -4 0 0];
A = [1 1 1 1 0;1 2 3 0 1 ]; b = [7 12]';
[xsol,ind,object] = barnes(A,b,c,0.00005);
```

```
fprintf('objective = %8.4f', object)
i = 1;
fprintf('\nSolution is:');
for j = ind
    fprintf('\nx(%1.0f) =%8.4f',j,xsol(i))
    i = i+1;
end;
fprintf('\nAll other variables are zero\n')
```

Running this script provides the result

```
objective = -19.0000
Solution is:
x(1) =   4.5000
x(3) =   2.5000
All other variables are zero
```

Since the original problem was to maximize objective function, its value is 19. This solution illustrates an important theorem of linear programming. The number of nonzero primal variables is at most equal to the number of independent constraints (excluding nonnegativity constraints). In this problem there are only two main constraints. Thus there are only two nonzero variables, x_1 and x_3. The slack variables x_4 and x_5 are zero and so is x_2.

The `lsqnonneg` function, discussed in Section 2.12, provides a method for finding a solution to an equation system in which all components of the solution are nonnegative. This corresponds to a basic feasible solution for the system but it is generally nonoptimum for a specific objective function.

Having examined the process for solving linear optimization problems, we now consider methods that are used to solve nonlinear optimization problems.

8.3 Optimizing Single-Variable Functions

We sometimes need to determine the maximum or minimum value of a one-variable nonlinear function. Throughout this discussion we assume we are seeking the minimum value of the function. If we require the maximum value, then we merely have to change the sign of the original function.

The most obvious way of determining the minimum of a function is to differentiate it and find the value of the independent variable that makes this derivative zero. However, there are situations in which it is not practical to find the derivative directly; see for example (8.4). A method is now described that provides an approximation to the minimum to any required accuracy.

Consider a function $y = f(x)$ and let us assume that in the range $[x_a \ \ x_b]$ there is a single minimum, as shown in Figure 8.2. Two additional points, x_1 and x_2, are chosen arbitrarily so that the range is divided into three intervals. Assuming that $x_a < x_1 < x_2 < x_b$, then

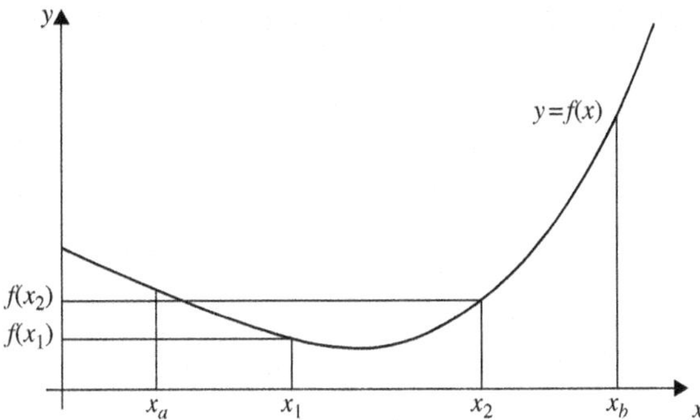

FIGURE 8.2 Graph of a function with a minimum in the range $[x_a\, x_b]$.

If $f(x_1) < f(x_2)$ then the minimum value must lie in the range $[x_a\; x_2]$.

If $f(x_1) > f(x_2)$ then the minimum value must lie in the range $[x_1\; x_b]$.

Either of these ranges must provide a smaller interval than $[x_a\;\; x_b]$ in which the minimum lies. This interval reduction process can be repeated continuously in successively smaller ranges until an acceptably small interval is found for the minimum.

It might be assumed that the most efficient procedure is to select x_1 and x_2 so that the range $[x_a\; x_b]$ is subdivided into three equal intervals. In fact, this is not so and for a more efficient procedure we take

$$x_1 = x_a + r(1-g), \quad x_2 = x_a + rg$$

where $r = x_b - x_a$ and

$$g = \frac{1}{2}\left(-1+\sqrt{5}\right) \approx 0.61803$$

The quantity g is called the *golden ratio*. This quantity has many interesting properties. For example, it is one of the roots of the equation

$$x^2 + x + 1 = 0$$

This golden ratio is also related to the famous Fibonacci series. This series is $1, 1, 2, 3, 5, 8, 13, \ldots$ and it is generated from

$$N_{k+1} = N_k + N_{k-1}, \quad k = 2, 3, 4, \ldots$$

where $N_2 = N_1 = 1$ and N_k is the kth term in the series. As k tends to infinity the ratio N_k/N_{k+1} tends to the golden ratio.

The algorithm described in the preceding is implemented in MATLAB as follows:

```
function [f,a,iter] = golden(func,p,tol)
% Golden search for finding min of one variable nonlinear function.
% Example call: [f,a] = golden(func,p,tol)
% func is the name of the user defined nonlinear function.
% p is a 2 element vector giving the search range.
% tol is the tolerance. a is the optimum value of the function.
% f is the minimum of the function. iter is the number of iterations
if p(1)<p(2)
    a = p(1);  b = p(2);
else
    a = p(2);  b = p(1);
end
g = (-1+sqrt(5))/2;
r = b-a;  iter = 0;
while r>tol
    x = [a+(1-g)*r a+g*r];
    y = feval(func,x);
    if y(1)<y(2)
        b = x(2);
    else
        a = x(1);
    end
    r = b-a; iter = iter+1;
end
f = feval(func,a);
```

We can use the function `golden` to search for the minimum value of the Bessel function of the second kind of order 2. The function `bessely(2,x)` is provided by MATLAB. The following command provides the output shown:

```
>> format long
>> [f,x,iter] = golden(@(x) bessely(2,x),[4 10],0.000001)

f =
  -0.279275263440711


x =
   8.350724427010965


iter =
    33
```

Note that if we had divided the search interval into three equal sections, rather than using the golden ratio, then 39 iterations would have been required.

The search algorithm has been developed assuming that there is only one minimum value of the function in the search range. If there are several minima in the search range, then the procedure locates one, but the one located is not necessarily the global minimum in the range. For example, a Bessel function of the second kind of order 2 has 3 minima in the range 4 to 25, as shown in Figure 8.3.

If we use the function `golden` and search in the ranges 4 to 24, 4 to 25, and 4 to 26, we obtain the results given in Table 8.1. In this table we see that a different minimum has been found when using different search ranges, even though all three minima are in each of the search ranges used. Ideally we require the search to determine the global minimum of the function, something that the function `golden` has failed to accomplish in two out of three tests. Obtaining a *global* solution is a major problem in minimization.

In this particular example we can verify the accuracy of these solutions by calculus. The derivative of the Bessel function of the second kind of order n is given by

$$\frac{d}{dx}\{Y_n(x)\} = \frac{1}{2}\{Y_{n-1}(x) - Y_{n+1}(x)\}$$

where $Y_n(x)$ is the Bessel function of the second kind of order n. (Sometimes $N_n(x)$ is used instead of $Y_n(x)$.)

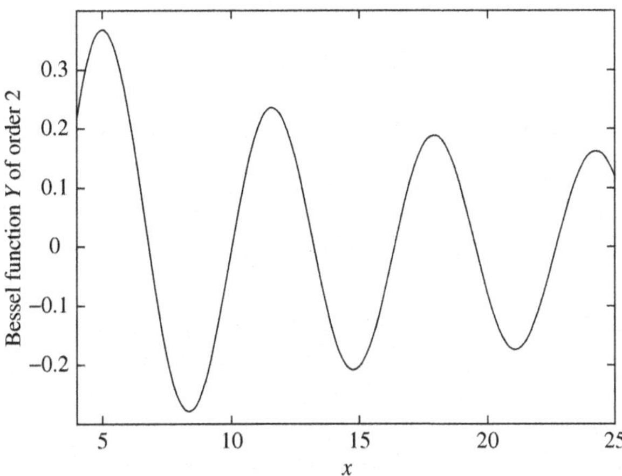

FIGURE 8.3 A plot of the Bessel function of the second kind showing three minima.

Table 8.1 Effect of Different Search Ranges

Search Range	Value of $f(x)$ at Min	Value of x at Min
4 to 24	−0.20844576503764	14.76085144779431
4 to 25	−0.17404548213116	21.09284729991696
4 to 26	−0.27927526323841	8.35068549680869

The minimum (or maximum) value of a function occurs when the derivative of the function is zero. Hence, with $n = 2$ we have a minimum (or maximum) occurring when

$$Y_1(x) - Y_3(x) = 0$$

We cannot escape from the need to use numerical methods because the only way to find the roots of this equation is to use a numerical procedure such as that implemented in the MATLAB function `fzero` (see Chapter 3). Thus, using this function to determine a root near 8 we have

```
>> format long
>> fzero(@(x) bessely(1,x)-bessely(3,x),8)

ans =
   8.350724701413078
```

We can also use `fzero` to find roots at 14.76090930620768 and 21.09289450441274. These results are in good agreement with the minima found using the function `golden`.

8.4 The Conjugate Gradient Method

Here we confine ourselves to solving the problem

$$\text{Minimize } f(\mathbf{x}) \quad \text{for all} \quad \mathbf{x} \in \mathrm{R}^{\mathrm{n}}$$

where $f(\mathbf{x})$ is a nonlinear function of $\mathbf{x}$ and $\mathbf{x}$ is an n-component column vector. This is called a nonlinear unconstrained optimization problem. These problems arise in many applications–for example, in neural network problems where an important aim is to find weights in a network that minimize the difference between the output of the network and the required output.

The standard approach for solving this problem is to assume an initial approximation $\mathbf{x}^0$ and then to proceed to an improved approximation by using an iterative formula of the form

$$\mathbf{x}^{k+1} = \mathbf{x}^k + s\mathbf{d}^k \quad \text{for} \quad k = 0, 1, 2, \ldots \tag{8.2}$$

Clearly, to use this formula we must determine values for the scalar s and the vector $\mathbf{d}^k$. The vector $\mathbf{d}^k$ represents a direction of search and the scalar s determines how far we should step in this direction. A vast literature has grown up that has examined the problem of choosing the best direction and the best step size to solve this problem efficiently. For example, see Adby and Dempster (1974). A simple choice for a direction of search is to take $\mathbf{d}^k$ as the negative gradient vector at the point $\mathbf{x}^k$. For a sufficiently small step value this can be shown to guarantee a reduction in the function value. This leads to an algorithm of the form

$$\mathbf{x}^{k+1} = \mathbf{x}^k - s\nabla f\left(\mathbf{x}^k\right) \quad \text{for} \quad k = 0,\ 1,\ 2,\ \ldots \tag{8.3}$$

where $\nabla f(\mathbf{x}) = (\partial f/\partial x_1, \partial f/\partial x_2, \ldots, \partial f/\partial x_n)$ and s is a small constant value. This is called the steepest descent algorithm. The minimum is reached when the gradient is zero, as in the ordinary calculus approach. We also assume that there exists only one local minimum that we wish to find in the range considered. The problem with this method is that although it reduces the function value, the step may be very small and therefore the algorithm is very slow. An alternative approach is to choose the step that gives the maximum reduction in the function value in the current direction. This may be described formally as

$$\text{For each } k \text{ find the value of } s \text{ that minimizes } f(\mathbf{x}^k - s\nabla f(\mathbf{x}^k)) \tag{8.4}$$

This procedure is known as a line search. The reader will note that this is also a minimization problem. However, since $\mathbf{x}^k$ is known, it is a *one-variable* minimization problem in the step size s. Although it is a difficult problem, numerical procedures are available to solve it, one of which is the search method given in Section 8.3. Equations (8.3) and (8.4) provide a workable algorithm but it is still slow. One reason for this poor performance lies in our choice of direction $-\nabla f(\mathbf{x}^k)$.

Consider the function we wish to minimize in (8.4). Clearly the value of s that minimizes $f(\mathbf{x}^k - s\nabla f(\mathbf{x}^k))$ is such that the derivative of $f(\mathbf{x}^k - s\nabla f(\mathbf{x}^k))$ with respect to s is zero. Now, differentiating $f(\mathbf{x}^k - s\nabla f(\mathbf{x}^k))$ with respect to s gives

$$\frac{df\left(\mathbf{x}^k - s\nabla f\left(\mathbf{x}^k\right)\right)}{ds} = -\left(\nabla f\left(\mathbf{x}^{k+1}\right)\right)^\top \nabla f\left(\mathbf{x}^k\right) = 0 \tag{8.5}$$

This shows that the successive directions of search are orthogonal. This is not the best way of getting from our original approximation to the optimum value since the changes in direction are so large.

The conjugate gradient method takes a combination of the previous direction and the new direction to approach the optimum more directly. It uses the same step size choice procedure given by (8.4), so we must now consider how the direction vector is chosen in the conjugate gradient method. Let $\mathbf{g}^{k+1} = \nabla f(\mathbf{x}^{k+1})$ so that the basic formula for the conjugate gradient direction is

$$\mathbf{d}^{k+1} = -\mathbf{g}^{k+1} + \beta\mathbf{d}^k \tag{8.6}$$

Thus the current direction of search is a combination of the current negative gradient plus a scalar β times the previous direction of search. The crucial question is: How is the value of β to be determined? The criterion used is that successive directions of search should be conjugate. This means that $(\mathbf{d}^{k+1})^\top \mathbf{A}\,\mathbf{d}^k = 0$ for some specified matrix $\mathbf{A}$.

This apparently obscure choice of requirement can be shown to lead to desirable convergence properties for the conjugate gradient method. In particular it has the property that the optimum of a positive definite quadratic function of n variables can be found in n or fewer steps. In the case of a quadratic, $\mathbf{A}$ is the matrix of coefficients of the squared and cross-product terms. It can be shown that the requirement of conjugacy leads to a value for β given by

$$\beta = \frac{\left(\mathbf{g}^{k+1}\right)^{\top} \mathbf{g}^{k+1}}{\left(\mathbf{g}^{k}\right)^{\top} \mathbf{g}^{k}} \tag{8.7}$$

Now (8.2), (8.4), (8.6), and (8.7) lead to the conjugate gradient algorithm given by Fletcher and Reeves (1964), which has the form

- *Step* 0: Input value for $\mathbf{x}^0$ and accuracy ε. Set $k = 0$ and compute $\mathbf{d}^k = -\nabla f(\mathbf{x}^k)$.
- *Step* 1: Determine s_k, which is the value of s that minimizes $f(\mathbf{x}^k + s\mathbf{d}^k)$.
  Calculate $\mathbf{x}^{k+1}$ where $\mathbf{x}^{k+1} = \mathbf{x}^k + s_k\mathbf{d}^k$ and compute $\mathbf{g}^{k+1} = \nabla f(\mathbf{x}^{k+1})$.
  If norm$(\mathbf{g}^{k+1}) < \varepsilon$, then terminate with solution $\mathbf{x}^{k+1}$, else go to step 2.
- *Step* 2: Calculate new conjugate direction $\mathbf{d}^{k+1}$ where

$$\mathbf{d}^{k+1} = -\mathbf{g}^{k+1} + \beta\mathbf{d}^k \quad \text{and} \quad \beta = (\mathbf{g}^{k+1})^{\top}\mathbf{g}^{k+1}/\{(\mathbf{g}^k)^{\top}\mathbf{g}^k\}$$

- *Step* 3: $k = k + 1$; go to step 1.

Note that in other forms of this algorithm steps 1, 2, and 3 are repeated n times and then restarted with a steepest descent step from step 0. The following is a MATLAB function for this method.

```
function [x1,df,noiter] = mincg(f,derf,ftau,x,tol)
% Finds local min of a multivariable nonlinear function in n variables
% using conjugate gradient method.
% Example call: res = mincg(f,derf,ftau,x,tol)
% f is a user defined multi-variable function,
% derf a user defined function of n first order partial derivatives.
% ftau is the line search function.
% x is a col vector of n starting values, tol gives required accuracy.
% x1 is solution, df is the gradient,
% noiter is the number of iterations required.
% WARNING. Not guaranteed to work with all functions. For difficult
% problems the linear search accuracy may have to be adjusted.
global p1 d1
n = size(x);  noiter = 0;
% Calculate initial gradient
df = feval(derf,x);
% main loop
```

```
while norm(df)>tol
    noiter = noiter+1;
    df = feval(derf,x);
    d1 = -df;
    %Inner loop
    for inner = 1:n
        p1 = x;  tau = fminbnd(ftau,-10,10);
        % calculate new x
        x1 = x+tau*d1;
        % Save previous gradient
        dfp = df;
        % Calculate new gradient
        df = feval(derf,x1);
        % Update x and d
        d = d1; x = x1;
        % Conjugate gradient method
        beta = (df'*df)/(dfp'*dfp);
        d1 = -df+beta*d;
    end
end
```

Notice that the MATLAB function fminbnd is used in the function mincg to perform the single-variable minimization to find the best step value. It is important to note that the function mincg requires three input functions, which must be supplied by the user. They are the function to be minimized, the partial derivatives of this function, and the line-search function. As implemented, the function mincg requires the input functions to be user-defined functions, not anonymous functions. An example of the use of mincg follows.

The function to be minimized, which is taken from Styblinski and Tang (1990), is

$$f(x_1,x_2) = \left(x_1^4 - 16x_1^2 + 5x_1\right)/2 + \left(x_2^4 - 16x_2^2 + 5x_2\right)/2$$

The function f01 and the derivative of this function, f01d, are defined as follows:

```
function f = f01(x)
f = 0.5*(x(1)^4-16*x(1)^2+5*x(1)) + 0.5*(x(2)^4-16*x(2)^2+5*x(2));

function f = f01d(x)
f = [0.5*(4*x(1)^3-32*x(1)+5); 0.5*(4*x(2)^3-32*x(2)+5)];
```

The MATLAB line-search function ftau2cg is defined as

```
function ftauv = ftau2cg(tau);
global p1 d1
q1 = p1+tau*d1;
ftauv = feval('f01',q1);
```

To test the `mincg` function we use the following simple MATLAB commands:

```
>> [sol,grad,iter] = mincg('f01','f01d','ftau2cg',[1 -1]', .000005)
```

The results of executing these statements are

```
sol =
   -2.9035
   -2.9035


grad =
  1.0e-006 *
    0.0156
   -0.2357


iter =
     3
```

Note that

```
>> f = f01(sol)

f =
  -78.3323
```

This is the minimum value of the function determined by `mincg`. It is interesting to see the function that has been optimized and we provide both a three-dimensional and a contour plot of the function in Figures 8.4 and 8.5. The latter includes a plot of the iterates and shows the path taken to reach the optimum solution from a particular starting point. The script used to obtain these graphs is

```
% e3s802.m
clf
[x,y] = meshgrid(-4.0:0.2:4.0,-4.0:0.2:4.0);
z = 0.5*(x.^4-16*x.^2+5*x)+0.5*(y.^4-16*y.^2+5*y);
figure(1)
surfl(x,y,z)
axis([-4 4 -4 4 -80 20])
xlabel('x1'), ylabel('x2'), zlabel('z')
x1=[1 2.8121 -2.8167 -2.9047 -2.9035];
y1=[0.5 -2.0304 -2.0295 -2.9080 -2.9035];
figure(2)
contour(-4.0:0.2:4.0,-4.0:0.2:4.0,z,15);
xlabel('x1'), ylabel('x2')
hold on
```

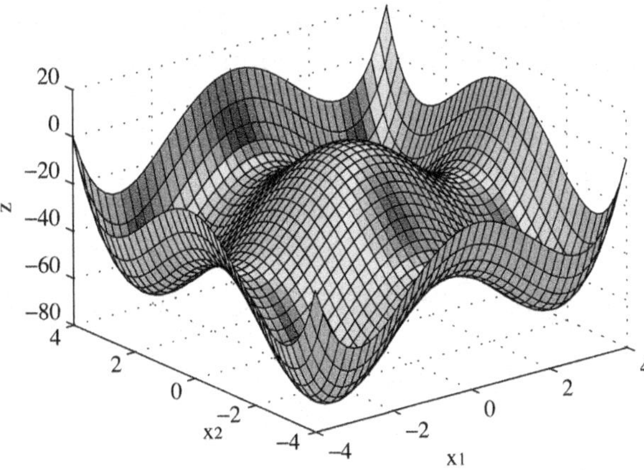

FIGURE 8.4 Three-dimensional plot of $f(x_1, x_2) = (x_1^4 - 16x_1^2 + 5x_1)/2 + (x_2^4 - 16x_2^2 + 5x_2)/2$.

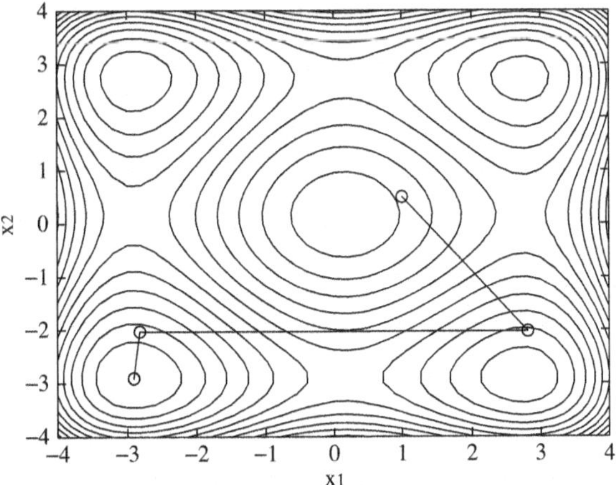

FIGURE 8.5 Contour plot of the function $f(x_1, x_2) = (x_1^4 - 16x_1^2 + 5x_1)/2 + (x_2^4 - 16x_2^2 + 5x_2)/2$ showing the location of four local minima. The conjugate gradient algorithm has found the one in the lower left corner. The search path taken by the algorithm is also shown.

```
plot(x1,y1,x1,y1,'o')
xlabel('x1'), ylabel('x2')
hold off
```

In this script the vectors `x1` and `y1` contain the iterates for the conjugate gradient solution of the given function. These values were obtained by running a modified version of the `mincg` function separately. The minimum we have obtained is in fact the smallest of the four local minima that exist for this function. However, this result was fortuitous; all that the conjugate gradient method is able to do is to find one of the four local minima and even this is not guaranteed for all problems. The conjugate gradient method, because of its small storage requirements, is one of the key algorithms used in neural network problems as part of the back propagation algorithm, but it has many other applications.

It should be noted that a MATLAB optimization toolbox is available and this provides a range of optimization procedures.

8.5 Moller's Scaled Conjugate Gradient Method

In 1993 Moller, when working on optimization methods for neural networks, introduced a much improved version of Fletcher's conjugate gradient method. Fletcher's conjugate gradient method uses a line-search procedure to solve a single-variable minimization problem, which is then used to find the optimum step to take in the chosen direction of search. The procedure used by Fletcher is a fragile, iterative, and computationally intensive process. In addition, the line search depends on a number of parameters that must be estimated by the user. Moller's paper (Moller 1993) introduced a method that allowed the line-search procedure to be replaced by a considerably simplified method for estimating an acceptable step size. However, using a simple estimation of the step size often fails and leads to nonstationary points. Moller noted that a simple approach to the problem fails because it only works for functions with positive definite matrices. Consequently, Moller suggested a method based on a combination of the Levenberg-Marquardt algorithm and the conjugate gradient algorithm. An outline of the algorithm is described in the following; for the details the reader is referred to the original paper.

Consider the n-variable nonlinear function $f(\mathbf{x})$. Moller introduces a scalar parameter λ_k, which is adjusted at each iteration k after considering the sign of δ_k where

$$\delta_k = \mathbf{p}_k^\top \mathbf{H}_k \mathbf{p}_k$$

where $\mathbf{p}_k$ for $k = 1, 2, \ldots, n$ are a set of conjugate directions and $\mathbf{H}_k$ is the Hessian matrix of the function $f(\mathbf{x})$. If $\delta_k \geq 0$ then $\mathbf{H}_k$ is positive definite. However, since only first-order derivitive information is known at each step of the conjugate gradient method, Moller suggests that the Hessian multiplied by $\mathbf{p}_k$ is approximated by

$$\mathbf{s}_k = \frac{f'(\mathbf{x}_k + \sigma_k \mathbf{p}_k) - f'(\mathbf{x}_k)}{\sigma_k} \quad \text{for} \quad 0 < \sigma_k < 1$$

In practice the value of σ_k should be kept as small as possible for a good approximation. This expression in the limit tends to the true Hessian matrix multiplied by $\mathbf{p}_k$. The scalar λ_k is now introduced to regulate the approximation to the Hessian to ensure it is positive definite, specifically by using the equation

$$\mathbf{s}_k = \frac{f'(\mathbf{x}_k + \sigma_k \mathbf{p}_k) - f'(\mathbf{x}_k)}{\sigma_k} + \lambda_k \mathbf{p}_k \quad \text{for} \quad 0 < \sigma_k < 1$$

Thus the value of λ_k is adjusted, and then we check the value of δ_k defined earlier using the approximation to the Hessian; if this is negative then the Hessian is no longer positive definite and the value of λ_k is increased and s_k is checked again. This is repeated until the current estimate of the Hessian is positive definite. The key question is how should λ_k be adjusted to ensure the Hessian estimate becomes positive definite. Let the λ_k be increased to $\bar{\lambda}_k$; then

$$\bar{\mathbf{s}}_k = \mathbf{s}_k + (\bar{\lambda}_k - \lambda_k)\mathbf{p}_k$$

Now at any iteration k a new δ_k, which we can denote as $\bar{\delta}_k$, can be computed from

$$\bar{\delta}_k = \mathbf{p}_k^{\top}\bar{\mathbf{s}}_k = \mathbf{p}_k^{\top}(\mathbf{s}_k + (\bar{\lambda}_k - \lambda_k)\mathbf{p}_k) = \mathbf{p}_k^{\top}\mathbf{s}_k + (\bar{\lambda}_k - \lambda_k)\mathbf{p}_k^{\top}\mathbf{p}_k$$

But $\mathbf{p}_k^{\top}\mathbf{s}_k$ is the original value of δ_k before λ_k was increased. So we have

$$\bar{\delta}^k = \delta^k + (\bar{\lambda}_k - \lambda_k)\mathbf{p}_k^{\top}\mathbf{p}_k$$

Clearly we now require that the new value of $\bar{\delta}^k$ be positive; hence we require that

$$\delta_k + (\bar{\lambda}_k - \lambda_k)\mathbf{p}_k^{\top}\mathbf{p}_k > 0$$

This will be true if

$$\bar{\lambda}_k > \lambda_k - \frac{\delta_k}{\mathbf{p}_k^{\top}\mathbf{p}_k}$$

Moller suggests a reasonable choice of $\bar{\lambda}_k$ is

$$\bar{\lambda}_k = 2\left(\lambda_k - \frac{\delta^k}{\mathbf{p}_k^{\top}\mathbf{p}_k}\right)$$

It is easily verified by back-substitution of this value for $\bar{\lambda}_k$ in our expression for $\bar{\delta}_k$ that

$$\bar{\delta}_k = -\delta_k + \lambda_k\mathbf{p}_k^{\top}\mathbf{p}_k$$

which, since δ_k is negative, λ_k is positive, and $\mathbf{p}_k^{\top}\mathbf{p}_k$ is a sum of squares, is clearly positive as required. The step size estimate is based on a quadratic approximation to the function being optimized at the current step and is calculated from

$$\alpha_k = \frac{\mu_k}{\delta_k} = \frac{\mu_k}{\mathbf{p}_k^{\top}\mathbf{s}_k + \lambda_k\mathbf{p}_k^{\top}\mathbf{p}_k}$$

Here μ_k is the current negative gradient times the current direction of search $\mathbf{p}_k$. This gives the basis of the algorithm. However, an important issue still to be decided, is how the value of λ_k can be safely and systematically varied. Moller provides a method based on a measure of how well the current quadratic approximation, defined as f_q, approximates the original function at the point considered. He does this by using the following definition:

$$\Delta_k = \frac{f(\mathbf{x}_k) - f(\mathbf{x}_k + \alpha_k \mathbf{p}_k)}{f(\mathbf{x}_k) - f_q(\alpha_k \mathbf{p}_k)}$$

By virtue of the fact that $f_q(\alpha_k\, \mathbf{p}_k)$ is a quadratic approximation at the current iteration, this can be shown to be equivalent to

$$\Delta_k = \frac{\delta_k^2 (f(\mathbf{x}_k) - f(\mathbf{x}_k + \alpha_k \mathbf{p}_k))}{\mu_k^2}$$

Now if Δ_k is close to 1 then the quadratic approximation $f_q(\alpha_k\, \mathbf{p}_k)$ must be close to $f(x_k + \alpha_k \mathbf{p}_k)$ and hence a good local approximation to the function. This leads to the following steps for the adjustment of λ_k. Use the definition of Δ_k described earlier as the quadratic approximation measure; more details can be found in Moller (1993). Then adjust λ_k as follows:

$$\begin{aligned} &\text{If } \Delta_k > 0.75 \quad \text{then} \quad \lambda_k = \lambda_k / 4 \\ &\text{If } \Delta_k < 0.25 \quad \text{then} \quad \lambda_k = \lambda_k + \frac{\delta_k (1 - \Delta_k)}{\mathbf{p}_k^\top \mathbf{p}_k} \end{aligned}$$

These steps, together with any of the methods for generating conjugate gradient directions of search, provide an algorithm with a simple line-search process. The outline of the major steps in Moller's algorithm are now given:

- *Step* 1: Choose the initial approximation $\mathbf{x}_0$ and initial values for $\sigma_i < 10^{-4}$, $\lambda_i < 10^{-4}$, and $\bar{\lambda}_i = 0$. These values were suggested by Moller. Calculate the initial negative gradient and assign it to $\mathbf{r}_1$ and assign $\mathbf{r}_1$ to the initial direction of search $\mathbf{p}_1$. Set $k = 1$.
- *Step* 2: Calculate second-order information. Specifically, calculate values for σ_k, $\bar{\mathbf{s}}_k$, and δ_k.
- *Step* 3: Scale δ_k using

$$\bar{\delta}^k = \delta^k + (\bar{\lambda}_k - \lambda_k) \mathbf{p}_k^\top \mathbf{p}_k$$

- *Step* 4: If $\delta_k < 0$ then make the Hessian approximation positive definite using

$$\bar{\delta}_k = -\delta_k + \lambda_k \mathbf{p}_k^\top \mathbf{p}_k$$

Set

$$\bar{\lambda}_k = 2\left(\lambda_k - \frac{\delta^k}{\mathbf{p}_k^\top \mathbf{p}_k}\right)$$

and

$$\bar{\lambda}_k = \lambda_k$$

- *Step* 5: Calculate step size from

$$\alpha_k = \frac{\mu_k}{\delta_k}$$

- *Step* 6: Calculate the factor to test goodness of quadratic fit Δ_k from

$$\Delta_k = \frac{\delta_k^2(f(\mathbf{x}_k) - f(\mathbf{x}_k + \alpha_k \mathbf{p}_k))}{\mu_k^2}$$

- *Step* 7: If $\Delta_k \geq 0$ then the function can be reduced toward the minimum, so use

$$\mathbf{x}_{k+1} = \mathbf{x}_k + \alpha_k \mathbf{p}_k$$

Calculate the new gradient

$$\mathbf{r}_{k+1} = -\nabla f(\mathbf{x}_{k+1})$$

Set $\bar{\lambda}_k = 0$. If k mod $N = 0$ then restart algorithm with

$$\mathbf{p}_{k+1} = \mathbf{r}_{k+1}$$

else calculate a new conjugate gradient direction.

Use some method to calculate the set of conjugate directions; see the Fletcher-Reeves (1964) for example. A number of other methods are available.

$$\text{If } \Delta_k \geq 0.75 \quad \text{then} \quad \lambda_k = 0.25\lambda_k$$

else

$$\bar{\lambda}_k = \lambda_k$$

- *Step* 8: If $\Delta_k < 0.25$ then increase the scale parameter:

$$\lambda_k = \lambda_k + (\delta_k(1 - \Delta_k)/\mathbf{p}_k^\top \mathbf{p}_k$$

- *Step* 9: If the gradient r_k is still not suffiently close to zero then set $k = k+1$ and go to step 2; otherwise terminate and return the optimum solution.

The following MATLAB function implements this method.

```
function [res, noiter] = minscg(f,derf,x,tol)
% Conjugate gradient optimization by Moller
% Finds local min of a multivariable nonlinear function in n variables
% Example call: [res, noiter] = minscg(f,derf,x,tol)
% f is a user defined multi-variable function,
```

```
% derf a user defined function of n first order partial derivatives.
% x is a col vector of n starting values, tol gives required accuracy.
% res is solution, noiter is the number of iterations required.
lambda = 1e-8; lambdabar = 0; sigmac = 1e-5; sucess = 1;
deltastep = 0; [n m] = size(x);
% Calculate initial gradient
noiter = 0;
pv = -feval(derf,x); rv = pv;
while norm(rv)>tol
    noiter = noiter+1;
    if deltastep==0
        df = feval(derf,x);
    else
        df = -rv;
    end
    deltastep = 0;
    if sucess==1
        sigma = sigmac/norm(pv);
        dfplus = feval(derf,x+sigma*pv);
        stilda = (dfplus-df)/sigma;
        delta = pv'*stilda;
    end
    % Scale
    delta = delta+(lambda-lambdabar)*norm(pv)^2;
    if delta<=0
        lambdabar = 2*(lambda-delta/norm(pv)^2);
        delta = -delta+lambda*norm(pv)^2;
        lambda = lambdabar;
    end
    % Step size
    mu = pv'*rv; alpha = mu/delta;
    fv = feval(f,x);
    fvplus = feval(f,x+alpha*pv);
    delta1 = 2*delta*(fv-fvplus)/mu^2;
    rvold = rv; pvold = pv;
    if delta1>=0
        deltastep = 1;
        x1 = x+alpha*pv;
        rv = -feval(derf,x1);
        lambdabar = 0; sucess = 1;
        if rem(noiter,n) == 0
            pv = rv;
```

```
        else
            %Alternative conj grad direction generators may be used here
            % beta = (rv'*rv)/(rvold'*rvold);
            rdiff = rv-rvold;
            beta = (rdiff'*rv)/(rvold'*rvold);
            pv = rv+beta*pvold;
        end
        if delta1>=0.75
            lambda = 0.25*lambda;
        end
    else
        lambdabar = lambda;
        sucess = 0;
        x1 = x+alpha*pv;
    end
    if delta1<0.25
        lambda = lambda+delta*(1-delta1)/norm(pvold)^2;
    end
    x = x1;
end
res = x1;
```

We now show the scaled conjugate gradient method applied to two problems:

$$\text{Minimize}\, f(x_1,x_2) = \left(x_1^4 - 16x_1^2 + 5x_1\right)/2 + \left(x_2^4 - 16x_2^2 + 5x_2\right)/2$$

and

$$\text{Minimize}\, f(x_1,x_2) = 100(x_2 - x_1^2)^2 + (1 - x_1)^2 \quad \text{(Rosenbrock's function)}$$

The first of these problems is also solved using `mincg`, and the user-defined functions `f01` and `f01d` are given in Section 8.4. Thus we have

```
>> [x, iterns] = minscg('f01','f01d',[1 -1]',.000005)

x =
    2.7468
   -2.9035

iterns =
     8
```

This is not the same solution as that determined by `mincg`. It is a local minimum value of the function but not the global minimum. Other initial values will lead to the global minimum.

To find a minimum of Rosenbrock's function we define the necessary anonymous functions, and solve the problem as follows:

```
>> fr = @(x) 100*(x(2)-x(1).^2).^2+(1-x(1)).^2;
>> frd = @(x) [-400*x(1).*(x(2)-x(1).^2)-2*(1-x(1)); 200*(x(2)-x(1).^2)];
>> [x, iterns] = minscg(fr,frd,[-1.2 1]',.0005)

x =
     1.0000
     1.0000

iterns =
   135
```

Note that a large number of iterations were required to solve this difficult problem.

8.6 Conjugate Gradient Method for Solving Linear Systems

We now apply the conjugate gradient algorithm to minimize a positive definite quadratic function, which has the standard form

$$f(\mathbf{x}) = \left(\mathbf{x}^\top \mathbf{A}\mathbf{x}\right)/2 + \mathbf{p}^\top \mathbf{x} + q \tag{8.8}$$

Here $\mathbf{x}$ and $\mathbf{p}$ are n-component column vectors, $\mathbf{A}$ is an $n \times n$ positive definite symmetric matrix, and q is a scalar. The minimum value of $f(\mathbf{x})$ is such that the gradient of $f(\mathbf{x})$ is zero. However, the gradient is easily found by direct differentiation as

$$\nabla f(\mathbf{x}) = \mathbf{A}\mathbf{x} + \mathbf{p} = \mathbf{0} \tag{8.9}$$

Thus finding the minimum is equivalent to solving this system of linear equations, which becomes, on letting $\mathbf{b} = -\mathbf{p}$,

$$\mathbf{A}\mathbf{x} = \mathbf{b} \tag{8.10}$$

Since we can use the conjugate gradient method to find the minimum of (8.8), we can use it to solve the equivalent system of linear equations (8.10). The conjugate gradient method provides a powerful method for solving linear equation systems with positive definite symmetric matrices, and it follows quite closely the algorithm we have described for solving nonlinear optimization problems. However, the line search is greatly simplified and the value of the gradient can be computed within the algorithm in this case. The algorithm takes the form

- *Step* 0: $k=0$: $\mathbf{x}^k=\mathbf{0}, \quad \mathbf{g}^k=\mathbf{b}, \quad \mu^k=\mathbf{b}^\top\mathbf{b}, \quad \mathbf{d}^k=-\mathbf{g}^k$
- *Step* 1: While system is not satisfied

$$\mathbf{q}^k=\mathbf{A}\mathbf{d}^k, \quad r^k=(\mathbf{d}^k)^\top\mathbf{q}^k, \quad s^k=\mu^k/r^k$$

$$\mathbf{x}^{k+1}=\mathbf{x}^k+s^k\mathbf{d}^k, \quad \mathbf{g}^{k+1}=\mathbf{g}^k+s^k\mathbf{q}^k$$

$$t^k=(\mathbf{g}^{k+1})^\top\mathbf{q}^k, \quad b^k=t^k/r^k$$

$$\mathbf{d}^{k+1}=-\mathbf{g}^{k+1}+\beta^k\mathbf{d}^k, \quad \mu^{k+1}=\beta^k\mu^k$$

$$k=k+1, \quad \text{end}$$

Notice that the values of the gradient $\mathbf{g}$ and the step s are calculated directly and no MATLAB function or user-defined function is required.

The MATLAB function `solvercg` implements this algorithm and utilizes the stopping procedure suggested by Karmarkar and Ramakrishnan (1991). See this paper and also Golub and Van Loan (1989) for more details.

```
function xdash = solvercg(a,b,n,tol)
% Solves linear system ax = b using conjugate gradient method.
% Example call: xdash = solvercg(a,b,n,tol)
% a is an n x n positive definite matrix, b is a vector of n
% coefficients. tol is accuracy to which system is satisfied.
% WARNING Large, ill-cond. systems will lead to reduced accuracy.
xdash = [ ];  gdash = [ ];
ddash = [ ];  qdash = [ ];
q=[ ];
mxitr = n*n;
xdash = zeros(n,1);  gdash = -b;
ddash = -gdash; muinit = b'*b;
stop_criterion1 = 1;
k = 0;
mu = muinit;
% main stage
while stop_criterion1==1
    qdash = a*ddash;
    q = qdash; r = ddash'*q;
    if r==0
        error('r=0, divide by 0!!!')
    end
    s = mu/r;
    xdash = xdash+s*ddash;
```

```
        gdash = gdash+s*q;
        t = gdash'*qdash;  beta = t/r;
        ddash = -gdash+beta*ddash;
        mu = beta*mu; k = k+1;
        val = a*xdash;
        if ((1-val'*b/(norm(val)*norm(b)))<=tol) & (mu/muinit<=tol)
            stop_criterion1 = 0;
        end
        if k>mxitr
            stop_criterion1 = 0;
        end
    end
```

The following script generates a system of 10 equations with randomly selected elements on which this algorithm can be tested:

```
% e3s803.m
n = 10; tol = 1e-8;
A = 10*rand(n); b = 10*rand(n,1);
ada = A*A';
% To ensure a symmetric positive definite matrix.
sol = solvercg(ada,b,n,tol);
disp('Solution of system is:')
disp(sol)
accuracy = norm(ada*sol-b);
fprintf('Norm of residuals =%12.9f\n',accuracy)
```

Running this script gives the following results:

```
Solution of system is:
    0.2527
   -0.2642
   -0.1706
    0.4284
    0.0017
   -0.1391
   -0.0231
   -0.0109
   -0.2310
    0.2928

Norm of residuals = 0.000000008
```

We note that the norm of the residuals is very small. For ill-conditioned matrices it is necessary to use some kind of preconditioner, which reduces the condition number of the matrix; otherwise the method becomes too slow. Karmarkar and Ramakrishnan (1991) used a preconditioned conjugate gradient method as part of an interior point algorithm to solve linear programming problems with 5000 rows and 333,000 columns.

MATLAB provides a range of iterative procedures based on conjugate gradient methods for solving $\mathbf{Ax} = \mathbf{b}$. These are the MATLAB functions `pcg`, `bicg`, and `cgs`.

8.7 Genetic Algorithms

In this section we introduce the ideas on which genetic algorithms are based and provide a group of MATLAB functions that implement the key features of a genetic algorithm. These are applied to the solution of some optimization problems. It is beyond the scope of this book to give a detailed account of this rapidly developing field of study and the reader is referred to the excellent text of Goldberg (1989).

Genetic algorithms have been the subject of considerable interest in recent years since they appear to provide a robust search procedure for solving difficult problems. The striking feature of these algorithms is that they are based on ideas from the science of genetics and the process of natural selection. This cross-fertilization from one field of science to another has led to stimulating and fruitful applications in many fields and particularly in computer science.

We will describe the genetic algorithm in the terminology used in the field and then explain how this relates to an optimization problem. The genetic algorithm works with an initial *population*, which may, for example, correspond to numerical values of a particular variable. The size of this population may vary and is generally related to the problem under consideration. The members of this population are usually strings of zeros and ones, that is, binary strings. For example, a small initial or first-generation population may take the form

1000010
1110000
1010101
1111001
1000001

In practice the population may be far larger than this and the strings longer. The strings themselves may be the encoded values of a variable or variables that we are examining. This initial population is generated randomly and we can use the terminology of genetics to characterize it. Each string in the population corresponds to a *chromosome* and each binary element of the string to a *gene*. A new population must now develop from this initial population; to do this we implement the analogue of specific fundamental genetic processes. These are

1. Selection based on fitness
2. Crossover
3. Mutation

A set of chromosomes is selected at the reproduction stage based on natural selection. Thus members of the population are chosen for reproduction on the basis of their fitness defined according to some specified criteria. The fittest are given a greater probability of reproducing in proportion to the value of their fitness.

The actual process of *mating* is implemented using the simple idea of crossover. This means that two members of the population exchange genes. There are many ways of implementing this crossover—for example having a single crossover point or many crossover points. These crossover points are selected randomly. A simple crossover is illustrated in the following for two chromosomes selected according to fitness. Here we have randomly selected a crossover point after the fourth digit.

1110|000
1010|101

After crossover this gives the new chromosomes

1110|101
1010|000

Applying this procedure to our original population, we produce a new generation. The final process is mutation. Here we randomly change a particular gene in a particular chromosome. Thus a 0 may be changed to a 1 or vice versa. The process of mutation in a genetic algorithm occurs very rarely and hence this probability of a change in a string is kept very low.

Having described the basic principles of a genetic algorithm, we now illustrate how it may be applied by considering a simple optimization problem and in so doing fill in some of the details to show how a genetic algorithm may be implemented. A manufacturer wishes to produce a container that consists of a hemisphere surmounted by a cylinder of fixed height. The height of the cylinder is fixed but the common radius of the cylinder and hemisphere may be varied between 2 and 4 units. The manufacturer wishes to find the radius value that maximizes the volume of the container. This is a simple problem and the optimum radius is 4 units. However, it serves to illustrate how the genetic algorithm may be applied.

We can formulate this as an optimization problem by taking r as the common radius of the cylinder and hemisphere and h as the height of the cylinder. Taking $h = 2$ units leads to the formula

$$\text{Maximize } v = 2\pi r^3/3 + 2\pi^2 \tag{8.11}$$

where $2 \leq r \leq 4$.

The first problem we must consider is how to transform this problem so that the genetic algorithm can be applied directly. First we must generate an initial set of strings to constitute the initial population. The number of bits in each string, that is, the string length, limits the accuracy with which we can find the solution to the problem so it must be chosen with care. In addition, we must select the size of the initial population; again this must be chosen with care since a large initial population increases the time taken to implement the steps of the algorithm. A large population may be unnecessary since the algorithm automatically generates new members of the population in the process of searching the region. The MATLAB function `genbin` is used to generate such an initial population and takes the form

```
function chromosome = genbin(bitl,numchrom)
% Example call: chromosome=genbin(bitl, numchrom)
% Generates numchrom chromosomes of bitlength bitl.
% Called by optga.m.
maxchros = 2^bitl;
if numchrom>=maxchros
  numchrom = maxchros;
end
for k = 1:numchrom
    for bd = 1:bitl
        if rand>=0.5
            chromosome(k,bd) = 1;
        else
            chromosome(k,bd) = 0;
        end
    end
end
```

This function can be defined more succinctly using the MATLAB `round` function as follows:

```
function chromosome = genbin(bitl,numchrom)
% Example call: chromosome = genbin(bitl,numchrom)
% Generates numchrom chromosomes of bitlength bitl.
% Called by optga.m
maxchros=2^bitl;
if numchrom>=maxchros
    numchrom = maxchros;
end
chromosome = round(rand(numchrom,bitl));
```

To generate an initial population of five chromosomes, each with six genes, we call this function as

```
>> chroms = genbin(6,5)

chroms =
     0     1     1     1     0     0       [Population member #1]
     1     1     1     1     0     1       [Population member #2]
     1     0     0     1     0     0       [Population member #3]
     0     0     0     0     1     1       [Population member #4]
     0     1     1     1     0     1       [Population member #5]
```

To aid the reader in the following discussion we have labeled the five members of the population #1 to #5. These labels are not, of course, part of the MATLAB output.

Since we are interested in values of r in the range 2 to 4, we must be able to transform these binary strings to values in the range 2 to 4. This is achieved using the MATLAB function `binvreal`, which converts a binary value to a real value in the required range.

```
function rval = binvreal(chrom,a,b)
% Converts binary string chrom to real value in range a to b.
% Example call rval=binvreal(chrom,a,b)
% Normally called from optga.
[pop bitlength] = size(chrom);
maxchrom = 2^bitlength-1;
realel = chrom.*((2*ones(1,bitlength)).^fliplr([0:bitlength-1]));
tot = sum(realel);
rval = a+tot*(b-a)/maxchrom;
```

We now call this function to convert the previously generated population:

```
>> for i = 1:5, rval(i) = binvreal(chroms(i,:),2,4); end
>> rval

rval =
    2.8889    3.9365    3.1429    2.0952    2.9206
```

As expected, these values are in the range 2 to 4 and provide the initial population of values for r. However, these values tell us nothing about their fitness and to discover this we must judge them against some fitness criterion. In this case the choice is easy since our objective is to maximize the value of the function (8.11). We simply find the values of our objective function (8.11) for these values of r. We must define our function as a MATLAB function and it takes the form

```
>> g = @(x) pi*(0.66667*x+2).*x.^2;
```

Now we use this to evaluate fitness by replacing `x` by the values `rval`:

```
>> fit = g(rval)
```

```
fit =
   102.9330   225.1246   127.0806    46.8480   105.7749
```

Notice at this stage that the total fitness is

```
>> sum(fit)

ans =
   607.7611
```

So the fittest is the value 3.9365, with a fitness value 225.1246, which corresponds to string or population member #2. Fortuitously this is a very good result. The function `fitness` implements the preceding process and is given as follows:

```
function [fit,fitot] = fitness(criteria,chrom,a,b)
% Example call: [fit,fitot] = fitness(criteria,chrom,a,b)
% Calculates fitness of set of chromosomes chrom in range a to b,
% using the fitness criterion given by the parameter criteria.
% Called by optga.
[pop bitl] = size(chrom);
for k = 1:pop
    v(k) = binvreal(chrom(k,:),a,b);
    fit(k) = feval(criteria,v(k));
end
fitot = sum(fit);
```

Thus, repeating the preceding calculations, we have

```
>> [fit, sum_fit] = fitness(g,chroms,2,4)

fit =
   102.9330   225.1246   127.0806    46.8480   105.7749

sum_fit =
   607.7611
```

as before.

The next stage is reproduction when the strings are copied according to their fitness. Thus there is a higher probability of more of the fittest chromosomes in the mating pool. This process of selection is more complex and is based on a process that simulates the use of a roulette wheel. The percentage of the roulette wheel that is allocated to a particular string is directly proportional to the fitness of the string. For the preceding fitness vector `fit` the percentages can be calculated from

```
>> percent = 100*fit/sum_fit

percent =
   16.9364   37.0416   20.9096    7.7083   17.4040

>> sum(percent)

ans =
  100.0000
```

Thus, conceptually, we spin a roulette wheel on which strings 1 to 5 have 16.9364, 37.0416, 20.9096, 7.7083, and 17.4040 percent of the area, respectively. These chromosomes or strings have this chance of being selected. This is implemented by the function `selectga` as follows:

```
function newchrom = selectga(criteria,chrom,a,b)
% Example call: newchrom = selectga(criteria,chrom,a,b)
% Selects best chromosomes from chrom for next generation
% using function criteria in range a to b.
% Called by function optga.
% Selects best chromosomes for next generation using criteria
[pop bitlength] = size(chrom);
fit = [ ];
% calculate fitness
[fit,fitot] = fitness(criteria,chrom,a,b);
for chromnum = 1:pop
    sval(chromnum) = sum(fit(1,1:chromnum));
end
% select according to fitness
parname = [ ];
for i = 1:pop
    rval = floor(fitot*rand);
    if rval<sval(1)
        parname = [parname 1];
    else
        for j = 1:pop-1
            sl = sval(j);  su = sval(j)+fit(j+1);
            if (rval>=sl) & (rval<=su)
                parname = [parname j+1];
            end
        end
    end
end
newchrom(1:pop,:) = chrom(parname,:);
```

We can now use this function to perform the selection stage as follows:

```
>> matepool = selectga(g,chroms,2,4)

matepool =
     1     1     1     1     0     1        [Population member #2]
     1     1     1     1     0     1        [Population member #2]
     0     1     1     1     0     0        [Population member #1]
     0     1     1     1     0     1        [Population member #5]
     0     1     1     1     0     0        [Population member #1]
```

Note that members #1 and #2 have been favored by the selection process and duplicated. Because of the random nature of the selection process, member #3 is not selected, even though it is the second fittest member. We can now use the `fitness` function to obtain the fitness of the new population:

```
>> fitness(g,matepool,2,4)

ans =
  225.1246   225.1246   102.9330   105.7749   102.9330

>> sum(ans)

ans =
  761.8902
```

Notice the substantial increase in overall fitness.

We can now mate the members of this population, but we mate only a proportion of them, in this case 60% or 0.6. In this example the population size is 5 and $0.6 \times 5 = 3$. This number is rounded down to an even number, that is, 2, since only an even number of members of the population can mate. Thus, 2 members of the population are randomly selected for mating. The function that carries this out is `matesome`, defined as follows:

```
function chrom1 = matesome(chrom,matenum)
% Example call: chrom1 = matesome(chrom,matenum)
% Mates a proportion, matenum, of chromosomes, chrom.
mateind = [ ]; chrom1 = chrom;
[pop bitlength] = size(chrom);
ind = 1:pop;
u = floor(pop*matenum);
if floor(u/2)~=u/2
    u = u-1;
end
```

```
% select percentage to mate randomly
while length(mateind)~=u
    i = round(rand*pop);
    if i==0
        i = 1;
    end
    if ind(i)~=-1
        mateind = [mateind i];
        ind(i) = -1;
    end
end
% perform single point crossover
for i = 1:2:u-1
    splitpos = floor(rand*bitlength);
    if splitpos==0
        splitpos = 1;
    end
    i1 = mateind(i); i2 = mateind(i+1);
    tempgene = chrom(i1,splitpos+1:bitlength);
    chrom1(i1,splitpos+1:bitlength) = chrom(i2,splitpos+1:bitlength);
    chrom1(i2,splitpos+1:bitlength) = tempgene;
end
```

We now use this function to mate the strings in the new population `matepool`:

```
>> newgen = matesome(matepool,0.6)

newgen =
     1     1     1     1     0     1     [Population member #2]
     1     1     1     1     0     0     [Created from #2 and #1]
     0     1     1     1     0     1     [Created from #1 and #2]
     0     1     1     1     0     1     [Population member #5]
     0     1     1     1     0     0     [Population member #1]
```

We see that two members of the original population, members #1 and #2, have mated by crossing over after the second digit to create two new members of the population.

Computing the new population fitness we have

```
>> fitness(g,newgen,2,4)

ans =
  225.1246  220.4945  105.7749  105.7749  102.9330

>> sum(ans)

ans =
  760.1018
```

Notice that the total fitness has not improved and indeed, at this stage, we cannot expect improvements every time.

Finally we perform a mutation before repeating this same cycle of steps. This is implemented by the function `mutate` as follows:

```
function chrom = mutate(chrom,mu)
% Example call: chrom = mutate(chrom,mu)
% mutates chrom at rate given by mu
% Called by optga
[pop bitlength] = size(chrom);
for i = 1:pop
    for j = 1:bitlength
        if rand<=mu
            if chrom(i,j)==1
                chrom(i,j) = 0;
            else
                chrom(i,j) = 1;
            end
        end
    end
end
```

This is called with a very small value for `mu` and a population of this size is unlikely to be changed in just one generation. This function is called in the following:

```
>> mutate(newgen,0.05)

ans =
     1     1     0     1     0     1
     1     1     1     1     0     0
     0     1     1     1     0     1
     0     1     1     1     0     1
     0     1     1     1     1     0
```

Notice that in this example two mutations have occurred; the third element of the first chromosome has changed from 1 to 0, and the fifth element of the last chromosome has changed from 0 to 1. Sometimes no mutation will occur. This completes the production of a new generation. The process of selection based on fitness, reproduction, and mutation is now repeated using the new generation and subsequently repeated for many generations.

The function `optga` includes all these steps in one function and is defined as follows:

```
function [xval,maxf] = optga(fun,range,bits,pop,gens,mu,matenum)
% Determines maximum of a function using the Genetic algorithm.
% Example call: [xval,maxf] = optga(fun,range,bits,pop,gens,mu,matenum)
% fun is name of a one variable user defined positive valued function.
```

```
% range is 2 element row vector giving lower and upper limits for x.
% bits is number of bits for the variable, pop is population size.
% gens is number of generations, mu is mutation rate,
% matenum is proportion mated in range 0 to 1.
% WARNING. Method is not guaranteed to find global optima.
newpop = [ ];
a = range(1); b = range(2);
newpop = genbin(bits,pop);
for i = 1:gens
    selpop = selectga(fun,newpop,a,b);
    newgen = matesome(selpop,matenum);
    newgen1 = mutate(newgen,mu);
    newpop = newgen1;
end
[fit,fitot] = fitness(fun,newpop,a,b);
[maxf,mostfit] = max(fit);
xval = binvreal(newpop(mostfit,:),a,b);
```

Now, applying this function to solve our original problem, we specify the range of x from 2 to 4, and use 8-bit chromosomes and an initial population of 10. The process is continued for 20 generations with a mutation probability of 0.005 and a mating proportion of 0.6. Note that `matenum` must be greater than zero and less than or equal to one. Thus

```
>> [x f] = optga(g,[2 4],8,10,20,0.005,0.6)

x =
    3.8980

f =
  219.5219
```

Since the exact solution is $x = 4$, this is a reasonable result. Figure 8.6 gives a graphical representation of the progress of the genetic algorithm. It should be noted that each run of the genetic algorithm can produce a different result because of the random nature of the process. In addition, the number of distinct values in the search space is limited by the chromosome length. In this example the chromosome length is 8 bits, giving 2^8 or 256 divisions. Thus the range of r from 2 to 4 is divided into 256 divisions, each equal to 0.0078125.

We now discuss the philosophy and theory behind this process and the real problems to which genetic algorithms may be applied. The reason why a genetic algorithm differs from a simple direct-search procedure is that it involves two special features: crossover and mutation. Thus, starting from an initial population, the algorithm develops new generations, which rapidly explore the region of interest. This is useful for difficult optimization problems and in particular for those where we wish to find the global

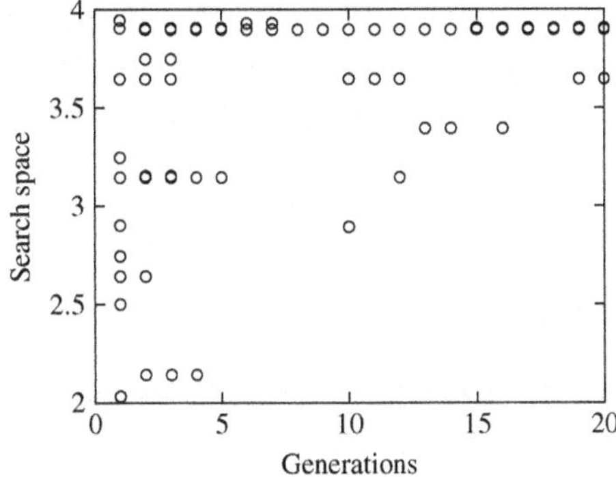

FIGURE 8.6 Each member of the population is represented by ∘. Successive generations of the population concentrate toward the value 4 approximately.

maximum or minimum of a function that has many local maxima and minima. In this case standard optimization methods such as the conjugate gradient method of Fletcher and Reeves can locate only the local optimum. However, a genetic algorithm may locate the global optimum, although this is not guaranteed. This is due to the way it explores the region of interest, avoiding getting stuck at a particular local minimum. We do not consider the theoretical justification in detail but describe the key result only.

We first introduce the concept of *schemata*. If we study the structure of the strings produced by a genetic algorithm, certain patterns of behavior begin to emerge. Strings that have high fitness values often have common features, such as a particular combination of binary elements. For example, the fittest strings may have the common feature that they start with 11 and end with 0 or always have the middle three elements 0. We can represent strings with this structure by 11*****0 and ***000** where the asterisks represent "wild card" elements, which may be either 0 or 1. These structures are called schemata and essentially they identify the common features of a set of strings. The reason why a particular schema is interesting is that we wish to study the propagation of such strings that have this structure and are associated with high values of fitness. The length of a schema is the distance between the outermost specified gene values. The order of a schemata is the number of positions specified by 0 or 1. For example,

String	Order	Length
***********1	1	1
******10*1**	3	4
10******	2	2
00******101	5	11
11**00	4	6

It is clear that schemata that are defined by substrings of short length are less likely to be affected by crossover and therefore propagate through the generations unchanged.

We can now state the *fundamental theorem of genetic algorithms*, due to Holland, in terms of these schemata. This states that schemata of short length and low order with above-average fitness are propagated in exponentially increasing numbers throughout the generations. The ones with below-average fitness die away exponentially. This key result explains some of the success of genetic algorithms.

We now provide some further examples that apply the MATLAB genetic algorithm function `optga` to a specific optimization routine.

Example 8.1

Determine the maximum of the following function in the range $x = 0$ to $x = 1$.

$$f(x) = e^x + \sin(3\pi x)$$

Let the function `h` be defined by

```
h = @(x) exp(x)+sin(3*pi*x);
```

Calling `optga` with this function we have

```
>> [x f] = optga(h,[0 1],8,40,50,0.005,0.6)

x =
      0.8627

f =
      3.3315
```

We now apply the supplied MATLAB function `fminsearch` to solve this problem. Note that for use with `fminsearch` the function `h(x)` has been modified by including a minus sign, thereby negating the function since `fminsearch` is performing minimization.

```
>> h1 = @(x) -(exp(x)+sin(3*pi*x));
>> fminsearch(h1,0,1)

ans =
      0.1802

>> h1(ans)

ans =
     -2.1893
```

Here the function `fminsearch` has found a optimal value of the function but it is only a local optimum. The genetic algorithm (GA) has found a good approximation to the global optimum.

Example 8.2

A more demanding problem is to maximize the function

$$f(x) = 10 + \left[\frac{1}{(x-0.16)^2 + 0.1}\right]\sin(1/x)$$

Calling `optga` with this function defined by the anonymous function `phi` gives the following:

```
>> phi = @(x) 10+(1./((x-0.16).^2+0.1)).*sin(1./x);
>> [x f] = optga(phi,[0.001 0.3],8,10,40,0.005,0.6)

x =
      0.1288

f =
     19.8631
```

Figure 8.7 illustrates the difficulty of this problem and shows that the result is a reasonable one.

The changing diversity of the population can be illustrated graphically; see Figures 8.8 and 8.9. These plots show the value of each bit (0 or 1) for each member of the population. If the rectangle corresponding to a particular bit in a particular population member is shaded, it indicates a unit value for the bit; if it is white, it indicates a zero value for the bit. Figure 8.8 shows the initial randomly selected population and it is seen that there is a random selection of white and shade rectangles. Figure 8.9 shows the final population after 50 generations and we see that many of the bits have identical values in each population member.

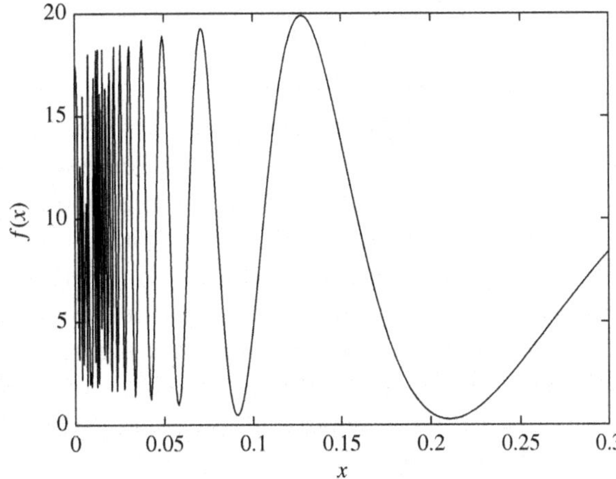

FIGURE 8.7 Plot of the function $10 + [1/\{(x-0.16)^2 + 0.1\}]\sin(1/x)$ showing many local maximum and minimum values.

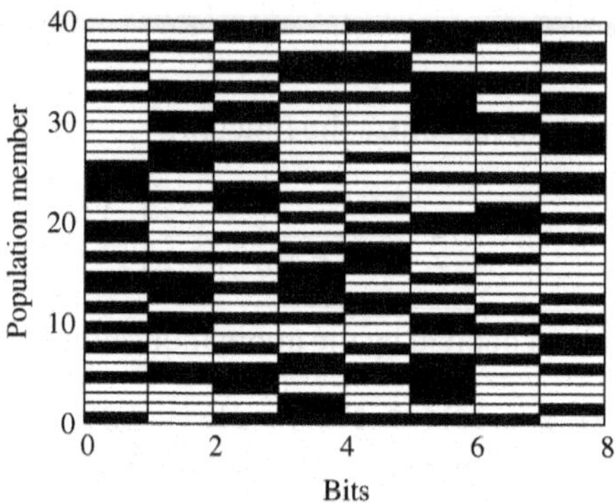

FIGURE 8.8 Initial random distribution of bits.

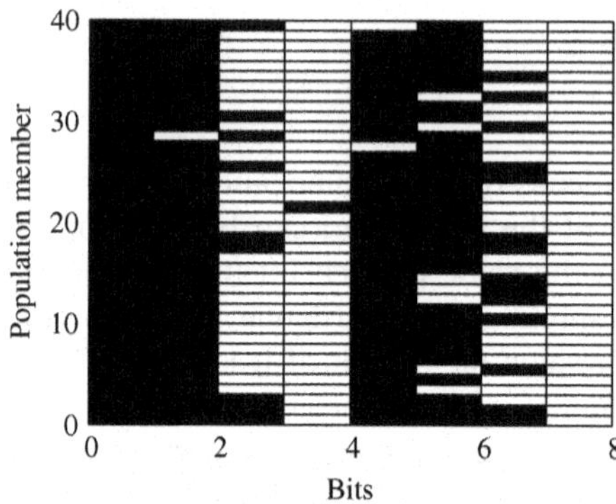

FIGURE 8.9 Distribution of bits after 50 generations.

Genetic algorithms are a developing area of research and many amendments could be made to the functions that we have supplied to implement a genetic algorithm. For example, Gray code rather than binary code can be used; the roulette wheel selection can be implemented in many different ways; crossover can be changed to multipoint crossover or other alternatives. It is often noticed that a genetic algorithm is slow in execution, but it should be remembered that it is best applied to difficult problems, such as those that have multiple optima and where the global optimum is required. Since standard algorithms often fail in these cases, the extra time taken by the genetic algorithm is worthwhile. There are many applications of genetic algorithms that we have not considered and the function `optga` only works for positive-valued functions in one independent variable. It would be easy to extend it to deal with two variable functions and this task is given as exercise for the reader (Problem 8.8)

We now consider using Gray code as an alternative strategy to the standard binary genetic algorithm. Here we interpret each of the strings as a number in Gray code. Gray code is a binary number system where two successive numbers differ by only one bit.

The code was originally developed by Gray to make the operation of systems of mechanical switches more reliable. It is useful at this point to define the Hamming distance. The Hamming distance is the count of the number of bits that are different between two binary vectors. Thus it follows that the Hamming distance between two succesive Gray code numbers is generally smaller than that of two succesive binary code numbers.

The following shows a comparison of the three-bit Gray code and the three-bit binary code

Decimal	0	1	2	3	4	5	6	7
Binary	000	001	010	011	100	101	110	111
Gray	000	001	011	010	110	111	101	100

For the genetic algorithm we must convert from Gray code to decimal. To do this we use two stages: convert Gray code to binary and then convert binary to decimal. To convert from Gray code to binary, the simple algorithm is

$$\begin{array}{l} \text{For bit 1 (the most significant bit)},\ b(1) = g(1) \\ \text{For bit } i, \text{ where } i = 2, \ldots, n, \\ \quad \text{if } b(i-1) = g(i) \text{ then } b(i) = 0, \text{ else } b(i) = 1 \end{array}$$

where $b(i)$ is a binary digit and $g(i)$ is the equivalent Gray code digit. This algorithm is implemented in the following function `grayvreal`

```
function rval = grayvreal(gray,a,b)
% Converts gray string to real value in range a to b.
% Example call rval = grayvreal(gray,a,b)
% Normally called from optga.
[pop bitlength] = size(gray);
maxchrom = 2^bitlength-1;
% Converts gray to binary
bin(1) = gray(1);
for i = 2:bitlength
    if bin(i-1) == gray(i)
        bin(i) = 0;
    else
        bin(i) = 1;
    end
end
% Converts binary to real
realel = bin.*((2*ones(1,bitlength)).^fliplr([0:bitlength-1]));
tot = sum(realel);
rval = a+tot*(b-a)/maxchrom;
```

This function can be used instead of `binvreal` in the MATLAB function `fitness` (renamed `fitness_g`) and this function is then used in `selectga` (renamed `selectga_g`). Finally, both of these functions are used in the function optga_g. The following example ilustrates the use of the function optga_g.

```
>> g = @(x) exp(x)+sin(3*pi*x);
>> [x f] = optga_g(g,[0 1],8,40,50,0.005,0.6)

x =
     0.8588

f =
     3.3317
```

This provides a result with similar accuracy to that achieved using the ordinary binary algorithm.

While some researchers have found no benefit in using Gray codes, others, such as Caruana and Schaffer (1988), claim that the Gray code GA can sometimes be of significant value.

The genetic algorithm can be used to solve optimization problems in which the solutions are constrained to members of a set of discrete values, rather than being continuous over a defined range. For example, steel sections are rolled in specific sizes, and it would be uneconomical to have special sections rolled. Thus a framework, for example, will be constructed from standard beam sizes and sections. The same applies to electronic circuit components such as resistors, which are manufactured in a set of standard values. Suppose we wish to optimize a design that only includes components that are available in eight sizes. Let us assume the vector of discrete values [10 15 24 36 50 75 90 120] gives some performance values for each of the eight possible components. We can use the binary numbers 000 to 111 to represent the indices of these eight performance values. The GA optimization proceeds in the normal way; however, to calculate the fitness corresponding to a particular binary number, this number is used as the index to the vector of performance values to obtain the corresponding performance value. For example, suppose we require the fitness corresponding to the binary number 100 (i.e., decimal 4). The performance value corresponding to this number is 36 and this number is used in the fitness calculation. A difficulty arises if the number of possible components, and hence the number of members in the set of performance values, is not an integer power of 2. For example, suppose we only have six possible component sizes with performance values of, say, [10 15 24 36 50 75]. To represent the six indices of this vector in binary code requires a minimum of three digits. However, the process of crossover and mutation may generate any of eight binary numbers, but only six have corresponding properties. To overcome this difficulty, two of the performance values are duplicated so that, for example, the properties corresponding to the eight binary numbers 000 to 111 are now [10 10 15 24 36 50 75 75]. This adjustment slightly affects the statistics of the process but it generally works satisfactorily. Although in this discussion we have assumed sets of six or eight component sizes, in most practical problems the component set is likely to have as many as 32 or 64 members.

8.8 Continuous Genetic Algorithm

The continuous genetic algorithm is similar in structure to the binary form of the genetic algorithm we have described in Section 8.7, in that an initial population in the region of interest is generated randomly, pairs are selected from the current population and are mated according to fitness, crossover occurs between chromosomes, and mutation of chromosomes occurs with a specified probabilty. However, these steps have significant differences in their implementation in the continuous GA. Basing our description on an optimization problem we randomly generate a set of chromosomes. This initial population is a set of random real numbers rather than binary digits. The key feature here is that the values can be any of the continuous set of values in the region of interest and not a discrete set of binary values that we have used in the binary form of the algorithm.

Suppose we assume that the function to be optimized has four variables. Then initially each chromosome is a vector of four randomly generated decimal numbers, each lying in the search range for the variable. If we choose to have a population of 20 chromosomes, then each has its fitness assessed according to a fitness criteria and a number of the fittest are chosen for the mating process. For example, the most fit 8 chromosomes from a group of 20 chromosomes may be chosen to constitute the mating pool. From this group, random pairs are chosen for crossover and mating.

The mating process is again broadly similar to that of the binary form of the GA in that a random point is chosen for crossover so that the parental chromosomes are intermixed about this point by simply interchanging the real variable values within the chromosomes. However at this point a crucial difference is introduced since crossover in this form simply interchanges the original set of randomly generated real values without producing new values in the region. So to help explore the region we need to introduce new values. For example, suppose the function to be minimized is a function of four variables, u, v, w, and x, and the two chromosomes to be mated, r_1 and r_2, are given by

$$r_1 = [u_1\ v_1\ w_1\ x_1],\quad r_2 = [u_2\ v_2\ w_2\ x_2]$$

Of course, these two chromosomes are chosen at random from the mating pool according to fitness. At a random point in the chromosome a new value in the region is created by forming two new elements from a linear random combination of the pair of chromosome elements at this point. These new values then replace the original chromosome values at the selected crossover point. The suggested formulae for generating new data values take the form

$$x_a = x_1 - \beta(x_1 - x_2),\quad x_b = x_2 + \beta(x_1 - x_2)$$

Similar equations can be applied to the variables u, v, and w. At each generation, β, the crossover point and the pairing of the fittest four members of the previous population will all be rechosen.

The crossover point can occur at random at points 1, 2, 3, or 4. Depending on which crossover point is chosen, after mating the new chromosomes are

Crossover at 1: $r_1 = [u_a\ v_2\ w_2\ x_2],\quad r_2 = [u_b\ v_1\ w_1\ x_1]$

Crossover at 2: $r_1 = [u_1\ v_a\ w_2\ x_2],\quad r_2 = [u_2\ v_b\ w_1\ x_1]$

Crossover at 3: $r_1 = [u_1\ v_1\ w_a\ x_2],\quad r_2 = [u_2\ v_2\ w_b\ x_1]$

Crossover at 4: $r_1 = [u_1\ v_1\ w_1\ x_a],\quad r_2 = [u_2\ v_2\ w_2\ x_b]$

Other crossover rules may be applied. Note that a one-dimensional problem cannot be solved using this particular mating algorithm.

The process of mutation is implemented in a very similar way to that of the binary genetic algorithm. A mutation rate is chosen; then the number of mutations can be calculated from the number of chromosomes and the number of components in the chromosome. Then positions are randomly selected in the chromosomes and these chromosome values are replaced by random values selected within the region. This is another way of helping the algorithm to explore the region, which increases the chance of finding the global minimum for the whole region. The `contgaf` function implements a continuous genetic algorithm. This particular implementation is arranged to find the *minimum* of a function.

```
function [x,f] = contgaf(func,nv,range,pop,gens,mu,matenum)
% function for continuous genetic algorithm
% func is the multivariable function to be optimised
% nv is the number of variables in the function (minimum = 2)
% range is row vector with 2 elements. i.e [lower bound upper bound]
% pop is the number of chromosomes, gens is the number of generations
% mu is the mutation rate in range 0 to 1.
% matenum is the proportion of the population mated in range 0 to 1.
pops = [ ];  fitv = [ ]; nc = pop;
% Generate chromosomes as uniformly distributed sets of random decimal
% numbers in the range 0 to 1
chrom = rand(nc,nv);
% Generate the initial population in the range a to b
a = range(1); b = range(2);
pops = (b-a)*chrom+a;
for MainIter = 1:gens
    % Calculate fitness values
    for i = 1:nc
        fitv(i) = feval(func, pops(i,:));
    end
    % Sort fitness values
    [sfit,indexf] = sort(fitv);
```

```
% Select only the best matnum values for mating
% ensure an even number of pairs is produced
nb = round(matenum*nc);
if nb/2~=round(nb/2)
    nb = round(matenum*nc)+1;
end
fitbest = sfit(1:nb);
% Choose mating pairs use rank weighting
prob = @(n) (nb-n+1)/sum(1:nb);
rankv = prob([1:nb]);
for i = 1:nb
    cumprob(i) = sum(rankv(1:i));
end
% Choose two sets of mating pairs
mp = round(nb/2);
randpm = rand(1,mp);  randpd = rand(1,mp);
mm = [ ];
for j = 1:mp
    if randpm(j)<cumprob(1)
        mm = [mm,1];
    else
        for i = 1:nb-1
            if (randpm(j)>cumprob(i)) && (randpm(j)<cumprob(i+1))
                mm = [mm i+1];
            end
        end
    end
end
% The remaining elements of nb = [1 2 3,...] are the other ptnrs
md = [ ];
md = setdiff([1:nb],mm);
% Mating between mm and md. Choose crossover
xp = ceil(rand*nv);
addpops = [ ];
for i = 1:mp
    % Generate new value
    pd = pops(indexf(md(i)),:);
    pm = pops(indexf(mm(i)),:);
    % Generate random beta
    beta = rand;
    popm(xp) = pm(xp)-beta*(pm(xp)-pd(xp));
    popd(xp) = pd(xp)+beta*(pm(xp)-pd(xp));
```

```
        if xp==nv
            % Swap only to left
            ch1 = [pm(1:nv-1),pd(nv)];
            ch2 = [pd(1:nv-1),pm(nv)];
        else
            ch1 = [pd(1:xp),pm(xp+1:nv)];
            ch2 = [pm(1:xp),pd(xp+1:nv)];
        end
        % New values introduced
        ch1(xp) = popm(xp);
        ch2(xp) = popd(xp);
        addpops = [addpops;ch1;ch2];
    end
    % Add these ofspring to the best to obtain a new population
    newpops = [ ];   newpops = [pops(indexf(1:nc-nb),:); addpops];
    % Calculate number of mutations, mutation rate mu
    Nmut = ceil(mu*nv*(nc-1));
    % Choose location of variables to mutate
    for k = 1:Nmut
        mui = ceil(rand*nc);  muj = ceil(rand*nv);
        if mui~=indexf(1)
            newpops(mui,muj) = (b-a)*rand+a;
        end
    end
    pops = newpops;
end
f = sfit(1);  x = pops(indexf(1),:);
```

We can test this function on an example already discussed in this chapter: a two-variable function that is taken from Styblinski and Tang (1990). This function is

$$f(x_1,x_2) = \left(x_1^4 - 16x_1^2 + 5x_1\right)/2 + \left(x_2^4 - 16x_2^2 + 5x_2\right)/2$$

We may define this function in MATLAB using an anonymous function as follows:

```
>> tf=@(x) 0.5*(x(1).^4-16*x(1).^2+5*x(1))+0.5*(x(2).^4- ...
16*x(2).^2+5*x(2));
```

This function has several local minima but the global optima is at $(-2.9035,\ -2.9035)$. Here we execute three runs of the continuous genetic algorithm:

```
>> [x,f] = contgaf(tf,2,[-4 4],50,50,0.2,0.6)

x =
   -2.9036   -2.9032
```

```
f =
  -78.3323

>> [x,f] = contgaf(tf,2,[-4 4],50,50,0.2,0.6)

x =
   -2.9035   -2.9037

f =
  -78.3323

>> [x,f] = contgaf(tf,2,[-4 4],50,50,0.2,0.6)

x =
   -2.9035   -2.8996

f =
  -78.3321
```

Notice the difference in the **x** values. The process involves a random element and will not produce the same result every time.

As a further example, determine the minimim value of the function

$$f(x) = \sum_{n=1}^{4}\left[100\left(x_{n+1} - x_n^2\right)^2 + (1 - x_n)^2\right]$$

Obviously, the minimum of this function is zero. Thus we have

```
ff = @(x)(1-x(4))^2+(1-x(3))^2+(1-x(2))^2+(1-x(1))^2+ ...
     100*((x(5)-x(4)^2)^2+(x(4)-x(3)^2)^2+(x(3)-x(2)^2)^2+(x(2)-x(1)^2)^2);
>> [x,f] = contgaf(ff,5,[-5 5],20,100,0.15,0.6)

x =
     0.7617     0.6677     0.6392     0.5876     0.3435

f =
     8.1752
```

This is a good result. The actual minimum is zero at $[1\ 1\ 1\ 1\ 1]$. However, the function value at $[-5\ -5\ -5\ -5\ -5]$ is 360,144. If we sought the solution for the minimum value to the nearest integer by evaluating the function in the range -5 to 5 in each dimension, the function would need to be evaluated 161,051 times. To find the solution to an accuracy of 0.1, we would require 1.051×10^{10} function evaluations—not a realistic approach.

For the discussion of the efficiency of the continuous GA, see Chelouah and Siarry (2000). Comparisons between binary and continous genetic algorithms have been carried out by several authors and the continuous GA has been found to have the advantage of greater consistency from run to run and higher precision (Michalewicz, 1996).

8.9 Simulated Annealing

Here we provide a brief introduction to the ideas on which optimization using simulated annealing is based. The technique should be applied to large and difficult problems where we require the global optima and where other techniques are inadequate. Even for relatively simple problems the technique can be slow.

If a metal is allowed to cool sufficiently slowly (metallurgically called postannealing), its metallurgical structure is naturally able to find a minimum energy state for the system. If, however, the metal is cooled quickly, say by quenching in water, then this minimum energy state is not found. This concept of the natural process of finding a minimum energy state can be used to find the global optima of given nonlinear functions. This optimization method is called simulated annealing.

The analogy is not perfect but the fast cooling process may be viewed as equivalent to finding a local minimum of a given nonlinear function corresponding to the energy level, while the slow cooling corresponds to finding the ideal energy state or a global minimum of the function. This slow cooling process may be implemented using the Boltzmann probability distribution of energy states, which plays a prominent part in thermodynamics and has the form

$$P(E) = \exp(-E/kT)$$

where $P(E)$ is the probability of E, a particular energy state, k is Boltzmann's constant, and T is the temperature. This function is used to reflect the cooling process where a change in the energy level, which may be initially unfavorable, ultimately leads to a final minimum global energy state.

This corresponds to the concept of moving out of the region of a local minimum of a nonlinear function in the search for a global solution for the problem. This may require a temporary increase in the value of the objective function, that is, climbing out of the valley of a local minimum, although convergence to the global optimum may still occur if the adjustment to the temperature is slow enough. These ideas lead to an optimization algorithm used by Kirkpatrick et al. (1983), which has the following general structure.

Let $f(\mathbf{x})$ be the nonlinear function to be minimized, where $\mathbf{x}$ is an n-component vector. Then

- *Step* 1: Set $k = 0$, $p = 0$. Chose a starting solution $\mathbf{x}^k$ and an initial, arbitrary temperature T_p.
- *Step* 2: Let a new value of $\mathbf{x}$, $\mathbf{x}^{k+1}$ cause a change, $\Delta f = f(\mathbf{x}^{k+1}) - f(\mathbf{x}^k)$; then

if $\Delta f < 0$, accept the change with probability 1 and $\mathbf{x}^{k+1}$ replaces $\mathbf{x}^k$, $k = k+1$.
If $\Delta f > 0$, accept the change with probability $\exp(-\Delta f/T_p)$ and $\mathbf{x}^{k+1}$ replaces $\mathbf{x}^k$, $k = k+1$.

- *Step* 3: Repeat from step 2 until there is no significant change of function value.
- *Step* 4: Lower the temperature using an appropriate reduction process $T_{p+1} = g(T_p)$, set $p = p+1$, and repeat from step 2 until there is no further significant change in the function value from temperature reduction.

The key difficulties with this algorithm are choosing an initial temperature and a temperature reduction regime. This has generated many research papers and the details are not discussed here.

The MATLAB function `asaq` is an improved implementation of the preceding algorithm. It is based on a modified and simplified version of an algorithm described by Lester Ingber (1993). This uses a exponential cooling regime with some quenching to accelerate the convergence of the algorithm. The key parameters, such as the values of `qf`, `tinit`, and `maxstep`, and the upper and lower bounds on the variables can be adjusted and may lead to some improvements in the convergence rate. A major change would be to use a different temperature adjustment regime and many alternatives have been suggested. The reader should view these parameter variations as an opportunity to experiment with simulated annealing.

```
function [fnew,xnew] = asaq(func,x,maxstep,qf,lb,ub,tinit)
% Determines optimum of a function using simulated annealing.
% Example call: [fnew,xnew]=asaq(func,x,maxstep,qf,lb,ub,tinit)
% func is the function to be minimized, x the initial approx.
% given as a column vector, maxstep the maximum number of main
% iterations, qf the quenching factor in range 0 to 1.
% Note: small value gives slow convergence, value close to 1 gives
% fast convergence, but may not supply global optimum.
% lb and ub are lower and upper bounds for the variables,
% tinit is the intial temperature value
% Suggested values for maxstep = 200, tinit = 100, qf = 0.9
% Initialisation
xold = x;  fold = feval(func,x);
n = length(x);  lk = n*10;
% Quenching factor q
q = qf*n;
% c values estimated
nv = log(maxstep*ones(n,1));
mv = 2*ones(n,1);
c = mv.*exp(-nv/n);
% Set values for tk
t0 = tinit*ones(n,1);  tk = t0;
```

```
% upper and lower bounds on x variables
% variables assumed to lie between -100 and 100
a = lb*ones(n,1);  b = ub*ones(n,1);
k = 1;
% Main loop
for mloop = 1:maxstep
    for tempkloop = 1:lk
        % Choose xnew as random neighbour
        fold = feval(func,xold);
        u = rand(n,1);
        y = sign(u-0.5).*tk.*((1+ones(n,1)./tk).^(abs((2*u-1))-1));
        xnew = xold+y.*(b-a);
        fnew = feval(func,xnew);
        % Test for improvement
        if fnew <= fold
            xold = xnew;
        elseif exp((fold-fnew)/norm(tk))>rand
            xold = xnew;
        end
    end
    % Update tk values
    tk = t0.*exp(-c.*k^(q/n));
    k = k+1;
end
tf = tk;
```

We will run this script to optimize the following function, which is taken from Styblinski and Tang (1990), and solve it by the conjugate gradient method in Section 8.4:

$$f(x_1, x_2) = \left(x_1^4 - 16x_1^2 + 5x_1\right)/2 + \left(x_2^4 - 16x_2^2 + 5x_2\right)/2$$

The results are as follows:

```
>> fv = @(x) 0.5*(x(1)^4-16*x(1)^2+5*x(1)) +...
          0.5*(x(2)^4-16*x(2)^2+5*x(2));
>> [fnew,xnew] = asaq(fv,[0 0].',200,0.9,-10,10,100)

fnew =
  -78.3323

xnew =
   -2.9018
   -2.9038
```

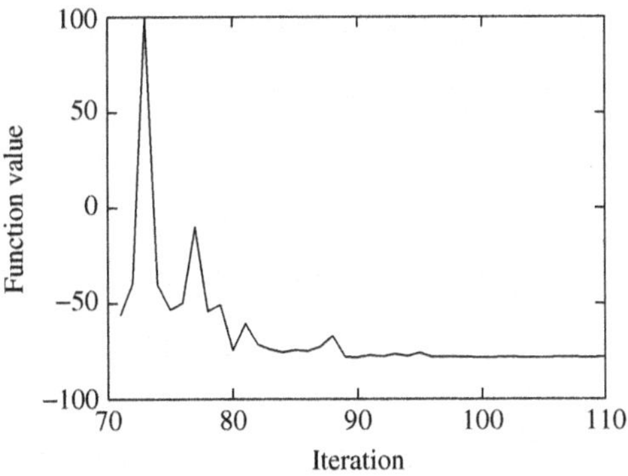

FIGURE 8.10 Graph showing the value of function $f(x_1, x_2) = (x_1^4 - 16x_1^2 + 5x_1)/2 + (x_2^4 - 16x_2^2 + 5x_2)/2$ for the final 40 iterations.

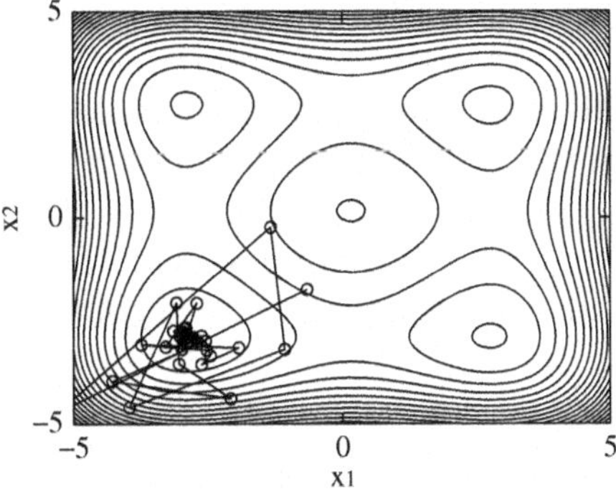

FIGURE 8.11 Contour plot of function $f(x_1, x_2) = (x_1^4 - 16x_1^2 + 5x_1)/2 + (x_2^4 - 16x_2^2 + 5x_2)/2$. The final stages in the simulated annealing process are shown. Note how these values are concentrated in the lower left corner, close to the global optimum.

Note that each run provides a different result and is not guaranteed to provide a global optimum unless the parameters are adjusted appropriately for the particular problem. Figure 8.10 provides a plot of the variation in the function value for the final 40 iterations. It illustrates the behavior of the algorithm, which allows both increases and decreases in the function value.

As a further illustration, a contour plot showing only the final stages of the iteration is given in Figure 8.11.

8.10 Constrained Nonlinear Optimization

In this section we consider the problem of optimizing a nonlinear function, subject to one or more nonlinear constraints. This problem can be expressed mathematically as follows:

$$\text{Minimize} f = f(\mathbf{x}) \quad \text{where} \quad \mathbf{x}^\top = [x_1 \; x_2 \; \dots \; x_n] \tag{8.12}$$

subject to the constraints

$$h_i(\mathbf{x}) = 0 \quad \text{where} \quad i = 1, \dots, p \tag{8.13}$$

Sometimes the minimization problem may have additional or alternative constraints that are of the form

$$g_j(\mathbf{x}) \geq b_j \quad \text{where} \quad j = 1, \dots, q \tag{8.14}$$

To solve this problem we can use the Lagrange multiplier method. This method is not a purely numerical method; it requires the user to apply calculus, and the resulting equations are solved numerically. For large problems this is too onerous on the user, and for this reason it is not a practical method for solving this type of problem. However, it is theoretically important in the development of other, more practical, methods.

If constraints of the form of (8.14) are present, they must be converted to the form of (8.13) as follows. Let $\theta_j^2 = g_j(\mathbf{x}) - b_j$. If the constraint $g_j(\mathbf{x}) \geq b_j$ is violated, then θ_j^2 is negative and θ_j is imaginary. Thus, we have a requirement that for the constraints to be satisfied θ_j must be real. Thus the constraint equation (8.14) becomes

$$\theta_j^2 - g_j(\mathbf{x}) + b_j = 0 \quad \text{where} \quad j = 1, \dots, q \tag{8.15}$$

This constraint equation is of the same general form as the constraint in (8.13).

To solve (8.12) we begin by forming the expression

$$L(\mathbf{x}, \boldsymbol{\theta}, \boldsymbol{\lambda}) = f(\mathbf{x}) + \sum_{i=1}^{p} \lambda_i h_i(\mathbf{x}) + \sum_{j=1}^{q} \lambda_{p+j}[\theta_j^2 - g_j(\mathbf{x}) + b_j] = 0 \tag{8.16}$$

The function L is called the Lagrange function and the scalar quantities λ_i are called the Lagrange multipliers. We now minimize this function using calculus; that is, we take the following partial derivatives and set them to zero.

$$\partial L/\partial x_k = 0, \quad k = 1, \dots, n$$

$$\partial L/\partial \lambda_r = 0, \quad r = 1, \dots, p+q$$

$$\partial L/\partial \theta_s = 0, \quad s = 1, \dots, q$$

We will find that when we set the differentials with respect to λ_r to zero, we force both $h_i(\mathbf{x})$ $(i = 1, 2, \dots, n)$ and $\theta_j^2 - g_j(\mathbf{x}) + b_j$ $(j = 1, 2, \dots, q)$ to be zero. Thus the constraints to be satisfied. If these terms are zero then minimizing (8.16) is equivalent to minimizing (8.12) subject to (8.13) and (8.14). If we are dealing with a quadratic function with linear constraints then the resulting equations are all linear and relative easy to solve.

Example 8.3

Consider the solution of a problem with a cubic function and quadratic constraints.

$$\text{Minimize } f = 2x + 3y - x^3 - 2y^2$$

subject to

$$x + 3y - x^2/2 \le 5.5$$
$$5x + 2y + x^2/10 \le 10$$
$$x \ge 0,\ y \ge 0$$

To use the Lagrange method we change the form of the constraints to be equality constraints, as follows:

$$\text{Minimize } f = 2x + 3y - x^3 - 2y^2$$

subject to

$$\theta_1^2 + x + 3y - x^2/2 - 5.5 = 0$$
$$\theta_2^2 + 5x + 2y + x^2/10 - 10 = 0$$
$$x \ge 0,\ y \ge 0$$

Hence, forming L we have

$$L = 2x + 3y - x^3 - 2y^2 + \lambda_1(\theta_1^2 + x + 3y - x^2/2 - 5.5) + \lambda_2(\theta_2^2 + 5x + 2y + x^2/10 - 10)$$

Taking partial derivaties of L and setting them to zero gives

$$\partial L/\partial x = 2 - 3x^2 + \lambda_1(1 - x) + \lambda_2(5 + x/5) = 0 \tag{8.17}$$

$$\partial L/\partial y = 3 - 4y + 3\lambda_1 + 2\lambda_2 = 0 \tag{8.18}$$

$$\partial L/\partial \lambda_1 = \theta_1^2 + x + 3y - x^2/2 - 5.5 = 0 \tag{8.19}$$

$$\partial L/\partial \lambda_2 = \theta_2^2 + 5x + 2y + x^2/10 - 10 = 0 \tag{8.20}$$

$$\partial L/\partial \theta_1 = 2\lambda_1\theta_1 = 0 \tag{8.21}$$

$$\partial L/\partial \theta_2 = 2\lambda_2\theta_2 = 0 \tag{8.22}$$

If (8.21) and (8.22) are to be satisfied then there are four cases to consider:

Case 1: $\theta_1^2 = \theta_2^2 = 0$. Then (8.17) through (8.20) become, with some rearrangement,

$$2 - 3x^2 + \lambda_1(1 - x) + \lambda_2(5 + x/5) = 0$$
$$3 - 4y + 3\lambda_1 + 2\lambda_2 = 0$$
$$x + 3y - x^2/2 - 5.5 = 0$$
$$5x + 2y + x^2/10 - 10 = 0$$

Case 2: $\lambda_1 = \theta_2^2 = 0$. Then (8.17) through (8.20) become, with some rearrangement,

$$2 - 3x^2 + \lambda_2(5 + x/5) = 0$$
$$3 - 4y + 2\lambda_2 = 0$$
$$\theta_1^2 + x + 3y - x^2/2 - 5.5 = 0$$
$$5x + 2y + x^2/10 - 10 = 0$$

Case 3: $\theta_1^2 = \lambda_2 = 0$. Then (8.17) through (8.20) become, with some rearrangement,

$$2 - 3x^2 + \lambda_1(1 - x) = 0$$
$$3 - 4y + 3\lambda_1 = 0$$
$$x + 3y - x^2/2 - 5.5 = 0$$
$$\theta_2^2 + 5x + 2y + x^2/10 - 10 = 0$$

Case 4: $\lambda_1 = \lambda_2 = 0$. Then (8.17) through (8.20) become, with some rearrangement,

$$2 - 3x^2 = 0$$
$$3 - 4y = 0$$
$$\theta_1^2 + x + 3y - x^2/2 - 5.5 = 0$$
$$\theta_2^2 + 5x + 2y + x^2/10 - 10 = 0$$

The solution of these sets of nonlinear equations requires some iterative procedure. The MATLAB function `fminsearch` finds the minimum of a scalar function of several variables, given an initial estimate. The application of this function to this problem is illustrated for *Case* 1 in the following MATLAB script. Since the right side of each equation is zero, when the solution is found, the function

$$[2 - 3x^2 + \lambda_1(1 - x) + \lambda_2(5 + x/5)]^2 + [3 - 4y + 3\lambda_1 + 2\lambda_2]^2 + \cdots$$
$$[x + 3y - x^2/2 - 5.5]^2 + [5x + 2y + x^2/10 - 10]^2$$

should equal zero. The function `fminsearch` will choose values of x, y, λ_1, and λ_2 to minimize this expression and bring it very close to zero. This is generally referred to as unconstrained nonlinear optimization. Thus we have converted a constrained optimization to an unconstrained one.

```
% e3s820.m
g = @(X) sqrt((2-3*X(1).^2+X(3).*(1-X(1))+X(4).*(5+X(1)/5)).^2 ...
    +(3-4*X(2)+3*X(3)+2*X(4)).^2+(X(1)+3*X(2)-X(1).^2/2-5.5).^2 ...
    +(5*X(1)+2*X(2)+X(1).^2/10-10).^2);
X = fminsearch(g, [1 1 1 1]);
x = X(1); y = X(2); f = 2*x+3*y-x^3-2*y^2;
lambda_1 = X(3); lambda_2 = X(4);
```

```
disp('Case 1')
disp(['x = ' num2str(x) ', y = ' num2str(y) ', f = ' num2str(f)])
disp(['lambda_1 = ' num2str(lambda_1) ...
    ', lambda_2 = ' num2str(lambda_2)])
[xx,yy] = meshgrid(0:0.1:2,0:0.1:2);
ff = 2*xx+3*yy-xx.^3-2*yy.^2;
contour(xx,yy,ff,20,'k'), hold on
x1 = 0:0.1:2;
y1 = (5.5-(x1-x1.^2/2))/3;
y2 = (10-(5*x1+x1.^2/10))/2;
plot(x1,y1,'k',x1,y2,'k')
plot(xp1,yp1,'ok',xp2,yp2,'ok',xp3,yp3,'ok',xp4,yp4,'ok')
hold off
xlabel('x'), ylabel('y')
```

Executing this script gives

```
Case 1
x = 1.2941, y = 1.6811, f = -0.18773
lambda_1 = 0.82718, lambda_2 = 0.62128
```

Comparing the values of f computed for each case (see Table 8.2), it is clear that Case 1 gives the minimum solution. The script also provides Figure 8.12. This graph shows the function,

Table 8.2 Possible Solutions of a Minimization Problem

Case	θ_1^2	θ_1^2	λ_1	λ_2	x	y	f
1	0	0	0.8272	0.6213	1.2941	1.6811	−0.1877
2	1.5654	0	0	0.8674	1.4826	1.1837	0.4552
3	0	2.3236	1.2270	0	0.8526	1.6703	0.5166
4	2.7669	4.3508	0	0	0.8165	0.7500	2.2137

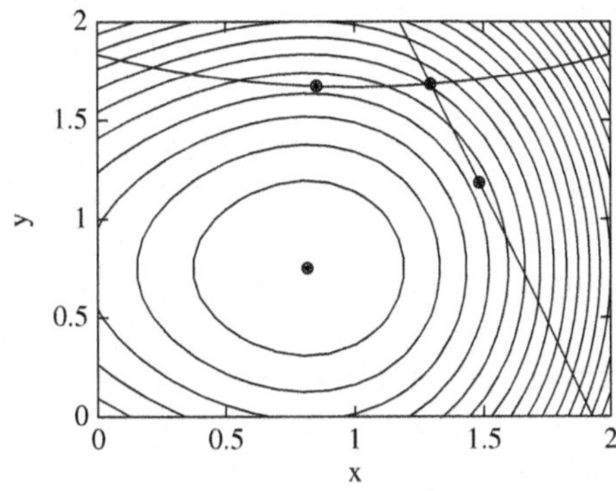

FIGURE 8.12 Function and constraints. The four solutions are also indicated.

the constraints, and the four possible solutions. The optimal solution is not necessarily at the intersection of the constraint boundaries as it is in a linear system. All solutions are feasible but only Case 1 is the global minimum. Case 2 and Case 3 are local minima at the constraint boundary and Case 4 is a local maximum. It should be noted that often one or more solutions will not be feasible; that is, they will not satisfy the constraints.

8.11 The Sequential Unconstrained Minimization Technique

We now give a brief introduction to a standard method for constrained optimization. The sequential unconstrained minimization technique (SUMT) for constrained optimization converts the solution of a constrained optimization problem to the solution of a sequence of unconstrained problems. This method was developed by Fiacco and McCormicks and others in the 1960s. See Fiacco and McCormicks (1964, 1990).

Consider the following optimization problem:

$$\begin{aligned} &\text{Minimize } f(\mathbf{x}), \text{ subject to} \\ &g_i(\mathbf{x}) \geq 0 \quad \text{for} \quad i = 1, 2, \ldots, p \\ &h_j(\mathbf{x}) = 0 \quad \text{for} \quad j = 1, 2, \ldots, s \end{aligned}$$

where $\mathbf{x}$ is a component vector. By using barrier and penalty functions the requirements of the constraints can be included with the function to be minimized so that the problem is converted to the unconstrained problem:

$$\text{Minimize } f(\mathbf{x}) - r_k \sum_{i=1}^{p} \log_e(g_i(\mathbf{x})) + \frac{1}{r_k} \sum_{j=1}^{s} h_j(\mathbf{x})^2$$

Notice the effect of the added terms. The first term imposes a barrier at zero on the inequality constraints in that as the $g_i(\mathbf{x})$ approaches zero the function approaches minus infinity, thus imposing a substantial penalty. Figure 8.13 illustrates this. The last term encourages the satisfaction of the equality constraints $h_j(\mathbf{x}) = 0$ since the smallest amount is added when all the constraints are zero; otherwise, a substantial penalty is imposed. This means that this approach encourages the maintenance of the feasibility of the solution assuming we start with an initial solution that is within the feasible region of the inequality constraints. These methods are sometimes called interior point methods.

A sequence of problems are generated by starting with an arbitrarily large value for r_0 and then using $r_{k+1} = r_k/c$ where $c > 1$ and solving the resulting sequence of unconstrained optimization problems. The unconstrained minimization steps may of course present formidable difficulties for some problems. A simple stopping criteria is to examine

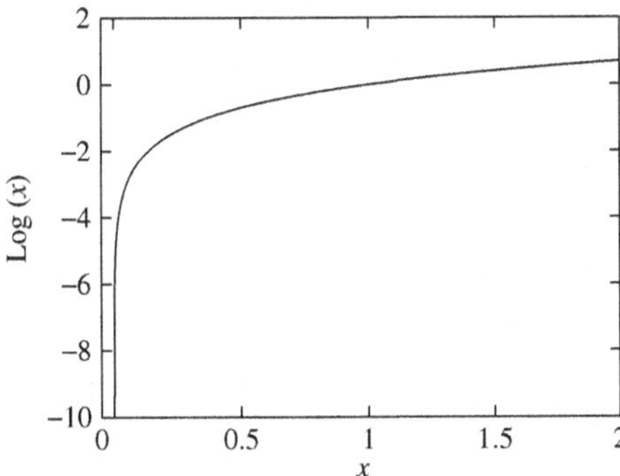

FIGURE 8.13 Graph of $\log_e(x)$.

the difference between the value of $f(\mathbf{x})$ between successive unconstrained optimizations. If the difference is below a specified tolerance then stop the procedure.

There are various alternatives to this algorithm. For example, a reciprocal barrier function can be used instead of the preceding logarithmic function. The barrier term can be replaced by a penalty function term of the form

$$\sum_{i=1}^{p} \max(0, g_i(\mathbf{x}))^2$$

This term will add a substantial penalty if $g_i(\mathbf{x}) < 0$; otherwise, no penalty is applied. This method has the advantage that feasibility is not required and is called an exterior point method. However, the resulting unconstrained problems may present additional problems for the unconstrained minimization procedure. For more details on these methods see Lesdon et al. (1996).

Although purpose-built software is available for this method, a very simple illustration of its operation is given. The program shows some steps of the method for solving a constrained minimization problem; notice that care must be taken in the choice of the initial r value and its reduction factor. To avoid the need for the derivative, the MATLAB function `fminsearch` is used to solve the following unconstrained problem first using the interior point method.

$$\text{Minimize } x_1^2 + 100x_2^2$$

subject to

$$4x_1 + x_2 \geq 6 \tag{8.23}$$

$$x_1 + x_2 = 3 \tag{8.24}$$

$$x_1, x_2 \geq 0 \tag{8.25}$$

```
% e3s810.m
r = 10; x0 = [5 5]
while r>0.01
    fm = @(x) x(1).^2+100*x(2).^2-r*log(-6+4*x(1)+x(2)) ...
        +1/r*(x(1)+x(2)-3).^2-r*log(x(1))-r*log(x(2));
    x1 = fminsearch(fm,x0);
    r = r/5;
    x0 = x1
end
optval = x1(1).^2+100*x1(2).^2

x0 =
     5     5

x0 =
    3.6097    0.2261

x0 =
    2.1946    0.1035

x0 =
    2.2084    0.0553

x0 =
    2.7463    0.0377

x0 =
    2.9217    0.0317

optval =
    8.6366
```

This shows convergence to the optimum solution (3,0), which satisfies the constraints. Using the exterior point method to solve the same problem we have:

```
% e3s811.m
r = 10; x0 = [5 5]
while r>0.01
    fm = @(x) x(1).^2+100*x(2).^2+1/r*min(0,(6+4*x(1)+x(2))).^2 ...
     +1/r*(x(1)+x(2)-3).^2+1/r*min(0,x(1)).^2+1/r*min(0,x(2)).^2;
    x1 = fminsearch(fm,x0);
    r = r/5;
    x0 = x1
end
optval = x1(1).^2+100*x1(2).^2
```

This produces the results:

```
x0 =
     5     5

x0 =
    0.2725    0.0027

x0 =
    0.9967    0.0100

x0 =
    2.1276    0.0213

x0 =
    2.7523    0.0275

x0 =
    2.9240    0.0292

optval =
    8.6352
```

Clearly the results are very similar.

8.12 Summary

In this chapter we have introduced a number of more advanced areas of numerical analysis. The genetic algorithm and simulated annealing are still topics for active research and development, and are mainly used to tackle difficult optimization problems. The congugate gradient method is well established and widely used for problems that present a range of difficulties. MATLAB functions have been provided to allow the reader to experiment and explore the problems more deeply. However, it must be remembered that no optimization technique is guaranteed to solve all optimization problems. The structure of the algorithms is well reflected by the structure of the MATLAB functions. The MathWorks Optimization Toolbox provides a useful selection of optimization functions, which may be used in both education and research.

Problems

8.1. Use the function `barnes` to minimize $z = 5x_1 + 7x_2 + 10x_3$ subject to $x_1 + x_2 + x_3 \geq 4$, $x_1 + 2x_2 + 4x_3 \geq 5$, and $x_1, x_2, x_3 \geq 0$.

8.2. Maximize $p = 4y_1 + 5y_2$ subject to $y_1 + y_2 \leq 5$, $y_1 + 2y_2 \leq 7$, $y_1 + 4y_2 \leq 10$, and $y_1, y_2 \geq 0$.

By introducing slack variables and subtracting one from each equality, write the constraints as equalities. Then apply the function `barnes` to solve this problem. Notice that the optimum value of p for this problem is equal to the optimum value of z in Problem 8.1. Problem 8.2 is called the dual of Problem 8.1. This is an example of an important theorem that the optima of the objective function of a problem and its dual are equal.

8.3. Maximize $z = 2u_1 - 4u_2 + 4u_3$ subject to $u_1 + 2u_2 + u_3 \leq 30$, $u_1 + u_2 = 10$, $u_1 + u_2 + u_3 \geq 8$, and $u_1, u_2, u_3 \geq 0$.
Hint: Remember to use slack variables to ensure that the main constraints are equalities.

8.4. Use the function `mincg`, with tolerance 0.005, to minimize Rosenbrock's function

$$f(x,y) = 100\left(x^2 - y^2\right) + (1-x)^2$$

starting with the initial approximation $x = 0.5$, $y = 0.5$ and using a line-search accuracy 10 times the machine precision in the MATLAB function `fminsearch`. To obtain an impression of how this function varies, plot it in the range $0 \leq x \leq 2$, $0 \leq y \leq 2$.

8.5. Use the function `mincg`, with tolerance 0.00005, to minimize the five-variable function

$$z = 0.5\left(x_1^4 - 16x_1^2 + 5x_1\right) + 0.5\left(x_2^4 - 16x_2^2 + 5x_2\right) + (x_3 - 1)^2 + (x_4 - 1)^2 + (x_5 - 1)^2$$

Use $x_1 = 1$, $x_2 = 2$, $x_3 = 0$, $x_4 = 2$, and $x_5 = 3$ for the starting values in `mincg`. Experiment further with other starting values.

8.6. Use the function `solvercg` to solve the matrix equation $\mathbf{Ax} = \mathbf{b}$ where

$$\mathbf{A} = \begin{bmatrix} 5 & 4 & 1 & 1 \\ 4 & 5 & 1 & 1 \\ 1 & 1 & 4 & 2 \\ 1 & 1 & 2 & 4 \end{bmatrix} \quad \mathbf{b} = \begin{bmatrix} 1 \\ 2 \\ 3 \\ 4 \end{bmatrix}$$

Check the accuracy of the solution by finding the value of `norm(b-Ax)`.

8.7. Maximize the function $y = 1/\{(x-1)^2 + 2\}$ in the range $x = 0$ to 2 using the function `optga`. Use different initial population sizes, mutation rates, and numbers of generations. Notice that this is not a simple exercise since for each set of conditions it is necessary to solve the problem several times to take account of the random nature of the process. Given that the optimum value of the function is 0.5, plot the error in the optimum value of the function for each run under a particular set of parameters. Then change one of the parameters and repeat the process. Differences in the plots may or may not be discernible.

8.8. Plot the function $z = x^2 + y^2$ in the range $0 \le x \le 2$ and $0 \le y \le 2$. The genetic algorithm given in Section 8.6 may be applied to maximize functions in more than one variable. Use the MATLAB function `optga` to determine the maximum value of the preceding function. In order to do this you must modify the `fitness` function so that the first half of the chromosome corresponds to values of x and the second half to values of y and these chromosomes must map to the values of x and y. For example, if an 8-bit chromosome 10010111 is split into two parts, 1001 and 0111, it will convert to $x = 9$ and $y = 7$.

8.9. Use the function `golden`, given in Section 8.3, to minimize the single-variable function $y = e^{-x}\cos(3x)$ for x in the range 0 to 2. Use a tolerance of 0.00001. You should check your result using the MATLAB function `fminsearch` to minimize the same function in the same range. As a further confirmation you might plot the function in the range 0 to 4 using the MATLAB function `fplot`.

8.10. Use the simulated annealing function `asaq` (with the same values of parameters used in the call of the function in Section 8.9) to minimize the two-variable function f where

$$f = (x_1 - 1)^2 + 4\,(x_2 + 3)^2$$

Compare your result with the exact answer, which is clearly $x_1 = 1$ and $x_2 = 3$. This function has only one minimum in the region considered. As a more demanding test, use the same call as used for the preceding function to minimize the function f where

$$f = 0.5\left(x_1^4 - 16x_1^2 + 5x_1\right) + 0.5\left(x_2^4 - 16x_2^2 + 5x_2\right) - 10\cos\{4\,(x_1 + 2.9035)\}\cos\{4\,(x_2 + 2.9035)\}$$

The global optimum for this problem is $x_1 = -2.9035$ and $x_2 = -2.9035$. Try several runs of the function `asaq` for this problem. All runs may not provide the global optimum since this problem has many local optima.

8.11. A method for solving a system of nonlinear equations is to re-express them as an optimization problem. Consider the system of equations

$$2x - \sin((x + y)/2) = 0$$
$$2y - \cos((x - y)/2) = 0$$

These can be rewritten as

$$\text{minimize } z = (2x - \sin((x + y)/2))^2 + (2y - \cos((x - y)/2))^2$$

Use the MATLAB function `minscg` to minimize this function using the starting point $x = 10$ and $y = -10$.

8.12. Write a MATLAB script that provides three-dimensional plots for the function $z = f(x, y)$ defined as follows:

$$z = f(x, y) = (1 - x)^2 e^{-p} - pe^{-p} - e^{(-(x+1)^2 - y^2)}$$

where p is defined by

$$p = x^2 + y^2$$

for x and y in the range $x = -4 : 0.1 : 4$ and $y = -4 : 0.1 : 4$. The script should use the MATLAB `surf` and `contour` functions to provide separate three-dimensional and contour plots. Use the MATLAB function `ginput` to select and assign to an appropriate matrix three points that appear to be optimal on the contour plot. Use the function $z = f(x, y)$ defined earlier to find the values of z for these points. Then find the maximum and minimum of these z values using the MATLAB functions `max` and `min`. Finally, approximate the global minimum and maximum of the function.

An alternative method of finding the minimum is to use the MATLAB function `fminsearch` in the form `x = fminsearch(funxy,xv)`, where `funxy` is an anonymous function or user-defined function given by the user and `xv = [-4 4]` is a vector of initial approximations to the location of the minimum. Experiment with different intial approximations to see if your results vary.

8.13. Solve the minimization problem described in Problem 8.12 using the continuous genetic algorithm. Use the MATLAB function `contgaf`.

8.14. Write a MATLAB script to minimize $f(x) = x_1^4 + x_2^2 + x_1$ subject to $4x_1^3 + x_2 > 6$, $x_1 + x_2 = 3$, and $x_1, x_2 > 0$. Use the sequential unconstrained minimization technique with a logarithmic barrier function and an initial approximation vector of $x = [5,\ 5]$. Use an initial value for the parameter r_0 of 10, a reduction parameter $c = 5$, and $r_{k+1} = r_k/c$. Continue iterations while r_k is greater than 0.0001.

8.15. Write a MATLAB script to solve the constrained optimization problem described in Problem 8.14. However, use the penalty function of the form $[\min(0,\ g_i(\mathbf{x}))]^2$ instead of the logarithmic barrier function, where the $g_i(\mathbf{x})$ are greater than or equal to the constraints. Use the same initial starting point and values for r_0 and c. Compare the solution you find with this method to that achieved with Problem 8.14.

8.16. Solve the Rosenbrooks two-variable optimization problem:

$$\text{Minimize } f(x) = 100(x_2 - x_1^2)^2 + (1 - x_1)^2$$

using the MATLAB function `asaq` with initial approximation $[-1.2\ 1]$. With quenching factor 0.9, the upper and lower bounds for the variables given by -10 and 10, respectively, initial temperature value of 100 and the maximum number of main iterations equal to 800. The solution to this problem is $[1\ 1]$.

9 Applications of the Symbolic Toolbox

The Symbolic Toolbox provides an extensive list of functions for the symbolic manipulation of symbolic expressions and equations. The use of symbolic functions and their analytic manipulation can often play a useful role in association with a numerical algorithm. In such algorithms the combination of the standard numerical functions with the facilities of the Symbolic Toolbox can be particularly beneficial, relieving the user of the tedious and error-prone task of symbolic manipulation. This allows the designer of the algorithm to provide a more user-friendly and complete function.

Originally the MATLAB Symbolic Toolbox used the Maplesoft symbolic software to carry out symbolic operations and pass the results to MATLAB. However, since late 2008 MATLAB has used Mupad for its symbolic engine. This change has caused minor changes in the way results are presented, but the results are generally equivalent.

9.1 Introduction to the Symbolic Toolbox

Since we are using the Symbolic Toolbox in the field of numerical analysis we begin by giving some examples of the beneficial application of this toolbox. It provides

1. The symbolic first derivative of a given single-variable nonlinear function, which is required by Newton's method for the solution of single-variable nonlinear equations (see Chapter 3)
2. The Jacobian for a system of nonlinear simultaneous equations (see Chapter 3)
3. The symbolic gradient vector of a given nonlinear function, which is required for the conjugate gradient method for minimizing a nonlinear function (see Chapter 8)

An important feature of the Symbolic Toolbox is that it allows an extra dimension of experimentation. For example, a user can test a given numerical algorithm by solving a test problem symbolically, providing it has a solution in the closed form, and compare this exact solution with a numerical solution. In addition, the study of the accuracy of computations can be enhanced using the Symbolic Toolbox variable-precision arithmetic feature. This feature allows the user to perform certain computations to an unlimited precision.

In Sections 9.2 through 9.14 we provide an introduction to some of the Symbolic Toolbox features but it is not our intention to provide details of all the features. In Section 9.15 we describe applications of the Symbolic Toolbox to specific numerical algorithms.

9.2 Symbolic Variables and Expressions

The first key point to note is that symbolic variables and expressions are different from the standard variables and expressions of MATLAB, and we must distinguish clearly between them. Symbolic variables and symbolic expressions do not have to have numerical values but rather define a structural relationship between symbolic variables, that is, an algebraic expresssion.

To define any variable as a symbolic variable the `sym` function must be used as follows:

```
>> x = sym('x')

x =
x

>> d1 = sym('d1')

d1 =
d1
```

Alternatively, we can use the statement `syms` to define any number of symbolic variables. Thus

```
>> syms a b c d3
```

provides four symbolic variables `a`, `b`, `c`, and `d3`. Note that there is no output to the screen. This is a useful shortcut for the definition of variables and we have used this approach in this text. To check which variables have been declared as symbolic we can use the standard `whos` command. Thus if we use this command after the preceding `syms` declaration, we obtain

```
>> whos
  Name       Size            Bytes  Class     Attributes

  a          1x1                60  sym
  ans        1x19               38  char
  b          1x1                60  sym
  c          1x1                60  sym
  d1         1x1                60  sym
  d3         1x1                60  sym
  x          1x1                60  sym
```

Once variables have been defined as symbolic, expressions can be written using them directly in MATLAB and they will be treated as symbolic expressions. For example, once `x` has been defined as a symbolic variable, the statement

```
>> syms x
>> 1/(1+x)
```

produces the symbolic expression

```
ans =
1/(x + 1)
```

To set up a symbolic matrix we first define any symbolic variables involved in the matrix. Then we enter the statement that defines the matrix in terms of these symbolic variables in the usual manner. On execution the matrix will be displayed as follows:

```
>> syms x y
>> d = [x+1 x^2 x-y;1/x 3*y/x 1/(1+x);2-x x/4 3/2]

d =
[ x + 1,      x^2,      x - y]
[   1/x, (3*y)/x, 1/(x + 1)]
[ 2 - x,      x/4,        3/2]
```

Note that `d` is automatically made symbolic by the assignment of a symoblic expression. We can address individual elements or specific rows and columns as follows:

```
>> d(2,2)

ans =
(3*y)/x

>> c = d(2,:)

c =
[ 1/x, (3*y)/x, 1/(x + 1)]
```

We now consider the manipulation of a symbolic expression. First we set up a symbolic expression as follows:

```
>> e = (1+x)^4/(1+x^2)+4/(1+x^2)

e =
(x + 1)^4/(x^2 + 1) + 4/(x^2 + 1)
```

To see more clearly what this expression represents we can use the function `pretty` to get a more conventional layout of the function:

```
>> pretty(e)

          4
  (x + 1)        4
  -------- + ------
    2          2
   x  + 1     x  + 1
```

Not very pretty! We can simplify the symbolic expression `e` using the function `simplify`:

```
>> simplify(e)

ans =
x^2 + 4*x + 5
```

We can expand expressions using `expand`:

```
>> p = expand((1+x)^4)

p =
x^4 + 4*x^3 + 6*x^2 + 4*x + 1
```

Note the layout of this and other expressions may vary slightly from one computer platform to another. The expression for `p`, in turn, can be rearranged into a nested form using the function `horner`:

```
>> horner(p)

ans =
x*(x*(x*(x + 4) + 6) + 4) + 1
```

We may factorize expressions using the function `factor`. Assuming `a`, `b`, and `c` have been declared as symbolic variables, then

```
>> syms a b c
>> factor(a^3+b^3+c^3-3*a*b*c)

ans =
(a + b + c)*(a^2 - a*b - a*c + b^2 - b*c + c^2)
```

When dealing with complicated expressions, it is useful to simplify the expression as far as possible and as soon as possible. However, it is not always immediately obvious which route should be taken in the simplification process. The function `simple` attempts to simplify the expression using a variety of methods and displays the various results to inform the user. Some of the methods used by the function are not available as separate functions, for example, `radsimp` and `combine(trig)`. The following illustrates the use of the function `simple`:

```
>> syms x; y = sqrt(cos(x)+i*sin(x));
>> simple(y)

simplify:
(cos(x) + sin(x)*i)^(1/2)

radsimp:
(cos(x) + sin(x)*i)^(1/2)

simplify(100):
exp(x*i)^(1/2)

.................
.................

rewrite(exp):
exp(x*i)^(1/2)

rewrite(sincos):
(cos(x) + sin(x)*i)^(1/2)

rewrite(sinhcosh):
(cosh(x*i) + sinh(x*i))^(1/2)

rewrite(tan):
((tan(x/2)*2*i)/(tan(x/2)^2 + 1)
                    - (tan(x/2)^2 - 1)/(tan(x/2)^2 + 1))^(1/2)

mwcos2sin:
(sin(x)*i - 2*sin(x/2)^2 + 1)^(1/2)

collect(x):
(cos(x) + sin(x)*i)^(1/2)

ans =
exp(x*i)^(1/2)
```

In this example the symbolic engine has tried no fewer than fifteen methods to simplify the original expression (not all displayed here)—many to little effect. However, the multiplicity of methods provided allows problems of differing algebraic and transcendental functions to be simplified. The final answer is the shortest, and we would judge it the most acceptable. A compact version of this result can be determined using

```
>> [r,how] = simple(y)

r =
exp(x*i)^(1/2)

how =
simplify(100)
```

When manipulating algebraic, trigonometric, and other expressions, it is important to be able to substitute an expression or constant for any given variable. For example,

```
>> syms u v w
>> fmv = pi*v*w/(u+v+w)

fmv =
(pi*v*w)/(u + v + w)
```

Now we substitute for various variables in this expression. The following statement substitutes the symbolic expression `2*v` for the variable `u`:

```
>> subs(fmv,u,2*v)

ans =
(pi*v*w)/(3*v + w)
```

This next statement substitutes the symbolic constant `1` for the variable `v` in the previous result held in `ans`:

```
>> subs(ans,v,1)

ans =
(pi*w)/(w + 3)
```

Finally, we substitute the symbolic constant `1` for `w` to give

```
>> subs(ans,w,1)

ans =
    0.7854
```

As a further example of the use of the subs function, consider the statements

```
>> syms y
>> f = 8019+20412*y+22842*y^2+14688*y^3+5940*y^4 ...
                                   +1548*y^5+254*y^6+24*y^7+y^8;
```

Now to substitute x-3 for y we use

```
>> subs(f,y,x-3)

ans =
20412*x + 22842*(x - 3)^2 + 14688*(x - 3)^3 + 5940*(x - 3)^4
   + 1548*(x - 3)^5 + 254*(x - 3)^6 + 24*(x - 3)^7 + (x - 3)^8 - 53217
```

Using the collect function to collect terms having the same power of x, we obtain a major simplification as follows:

```
>> collect(ans)

ans =
x^8 + 2*x^6
```

The process of rearranging and simplifying algebraic and transcendental expressions is difficult and the Symbolic Toolbox can be both powerful and frustrating. Powerful because, as we have shown, it is capable of simplifying complex expressions; frustrating because it sometimes fails on relatively simple problems.

Now that we have seen how to manipulate symbolic expressions, we may require a graphical representation. A simple approach to plotting symbolic functions is to use the MATLAB function ezplot, although it must be emphasized that this is restricted to single-variable functions. The following function provides the plot of the normal curve between the values −5 and 5, as shown in Figure 9.1.

```
>> syms x
>> ezplot(exp(-x*x/2),-5,5); grid
```

The alternative to using ezplot is to substitute numerical values into the symbolic expression using the subs function and then use conventional plotting functions.

9.3 Variable-Precision Arithmetic in Symbolic Calculations

In symbolic calculations involving numerical values, the function vpa may be used to obtain any number of decimal places. It should be noted that the result of using this function is a symbolic constant, not a numerical value. Thus to provide $\sqrt{6}$ to 100 places we write vpa(sqrt(6),100). The accuracy here is not restricted to 16 decimal places as in

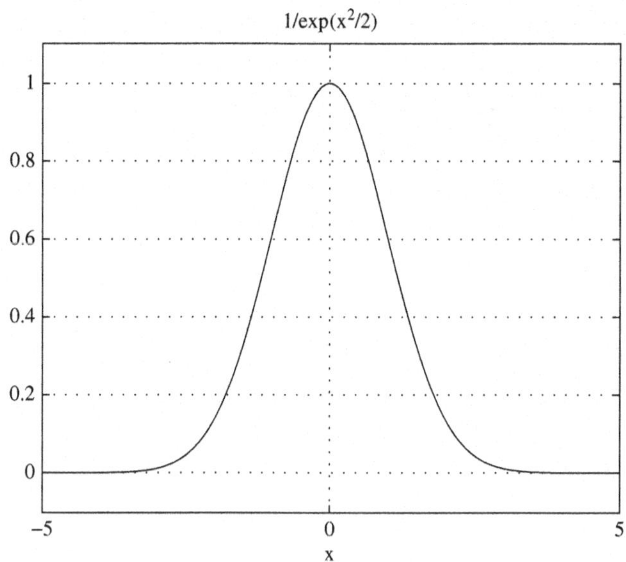

FIGURE 9.1 A plot of the normal curve using the function `ezplot`.

ordinary arithmetic calculations. A nice illustration of this feature is given in the following where we provide an implementation of the famous algorithm of the Borweins, which amazingly quadruples the number of accurate decimal places of π at each iteration! We include this script for illustration only; MATLAB can give π to as many digits as required by writing, for example, `vpa(pi,100)`.

```
% Script e3s901.m  Borwein iteration for pi
n = input('enter n')
y0 = sqrt(2)-1; a0 = 6-4*sqrt(2);
np = 4;
for k = 0:n
    yv = (1-y0^4)^0.25; y1 = (1-yv)/(1+yv);
    a1 = a0*(1+y1)^4-2.0^(2*k+3)*y1*(1+y1+y1^2);
    rpval = a1;  pval = vpa(1/rpval,np)
    a0 = a1; y0 = y1; np = 4*np;
end
```

The results of three iterations follow:

```
enter n 3

n =

     3


pval =
3.142
```

```
pval =
3.141592653589793

pval =
3.14159265358979323846264338327950288419716939937510582097494459

pval =
3.14159265358979323846264338327950288419716939937510582097494459
30781640628620899862803482534211706798214808651328230664709384460
95505822317253594081284811174502841027019385211055596446229489549
303819644288109756659334461284756482337867831652712019091456449
```

Theoretically, the function `vpa` can be utilized to give results to any number of decimal places. An excellent introduction to the computation of π is given by Bailey (1988).

9.4 Series Expansion and Summation

In this section we consider both the development of series approximations for functions and the summation of series.

We begin by showing how a symbolic function can be expanded in the form of a Taylor series using the MATLAB function `taylor(f,n)`. This provides the $(n-1)$th-degree polynomial approximation for the symbolically defined function `f`. If the `taylor` function has only one parameter, then it provides the fifth-degree polynomial approximation for that function.

Consider the following examples:

```
>> syms x
>> taylor(cos(exp(x)),4)

ans =
- (cos(1)*x^3)/2 + (- cos(1)/2 - sin(1)/2)*x^2 - sin(1)*x + cos(1)

>> s = taylor(exp(x),8)

s =
x^7/5040 + x^6/720 + x^5/120 + x^4/24 + x^3/6 + x^2/2 + x + 1
```

The series expansion for the exponential function can be summed using the function `symsum` with $x = 0.1$. To use this function we must know the form of the general term. In this case the series definition is

$$e^{0.1} = \sum_{r=1}^{\infty} \frac{0.1^{r-1}}{(r-1)!} \quad \text{or} \quad \sum_{r=1}^{\infty} \frac{0.1^{r-1}}{\Gamma(r)}$$

Thus, to sum the first eight terms we have

```
>> syms r
>> symsum((0.1)^(r-1)/gamma(r),1,8)

ans =
55700614271/50400000000
```

This can then be evaluated using the function `double`:

```
>> double(ans)

ans =
     1.1052
```

In this example a simple alternative is to use the function `subs` to replace `x` by 0.1 in the symbolic function represented by `s`, and then use `double` to evaluate it, as follows:

```
>> double(subs(s,x,0.1))

ans =
     1.1052
```

This gives a good approximation since we can see that

```
>> exp(0.1)

ans =
     1.1052
```

The function `symsum` can be used to perform many different summations using different combinations of parameters. The following examples illustrate the different cases. To sum the series

$$S = 1 + 2^2 + 3^2 + 4^2 + \cdots + n^2 \tag{9.1}$$

we proceed as follows:

```
>> syms r n
>> symsum(r*r,1,n)

ans =
(n*(2*n + 1)*(n + 1))/6
```

Another example is to sum the series

$$S = 1 + 2^3 + 3^3 + 4^3 + \cdots + n^3 \tag{9.2}$$

Here we use

```
>> symsum(r^3,1,n)

ans =
(n^2*(n + 1)^2)/4
```

Infinity can be used as an upper limit. As an example of this, consider the following infinite sum:

$$S = 1 + \frac{1}{2^2} + \frac{1}{3^2} + \frac{1}{4^2} + \cdots \frac{1}{r^2} + \cdots$$

Summing an infinity of terms, we have

```
>> symsum(1/r^2,1,inf)

ans =
pi^2/6
```

This is an interesting series and is a particular case of the Riemann zeta function (implemented in MATLAB by `zeta(k)`), which gives the sum of the following series:

$$\zeta(k) = 1 + \frac{1}{2^k} + \frac{1}{3^k} + \frac{1}{4^k} + \cdots + \frac{1}{r^k} + \cdots \tag{9.3}$$

For example:

```
>> zeta(2)

ans =
    1.6449

>> zeta(3)

ans =
    1.2021
```

A further interesting example is a summation involving the gamma function (Γ), where $\Gamma(r) = 1.2.3\ldots(r-2)(r-1) = (r-1)!$ for integer values of r. This function is implemented in MATLAB by `gamma(r)`. For example, to sum the series

$$S = 1 + \frac{1}{1} + \frac{1}{2!} + \frac{1}{3!} + \cdots + \frac{1}{r!} + \cdots$$

to infinity, we use the MATLAB statement

```
>> symsum(1/gamma(r),1,inf)

ans =
exp(1)

>> vpa(ans,100)

ans =
2.718281828459045235360287471352662497757247093699959574966967627
7724076630353547594571382178525166427
```

Notice that the use of the vpa function leads to an interesting evaluation of e to a large number of decimal places.

A further example is given by

$$S = 1 + \frac{1}{1!} + \frac{1}{(2!)^2} + \frac{1}{(3!)^2} + \frac{1}{(4!)^2} + \cdots$$

In MATLAB, this becomes

```
>> symsum(1/gamma(r)^2,1,inf)

ans =
sum(1/gamma(r)^2, r = 1..Inf)
```

This is an example of a case where symsum has not worked.

9.5 Manipulation of Symbolic Matrices

Some of the MATLAB functions, such as eig, that can be applied to numerical matrices can also be applied directly to symbolic matrices. However, these features must be used with care for two reasons. First, manipulating large symbolic matrices can be a very slow process. Second, the symbolic results derived from such operations can be of such algebraic complexity that it is difficult or almost impossible to obtain insight into the meaning of the equation.

We will begin by finding the eigenvalues of a simple 4×4 matrix expressed in terms of two symbolic variables.

```
>> syms a b
>> Sm = [a b 0 0;b a b 0;0 b a b;0 0 b a]

Sm =
[ a, b, 0, 0]
[ b, a, b, 0]
[ 0, b, a, b]
[ 0, 0, b, a]
```

```
>> eig(Sm)

ans =
 a - b/2 - (5^(1/2)*b)/2
 a - b/2 + (5^(1/2)*b)/2
 a + b/2 - (5^(1/2)*b)/2
 a + b/2 + (5^(1/2)*b)/2
```

In this problem the expressions for the eigenvalues are quite simple. In contrast, we will now consider an example that appears to be equally simple but develops into a problem with eigenvalues that are much more complicated.

```
>> syms A p
>> A = [1 2 3;4 5 6;5 7 9+p]

A =
[ 1, 2,     3]
[ 4, 5,     6]
[ 5, 7, p + 9]
```

Inspection of this matrix reveals that if $p = 0$, the matrix is singular. Evaluating the determinant of the matrix, we have

```
>> det(A)

ans =
-3*p
```

From this simple result it can immediately be seen that as p tends to zero, the determinant of the matrix tends to zero, indicating that the matrix is singular. Inverting the matrix gives

```
>> B = inv(A)

B =
[    -(5*p + 3)/(3*p), (2*p - 3)/(3*p),  1/p]
[ (2*(2*p + 3))/(3*p),  -(p - 6)/(3*p), -2/p]
[                -1/p,            -1/p,  1/p]
```

This is much more difficult to interpret, although it can be seen that as p tends to zero, each element of the inverse matrix tends to infinity; that is, the inverse does not exist. Note that every element of this inverse matrix is a function of p whereas only one element of the original matrix is a function of p. Finally, we can compute the eigenvalues of the original matrix using the statement `v = eig(A)`. The value of the symbolic object `v` is not shown here because it is so long and complicated. We can find out how many characters (including spaces) are required to express the three eigenvalues symbolically as follows:

```
>>n = length(char(v))

n =
1720
```

This output is very difficult to read, let alone understand. Using the pretty print facility (`pretty`) improves the situation, but the ouput still requires 106 character spaces per line—far more than can be displayed on this page.

The following scripts compute the eigenvalues both symbolically and numerically. In each case the parameter p is varied from 0 to 2 in steps of 0.1. However, only the eigenvalues that correspond to $p = 0.9$ and 1.9 are displayed. The following script determines the eigenvalues symbolically, and then the values of p are substituted into the symbolic eigenvalue expressions using the function `subs`; the function `double` is then used to provide the numerical results.

```
% e3s902.m
disp('Script 1; Symbolic - numerical solution')
c = 1; v = zeros(3,21);
tic
syms a p u w
a = [1 2 3;4 5 6;5 7 9+p];
w = eig(a);  u = [ ];
for s = 0:0.1:2
    u = [u,subs(w,p,s)];
end
v = sort(real(double(u)));
toc
v(:,[10 20])
```

Running this script gives

```
Script 1; Symbolic - numerical solution
Elapsed time is 2.108940 seconds.

ans =
   -0.4255   -0.4384
    0.3984    0.7854
   15.9270   16.5530
```

An alternative approach is to find the eigenvalues of the same matrix by substituting numerical values of p into the numerical matrix and thus determine the eigenvalues. A script to carry out this process follows. Again, only the eigenvalues for $p = 0.9$ and 1.9 are displayed.

```
% e3s903
disp('Script 2: Numerical solution')
c = 1; v = zeros(3,21);
tic
for p = 0:.1:2
    a = [1 2 3;4 5 6;5 7 9+p];
    v(:,c) = sort(eig(a));
    c = c+1;
end
toc
v(:,[10 20])
```

Running this script gives

```
Script 2: Numerical solution
Elapsed time is 0.000934 seconds.

ans =
   -0.4255   -0.4384
    0.3984    0.7854
   15.9270   16.5530
```

As expected, the methods give identical results and show that the eigenvalues are real. Note that the symbolic approach is much slower.

We conclude this section with an example that illustrates the advantage of the symbolic approach for certain problems. We wish to find the eigenvalues of a matrix that can be generated using the MATLAB statement `gallery(5)`. We will begin by finding the eigenvalues of this matrix in a nonsymbolic way.

```
>> B = gallery(5)

B =
          -9          11         -21          63        -252
          70         -69         141        -421        1684
        -575         575       -1149        3451      -13801
        3891       -3891        7782      -23345       93365
        1024       -1024        2048       -6144       24572
```

```
>> format long e
>> eig(B)

ans =
    -4.052036755439267e-002
    -1.177933343414123e-002 +3.828611372186529e-002i
    -1.177933343414123e-002 -3.828611372186529e-002i
     3.203951721060507e-002 +2.281159217067240e-002i
     3.203951721060507e-002 -2.281159217067240e-002i
```

The eigenvalues appear to be small with one real value and the remaining values forming complex conjugate pairs. However, using the symbolic approach we have

```
>> A = sym(gallery(5))

A =
[   -9,     11,    -21,     63,    -252]
[   70,    -69,    141,   -421,    1684]
[ -575,    575,  -1149,   3451,  -13801]
[ 3891,  -3891,   7782, -23345,   93365]
[ 1024,  -1024,   2048,  -6144,   24572]

>> eig(A)

ans =
 0
 0
 0
 0
 0
```

How do we verify which of these two solutions is correct? If we rearrange the eigenvalue problem into the form given by (2.38), that is,

$$(\mathbf{A} - \lambda\mathbf{I})\mathbf{x} = 0$$

then the eigenvalues are the roots of $|\mathbf{A} - \lambda\mathbf{I}| = 0$. We can find these roots symbolically in MATLAB as follows:

```
>> syms lambda
>> D = A-lambda*sym(eye(5));
>> det(D)

ans =
-lambda^5
```

We have shown that $|\mathbf{A} - \lambda\mathbf{I}| = -\lambda^5$ and hence the eigenvalues are the roots of $-\lambda^5 = 0$, that is, zero. Here the Symbolic Toolbox has revealed the true solution.

9.6 Symbolic Methods for the Solution of Equations

The function available in MATLAB to solve symbolic equations is `solve`. This function is most useful for solving polynomials since it provides expressions for all roots. To use `solve` we must set up the expression for the equation we wish to solve in terms of a symbolic variable. For example,

```
>> syms x
>> f = x^3-7/2*x^2-17/2*x+5

f =
x^3 - (7*x^2)/2 - (17*x)/2 + 5

>> solve(f)

ans =
   5
  -2
 1/2
```

The following example illustrates how to solve a system of two equations in two variables. In this example we have chosen to enter the two equations directly in the function by placing them in quotes.

```
>> syms x y
>> [x y] = solve('x^2+y^2=a','x^2-y^2=b')
```

This gives four solutions:

```
x =
  (2^(1/2)*(a + b)^(1/2))/2
 -(2^(1/2)*(a + b)^(1/2))/2
  (2^(1/2)*(a + b)^(1/2))/2
 -(2^(1/2)*(a + b)^(1/2))/2

y =
  (2^(1/2)*(a - b)^(1/2))/2
  (2^(1/2)*(a - b)^(1/2))/2
 -(2^(1/2)*(a - b)^(1/2))/2
 -(2^(1/2)*(a - b)^(1/2))/2
```

A simple check of this solution may be obtained by inserting these solutions back in the original equations:

```
>> x.^2+y.^2, x.^2-y.^2

ans =
 a
 a
 a
 a

ans =
 b
 b
 b
 b
```

Hence all four solutions satisfy the equations simultaneously.

It should be noted that if `solve` fails to obtain a symbolic solution to a given equation or set of equations, it will attempt to use the standard numerical processes where appropriate. In practice it is rarely possible to determine the symbolic solutions of general single-variable or multivariable nonlinear equations.

9.7 Special Functions

The MATLAB Symbolic Toolbox provides the user with access to a range of over 50 special functions and polynomials that can be used symbolically. These functions are not m-files and the standard MATLAB `help` command cannot be used to gain information about them. We can obtain a list of these functions using the command `help mfunlist`. The function `mfun` allows these functions to be evaluated numerically.

One of these functions is the Fresnel sine integral. In `mfunlist` the function `FresnelS` defines the Fresnel sine integral of x. To evaluate it for $x = 4.2$ we enter

```
>> x = 4.2; y = mfun('FresnelS',x)

y =
     0.5632
```

Note that the first and last letters in `FresnelS` must be capitalized. We plot this function (Figure 9.2) using the following script:

```
>> x=1:.01:3; y = mfun('FresnelS',x);
>> plot(x,y)
>> xlabel('x'), ylabel('Fresnel sine integral')
```

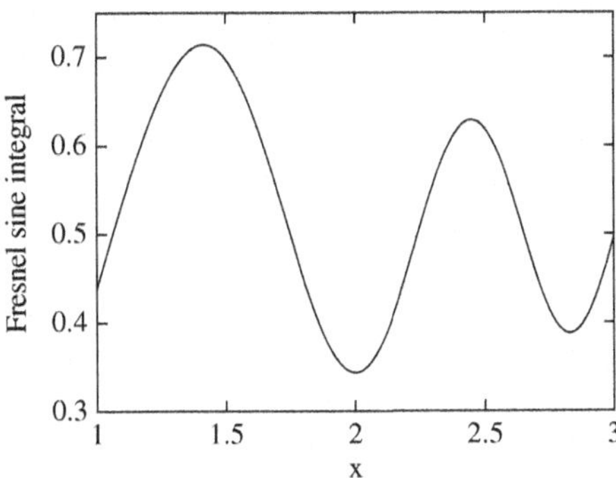

FIGURE 9.2 Plot of the Fresnel sine integral.

Another interesting function available in the `mfunlist` is the logarithmic integral, Li. For example,

```
>> y = round(mfun('Li',[1000 10000 100000]))

y =
         178          1246          9630
```

The logarithmic integral can be used to predict the number of primes below a particular value, and the prediction becomes more accurate as the number of primes increases. To find the number of primes below 1000, 10000, and 100000 we have

```
>> p1 = length(primes(1000)); p2 = length(primes(10000));
>> p3 = length(primes(100000)); p = [p1 p2 p3]

p =
         168          1229          9592
```

Taking the ratio between the values of the logarithmic integral `y` and the number of primes `p`, we have

```
>> p./y

ans =
    0.9438    0.9864    0.9961
```

Notice that the ratio of the value of the logarithmic integral and the number of primes less N tends to 1 as N tends to infinity.

Two important functions are the Dirac delta and Heaviside functions. These functions are not part of the `mfunlist`, and information about them can be obtained using the command `help` in the usual way.

The Dirac delta, or impulse, function $\delta(x)$, is defined as follows:

$$\delta(x - x_0) = 0 \quad x \neq x_0 \tag{9.4}$$

$$\int_{-\infty}^{\infty} f(x)\,\delta(x - x_0)\,dx = f(x_0) \tag{9.5}$$

If $f(x) = 1$, then

$$\int_{-\infty}^{\infty} \delta(x - x_0)\,dx = 1 \tag{9.6}$$

By (9.4), the Dirac delta function only exists at $x = x_0$. Furthermore, from (9.6), the area under the Dirac delta function is unity. This function is implemented in MATLAB by `dirac(x)`.

The Heaviside, or unit step, function is defined as follows:

$$u(x - x_0) = \begin{cases} 1, & \text{for} \quad x - x_0 > 0 \\ 0.5, & \text{for} \quad x - x_0 = 0 \\ 0, & \text{for} \quad x - x_0 < 0 \end{cases} \tag{9.7}$$

This function is implemented in MATLAB by `heaviside(x)`. Thus

```
heaviside(2)

ans =
     1
```

In Sections 9.8, 9.10, and 9.14, we illustrate some of the properties of these functions.

9.8 Symbolic Differentiation

Essentially any function can be differentiated symbolically, but the process is not always easy. For example, the following function given by Swift (1977) is tedious to differentiate:

$$f(x) = \sin^{-1}\left(\frac{e^x \tan x}{\sqrt{x^2 + 4}}\right)$$

Differentiating this function using the MATLAB function `diff` gives

```
>> syms x
>> diff(asin(exp(x)*tan(x)/sqrt(x^2+4)))

ans =
((exp(x)*(tan(x)^2 + 1))/(x^2 + 4)^(1/2) +
(exp(x)*tan(x))/(x^2 + 4)^(1/2) -
(x*exp(x)*tan(x))/(x^2 + 4)^(3/2))/
(1 - (exp(2*x)*tan(x)^2)/(x^2 + 4))^(1/2)

>> pretty(ans)
                     2
  exp(x) (tan(x)  + 1)    exp(x) tan(x)    x exp(x) tan(x)
  -------------------- + ------------- - ---------------
        2     1/2               2     1/2        2     3/2
      (x  + 4)                (x  + 4)         (x  + 4)
  ---------------------------------------------------------
                  /                     2 \1/2
                  |     exp(2 x) tan(x)   |
                  | 1 - ---------------   |
                  |           2           |
                  \          x  + 4       /
```

We now explain specifically how this is achieved. Differentiation is performed on a specified function with respect to a specific variable. When using the Symbolic Toolbox, the variables or variable of the expression must be defined as symbolic. Once this is done, the expression will be treated as symbolic and differentiation may be performed. Consider the following example:

```
>> syms z k
>> f = k*cos(z^4);
>> diff(f,z)

ans =
-4*k*z^3*sin(z^4)
```

Note that the differentiation is performed with respect to `z`, the variable indicated in the second parameter. Differentiation may be performed with respect to `k` as follows:

```
>> diff(f,k)

ans =
cos(z^4)
```

If the variable with respect to which the differentiation is being performed is not indicated, MATLAB chooses the variable name alphabetically closest to x.

Higher-order differentiation can also be performed by including an additional integer parameter that indicates the order of the differentiation as follows:

```
>> syms n
>> diff(k*z^n,4)

ans =
k*n*z^(n - 4)*(n - 1)*(n - 2)*(n - 3)
```

The following example illustrates how we can obtain a standard numerical value from our symbolic differentiation:

```
>> syms x
>> f = x^2*cos(x);
>> df = diff(f)

df =
2*x*cos(x) - x^2*sin(x)
```

We can now substitute a numerical value for x:

```
>> subs(df,x,0.5)

ans =
    0.7577
```

Finally, we consider the symbolic differentiation of the Heaviside, or unit step, function:

```
 diff(heaviside(x))

ans =
dirac(x)
```

This is as expected since the differention of the Heaviside function is zero for all x except when $x = 0$.

9.9 Symbolic Partial Differentiation

Partial derivatives of any multivariable function can be found by differentiating with respect to each variable in turn. As an example, we set up a symbolic function of three variables, assigned to fmv, as follows:

```
>> syms u v w
>> fmv =u*v*w/(u+v+w)

fmv =
(u*v*w)/(u + v + w)

>> pretty(fmv)
    u v w
  ---------
  u + v + w
```

Now we differentiate with respect to u, v, and w in turn:

```
>> d = [diff(fmv,u) diff(fmv,v) diff(fmv,w)]

d =
[ (v*w)/(u + v + w) - (u*v*w)/(u + v + w)^2,
        (u*w)/(u + v + w) - (u*v*w)/(u + v + w)^2,
                  (u*v)/(u + v + w) - (u*v*w)/(u + v + w)^2]
```

To obtain mixed partial derivatives we simply differentiate with respect to the second variable the answer obtained after differentiation with respect to the first. For example,

```
>> diff(d(3),u)

ans =
v/(u + v + w) - (u*v)/(u + v + w)^2
                     - (v*w)/(u + v + w)^2 + (2*u*v*w)/(u + v + w)^3
```

To clarify the structure of the preceding expression, we use the function pretty:

```
>> pretty(ans)
      v             u v             v w            2 u v w
  --------- - ------------ - ------------ + ------------
  u + v + w              2              2              3
              (u + v + w)    (u + v + w)    (u + v + w)
```

This expression provides the mixed second-order partial derivative with respect to w and then u.

9.10 Symbolic Integration

The integration process presents more difficulties than differentiation because not all functions can be integrated in the closed form to produce a symbolic algebraic expression. Even functions that can be integrated in the closed form often require considerable skill and experience for their evaluation.

To use the Symbolic Toolbox we begin by defining the symbolic expression `f`. Then the function `int(f,a,b)` performs symbolic integration where `a` and `b` are the lower and upper limits, respectively. This results in a symbolic constant. If the upper and lower limits are omitted, the result is an expression that is the formula for the indefinite integral. In either case, if the integral cannot be evaluated, which frequently happens, the original function is returned. It must be stressed that many integrals can only be evaluated numerically.

Consider the following indefinite integral:

$$I = \int u^2 \cos u du$$

In MATLAB this becomes

```
>> syms u
>> f = u^2*cos(u); int(f)

ans =
sin(u)*(u^2 - 2) + 2*u*cos(u)
```

We note that the result is a formula as expected and not a numerical value. However, if upper and lower limits are specified, we obtain a symbolic constant. For example, consider

$$y = \int_0^{2\pi} e^{-x/2} \cos(100x) dx$$

Thus we have

```
>> syms x, res = int(exp(-x/2)*cos(100*x),0,2*pi)

res =
2/40001 - 2/(40001*exp(pi))
```

We can obtain a numerical value for this using the `vpa` function as follows:

```
>> vpa(res)

ans =
0.000047838108134108034810408852920091
```

This result confirms the numerical solution given in Section 4.11.

The following examples require infinite limits. Limits of this type are easily accommodated using the symbols `inf` and `-inf`. Consider the following integrals:

$$y = \int_{0}^{\infty} \log_e(1+e^{-x})\,dx \quad \text{and} \quad y = \int_{-\infty}^{\infty} \frac{dx}{(1+x^2)^2}$$

These can be evaluated as follows:

```
>> syms x, int(log(1+exp(-x)),0,inf)

ans =
pi^2/12

>> syms x, f = 1/(1+x^2)^2;
>> int(f,-inf,inf)

ans =
pi/2
```

These results confirm those given by the numerical integration in Section 4.8.

We now consider what happens when an integral cannot be evaluated by symbolic means. In this case we must resort to numerical procedures to find an approximate numerical value for our integral.

```
>> p = sin(x^3);
>> int(p)
Warning: Explicit integral could not be found.

ans =
int(sin(x^3), x)
```

Note that the integral cannot be evaluated symbolically. If we now insert upper and lower limits for this integral, we have

```
>> int(p,0,1)

ans =
hypergeom([2/3], [3/2, 5/3], -1/4)/4
```

This is the hypergeometric function; see Abramowitz and Stegun (1965) or Olver et al. (2010). It can be evaluated under certain conditions; see MATLAB `help hypergeom`.

Clearly the result is not reduced to a numerical value, but in this case we can use a numerical method to solve it as follows:

```
>> fv = @(x) sin(x.^3);
```

```
>> quad(fv,0,1)

ans =
     0.2338
```

We now consider two interesting examples that raise new issues in symbolic processing. Consider the two integrals

$$\int_0^\infty e^{-x}\log_e(x)\,dx \quad \text{and} \quad \int_0^\infty \frac{\sin^4(mx)}{x^2}\,dx$$

We may evaluate the first integral as follows:

```
>> syms x; int(exp(-x)*log(x),0,inf)

ans =
-eulergamma

y = vpa('-eulergamma',10)

y =
-0.5772156649
```

In MATLAB, `eulergamma` is Euler's constant and is defined by

$$C = \lim_{p\to\infty}\left[-\log_e p + \frac{1}{2} + \frac{1}{3} + \cdots + \frac{1}{p}\right] = 0.577215\ldots$$

Thus MATLAB shows how Euler's constant arises and allows us to evaluate it to any specified number of significant figures.

Now, considering the second integral,

```
>> syms m, int(sin(m*x)^4/x^2,0,inf)
Warning: Explicit integral could not be found.

ans =
piecewise([0 < m, (pi*m)/4], [m in R_, (pi*abs(m))/4],
                [not m in R_, int(sin(x*m)^4/x^2, x = 0..Inf)])
```

This complicated MATLAB result attempts to provide, where possible, an evaluation of the integral for various ranges of m. Finally, we integrate the Dirac delta function from $-\infty$ to ∞:

```
>> int(dirac(x),-inf, inf)

ans =
1
```

This result accords with the definition of the Dirac delta function. We now consider the integral of the Heaviside function from −5 to 3:

```
>> int(heaviside(x),-5,3)

ans =
3
```

This result is as expected.

Symbolic integration can be performed for two variable functions by repeated application of the int function. Consider the double integral defined by (4.48) and repeated here for convenience:

$$\int_{x^2}^{x^4} dy \int_{1}^{2} x^2 y \, dx$$

This can be evaluated symbolically as follows:

```
>> syms x y; f = x^2*y;
>> int(int(f,y,x^2,x^4),x,1,2)

ans =
6466/77
```

which confirms the numerical result given in Section 4.14.2.

9.11 Symbolic Solution of Ordinary Differential Equations

The Symbolic Toolbox can be used to solve, symbolically, first-order differential equations, systems of first-order differential equations, or higher-order differential equations, together with any initial conditions provided. This symbolic solution of differential equations is implemented in MATLAB using the function dsolve, and its use is illustrated by a range of examples.

It is important to note that this approach only provides a symbolic solution if one exists. If no solution exists, then the user should apply one of the numerical techniques provided in MATLAB, such as ode45.

The general form of a call of the function dsolve to solve a differential equation system is

```
sol = dsolve('de1, de2, de3, ... , den, in1, in2, in3, ... , inn');
```

The independent variable is assumed to be t unless given by an optional final parameter of dsolve. The parameters de1, de2, de3 up to den stand for the individual differential equations. These must be written in symbolic form using symbolic variables, standard MATLAB operators, and the symbols `D`, `D2`, `D3`, and so on, which represent the first-, second-, third-, and higher-order differential operators, respectively. The parameters `in1`, `in2`, `in3`, `in4`, and so on, represent the initial conditions for the differential equations, if these conditions are required. An example of how these initial conditions should be written, assuming a dependent variable `y`, is

```
y(0) = 1,  Dy(0) = 0,  D2y(0) = 9.1
```

which means that the value of y is 1, $dy/dt = 0$, and $d^2y/dt^2 = 9.1$ when $t = 0$. It is important to note that `dsolve` accepts up to a maximum of 12 input parameters. If initial conditions are required, this is a significant restriction!

The solution is returned to `sol` as a MATLAB structure, and consequently the names of the dependent variables must be used to indicate the individual components. For example, if `g` and `y` are two dependent variables for the differential equation, `sol.y` gives the solution for the dependent variable `y` and `sol.g` gives the solution for the dependent variable `g`.

To illustrate these points we consider some examples. Consider the following first-order differential equation:

$$\left(1+t^2\right)\frac{dy}{dt} + 2ty = \cos t$$

We may solve this without initial conditions using `dsolve` as follows:

```
>> s = dsolve('(1+t^2)*Dy+2*t*y=cos(t)')

s =
-(C3 - sin(t))/(t^2 + 1)
```

Notice that the solution contains the arbitrary constant `C3`. If we now solve the same equation using an initial condition, we proceed as follows:

```
>> s = dsolve('(1+t^2)*Dy+2*t*y=cos(t),y(0)=0')

s =
sin(t)/(t^2 + 1)
```

Note that in this case there is no arbitrary constant.

We now solve a second-order system

$$\frac{d^2y}{dx^2} + y = \cos 2x$$

with the initial conditions $y = 0$ and $dy/dx = 1$ at $x = 0$. To solve this differential equation `dsolve` has the form

```
>> dsolve('D2y+y=cos(2*x), Dy(0)=1, y(0)=0','x')

ans =
(2*cos(x))/3 + sin(x) + sin(x)*(sin(3*x)/6 + sin(x)/2) -
  (2*cos(x)*(6*tan(x/2)^2 - 3*tan(x/2)^4 + 1))/(3*(tan(x/2)^2 + 1)^3)

>> simplify(ans)

ans =
sin(x) + (2*sin(x)^2)/3 - (2*sin(x/2)^2)/3
```

Notice that since the independent variable is x, this is indicated in `dsolve` by a final parameter `x` in the list of parameters.

We now solve the fourth-order differential equation

$$\frac{d^4y}{dt^4} = y$$

with initial conditions

$$y = 1,\ dy/dt = 0,\ d^2y/dt^2 = -1,\ d^3y/dt^3 = 0 \quad \text{when} \quad t = \pi/2$$

We again use `dsolve`. In this example `D4` stands for the fourth derivative operator, with respect to t, and so on.

```
>> dsolve('D4y=y, y(pi/2)=1, Dy(pi/2)=0, D2y(pi/2)=-1, D3y(pi/2)=0')

ans =
sin(t)
```

However, we note that if we try to solve the apparently simple problem

$$\frac{dy}{dx} = \frac{e^{-x}}{x}$$

with the initial condition $y = 1$ when $x = 1$, difficulties arise. Applying `dsolve`, we have

```
>> dsolve('Dy=exp(-x)/x, y(1)=1', 'x')

ans =
1 - Ei(1, x) - Ei(-1)
```

Note that `Ei(-1) = -Ei(1,1)`. Clearly this result is not an explicit solution. The function `Ei(1,x)` is the exponential integral and can be found in `mfunlist`. Details of this mathematical function can be found in Abramowitz and Stegun (1965) and Olver et al.

(2010). We can evaluate the function `Ei` using the `mfun` function for any parameters. For example

```
>> y = mfun('Ei',1,1)

y =
      0.2194
```

If we require more digits in the solution, we have

```
vpa('Ei(1,1)',20)

ans =
0.21938393439552027368
```

We will now attempt to solve the following apparently simple differential equation:

$$\frac{dy}{dx} = \cos(\sin x)$$

Applying `dsolve` we have

```
>> dsolve('Dy=cos(sin(x))','x')

ans =
C17 + int(cos(sin(x)), x, IgnoreAnalyticConstraints)
```

In this case `dsolve` fails to solve the equation.

The differential equation may also contain constants represented as symbols. For example, if we wish to solve the equation

$$\frac{d^2x}{dt^2} + \frac{a}{b}\sin t = 0$$

with initial conditions $x = 1$ and $dx/dt = 0$ when $t = 0$, we enter the following:

```
>> syms x t a b
>> x = dsolve('D2x+(a/b)*sin(t)=0,x(0)=1,Dx(0)=0')

x =
(a*sin(t))/b - (a*t)/b + 1
```

Note how the variables a and b appear in the solution as expected.

As an example of solving two simultaneous differential equations, we note that this differential equation may be rewritten as

$$\frac{du}{dt} = -\frac{a}{b}\sin t$$

$$\frac{dx}{dt} = u$$

Using the same initial conditions, dsolve may be applied to solve these equations by writing

```
>> syms u
>> [u x] = dsolve('Du+(a/b)*sin(t)=0,Dx=u,x(0)=1,u(0)=0')

u =
(a*cos(t))/b - a/b

x =
(a*sin(t))/b - (a*t)/b + 1
```

This gives the same solution as that obtained from dsolve applied directly to solve the second-order differential equation.

The following example provides an interesting comparison of the symbolic and numerical approach. It consists of a script and the output from the script. The script compares the use of dsolve for the symbolic solution of a differential equation and the use ode45 for the numerical solution of the same differential equation. Note that the symbolic solution is obtained in two ways: by solving the second-order equation directly and by separating it into two first-order simultaneous equations. Both approaches provide the same solution.

```
% e3s904.m  Simultaneous first order differential equations
% dx/dt = y, Dy = 3*t-4*x.
% Using dsolve this becomes
syms y t x
x = dsolve('D2x+4*x=3*t','x(0)=0', 'Dx(0)=1')
tt = 0:0.1:5; p = subs(x,t,tt); pp = double(p);
% Plot the symbolic solution to the differential equ'n
plot(tt,pp,'r')
hold on
xlabel('t'), ylabel('x')
sol = dsolve('Dx=y','Dy=3*t-4*x', 'x(0)=0', 'y(0)=1');
sol_x = sol.x, sol_y = sol.y
fv = @(t,x) [x(2); 3*t-4*x(1)];
options = odeset('reltol', 1e-5,'abstol',1e-5);
tspan = [0 5]; initx = [0 1];
[t,x] = ode45(fv,tspan,initx,options);
plot(t,x(:,1),'k+');
axis([0 5 0 4])
```

Executing the script gives the symbolic solution

```
x =
 (3*t)/4 + sin(2*t)/8
```

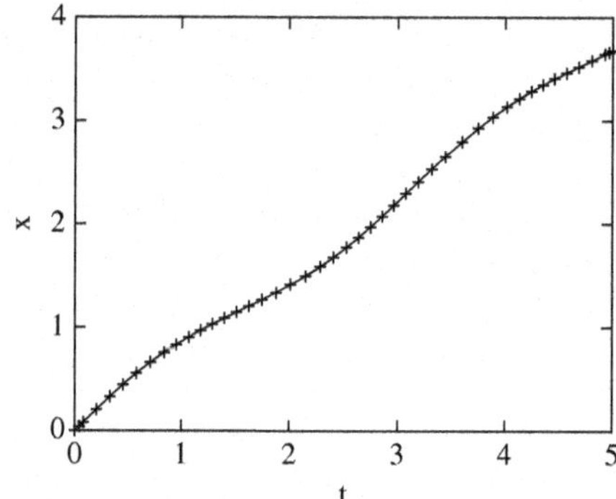

FIGURE 9.3 Symbolic solution and numerical solution indicated by +.

```
sol_x =
(3*t)/4 + sin(2*t)/8

sol_y =
cos(2*t)/4 + 3/4
```

This script also provides the graph of the symbolic solution with alternate numerical solution values also plotted (Figure 9.3). Note how the numerical solution is consistent with the symbolic one.

9.12 The Laplace Transform

The Symbolic Toolbox allows the symbolic determination of the Laplace transforms of many functions. The Laplace transform is used to transform a linear differential equation into an algebraic equation in order to simplify the process of obtaining the solution. It can also be used to transform a system of *differential* equations into a system of *algebraic* equations. The Laplace transform maps a continuous function $f(t)$ in the t domain into the function $F(s)$ in the s domain where $s = \sigma + \jmath\omega$; that is, s is complex. Let $f(t)$ have a finite origin, which can be assumed to be at $t = 0$. In this case we can write

$$F(s) = \int_0^\infty f(t)e^{-st}\,dt \tag{9.8}$$

where $f(t)$ is a given function defined for all positive values of t and $F(s)$ is the Laplace transform of $f(t)$. This transform is called the one-sided Laplace transform. The parameter

s is restricted so that the integral converges, and it should be noted that for many functions the Laplace transform does not exist. The inverse transform is given by

$$f(t) = \frac{1}{\jmath 2\pi} \int_{\sigma_0 - \jmath\infty}^{\sigma_0 + \jmath\infty} F(s) e^{st} ds \tag{9.9}$$

where $\jmath = \sqrt{-1}$ and σ_0 is any real value such that the contour $\sigma_0 - \jmath\omega$, for $-\infty < \omega < \infty$, is in the region of convergence of $F(s)$. In practice (9.9) is not used to compute the inverse transform. The Laplace transform of $f(t)$ may be denoted by the operator $\mathcal{L}$. Thus

$$F(s) = \mathcal{L}[f(t)] \quad \text{and} \quad f(t) = \mathcal{L}^{-1}[F(s)]$$

We now give examples of the use of the Symbolic Toolbox for finding Laplace transforms of certain functions.

```
>> syms t
>> laplace(t^4)

ans =
24/s^5
```

We can use a variable other than `t` as the independent variable as follows:

```
>> syms x; laplace(heaviside(x))

ans =
1/s
```

In this brief introduction to the Laplace transform we will not discuss its properties but merely state the following results:

$$\mathcal{L}\left[\frac{df}{dt}\right] = sF(s) - f(0)$$

$$\mathcal{L}\left[\frac{d^2f}{dt^2}\right] = s^2F(s) - sf(0) - f^{(1)}(0)$$

where $f(0)$ and $f^{(1)}(0)$ are the values of $f(t)$ and its first derivative when $t = 0$. This pattern is continued for higher derivatives.

Suppose we wish to solve the following differential equation:

$$\ddot{y} - 3\dot{y} + 2y = 4t + e^{3t}, \quad y(0) = 1, \quad \dot{y}(0) = -1 \tag{9.10}$$

where the dot notation denotes differentiation with respect to time. Taking the Laplace transform of (9.10), we have

$$s^2Y(s) - sy(0) - y^{(1)}(0) - 3\{sY(s) - y(0)\} + 2Y(s) = \mathcal{L}\left[4t + e^{3t}\right] \tag{9.11}$$

Now, from the definition of the Laplace transform or from tables, we can determine $\mathcal{L}[4t + e^{3t}]$. Here we use the Symbolic Toolbox to determine the required transform:

```
>> syms s t
>> laplace(4*t+exp(3*t))

ans =
1/(s - 3) + 4/s^2
```

Substituting this result in (9.11) and rearranging gives

$$(s^2 - 3s + 2)Y(s) = \frac{4}{s^2} + \frac{1}{s-3} - 3y(0) + sy(0) + y^{(1)}(0)$$

Applying the initial conditions and further rearranging, we have

$$Y(s) = \left(\frac{1}{s^2 - 3s + 2}\right)\left(\frac{4}{s^2} + \frac{1}{s-3} - 4 + s\right)$$

To obtain the solution $y(t)$, we must determine the inverse transform of this equation. It has already been stated that in practice (9.9) is not used to compute the inverse transform. The usual procedure is to rearrange the transform into one that can be recognized in tables of Laplace transforms, and typically the method of partial fractions is used for this purpose. However, the MATLAB Symbolic Toolbox allows us to avoid this task and determine inverse transforms using the `ilaplace` statement as follows:

```
>> ilaplace((4/s^2+1/(s-3)-4+s)/(s^2-3*s+2))

ans =
2*t - 2*exp(2*t) + exp(3*t)/2 - exp(t)/2 + 3
```

Thus $y(t) = 2t + 3 + 0.5(e^{3t} - e^t) - 2e^{2t}$.

9.13 The *Z*-Transform

The Z-transform plays a similar role to the Laplace transform in the solution of difference equations representing discrete systems. The Z-transform is defined by

$$F(z) = \sum_{n=0}^{\infty} f_n z^{-n} \tag{9.12}$$

where f_n is a sequence of data beginning at f_0. The function $F(z)$ is called the unilateral or single-sided Z-transform of f_n and is denoted $\mathcal{Z}[f_n]$. Thus

$$F(z) = \mathcal{Z}[f_n]$$

The inverse transform is denoted by $\mathcal{Z}^{-1}[F(z)]$. Thus

$$f_n = \mathcal{Z}^{-1}[F(z)]$$

Like the Laplace transform, the Z-transform has many important properties. These properties will not be discussed here, but we do provide the following important results:

$$\mathcal{Z}\left[f_{n+k}\right] = z^k F(z) - \sum_{m=0}^{k-1} z^{k-m} f_m \tag{9.13}$$

$$\mathcal{Z}\left[f_{n-k}\right] = z^{-k} F(z) + \sum_{m=1}^{k} z^{-(k-m)} f_{(-m)} \tag{9.14}$$

These are the left- and right-shifting properties, respectively.

We can use the Z-transform to solve *difference* equations in much the same way as we use the Laplace transform to solve *differential* equations. For example, consider the following difference equation:

$$6y_n - 5y_{n-1} + y_{n-2} = \frac{1}{4^n}, \quad n \geq 0 \tag{9.15}$$

Here y_n is a sequence of data values beginning at y_0. However, when $n = 0$ in (9.15), we require the values of y_{-1} and y_{-2} to be specified. These are *initial conditions* and play a similar role to the initial conditions in a differential equation. Let the initial conditions be $y_{-1} = 1$ and $y_{-2} = 0$. Taking $k = 1$ in (9.14), we have

$$\mathcal{Z}\left[y_{n-1}\right] = z^{-1} Y(z) + y_{-1}$$

and taking $k = 2$ in (9.14), we have

$$\mathcal{Z}\left[y_{n-2}\right] = z^{-2} Y(z) + z^{-1} y_{-1} + y_{-2}$$

Taking the Z-transform of (9.15), and substituting for the transformed values of y_{-1} and y_{-2}, we have

$$6Y(z) - 5\{z^{-1}Y(z) + y_{-1}\} + \{z^{-2}Y(z) + z^{-1}y_{-1} + y_{-2}\} = \mathcal{Z}\left[\frac{1}{4^n}\right] \tag{9.16}$$

We could use the basic definition of the Z-transform or tabulated relationships to determine the Z-transform of the right side of this equation. However, the MATLAB Symbolic

Toolbox gives the Z-transform of a function as follows:

```
>> syms z n
>> ztrans(1/4^n)

ans =
z/(z - 1/4)
```

Substituting this result in (9.16), we have

$$(6-5z^{-1}+z^{-2})Y(z)=\frac{4z}{4z-1}-z^{-1}y_{-1}-y_{-2}+5y_{-1}$$

Substituting for y_{-1} and y_{-2} gives

$$Y(z)=\left(\frac{1}{6-5z^{-1}+z^{-2}}\right)\left(\frac{4z}{z-1}-z^{-1}+5\right)$$

We can determine y_n by taking the inverse of the Z-transform. Using the MATLAB function `iztrans` we have

```
>> iztrans((4*z/(4*z-1)-z^(-1)+5)/(6-5*z^(-1)+z^(-2)))

ans =
(5*(1/2)^n)/2 - 2*(1/3)^n + (1/4)^n/2
```

Thus

$$y_n=\frac{5}{2}\left(\frac{1}{2}\right)^n-2\left(\frac{1}{3}\right)^n+\frac{1}{2}\left(\frac{1}{4}\right)^n$$

Evaluating this solution for $n=-2$ and -1 shows that it satisfies the initial conditions.

9.14 Fourier Transform Methods

Fourier analysis transforms data or functions from the time or spatial domain into the frequency domain. Here, we will transform from the x domain into the ω domain because MATLAB uses `x` and `w` (corresponding to ω) as the default parameters in the implementation of the Fourier and inverse Fourier transforms.

The Fourier series transforms a periodic function in the x domain into corresponding discrete values in the frequency domain; these are the Fourier coefficients. In contrast, the Fourier transform takes a nonperiodic and continuous function in the x domain and transforms it into a infinite and continuous function in the frequency domain.

The Fourier transform of a function $f(x)$ is given by

$$F(s)=\mathcal{F}[f(x)]=\int_{-\infty}^{\infty}f(x)e^{-sx}dx \tag{9.17}$$

where $s = \jmath\omega$; that is, s is imaginary. Thus we may write

$$F(\omega) = \mathcal{F}[f(x)] = \int_{-\infty}^{\infty} f(x)e^{-\jmath\omega x}dx \tag{9.18}$$

$F(\omega)$ is complex and is called the frequency spectrum of $f(x)$. The inverse Fourier transform is given by

$$f(x) = \mathcal{F}^{-1}[F(\omega)] = \frac{1}{2\pi}\int_{-\infty}^{\infty} F(\omega)e^{\jmath\omega x}d\omega \tag{9.19}$$

Here we use the operators $\mathcal{F}$ and $\mathcal{F}^{-1}$ to indicate the Fourier transform and the inverse Fourier transform, respectively. Not all functions have a Fourier transform, and certain conditions must be met if the Fourier transform is to exist (see Bracewell, 1978). The Fourier transform has important properties that are introduced as appropriate.

We will begin by using the MATLAB symbolic function `fourier` to determine the Fourier transform of $\cos(3x)$.

```
>> syms x, y = fourier(cos(3*x))

y =
pi*(dirac(w - 3) + dirac(w + 3))
```

This Fourier transform pair is shown diagrammatically in Figure 9.4. The Fourier transform tells us that the frequency spectrum of this cosine function consists of two infinitely narrow components, one at $\omega = 3$ and one at $\omega = -3$ (for a description of the Dirac delta function, see Section 9.7). The MATLAB function `ifourier` implements the symbolic inverse Fourier transform as follows:

```
>> z = ifourier(y)

z =
1/(2*exp(x*3*i)) + exp(x*3*i)/2

>> simplify(z)

ans =
cos(3*x)
```

As a second example of the use of the Fourier transform, consider the transform of the function shown in Figure 9.5, which has a unit value in the range $-2 < x < 2$ and zero elsewhere. Note how this has been constructed from two Heaviside functions (the Heaviside function is described in Section 9.7).

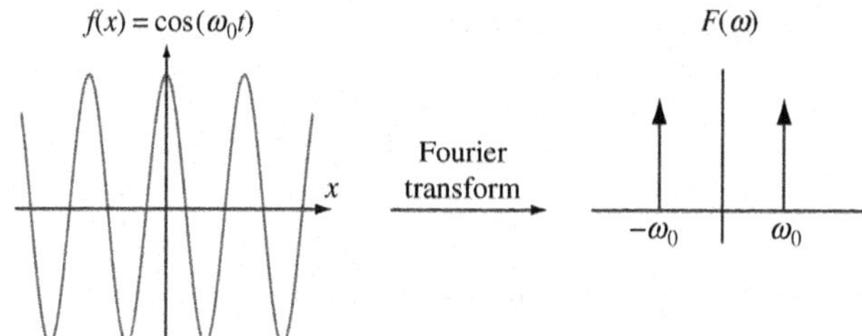

FIGURE 9.4 The Fourier transform of a cosine function.

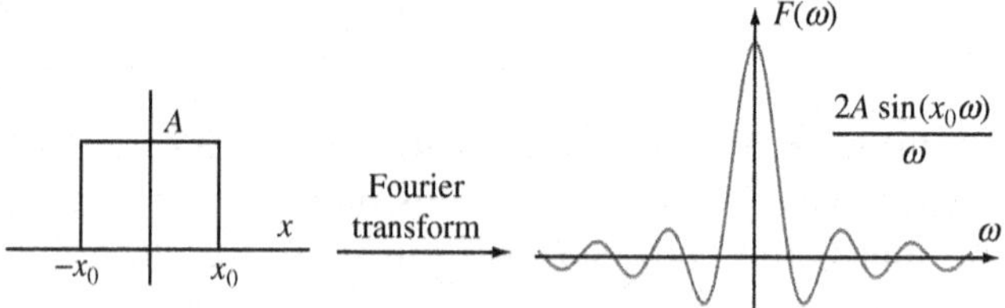

FIGURE 9.5 The Fourier transform of a "top-hat" function.

```
>> syms x
>> fourier(heaviside(x+2) - heaviside(x-2))

ans =
(cos(2*w)*i + sin(2*w))/w - (cos(2*w)*i - sin(2*w))/w
```

This expression can be simplified as follows:

```
simplify(ans)

ans =
(2*sin(2*w))/w
```

Note that the original function, which in the x domain is limited to the range $-2 < x < 2$, has a frequency spectrum that is continuous between $-\infty < \omega < \infty$. This is shown in Figure 9.5.

We now illustrate the use of the Fourier transform in the solution of a partial differential equation. Consider the equation

$$\frac{\partial u}{\partial t} = \frac{\partial^2 u}{\partial x^2} \quad (-\infty < x < \infty,\ t > 0) \tag{9.20}$$

subject to the initial condition

$$u(x, 0) = \exp(-a^2 x^2) \quad \text{where} \quad a = 0.1$$

It can be proved that

$$\mathcal{F}\left[\frac{\partial^2 u}{\partial x^2}\right] = -\omega^2 \mathcal{F}[u] \quad \text{and} \quad \mathcal{F}\left[\frac{\partial u}{\partial t}\right] = \frac{\partial}{\partial t}\{\mathcal{F}[u]\}$$

Thus, taking the Fourier transform of (9.20), we have

$$\frac{\partial}{\partial t}\{\mathcal{F}[u]\} + \omega^2 \mathcal{F}[u] = 0$$

Solving this first-order differential equation for $\mathcal{F}[u]$ gives

$$\mathcal{F}[u] = A\exp(-\omega^2 t) \tag{9.21}$$

To determine the constant A we must apply the initial conditions. We begin by using MATLAB to find the Fourier transform of the initial conditions:

```
>> syms x y z w
>> z = fourier(exp(-x^2/100))

z =
(10*pi^(1/2))/exp(25*w^2)
```

Hence

$$\mathcal{F}[u(x,0)] = \sqrt{100\pi}\exp(-25\omega^2)$$

Comparing this equation with (9.21) when $t = 0$, we see that

$$A = \sqrt{100\pi}\exp(-25\omega^2)$$

Substituting this result in (9.21), we have

$$\mathcal{F}[u] = \sqrt{100\pi}\exp(-25\omega^2)\exp(-\omega^2 t)$$

Taking the inverse transform of this equation gives

$$u(x,t) = \mathcal{F}^{-1}\left[\sqrt{100\pi}\exp(-25\omega^2)\exp(-\omega^2 t)\right]$$

Suppose we require a solution when $t = 4$. Using MATLAB to compute the inverse Fourier transform, we have

```
>> y = z*exp(-4*w^2)

y =
(10*pi^(1/2))/exp(29*w^2)
```

```
>> ifourier(y)

ans =
(5*29^(1/2))/(29*exp(x^2/116))
```

Thus, when $t = 4$, the solution of (9.20) is

$$u(x,4) = \frac{5\sqrt{29}}{29}\exp(-x^2/116)$$

Consider now the Fourier transform of the Heaviside or step function.

```
>> syms x
>> fourier(heaviside(x))

ans =
pi*dirac(w) - i/w
```

Thus the real part of the Fourier transform of the Heaviside function is π times the Dirac delta function at $\omega = 0$ and the imaginary part is $1/\omega$, which tends to plus and minus infinity when $\omega = 0$. The step function has many applications. For example, if we require the Fourier transform of the function

$$f(x) = \begin{cases} e^{-2x} & x \geq 0 \\ 0 & x < 0 \end{cases}$$

using the Heaviside or unit step function, we can rewrite this as

$$f(x) = u(x)e^{-2x} \quad \text{for all} \quad x$$

where $u(x)$ is the Heaviside function. The MATLAB implementation is

```
>> syms x
>> fourier(heaviside(x)*exp(-2*x))

ans =
1/(w*i + 2)
```

9.15 Linking Symbolic and Numerical Processes

Symbolic algebra can be used to ease the burden for the user in the numerical solution process. To illustrate this we show how the Symbolic Toolbox can be used in a version of Newton's method for solving a nonlinear equation that only requires the user to supply the function itself. The usual implementation of this algorithm (see Section 3.7) requires the user to supply the first-order derivative of the function as well as the function itself. The modified algorithm takes the following form in MATLAB:

```
function [res, it] = fnewtsym(func,x0,tol)
% Finds a root of f(x) = 0 using Newton's method
% using the symbolic toolbox.
% Example call: [res, it] = fnewtsym(func,x,tol)
% The user defined function func is the function f(x) which must
% be defined as a symbolic function.
% x is an initial starting value, tol is required accuracy.
it = 1; syms dfunc x
% Now perform the symbolic differentiation:
dfunc = diff(sym(func));
d = double(subs(func,x,x0)/subs(dfunc,x,x0));
while abs(d)>tol
    x1 = x0-d; x0 = x1;
    d = double(subs(func,x,x0)/subs(dfunc,x,x0));
    it = it+1;
end
res = x0;
```

Notice the use of the subs and double functions so that a numerical value is returned. To illustrate the use of fnewtsym we will solve $\cos x - x^3 = 0$ to find the root closest to 1 with an accuracy to four decimal places.

```
>> [r,iter] = fnewtsym('cos(x)-x^3',1,0.00005)

r =
    0.8655

iter =
     4
```

These results are identical to those obtained using function fnewton (see Section 3.7), but here the user is not required to provide the derivative of the function.

The following examples provide further illustrations of how the Symbolic Toolbox can help users perform the routine, tedious, and sometimes difficult tasks required by some numerical methods. We have seen how the single-variable Newton's method can be modified using symbolic differentiation; now we extend this to Newton's multivariable method to solve a system of equations. Here the use of symbolic functions provides an even greater savings for users. The equations solved in this example are

$$x_1 x_2 = 2$$
$$x_1^2 + x_2^2 = 4$$

The MATLAB function takes the form

```
function [x1,fr,it] = newtmvsym(x,f,n,tol)
% Newton's method for solving a system of n nonlinear equations
% in n variables. This version is restricted to two variables.
% Example call: [xv,it] = newtmvsym(x,f,n,tol)
% Requires an initial approximation column vector x. tol is
% required accuracy.
% User must define functions f, the system equations.
% xv is the solution vector, parameter it is number of iterations.
syms a b
xv = sym([a b]); it = 0;
fr = double(subs(f,xv,x));
while norm(fr)>tol
    Jr = double(subs(jacobian(f,xv),xv,x));
    x1 = x-(Jr\fr')'; x = x1;
    fr = double(subs(f,xv,x1));
    it = it+1;
end
```

Notice how this function uses the symbolic Jacobian function in the line

```
Jr = double(subs(jacobian(f,xv),xv,x));
```

This ensures that the user is not required to find the four partial derivatives required in this case. To use this function we run the script

```
% e3s905.m.  Script for running newtonmvsym.m
syms a b
x = sym([a b]);
format long
f = [x(1)*x(2)-2,x(1)^2+x(2)^2-4];
[x1,fr,it] = newtmvsym([1 0],f,2,.000000005)
```

Running this script gives

```
x1 =
   1.414244079950892   1.414183044795298

fr =
  1.0e-008 *
  -0.093132257461548   0.186264514923096

it =
    14
```

We note that this result provides an accurate solution for the given problem in two variables.

It is interesting to note that in a similar way we can write a script for the conjugate gradient method that enables the user to avoid having to provide the first-order partial derivatives of the function to be minimized. This script uses the statements

```
for i = 1:n, dfsymb(i) = diff(sym(f),xv(i)); end
df = double(subs(dfsymb,xv(1:n),x(1:n)'));
```

to obtain the gradient of the function where required. To run this modified function a script similar to that given earlier for the multivariable Newton's method must be written. This script defines the nonlinear function to be optimized and defines a symbolic vector x.

9.16 Summary

We have introduced a wide range of symbolic functions and shown how they can be applied to such standard mathematical processes as integration, differentiation, expansion, and simplification. We have also shown how symbolic methods can sometimes be directly linked to numerical procedures with good effect. For those with access to the Symbolic Toolbox, care must be taken to choose the appropriate method, symbolic or numerical, for the problem.

Problems

9.1. Using the appropriate MATLAB symbolic function, rearrange the following expression as a polynomial in x:

$$\left(x-\frac{1}{a}-\frac{1}{b}\right)\left(x-\frac{1}{b}-\frac{1}{c}\right)\left(x-\frac{1}{c}-\frac{1}{a}\right)$$

9.2. Using the appropriate MATLAB symbolic function, multiply the polynomials

$$f(x)=x^4+4x^3-17x^2+27x-19 \quad \text{and} \quad g(x)=x^2+12x-13$$

and simplify the resulting expression. Then arrange the solution in a nested form.

9.3. Using the appropriate MATLAB symbolic function, expand the following functions:

(a) $\tan(4x)$ in terms of powers of $\tan(x)$
(b) $\cos(x+y)$ in terms of $\cos(x)$, $\cos(y)$, $\sin(x)$, $\sin(y)$
(c) $\cos(3x)$ in terms of powers of $\cos(x)$
(d) $\cos(6x)$ in terms of powers of $\cos(x)$.

9.4. Using the appropriate MATLAB symbolic function, expand $\cos(x+y+z)$ in terms of $\cos(x)$, $\cos(y)$, $\cos(z)$, $\sin(x)$, $\sin(y)$, and $\sin(z)$.

9.5. Using the appropriate MATLAB symbolic function, expand the following functions in ascending powers of x up to x^7:

(a) $\sin^{-1}(x)$
(b) $\cos^{-1}(x)$
(c) $\tan^{-1}(x)$

9.6. Using the appropriate MATLAB symbolic function, expand $y = \log_e(\cos(x))$ in ascending powers of x up to x^{12}.

9.7. The first three terms of a series are

$$\frac{4}{1\cdot 2\cdot 3}\left(\frac{1}{3}\right)+\frac{5}{2\cdot 3\cdot 4}\left(\frac{1}{9}\right)+\frac{6}{3\cdot 4\cdot 5}\left(\frac{1}{27}\right)+\cdots$$

Sum this series to n terms using the MATLAB functions `symsum` and `simple`.

9.8. Using the appropriate MATLAB symbolic function, sum the first hundred terms of the series whose kth term is k^{10}.

9.9. Using the appropriate MATLAB symbolic function, verify that

$$\sum_{k=1}^{\infty} k^{-4} = \frac{\pi^4}{90}$$

9.10. For the matrix

$$\mathbf{A} = \begin{bmatrix} 1 & a & a^2 \\ 1 & b & b^2 \\ 1 & c & c^2 \end{bmatrix}$$

determine $\mathbf{A}^{-1}$ symbolically using the MATLAB function `inv` and, using the function `factor`, express it in the following form:

$$\begin{bmatrix} \dfrac{cb}{(a-c)(a-b)} & \dfrac{-ac}{(b-c)(a-b)} & \dfrac{ab}{(b-c)(a-c)} \\ \dfrac{-(b+c)}{(a-c)(a-b)} & \dfrac{(a+c)}{(b-c)(a-b)} & \dfrac{-(a+b)}{(b-c)(a-c)} \\ \dfrac{1}{(a-c)(a-b)} & \dfrac{-1}{(b-c)(a-b)} & \dfrac{1}{(b-c)(a-c)} \end{bmatrix}$$

9.11. Using the appropriate MATLAB symbolic function, show that the characteristic equation of the matrix

$$\mathbf{A} = \begin{bmatrix} a_1 & a_2 & a_3 & a_4 \\ 1 & 0 & 0 & 0 \\ 0 & 1 & 0 & 0 \\ 0 & 0 & 1 & 0 \end{bmatrix}$$

is

$$\lambda^4 - a_1\lambda^3 - a_2\lambda^2 - a_3\lambda - a_4 = 0$$

9.12. The rotation matrix **R** follows. Define it symbolically and find its second and fourth powers using MATLAB.

$$\mathbf{R} = \begin{bmatrix} \cos\theta & \sin\theta \\ -\sin\theta & \cos\theta \end{bmatrix}$$

9.13. Solve, symbolically, the general cubic equation having the form $x^3 + 3hx + g = 0$. *Hint*: Use the MATLAB function `subexpr` to simplify your result.

9.14. Using the appropriate MATLAB symbolic function, solve the cubic equation $x^3 - 9x + 28 = 0$.

9.15. Using the appropriate MATLAB symbolic functions, find the roots of $z^6 = 4\sqrt{2}(1 + \jmath)$ and plot these roots in the complex plane. *Hint*: Convert the answer using `double`.

9.16. Using the appropriate MATLAB symbolic function, differentiate the following function with respect to x:

$$y = \log_e \left\{ \frac{(1-x)(1+x^3)}{1+x^2} \right\}$$

9.17. Given that Laplace's equation is

$$\frac{\partial^2 z}{\partial x^2} + \frac{\partial^2 z}{\partial y^2} = 0$$

use the appropriate MATLAB symbolic function to verify that this equation is satisfied by the following functions:

(a) $z = \log_e(x^2 + y^2)$
(b) $z = e^{-2y}\cos(2x)$

9.18. Using the appropriate MATLAB symbolic function, verify that $z = x^3 \sin y$ satisfies the following conditions:

$$\frac{\partial^2 z}{\partial x \partial y} = \frac{\partial^2 z}{\partial y \partial x} \quad \text{and} \quad \frac{\partial^{10} z}{\partial x^4 \partial y^6} = \frac{\partial^{10} z}{\partial y^6 \partial x^4} = 0$$

9.19. Using the appropriate MATLAB symbolic functions, determine the integrals of the following functions and then differentiate the result to recover the original function:

(a) $\dfrac{1}{(a+fx)(c+gx)}$ **(b)** $\dfrac{1-x^2}{1+x^2}$

9.20. Using the appropriate MATLAB symbolic function, determine the following indefinite integrals:

$$\textbf{(a)} \quad \int \frac{1}{1+\cos x+\sin x}dx \qquad \textbf{(b)} \quad \int \frac{1}{a^4+x^4}dx$$

9.21. Using the appropriate MATLAB symbolic function, verify the following result:

$$\int_0^\infty \frac{x^3}{e^x-1}dx = \frac{\pi^4}{15}$$

9.22. Using the appropriate MATLAB symbolic function, evaluate the following integrals:

$$\textbf{(a)} \quad \int_0^\infty \frac{1}{1+x^6}dx \qquad \textbf{(b)} \quad \int_0^\infty \frac{1}{1+x^{10}}dx$$

9.23. Using the appropriate MATLAB symbolic function, evaluate the integral

$$\int_0^1 \exp(-x^2)dx$$

by expanding $\exp(-x^2)$ in an ascending series of powers of x and then integrating term by term to obtain a series approximation. Expand the series to both x^6 and x^{14} and hence find two approximations to the integral. Compare the accuracy of your results. The solution to 10 decimal places is 0.7468241328.

9.24. Using the appropriate MATLAB symbolic function, verify the following result:

$$\int_0^\infty \frac{\sin(x^2)}{x}dx = \frac{\pi}{4}$$

9.25. Using the appropriate MATLAB symbolic function, evaluate the integral

$$\int_0^1 \log_e(1+\cos x)dx$$

by expanding $\log_e(1+\cos x)$ in an ascending series of powers of x and then integrating term by term to obtain a series approximation. Expand the series up to the term in x^4. Compare the accuracy of your result. The solution to 10 decimal places is 0.6076250333.

9.26. Using the appropriate MATLAB symbolic function, evaluate the following repeated integral:

$$\int_0^1 dy \int_0^1 \frac{1}{1-xy}\,dx$$

9.27. Using the appropriate MATLAB symbolic function, solve the following differential equation, which arises in the study of consumer behavior:

$$\frac{d^2y}{dt^2} + (bp+aq)\frac{dy}{dt} + ab(pq-1)y = cA$$

Also find the solution in the case where $p=1$, $q=2$, $a=2$, $b=1$, $c=1$, and $A=20$. *Hint*: Use the MATLAB function `subs`.

9.28. Using the appropriate MATLAB symbolic function, solve the following pair of simultaneous differential equations:

$$2\frac{dx}{dt} + 4\frac{dy}{dt} = \cos t$$
$$4\frac{dx}{dt} - 3\frac{dy}{dt} = \sin t$$

9.29. Using the appropriate MATLAB symbolic function, solve the following differential equation:

$$(1-x^2)\frac{d^2y}{dx^2} - 2x\frac{dy}{dx} + 2y = 0$$

9.30. Using the appropriate MATLAB symbolic function, solve the following differential equations using the Laplace transform:

(a) $\dfrac{d^2y}{dt^2} + 2y = \cos(2t), \quad y=-2 \quad \text{and} \quad \dfrac{dy}{dt} = 0 \quad \text{when} \quad t=0$

(b) $\dfrac{dy}{dt} - 2y = t, \quad y=0 \quad \text{when} \quad t=0$

(c) $\dfrac{d^2y}{dt^2} - 3\dfrac{dy}{dt} + y = \exp(-2t), \; y=-3 \quad \text{and} \quad \dfrac{dy}{dt} = 0 \quad \text{when} \quad t=0$

(d) $\dfrac{dq}{dt} + \dfrac{q}{c} = 0, \quad q=V \quad \text{when} \quad t=0$

9.31. Solve the following difference equations symbolically using the Z-transform:

(a) $y_n + 2y_{n-1} = 0, \quad y_{-1} = 4$
(b) $y_n + y_{n-1} = n, \quad y_{-1} = 10$
(c) $y_n - 2y_{n-1} = 3, \quad y_{-1} = 1$
(d) $y_n - 3y_{n-1} + 2y_{n-2} = 3(4^n), \quad y_{-1} = -3, \quad y_{-2} = 5$

Matrix Algebra

The aim of this appendix is to give a brief overview of matrix algebra, which covers a number of issues referred to in the main text. It includes an introduction to matrix properties, operators, and classes.

A.1 Introduction

Since many MATLAB functions and operators act on matrices and arrays, it is important that MATLAB users feel at ease with matrix notation and matrix algebra. MATLAB is an ideal environment in which to experiment and learn matrix algebra. While it cannot provide a formal proof of any relationship, it does allow users to verify results and rapidly gain experience in matrix manipulation. In this appendix only definitions and results are provided. For proofs and further explanation, it is recommended that the reader consult Golub and Van Loan (1989).

A.2 Matrices and Vectors

A matrix is a rectangular array of elements that in itself cannot be evaluated. An element of a matrix can be a real or complex number, an algebraic expression, or another matrix. Normally matrices are enclosed in square brackets, parentheses, or braces. In this text square brackets are used. A complete matrix is denoted by an emboldened character. For example,

$$\mathbf{A} = \begin{bmatrix} 3 & -2 \\ -2 & 4 \end{bmatrix}, \quad \mathbf{B} = \begin{bmatrix} \mathbf{A} & \mathbf{A} & 2\mathbf{A} \\ \mathbf{A} & -\mathbf{A} & \mathbf{A} \end{bmatrix}$$

$$\mathbf{x} = \begin{bmatrix} 11 \\ -3 \\ 7 \end{bmatrix}, \quad \mathbf{e} = \begin{bmatrix} (2+3\imath) & \left(p^2+q\right) & (-4+7\imath) & (3-4\imath) \end{bmatrix}$$

and hence

$$\mathbf{B} = \begin{bmatrix} 3 & -2 & 3 & -2 & 6 & -4 \\ -2 & 4 & -2 & 4 & -4 & 8 \\ 3 & -2 & -3 & 2 & 3 & -2 \\ -2 & 4 & 2 & -4 & -2 & 4 \end{bmatrix}$$

where $\imath = \sqrt{(-1)}$. In the preceding examples **A** is a 2×2 square matrix with two rows and two columns of real coefficients. It also has the property of being a symmetric matrix (see Section A.7). The matrix **B** is built up from the matrix **A** and so **B** is a 4×6 real matrix. The matrix **x** is a 3×1 matrix and is usually called a column vector, and **e** is a 1×4 complex matrix, usually called a row vector. Note that **e** has the algebraic expression $p^2 + q$ for its second element. In this vector each element is enclosed in parentheses to clarify its structure. Enclosing an element in parentheses is not a requirement.

If we wish to refer to a particular element in a matrix, we use subscript notation: The first subscript denotes the row, the second the column. In the case of the row and column vectors it is conventional to use a single subscript. Thus, in the preceding examples,

$$a_{21} = -2, \; b_{25} = -4, \; x_2 = -3, \; e_4 = 3 - 4\imath$$

Note also that although **A** and **B** are uppercase letters it is conventional to refer to their elements by lowercase letters. In general the element in the ith row and jth column of **A** is denoted by a_{ij}.

A.3 Some Special Matrices

The identity matrix. The identity matrix, denoted by **I**, has unit values along the leading diagonal and zeros elsewhere. The leading diagonal is the diagonal of elements from the top left to the bottom right of the matrix. For example,

$$\mathbf{I}_2 = \begin{bmatrix} 1 & 0 \\ 0 & 1 \end{bmatrix}, \quad \mathbf{I}_3 = \begin{bmatrix} 1 & 0 & 0 \\ 0 & 1 & 0 \\ 0 & 0 & 1 \end{bmatrix}$$

The subscript indicating the size of the matrix is usually omitted. The identity matrix behaves rather like the scalar quantity 1. In particular, pre- or postmultiplying a matrix by **I** does not change it.

The diagonal matrix. This matrix is square and has nonzero elements *only* along the leading diagonal. Thus

$$\mathbf{A} = \begin{bmatrix} 4 & 0 & 0 & 0 \\ 0 & -2 & 0 & 0 \\ 0 & 0 & 0 & 0 \\ 0 & 0 & 0 & 9 \end{bmatrix}, \quad \mathbf{B} = \begin{bmatrix} 12 & 0 & 0 \\ 0 & -2 & 0 \\ 0 & 0 & -6 \end{bmatrix}$$

The tridiagonal matrix. This matrix is square and has nonzero elements along the leading diagonal and the diagonals immediately above and below it. Thus, using "x" to denote a nonzero element,

$$\mathbf{A} = \begin{bmatrix} x & x & 0 & 0 & 0 \\ x & x & x & 0 & 0 \\ 0 & x & x & x & 0 \\ 0 & 0 & x & x & x \\ 0 & 0 & 0 & x & x \end{bmatrix}$$

Triangular and Hessenberg matrices. A lower triangular matrix has nonzero elements only on and below the leading diagonal. An upper triangular matrix has nonzero elements on and above the leading diagonal. The Hessenberg matrix is similar to the triangular matrix except that in addition it has nonzero elements on the diagonals adjacent to the leading diagonal.

$$\begin{bmatrix} x & x & x & x & x \\ 0 & x & x & x & x \\ 0 & 0 & x & x & x \\ 0 & 0 & 0 & x & x \\ 0 & 0 & 0 & 0 & x \end{bmatrix}, \quad \begin{bmatrix} x & 0 & 0 & 0 & 0 \\ x & x & 0 & 0 & 0 \\ x & x & x & 0 & 0 \\ x & x & x & x & 0 \\ x & x & x & x & x \end{bmatrix}, \quad \begin{bmatrix} x & x & x & x & x \\ x & x & x & x & x \\ 0 & x & x & x & x \\ 0 & 0 & x & x & x \\ 0 & 0 & 0 & x & x \end{bmatrix}$$

The first matrix is upper triangular, the second is lower triangular, and the last is upper Hessenberg.

A.4 Determinants

The determinant of **A** is written $|\mathbf{A}|$ or $\det(\mathbf{A})$. For a 2×2 array we define its determinant as follows:

$$\text{If } \mathbf{A} = \begin{bmatrix} a_{11} & a_{12} \\ a_{21} & a_{22} \end{bmatrix} \text{ then } \det(\mathbf{A}) = \begin{vmatrix} a_{11} & a_{12} \\ a_{21} & a_{22} \end{vmatrix} = a_{11}a_{22} - a_{21}a_{12} \tag{A.1}$$

In general for an $n \times n$ array **A**, cofactors $C_{ij} = (-1)^{i+j}\,\Delta_{ij}$ can be defined. In this definition Δ_{ij} is the determinant formed from **A** when the elements of the ith row and jth column are deleted. Δ_{ij} is called the minor of **A**. Then

$$\det(\mathbf{A}) = \sum_{k=1}^{n} a_{ik}C_{ik} \quad \text{for any} \quad i = 1, 2, \ldots, n \tag{A.2}$$

This is known as an expansion along the ith row. Frequently the first row is used. This equation replaces the problem of evaluating one $n \times n$ determinant **A** by the evaluation of n, $(n-1) \times (n-1)$ determinants. The process can be continued until the cofactors are reduced to 2×2 determinants. Then the formula (A.1) is used. This is the formal definition for the determinant of **A** but it is not a computationally efficient procedure.

A.5 Matrix Operations

Matrix transposition. In this operation the rows and columns of a matrix are interchanged or transposed. The transposition of a real matrix $\mathbf{A}$ is denoted by $\mathbf{A}^\top$. For example,

$$\mathbf{A} = \begin{bmatrix} 1 & -2 & 4 \\ 2 & 1 & 7 \end{bmatrix}, \quad \mathbf{A}^\top = \begin{bmatrix} 1 & 2 \\ -1 & 1 \\ 4 & 7 \end{bmatrix}, \quad \mathbf{x} = \begin{bmatrix} 1 \\ 2 \\ 3 \end{bmatrix}, \quad \mathbf{x}^\top = \begin{bmatrix} 1 & 2 & 3 \end{bmatrix}$$

Note that a square matrix remains square when it is transposed and a column vector transposes into a row vector and vice versa.

Matrix addition and subtraction. This is done by adding or subtracting corresponding elements in the matrices. Thus

$$\begin{bmatrix} 1 & 3 \\ -4 & 5 \end{bmatrix} + \begin{bmatrix} 5 & -4 \\ 6 & 6 \end{bmatrix} = \begin{bmatrix} 6 & -1 \\ 2 & 11 \end{bmatrix}, \quad \begin{bmatrix} -4 \\ 6 \\ 11 \end{bmatrix} - \begin{bmatrix} 3 \\ -3 \\ 2 \end{bmatrix} = \begin{bmatrix} -7 \\ 9 \\ 9 \end{bmatrix}$$

It is apparent that only matrices with the same number of rows and the same number of columns can be added and subtracted. In general, if $\mathbf{A} = \mathbf{B} + \mathbf{C}$ then $a_{ij} = b_{ij} + c_{ij}$.

Scalar multiplication. Every element of a matrix is multiplied by a scalar quantity. Thus if $\mathbf{A} = s\mathbf{B}$, where s is a scalar, then $a_{ij} = sb_{ij}$.

Matrix multiplication. We can only multiply two matrices $\mathbf{B}$ and $\mathbf{C}$ together if the number of columns in $\mathbf{B}$ is equal to the number of rows in $\mathbf{C}$. Such matrices are said to be conformable. If $\mathbf{B}$ is a $p \times q$ matrix and $\mathbf{C}$ is a $q \times r$ matrix, then we can determine the product $\mathbf{A} = \mathbf{BC}$ and the result will be a $p \times r$ matrix. Because the order of matrix multiplication is important, we say that $\mathbf{B}$ premultiplies $\mathbf{C}$ or $\mathbf{C}$ postmultiplies $\mathbf{B}$. If $\mathbf{A} = \mathbf{BC}$, the elements of $\mathbf{A}$ are determined from the following relationship:

$$a_{ij} = \sum_{k=1}^{q} b_{ik} c_{kj} \quad \text{for} \quad i = 1, 2, \ldots, p; \quad j = 1, 2, \ldots, r$$

For example,

$$\begin{aligned} &\begin{bmatrix} 2 & -3 & 1 \\ -5 & 4 & 3 \end{bmatrix} \begin{bmatrix} -6 & 4 & 1 \\ -4 & 2 & 3 \\ 3 & -7 & -1 \end{bmatrix} \\ &= \begin{bmatrix} 2(-6) + (-3)(-4) + 1(3) & 2(4) + (-3)2 + 1(-7) & 2(1) + (-3)3 + 1(-1) \\ (-5)(-6) + 4(-4) + 3(3) & (-5)4 + 4(2) + 3(-7) & (-5)1 + 4(3) + 3(-1) \end{bmatrix} \\ &= \begin{bmatrix} 3 & -5 & -8 \\ 23 & -33 & 4 \end{bmatrix} \end{aligned}$$

Note that the product of a 2×3 and a 3×3 matrix is a 2×3 matrix. Consider four further examples of matrix multiplication:

$$\begin{bmatrix} 1 & 2 \\ 3 & 4 \end{bmatrix} \begin{bmatrix} 5 & 6 \\ 3 & 2 \end{bmatrix} = \begin{bmatrix} 11 & 10 \\ 27 & 26 \end{bmatrix}, \quad \begin{bmatrix} 5 & 6 \\ 3 & 2 \end{bmatrix} \begin{bmatrix} 1 & 2 \\ 3 & 4 \end{bmatrix} = \begin{bmatrix} 23 & 34 \\ 9 & 14 \end{bmatrix}$$

$$\begin{bmatrix} 1 & 2 & 3 \end{bmatrix} \begin{bmatrix} -4 \\ 3 \\ 3 \end{bmatrix} = 11, \quad \begin{bmatrix} -4 \\ 3 \\ 3 \end{bmatrix} \begin{bmatrix} 1 & 2 & 3 \end{bmatrix} = \begin{bmatrix} -4 & -8 & -12 \\ 3 & 6 & 9 \\ 3 & 6 & 9 \end{bmatrix}$$

In the preceding examples note that while the 2×2 matrices can be multiplied in either order, the product is different. This is an important observation and in general $\mathbf{BC} \neq \mathbf{CB}$. Note also that multiplying a row by a column vector gives a scalar whereas multiplying a column by a row results in a matrix.

Matrix inversion. The inverse of a square matrix $\mathbf{A}$ is written $\mathbf{A}^{-1}$ and is defined by

$$\mathbf{AA}^{-1} = \mathbf{A}^{-1}\mathbf{A} = \mathbf{I}$$

The formal definition of $\mathbf{A}^{-1}$ is

$$\mathbf{A}^{-1} = \text{adj}(\mathbf{A}) / \det(\mathbf{A}) \tag{A.3}$$

where adj($\mathbf{A}$) is the adjoint of $\mathbf{A}$. The adjoint of $\mathbf{A}$ is given by

$$\text{adj}(\mathbf{A}) = \mathbf{C}^{\top}$$

where $\mathbf{C}$ is a matrix composed of the cofactors of $\mathbf{A}$. Using (A.3) is not an efficient way to compute an inverse.

A.6 Complex Matrices

A matrix can have elements that are complex and such a matrix can be expressed in terms of two real matrices. Thus

$$\mathbf{A} = \mathbf{B} + \imath\mathbf{C} \quad \text{where} \quad \imath = \sqrt{(-1)}$$

Here $\mathbf{A}$ is complex and $\mathbf{B}$ and $\mathbf{C}$ are real matrices. The complex conjugate of $\mathbf{A}$ is normally denoted by $\mathbf{A}^*$ and is equal to

$$\mathbf{A}^* = \mathbf{B} - \imath\mathbf{C}$$

Matrix $\mathbf{A}$ can be transposed so that

$$\mathbf{A}^{\top} = \mathbf{B}^{\top} + \imath\mathbf{C}^{\top}$$

Matrix **A** can be transposed *and* conjugated at the same time and this is denoted by $\mathbf{A}^{\mathrm{H}}$ and called the Hermitian transpose. Thus

$$\mathbf{A}^{\mathrm{H}} = \mathbf{B}^{\top} - \imath\mathbf{C}^{\top}$$

For example,

$$\mathbf{A} = \begin{bmatrix} 1-\imath & -2-3\imath & 4\imath \\ 2 & 1+2\imath & 7+5\imath \end{bmatrix}, \quad \mathbf{A}^* = \begin{bmatrix} 1+\imath & -2+3\imath & -4\imath \\ 2 & 1-2\imath & 7-5\imath \end{bmatrix}$$

$$\mathbf{A}^{\top} = \begin{bmatrix} 1-\imath & 2 \\ -2-3\imath & 1+2\imath \\ 4\imath & 7+5\imath \end{bmatrix}, \quad \mathbf{A}^{\mathrm{H}} = \begin{bmatrix} 1+\imath & 2 \\ -2+3\imath & 1-2\imath \\ -4\imath & 7-5\imath \end{bmatrix}$$

It is important to note that the MATLAB expression `A'` gives the conjugation and transposition of `A` when applied to a complex matrix; that is, it is equivalent to $\mathbf{A}^{\mathrm{H}}$. However, `A.'` gives ordinary transposition, which corresponds to $\mathbf{A}^{\top}$.

A.7 Matrix Properties

The real square matrix **A** is

$$\text{symmetric if } \mathbf{A}^{\top} = \mathbf{A}$$

$$\text{skew-symmetric if } \mathbf{A}^{\top} = -\mathbf{A}$$

$$\text{orthogonal if } \mathbf{A}^{\top} = \mathbf{A}^{-1}$$

$$\text{nilpotent if } \mathbf{A}^{p} = \mathbf{0}, \quad \text{where } p \text{ is a positive integer and } \mathbf{0} \text{ is the matrix of zeros}$$

$$\text{idempotent if } \mathbf{A}^{2} = \mathbf{A}$$

The complex square matrix $\mathbf{A} = \mathbf{B} + \imath\mathbf{C}$ is

$$\text{Hermitian if } \mathbf{A}^{\mathrm{H}} = \mathbf{A}$$

$$\text{unitary if } \mathbf{A}^{\mathrm{H}} = \mathbf{A}^{-1}$$

A.8 Some Matrix Relationships

If **P**, **Q**, and **R** are matrices such that

$$\mathbf{W} = \mathbf{P}\mathbf{Q}\mathbf{R}$$

then

$$\mathbf{W}^{\top} = \mathbf{R}^{\top}\mathbf{Q}^{\top}\mathbf{P}^{\top} \tag{A.4}$$

and

$$\mathbf{W}^{-1} = \mathbf{R}^{-1}\mathbf{Q}^{-1}\mathbf{P}^{-1} \tag{A.5}$$

If **P**, **Q**, and **R** are complex, then (A.5) is still valid and (A.4) becomes

$$\mathbf{W}^{\mathrm{H}} = \mathbf{R}^{\mathrm{H}}\mathbf{Q}^{\mathrm{H}}\mathbf{P}^{\mathrm{H}} \tag{A.6}$$

A.9 Eigenvalues

Consider the eigenvalue problem

$$\mathbf{A}\mathbf{x} = \lambda\mathbf{x}$$

If **A** is an $n \times n$ symmetric matrix, then there are n real eigenvalues, λ_i, and n real eigenvectors, $\mathbf{x}_i$, that satisfy this equation. If **A** is an $n \times n$ Hermitian matrix, then there are n real eigenvalues, λ_i, and n complex eigenvectors, $\mathbf{x}_i$, that satisfy the eigenvalue problem. The polynomial in λ given by $\det(\mathbf{A} - \lambda\mathbf{I}) = 0$ is called the characteristic equation. The roots of this polynomial are the eigenvalues of **A**. The sum of the eigenvalues of **A** equals trace(**A**) where trace(**A**) is defined as the sum of the elements on the leading diagonal of **A**. The product of the eigenvalues of **A** equals det(**A**).

It is interesting to note that if we define **C** as

$$\mathbf{C} = \begin{bmatrix} -p_1/p_0 & -p_2/p_0 & \dots & -p_{n-1}/p_0 & -p_n/p_0 \\ 1 & 0 & \dots & 0 & 0 \\ 0 & 1 & \dots & 0 & 0 \\ \vdots & \vdots & & \vdots & \vdots \\ 0 & 0 & \dots & 1 & 0 \end{bmatrix}$$

then the eigenvalues of **C** are the roots of the polynomial

$$p_0x^n + p_1x^{n-1} + \dots + p_{n-1}x + p_n = 0$$

The matrix **C** is called the companion matrix.

A.10 Definition of Norms

The p-norm for the vector **v** is defined as follows:

$$||\mathbf{v}||_p = \left(|v_1|^p + |v_2|^p + \dots + |v_n|^p\right)^{1/p} \tag{A.7}$$

The parameter p can take any value but only three values are commonly used. If $p = 1$ in (A.7), we have the 1-norm, $||\mathbf{v}||_1$:

$$||\mathbf{v}||_1 = |v_1| + |v_2| + \cdots + |v_n| \tag{A.8}$$

If $p = 2$ in (A.7), we have the 2-norm or Euclidean norm of the vector $\mathbf{v}$, which is written $||\mathbf{v}||$ or $||\mathbf{v}||_2$ and is defined as follows:

$$||\mathbf{v}||_2 = \sqrt{v_1^2 + v_2^2 + \cdots + v_n^2} \tag{A.9}$$

Note that it is not necessary to take the modulus of the elements because in this case each element value is squared. The Euclidean norm is also called the length of the vector. These names arise from the fact that in two- or three-dimensional Euclidean space a vector of two or three elements is used to specify a position in space. The distance from the origin to the specified position is identical to the Euclidean norm of the vector.

If p tends to infinity in (A.7), we have $||\mathbf{v}||_\infty = \max(|v_1|, |v_2|, \ldots, |v_n|)$, the infinity norm. At first sight this might appear inconsistent with (A.7). However, when p tends to infinity, the modulus of each element is raised to a very large power and the largest element will dominate the summation.

These functions are implemented in MATLAB; `norm(v,1)`, `norm(v,2)` (or `norm(v)`), and `norm(v,inf)` return the 1, 2, and infinity norms of the vector `v`, respectively.

A.11 Reduced Row Echelon Form

The reduced row echelon form (RREF) of a matrix also has an important role to play in the theoretical understanding of linear algebra. A matrix is transformed into its RREF when the following conditions have been met:

1. All zero rows, if they exist, are at the bottom of the matrix.
2. The first nonzero element in every nonzero row is unity.
3. For each nonzero row, the first nonzero element appears to the right of the first nonzero element of the preceding row.
4. For any column in which the first nonzero element of a row appears, all other elements are zero.

The RREF is determined by using a finite sequence of elementary row operations. It is a standard form and the most fundamental form of a matrix that can be achieved using elementary row operations alone.

For a system of equations $\mathbf{Ax} = \mathbf{b}$ we can define the augmented matrix $[\mathbf{A}\,\mathbf{b}]$. If this matrix is transformed into its RREF, the following may be deduced:

1. If $[\mathbf{A}\,\mathbf{b}]$ is derived from an inconsistent system (i.e., no solution exists) the RREF has a row of the form $[0 \ldots 0\ 1]$.

2. If [**A b**] is derived from a consistent system with an infinity of solutions, then the number of columns of the coefficient matrix is greater than the number of nonzero rows in the RREF; otherwise there is a unique solution and it appears in the last (augmented) column of the RREF.
3. A zero row in the RREF indicates that the original set of equations contained equations with redundant information, that is, information contained in other equations of the system.

In computing the RREF, numerical problems can arise that are common to other procedures that use elementary row operations (see Section 2.6).

A.12 Differentiating Matrices

The rules for matrix differentiation are essentially the same as those for scalars, but care must be taken to ensure that the order of the matrix operations is maintained. The process is illustrated by the following example: differentiate $f(\mathbf{x}) = \mathbf{x}^\top \mathbf{A}\mathbf{x}$ with respect to each element of $\mathbf{x}$, where $\mathbf{x}$ is a column vector with n elements, $(x_1, x_2, x_3, \ldots, x_n)^\top$, and $\mathbf{A}$ has elements a_{ij} for $i,j = 1,2,\ldots, n$. We note first that any matrix associated with a quadratic form must be symmetric. Hence the matrix $\mathbf{A}$ is symmetric. We require the gradient of $f(\mathbf{x})$ (i.e., $\nabla f(\mathbf{x})$). The gradient consists of all the first-order partial derivatives of $f(\mathbf{x})$ with respect to each component of the vector $\mathbf{x}$. Now multiplying out the terms of $f(\mathbf{x})$ we have $f(\mathbf{x})$ expressed in component terms as

$$f(\mathbf{x}) = \sum_{i=1}^{n} a_{ii}x_i^2 + \sum_{i=1}^{n} \sum_{\substack{j=1,\\ j\neq i}}^{n} a_{ij}x_i x_j$$

However, we note that since $\mathbf{A}$ is symmetric $a_{ij} = a_{ji}$ and consequently the terms $a_{ij}x_i x_j + a_{ji}x_i x_j$ can be written as $2a_{ij}x_i x_j$. Hence

$$\frac{\partial f(\mathbf{x})}{\partial x_k} = 2a_{kk}x_k + 2\sum_{\substack{j=1,\\ j\neq i}}^{n} a_{kj}x_j \quad \text{for } k = 1, 2, \ldots, n$$

This is of course equivalent to the matrix form

$$\nabla f(\mathbf{x}) = 2\mathbf{A}\mathbf{x}$$

and this provides the standard matrix result where $\mathbf{x}$ is a column vector.

A.13 Square Root of a Matrix

In order to have a square root, a matrix must be square. If **A** is a square matrix and $\mathbf{BB} = \mathbf{A}$, then **B** is the square root of **A**. If **A** is singular, it may not have a square root.

The square matrix **A** can be factorized to give $\mathbf{A} = \mathbf{XDX}^{-1}$ where **D** is a diagonal matrix comprising the n eigenvalues of **A**, and **X** is an $n \times n$ array of the eigenvectors of **A**. We can expand this expression for **A** to give

$$\mathbf{A} = (\mathbf{XD}^{1/2}\mathbf{X}^{-1})(\mathbf{XD}^{1/2}\mathbf{X}^{-1})$$

Since

$$\mathbf{A} = \mathbf{BB}$$

then

$$\mathbf{B} = \mathbf{XD}^{1/2}\mathbf{X}^{-1}$$

The square root of the diagonal matrix of eigenvalues, **D**, is determined by taking the square root of each diagonal element, that is, each eigenvalue. Any number, real or complex, will have one positive and one negative square root. Thus to determine the square root of **D** (and hence **A**) we must consider every combination of the positive and negative square roots of the eigenvalues. This gives 2^n possible combinations and hence there are 2^n expressions for $\mathbf{D}^{1/2}$. This will lead to 2^n different square root matrices, **B**. If $\mathbf{D}^{1/2}$ comprises all the positive roots then the resulting square root matrix is called the principal square root. This matrix is unique.

Consider the following example. If

$$\mathbf{A} = \begin{bmatrix} 31 & 37 & 34 \\ 55 & 67 & 64 \\ 91 & 115 & 118 \end{bmatrix}$$

then, taking the $2^3 = 8$ combinations of square roots, we obtain the following square roots of **A**. Note that $\mathbf{B}_0$ is the principal square root.

$$\mathbf{B}_0 = \begin{bmatrix} 2.9798 & 2.9296 & 1.8721 \\ 4.3357 & 5.0865 & 3.9804 \\ 5.0313 & 7.1413 & 8.9530 \end{bmatrix} \quad \mathbf{B}_1 = \begin{bmatrix} 1.0000 & 2.0000 & 3.0000 \\ 3.0000 & 4.0000 & 5.0000 \\ 8.0000 & 9.0000 & 7.0000 \end{bmatrix}$$

$$\mathbf{B}_2 = \begin{bmatrix} 2.8115 & 3.0713 & 1.8437 \\ 4.5426 & 4.9123 & 4.0153 \\ 4.9594 & 7.2019 & 8.9408 \end{bmatrix} \quad \mathbf{B}_3 = \begin{bmatrix} 1.1683 & 1.8583 & 3.0284 \\ 2.7931 & 4.1742 & 4.9651 \\ 8.0719 & 8.9395 & 7.0121 \end{bmatrix}$$

The negative of these matrices give a further four square roots of **A**. Multiplying any one of these matrices by itself will result in the original matrix **A**.

Error Analysis

All numerical processes are subject to error. Errors may be of the following types:

1. Truncation errors that are inherent in the numerical algorithm
2. Rounding errors due to the necessity to work to a finite number of significant figures
3. Errors due to inaccurate input data
4. Simple human errors in coding, which should not happen but does!

Examples of the errors described in (1) can be found throughout this text—see, for example, Chapters 3, 4, and 5. Here we consider the implications of the errors described in (2) and (3). Errors of the type described in (4) are outside the scope of this text.

B.1 Introduction

Error analysis estimates the error in some computation caused by errors in some previous process. The previous process may be some experimentation, observation, or rounding in a calculation. Generally we require an upper estimate of the error that can arise when circumstances conspire to be at their worst! We now illustrate this by a specific example. Suppose that $a = 4 \pm 0.02$ (which implies an error of $\pm 0.5\%$) and $b = 2 \pm 0.03$ (which implies an error of $\pm 1.5\%$); then the highest value of a/b results when we divide 4.02 by 1.97 (to give 2.041) and the lowest value of a/b results when we divide 3.98 by 2.03 (to give 1.960). Thus compared with the nominal value of a/b $(= 2)$ we see that the extremes are 2.05% above and 2.0% below the nominal value.

A particular aspect of error analysis is to determine how sensitive a particular calculation is to an error in a specific parameter. Thus we deliberately modify the value of a parameter to determine how sensitive the final answer is to changes in that parameter. For example, consider the following equation:

$$a = 100\frac{\sin\theta}{x^3}$$

If a is evaluated for $\theta = 70^\circ$ and $x = 3$, then $a = 3.4803$. If θ is increased by 10%, then $a = 3.6088$, an increase of 3.69%. If θ is decreased by 10%, then $a = 3.3$, a decrease of 5.18%, Similarly, if we independently increase x by 10%, then $a = 2.6148$, which is a decrease of 24.8%. If we decrease x by 10%, then $a = 4.7741$, an increase of 37.17%. Clearly the value of a is much more sensitive to small changes in x than in θ.

B.2 Errors in Arithmetic Operations

More usually, each of the independent variables has a specified error and we wish to find the overall error in a calculation. We now consider how we can estimate the errors that arise from the standard arithmetic operations. Let x_a, y_a, and z_a be approximations to the exact values x, y, and z, respectively. Let the errors in x, y, and z be x_ε, y_ε, and z_ε, respectively. Then these are given by

$$x_\varepsilon = x - x_a,\ \ y_\varepsilon = y - y_a,\ \ z_\varepsilon = z - z_a$$

Thus

$$x = x_\varepsilon + x_a,\ \ y = y_\varepsilon + y_a,\ \ z = z_\varepsilon + z_a$$

If $z = x \pm y$, then

$$z = (x_a + x_\varepsilon) \pm (y_a + y_\varepsilon) = (x_a \pm y_a) + (x_\varepsilon \pm y_\varepsilon)$$

Now $z_a = x_a \pm y_a$ and hence from the preceding definitions $z_\varepsilon = x_\varepsilon \pm y_\varepsilon$. Normally we are concerned with the maximum possible error and since x_ε and y_ε may be positive or negative quantities, then

$$\max(|z_\varepsilon|) = |x_\varepsilon| + |y_\varepsilon|$$

Consider now the process of multiplication. If $z = xy$, then

$$z = (x_a + x_\varepsilon)(y_a + y_\varepsilon) = x_a y_a + x_\varepsilon y_a + y_\varepsilon x_a + x_\varepsilon y_\varepsilon \tag{B.1}$$

Assuming the errors are small, we can neglect the product of errors in the preceding equation. It is convenient to work in terms of relative error, where the relative error in x, x_ε^R is given by

$$x_\varepsilon^R = x_\varepsilon / x \approx x_\varepsilon / x_a$$

Thus, dividing (B.1) by $z_a = x_a y_a$ we have

$$\frac{(z_a + z_\varepsilon)}{z_a} = 1 + \frac{x_\varepsilon}{x_a} + \frac{y_\varepsilon}{y_a}$$

or

$$\frac{z_\varepsilon}{z_a} = \frac{x_\varepsilon}{x_a} + \frac{y_\varepsilon}{y_a} \tag{B.2}$$

(B.2) can be written

$$z_\varepsilon^R = x_\varepsilon^R + y_\varepsilon^R$$

Again, we want to estimate the worst-case error in z and since the error in x and y may be positive or negative, we have

$$\max\left(\left|z_\varepsilon^R\right|\right) = \left|x_\varepsilon^R\right| + \left|y_\varepsilon^R\right| \tag{B.3}$$

It can easily be shown that if $z = x/y$, the maximum relative error in z is also given by (B.3). This proof is left as an exercise for the reader.

A more general approach to error analysis is to use a Taylor series. Thus if $y = f(x)$ and $y_a = f(x_a)$, then we can write

$$y = f(x) = f(x_a + x_\varepsilon) = f(x_a) + x_\varepsilon f'(x_a) + \cdots$$

Now

$$y_\varepsilon = y - y_a = f(x) - f(x_a)$$

Therefore

$$y_\varepsilon \approx x_\varepsilon f'(x_a)$$

For example, consider $y = \sin\theta$ where $\theta = \pi/3 \pm 0.08$. Thus $\theta_\varepsilon = \pm 0.08$. Hence

$$y_\varepsilon \approx \theta_\varepsilon \frac{d}{d\theta}\{\sin(\theta)\} = \theta_\varepsilon \cos(\pi/3) = 0.08 \times 0.5 = 0.04$$

B.3 Errors in the Solution of Linear Equation Systems

We now consider the problem of estimating the error in the solution of a set of linear equations, $\mathbf{Ax} = \mathbf{b}$. For this analysis we must introduce the concept of a matrix norm.

The formal definition of a matrix p-norm is

$$\|\mathbf{A}\|_p = \max \frac{\|\mathbf{Ax}\|_p}{\|\mathbf{x}\|_p} \quad \text{if} \quad x \neq 0$$

where $\|\mathbf{x}\|_p$ is the vector norm defined in Section A.10. In practice matrix norms are not computed using this definition directly. For example, the 1-norm, 2-norm, and the infinity-norm are computed as follows:

$||\mathbf{A}||_1$ = maximum absolute column sum of **A**

$||\mathbf{A}||_2$ = maximum singular value of **A**

$||\mathbf{A}||_\infty$ = maximum absolute row sum of **A**

Having defined the matrix norm we now consider the solution of the equation system

$$\mathbf{Ax} = \mathbf{b}$$

Let the exact solution of this system be $\mathbf{x}$ and the computed solution be $\mathbf{x}_c$. Then we may define the error as

$$\mathbf{x}_e = \mathbf{x} - \mathbf{x}_c$$

We can also define the residual $\mathbf{r}$ as

$$\mathbf{r} = \mathbf{b} - \mathbf{Ax}_c$$

We note that large residuals are indicative of inaccuracies but small residuals do not guarantee accuracy. For example, consider the case where

$$\mathbf{A} = \begin{bmatrix} 2 & 1 \\ 2+\varepsilon & 1 \end{bmatrix}, \quad \mathbf{b} = \begin{bmatrix} 3 \\ 3+\varepsilon \end{bmatrix}$$

The exact solution of $\mathbf{Ax} = \mathbf{b}$ (with a residual $\mathbf{r} = \mathbf{0}$) is

$$\mathbf{x} = \begin{bmatrix} 1 \\ 1 \end{bmatrix}$$

However, if we consider the very poor approximation

$$\mathbf{x}_c = \begin{bmatrix} 1.5 \\ 0 \end{bmatrix}$$

then the residual is

$$\mathbf{b} - \mathbf{Ax}_c = \begin{bmatrix} 0 \\ -0.5\varepsilon \end{bmatrix}$$

If $\varepsilon = 0.00001$, then the residual is very small even though the solution is very inaccurate.

To obtain a formula that provides bounds on the relative error of the computed value, $\mathbf{x}_c$, we proceed as follows:

$$\mathbf{r} = \mathbf{b} - \mathbf{Ax}_c = \mathbf{Ax} - \mathbf{Ax}_c = \mathbf{Ax}_\varepsilon \tag{B.4}$$

From (B.4) we have

$$\mathbf{x}_\varepsilon = \mathbf{A}^{-1}\mathbf{r}$$

Taking the norms of this equation we have

$$\|\mathbf{x}_\varepsilon\| = \left\|\mathbf{A}^{-1}\mathbf{r}\right\| \tag{B.5}$$

We can choose to use any p-norm and in the analysis that follows the subscript p is omitted. A property of norms is that $||\mathbf{AB}|| \leq ||\mathbf{A}||\,||\mathbf{B}||$. Thus we have, from (B.5),

$$\|\mathbf{x}_\varepsilon\| \leq \left\|\mathbf{A}^{-1}\right\| \|\mathbf{r}\| \tag{B.6}$$

But $\mathbf{r} = \mathbf{A}\mathbf{x}_\varepsilon$ and so

$$\|\mathbf{r}\| \leq \|\mathbf{A}\|\ \|\mathbf{x}_\varepsilon\|$$

Therefore

$$\frac{\|\mathbf{r}\|}{\|\mathbf{A}\|} \leq \|\mathbf{x}_\varepsilon\|$$

Combining this equation with (B.6) we have

$$\frac{\|\mathbf{r}\|}{\|\mathbf{A}\|} \leq \|\mathbf{x}_\varepsilon\| \leq \left\|\mathbf{A}^{-1}\right\|\ \|\mathbf{r}\| \tag{B.7}$$

Now since $\mathbf{x} = \mathbf{A}^{-1}\mathbf{b}$ we have, similarly,

$$\frac{\|\mathbf{b}\|}{\|\mathbf{A}\|} \leq \|\mathbf{x}\| \leq \left\|\mathbf{A}^{-1}\right\| \|\mathbf{b}\| \tag{B.8}$$

If none of the terms in the preceding equation are zero, we can take reciprocals to give

$$\frac{1}{\left\|\mathbf{A}^{-1}\right\| \|\mathbf{b}\|} \leq \frac{1}{\|\mathbf{x}\|} \leq \frac{\|\mathbf{A}\|}{\|\mathbf{b}\|} \tag{B.9}$$

Multiplying the corresponding terms in (B.7) and (B.9) gives

$$\frac{1}{\|\mathbf{A}\|\ \left\|\mathbf{A}^{-1}\right\|}\frac{\|\mathbf{r}\|}{\|\mathbf{b}\|} \leq \frac{\|\mathbf{x}_\varepsilon\|}{\|\mathbf{x}\|} \leq \|\mathbf{A}\|\ \left\|\mathbf{A}^{-1}\right\| \frac{\|\mathbf{r}\|}{\|\mathbf{b}\|} \tag{B.10}$$

This equation gives error bounds for the relative error in the computation that are directly computable. The condition number of $\mathbf{A}$ is given by $\text{cond}(\mathbf{A}, p) = ||\mathbf{A}||_p||\mathbf{A}^{-1}||_p$. Hence (B.10) can be rewritten in terms of $\text{cond}(\mathbf{A}, p)$. When $p = 2$, $\text{cond}(\mathbf{A})$ is the ratio of the largest singular value of $\mathbf{A}$ to the smallest.

We now show how (B.10) can be used to estimate the relative error in the solution of $\mathbf{Ax} = \mathbf{b}$ when $\mathbf{A}$ is the Hilbert matrix. We have chosen the Hilbert matrix because its condition number is large and its inverse is known and so we can compute the actual error in the computation of $\mathbf{x}$. The following MATLAB script evaluates (B.10) for a specific Hilbert matrix using the 2-norm.

```
n = 6, format long
a = hilb(n); b = ones(n,1);
xc = a\b;
x = invhilb(n)*b;
```

```
exact_x = x';
err = abs((xc-x)./x);
nrm_err = norm(xc-x)/norm(x)
r = b-a*xc;
L_Lim = (1/cond(a))*norm(r)/norm(b)
U_Lim = cond(a)*norm(r)/norm(b)
```

Running this script gives

```
n =
     6

nrm_err =
3.316798106133016e-11

L_Lim =
3.351828310510846e-21

U_Lim =
7.492481073232495e-07
```

We see that the norm of the actual relative error, 3.316×10^{-11}, lies between the bounds 3.35×10^{-21} and 7.49×10^{-7}.

Solutions to Selected Problems

Chapter 1

1.1. **(a)** Since some x are negative, the corresponding square roots are imaginary and $\imath = \sqrt{-1}$ is used.
(b) In executing `x./y`, the divide by zero produces the symbol ∞ and a warning.

1.2. **(b)** Note that `t2` is identical to `c` but `t1` is not since the `sqrt` function gives the square root of the individual elements of `c`.

1.4. $x = 2.4545$, $y = 1.4545$, $z = -0.2727$. Note that when using the / operator the solution is given by `x=b'/a'`.

1.8. The plot does not truly represent the function $\cos(x^3)$ because there are insufficient plotting points.

1.9. The function `fplot` automatically adjusts to provide a smoother plot. However, changing `x` to `-2:0.01:2` gives a similar quality graph using the function `plot`.

1.12. $x = 1.6180$.

1.14. Using $x_1 = 1, x_2 = 2, \ldots, x_6 = 6$, a suitable script is

```
n = 6; x = 1:n;
for j = 1:n,
    p(j) = 1;
    for i = 1:n
        if i~=j
            p(j) = p(j)*x(i);
        end
    end
end
p
```

1.15. A suitable script is

```
x = 0.82; tol = 0.005; s = x; i = 2; term = x;
while abs(term)>tol
    term = -term*x; s = s+term/i; i = i+1;
end
s, log(1+x)
```

Note: The scripts may have been compressed to save space.

1.17. The form of the function is

```
function [x1,x2] = funct1(a,b,c)
d = b*b-4*a*c;
if d==0
    x1 = -b/(2*a); x2 = x1;
else
    x1 = (-b+sqrt(d))/(2*a); x2 = (-b-sqrt(d))/(2*a);
end
```

1.18. A possible script is

```
function [x1,x2] = funct2(a,b,c)
if a~= 0
    %as in problem 1.17
else
    disp('warning only one root'); x1 = -c/b; x2 = x1;
end
```

1.19. The graph provides an initial approximation of 1.5. Use the function call `fzero('funct3',1.5)` to obtain the root as 1.2512.

1.20. A possible script is

```
x=[ ]; x(1) = 1873;
c = 1; xc = x(1);
while xc>1
    if (x(c)/2)==floor(x(c)/2)
        x(c+1) = (x(c))/2;
    else
        x(c+1) = 3*x(c)+1;
    end
    xc = x(c+1); c = c+1;
    if c>1000
        break
    end
end
plot(x)
```

Try different values for `x(1)`. For example, 1173, 1409, and so on.

1.21. A possible script is

```
x = -4:0.1:4; y = -4:0.1:4;
[x,y] = meshgrid(-4:0.1:4,-4:0.1:4);
p = x.^2+y.^2;
z = (1-x.^2).*exp(-p)-p.*exp(-p)-exp(-(x+1).^2-y.^2);
```

```
subplot(3,1,1)
mesh(x,y,z)
xlabel('x'), ylabel('y'), zlabel('z')
title('mesh')
subplot(3,1,2)
surf(x,y,z)
xlabel('x'), ylabel('y'), zlabel('z')
title('surf')
subplot(3,1,3)
mesh(x,y,z)
xlabel('x'), ylabel('y'), zlabel('z')
title('contour')
```

1.22. A possible script is

```
clf
a = 11; b= 6;
t = -20:0.1:20;
% Cycloid
x = a*(t-sin(t));y=a*(1-cos(t));
subplot(3,1,1), plot(x,y)
xlabel('x-xis'), ylabel('y-xis'), title('Cycloid')
% witch of agnesi
x1 = 2*a*t;y1=2*a./(1+t.^2);
subplot(3,1,2), plot(x1,y1)
xlabel('x-xis'), ylabel('y-xis')
title('witch of agnesi')
% Complex structure
x2 = a*cos(t)-b*cos(a/b*t);
y2 = a*sin(t)-b*sin(a/b*t);
subplot(3,1,3), plot(x2,y2)
xlabel('x-xis'), ylabel('y-xis')
title('Complex structure')
```

1.23. A possible function is

```
function r = zetainf(s,acc)
sum = 0; n = 1; term = 1+acc;
while abs(term)>acc
    term = 1/n.^s;
    sum = sum +term;
    n = n+1;
end
r = sum;
```

1.24. A possible function is

```
function res = sumfac(n)
sum = 0;
for i = 1:n
    sum = sum+i^2/factorial(i);
end
res = sum;
```

1.26. A possible script is

```
rho1 = [zeros(2), eye(2); eye(2), zeros(2)]
rho2 = [zeros(2), i*eye(2); -i*eye(2), zeros(2)]
rho3 = [eye(2), zeros(2); zeros(2), -eye(2)]
q1 = [zeros(4)  rho1;-rho1  zeros(4)]
q1 = [zeros(4)  rho2;-rho2  zeros(4)]
q1 = [zeros(4)  rho3;-rho3  zeros(4)]
```

1.27. A possible script is

```
x = -4:0.001:4;
y = 1./(((x+2.5).^2).*((x-3.5).^2));
plot(x,y)
ylim([0,20])
xlim([-3,-2])
```

1.28. A possible script is

```
y = @(x)x.^2.*cos(1+x.^2);
y1 = @(x) (1+exp(x))./(cos(x)+sin(x));
x = 0:0.1:2;
subplot(1,2,1), plot(x,y(x))
xlabel('x'), ylabel('y')
subplot(1,2,2), plot(x,y1(x))
xlabel('x'), ylabel('y')
```

Chapter 2

2.1.

n	norm(p-r)	norm(q-r)
3	0.0000	0.0000
4	0.0849	0.0000
5	84.1182	0.1473
6	4.7405e10	6.7767e3

Note the large error in the inverse of the square of the Hilbert matrix when $n = 6$.

2.2. For $n = 3, 4, 5$, and 6, the answers are 2.7464×10^5, 2.4068×10^8, 2.2715×10^{11}, and 2.2341×10^{14}, respectively. The large errors in Problem 2.1 arise from the fact that the Hilbert matrix is very ill-conditioned, as shown by these results.

2.3. For example, taking $n = 5$, $a = 0.2$, and $b = 0.1, a + 2b < 1$ and maximum error in the matrix coefficients is 1.0412×10^5. Taking $n = 5$, $a = 0.3$, $b = 0.5$, $a + 2b > 1$ and after 10 terms, maximum error in the matrix coefficients is 10.8770. After 20 terms, maximum error is 50.5327, clearly diverging.

2.4. The eigenvalues are 5, $2 + 2\imath$, and $2 - 2\imath$. Thus taking $\lambda = 5$ in the matrix $(\mathbf{A} - \lambda\mathbf{I})$ and finding the RREF gives

$$\mathbf{p} = \begin{bmatrix} 1 & 0 & -1.3529 \\ 0 & 1 & 0.6471 \\ 0 & 0 & 0 \end{bmatrix}$$

Hence $\mathbf{px} = \mathbf{0}$. Solving this gives $x_1 = 1.3529x_3$, $x_2 = -0.6471x_3$, and x_3 is arbitrary.

2.6. $\mathrm{x}^\top = [0.95000.98110.9727]$. All methods give the identical solution. Note that if `[q,r] = qr(a)` and `y = q'*b;`, then `x = r(1:3,1:3)\y(1:3)`.

2.7. The solution is $[0\ 0\ 0\ 0\ \ldots\ n+1]$.

2.10. For $n = 20$ the condition number is 178.0643; the theoretical condition number is 162.1139. For $n = 50$ the condition number is 1053.5; the theoretical condition number is 1013.2.

2.11. The right vectors are

$$\begin{bmatrix} 0.0484 + 0.4447\imath \\ -0.3962 + 0.4930\imath \\ 0.4930 + 0.3962\imath \end{bmatrix} \quad \begin{bmatrix} 0.0484 - 0.4447\imath \\ -0.3962 - 0.4930\imath \\ 0.4930 - 0.3962\imath \end{bmatrix} \quad \begin{bmatrix} 0.4082 \\ 0.8165 \\ 0.4082 \end{bmatrix}$$

The corresponding eigenvalues are $2 + 4\imath$, $2 - 4\imath$, and 1. The left vectors are obtained by using the function `eig` on the transposed matrix.

2.12. **(a)** The largest eigenvalue is 242.9773.
(b) The eigenvalue nearest 100 is 112.1542.
(c) The smallest eigenvalue is 77.6972.

2.14. For $n = 5$, the largest eigenvalue is 12.3435 and the smallest eigenvalue is 0.2716. For $n = 50$, the largest eigenvalue is 1.0337×10^3 and the smallest eigenvalue is 0.2502.

2.15. Using the function `roots` we compute the eigenvalues 22.9714, -11.9714, $1.0206 \pm 0.0086\imath$, $1.0083 \pm 0.0206\imath$, $0.9914 \pm 0.0202\imath$, and $0.9798 \pm 0.0083\imath$. Using the function `eig` we have 22.9714, -11.9714, 1, 1, 1, 1, 1, 1, 1, and 1. This is a more accurate solution.

2.16. Both eig and roots give results that only differ by less than 1×10^{-10}. The eigenvalues are 242.9773, 77.6972, 112.1542, 167.4849, and 134.6865.

2.17. The sum of eigenvalues is 55; the product of eigenvalues is 1.

2.18. $c = 0.641n^{1.8863}$.

2.19. A suitable function is

```
function appinv = invapprox(A,k)
ev = eig(A);
evm = max(ev);
if abs(evm)>1
    disp('Method fails')
    appinv = eye(size(A));
else
    appinv = eye(size(A));
    for i = 1:k
        appinv = appinv+A^i;
    end
end
```

2.20. The MATLAB operator gives a much better result. A suitable function is

```
function [res1,res2, nv1,nv2] = udsys(A,b)
newA = A'*A; newb = A'*b;
x1 = inv(newA)*newb;
nv1 = norm(A*x1-b);
x2 = A\b;
nv2 = norm(A*x2-b);
res1 = x1; res2 = x2;
```

2.22. The exact solution is

$$\mathbf{x} = [-12.5 \;\; -24 \;\; -34 \;\; -42 \;\; -47.5 \;\; -50 \;\; -49 \;\; -44 \;\; -34.5 \;\; -20]^\top.$$

The Gauss–Seidel method requires 149 iterations and the Jacobi method requires 283 iterations to give the result to the required accuracy.

Chapter 3

3.2. The solution is 27.8235.

3.3. The solutions are −2 and 1.6344.

3.4. For $c = 5$, with the initial approximation 1.3 or 1.4, the root 1.3735 is obtained after two or three iterations. When $c = 10$, with the initial approximation 1.4, the root 1.4711 is obtained after five iterations. With initial approximation 1.3, convergence is to 193.1083 after 41 iterations. This is a root, but the discontinuity in the function has degraded the performance of the Newton algorithm.

3.5. Schroder's method provides the solution $x = 1.0285$ in 62 iterations, but Newton's method gives $x = 1.0624$ and requires 161 iterations. The solution obtained by Schroder's method is more accurate.

3.6. The equation can be rearranged into the form $x = \exp(x/10)$. Iteration gives $x = 1.1183$. There may be other successful rearrangements.

3.7. The solution is $E = 0.1280$.

3.8. The answers are -3.019×10^{-6} and -6.707×10^{-6} for initial values 1 and -1.5, respectively. The exact solution is clearly 0, but this is a difficult problem.

3.9. The three answers are 1.4299, 1.4468, and 1.4458, which are obtained for four, five, and six terms, respectively. These answers are converging to the correct answer.

3.10. Both approaches give identical results, $x = 8.2183$, $y = 2.2747$. The single variable function is $x/5 - \cos x = 2$. Alternatively, the following call can be used:

```
newtonmv([1 1]','p310','p310d',2,1e-4)
```

It requires the functions and derivatives to be defined as follows:

```
function v = p310(x)
v = zeros(2,1);
v(1) = exp(x(1)/10)-x(2);
v(2) = 2*log(x(2))-cos(x(1))-2;

function vd = p310d(x)
vd = zeros(2,2);
vd(1,:) = [exp(x(1)/10)/10 -1];
vd(2,:) = [sin(x(1)) 2/x(2)];
```

3.11. The solution given by `broyden` is $x = 0.1605$, $y = 0.4931$.

3.12. A solution is $x = 0.9397$, $y = 0.3420$. The MATLAB function `newtonmv` requires 7 iterations; `broyden` requires 33.

3.14. The five roots are 1, $-\iota$, ι, $-\sqrt{2}$, $\sqrt{2}$.

3.15. The solution is $x = -0.1737$, -0.9848ι, $0.9397 + 0.3420\iota$, and $-0.7660 + 0.6428\iota$. This is identical to the exact answer.

3.16. The MATLAB function required is

```
function v = jarrett(f,x1,x2,tol)
gamma = 0.5; d = 1;
while abs(d)>to
    f2 = feval(f,x2);f1=feval(f,x1);
    df = (f2-f1)/(x2-x1); x3 = x2-f2/df; d = x2-x3;
    if f1*f2>0
        x2 = x1; f2 = gamma*f1;
    end
    x2 = x3
end
```

3.17. The third-order method provides the required accuracy after seven iterations. The second-order method requires ten iterations.

3.18. The graphs show that for $c = 2.8$ there is convergence to a single solution, for $c = 3.25$ the iteration oscillates between two values, for $c = 3.5$ the iteration oscillates between four values, and for $c = 3.8$ there is chaotic oscillation between many values.

3.20. Here is an example with p and q chosen to give real roots:

```
>> p=2.5; q = -1; if p^3/q^2>27/4, r = roots([1 0 -p -q]), end

r =
   -1.7523
    1.3200
    0.4323
```

3.21. The commands to solve this problem are

```
>> y1 = roots([1 0 6 -60 36])

y1 =
  -1.8721 + 3.8101i
  -1.8721 - 3.8101i
   3.0999
   0.6444

>> y = y1(3:4)'

y =
    3.0999    0.6444
```

```
>> x = 6./y

x =
     1.9356    9.3110

>> z = 10-x-y

z =
     4.9646    0.0446
```

3.22. A script to solve this problem is

```
c1=(sinh(x)+sin(x))./(2*x);
c3=(sinh(x)-sin(x))./(2*x.^3);
fzero(@ (x) c1^2-x.^4*c3.^2,5)
fzero(@ (x) c1^2-x.^4*c3.^2,30)
```

Chapter 4

4.1. The first derivative is 0.2391, the second derivative is −2.8256. The function `diffgen` gives accurate answers using either $h = 0.1$ or 0.01. The function changes slowly over this range of values.

4.2. When $x = 1$, the computed and exact derivative is −5.0488; when $x = 2$, the computed derivative is −176.6375 (exact = −176.6450), and when $x = 3$, the computed derivative is −194.4680 (exact = −218.6079).

4.3. Using the new formula for Problem 4.1, the first derivative estimate is 0.2267 and 0.2390 for $h = 0.1$ and 0.01, respectively. The second derivative is −2.8249 and −2.8256 for $h = 0.1$ and 0.01, respectively. In Problem 4.2 for $x = 1$, 2, and 3 the first derivative estimates are −5.0489, −175.5798, and −150.1775, respectively. Note that these are less accurate than using `diffgen`.

4.4. The approximate derivatives are −1367.2, −979.4472, −1287.7, and −194.4680. If h is decreased to 0.0001, then the values are the same as the exact derivatives to the given number of decimal places.

4.5. The exact partial derivatives with respect to x and y are 593.652 and 445.2395, respectively. The corresponding approximate values are 593.7071 and 445.2933.

4.6. The integral method estimates 6.3470 primes in the range 1 to 10, 9.633 primes in the range 1 to 17, and 15.1851 primes in the range 1 to 30. The actual numbers are 7, 10, and 15.

4.7. The exact values are 1.5708, 0.5236, and 0.1428 for $r = 0$, 1, and 2, respectively. Approximations provided by integral are 1.5338, 0.5820, and 0.2700.

4.8. The exact values are -0.0811 for $a = 1$ and 0.3052 for $a = 2$. Using `simp1` with 512 points gives agreement to 12 decimal places.

4.9. The exact answer is -0.915965591, and the answer given by `fgauss` is -0.9136. Function `simp1` cannot be used because of the singularity at $x = 0$.

4.10. The exact answer is 0.915965591; `fgauss` gives 0.9159655938. Note that the integrals of Problems 4.9 and 4.10 have the same value apart from the sign.

4.11. **(a)** Using (4.32) with 10 points gives 3.97746326050642; 16-point Gauss gives 3.8145. **(b)** Using (4.33) gives 1.77549968921218; 16-point Gauss gives 1.7758.

4.13. The function `filon` gives

$$2.00000000000098, -0.13333333344440, \text{ and } -2.000199980281494 \times 10^{-4}$$

4.14. Using Romberg's method with nine divisions gives $-2.000222004003794 \times 10^{-4}$. Using Simpson's rule with 1024 intervals gives $-1.999899106566088 \times 10^{-4}$.

4.18. The solution for (a) is 48.96321182552904 and for (b) is 9.726564917628732^3. These compare well with the exact solution, which can be computed from the formula $4\pi^{(n+1)}/(n+1)^2$ where n is the power of x and y.

4.19. **(a)** To fix limits, substitute $y = \surd(x/3) - 1z + 1$. Answer: -1.71962748468952. **(b)** To fix limits, substitute $y = (2 - x)z$. Answer: 0.22222388780205.

4.20. The answers are **(a)** -1.71821293254848 and **(b)** 0.22222222200993.

4.21. Values of the integral are given in the following table:

z	Exact	16-point Gauss
0.5	0.493107418	0.49310741784618
1.0	0.946083070	0.94608306999140
2.0	1.605412977	1.60541297617644

4.22. Use `gauss2v` and define the following function:

```
z = @(x,y) 1./(1-x.*y);
```

4.23. The folowing will provide the solution of this problem:

```
% Probability of engine failure
p = [ ];
a = 3.5; b = 8200;
i = 1;
```

```
for T = 200:100:4000
P(i) = quad(@(x) a*b^a./((x+b).^(a+1)),0.001,T);
i = i+1;
end
figure(1)
plot(200:100:4000,P)
xlabel('Time in hours'), ylabel('Probability of failure')
title('plot of probaility of failure against time')
grid
```

4.24. The folowing will provide the solution of this problem; the value of the integral is −0.15415 correct to 5 places.

```
p = 3; q = 4; r = 2;
f = @(x) (x.^p-x.^q).*x.^r ./log(x);
val = quad(f,0,1);
fprintf('\n value of integral = %6.5f\n',val)
check = log((p+r+1)/(q+r+1))
fprintf('\n value of integral = %6.5f\n',check)
```

4.25. The three integrals are approximately equal to 0.91597.

4.26. The integral equals zero to five decimal places.

4.27. A fairly low accuracy result is obtained.

4.28. Accuracy to two decimal places is obtained.

4.29. There is good agreement between values. The script is

```
f = @(x) -log(x).^3 .* exp(-x)
val = quadgk(f,0,Inf);
fprintf('\n value of integral = %6.5f\n',val)
gam = 0.57722;
S3 = gam^3+0.5*gam*pi^2+2*zeta(3)
fprintf('\n Approximate sum of series = %6.5f\n',S3)
```

4.30. The best result is given by `dblquad` and is 2.01131. The script is

```
R = dblquad(@(x,y) (1-cos(50*x).*cos(100*y))./(2-cos(x)-cos(y))...
,0.0001,pi,0.0001,pi);R=R/pi^2;
fprintf('\nValue of integral using dblquad = %6.5f\n',R)
R1 = simp2v(@(x,y) (1-cos(50*x).*cos(100*y))./(2-cos(x)-cos(y))...
    ,.00001,pi,0.00001,pi,64);R1=R1/pi^2;
```

```
gamma = -psi(1);
R = (gamma+3*log(2)/2+log(50^2+100^2)/2)/pi;
fprintf('\nValue of integral using simp2v = %6.5f\n',R1)
fprintf('\n Approximate value check = %6.5f\n',R)
```

4.31. Value of integral = 0.46306.

Chapter 5

5.1. When $t = 10$, the exact value is 30.326533. `feuler`: 29.9368, 30.2885, 30.3227 with $h = 1$, 0.1, and 0.01, respectively. `eulertp`: 30.3281, 30.3266 with $h = 1$ and 0.1, respectively. `rkgen`: 30.3265 with $h = 1$.

5.2. The classical method gives 108.9077, the Butcher method gives 109.1924, and the Merson method gives 109.0706. The exact answer is $2\exp(x^2) = 109.1963$.

5.3. The Adams–Bashforth–Moulton method gives 4.1042, and Hamming's method gives 4.1043. The exact answer is 4.1042499.

5.4. Using `ode23` gives 0.0456; using `ode45` gives 0.0588.

5.5. The solution of Problem 5.1 with $h = 1$ is 30.3265. The solution of Problem 5.2 with $h = 0.2$ is 108.8906. The solution of Problem 5.2 with $h = 0.02$ is 109.1963.

5.6. (a) 7998.6, exact = 8000. (b) 109.1963.

5.9. The method is stable for $h = 0.1$ and 0.2, and unstable for $h = 0.4$.

5.11. Define the right sides using the following function:

```
function = p511(t,x)
v = ones(2,1);
v(1) = x(1)*(1-0.001*x(1)-1.8*x(2));
v(2) = x(2)*(.3-.5*x(2)/x(1));
```

5.12. Define the right sides using the following function:

```
function v = p512(t,x)
v = ones(2,1);
v(1) = -20*x(1); v(2) = x(1);
```

5.13. Define the right sides using the following function:

```
function v = p513(t,x)
v = ones(2,1);
v(1) = -30*x(2);
v(2) = -.01*x(1)*x(2);
```

5.14. Define the right sides using the following function:

```
function v = p514(t,x)
global c
k = 4; m = 1; F = 1;
v = ones(2,1);
v(1) = (F-c*x(1)-k*x(2))/m;
v(2) = x(1);
```

The script to solve this equation is

```
global c
i = 0;
for c = [0,2,1]
    i = i+1;
    c
    [t,x] = ode45('q514',[0 10],[0 0]');
    figure(i)
    plot(t,x(:,2))
end
```

5.16. The function to solve this problem is

```
function prhs = planetrhs(t,x)
% global x0
% NB global is used if initial values x0
% are used to calculate impact probabilities
% rather than x the variable values
for i=1:3
    for j=1:3
        A(i,j)=x(i).*x(j)./(x(i)+x(j))/1000;
    end
end
prhs = zeros(3,1);
prhs(1) = -x(1).*(A(1,2).*x(2)+A(1,3).*x(3));
prhs(2) = 0.5*A(1,1)*x(1).*x(1)-x(2).*(A(2,2).*x(2)+A(2,3).*x(3));
prhs(3) = 0.5*A(1,2)*x(1).*x(2);
```

The script is as follows:

```
% Solution of planetary growth
% The coagulation equation three size model
% Let x(1), x(2) and x(3) represent the
% number of planetesimals of the three sizes
global x0
% Initially
```

```
x0 = [200,25,1];
tspan = [0,2];
[t,x] = ode45('planetrhs', tspan,x0);
fprintf('\n number of smallest planets= %3.0f',x(end,1))
fprintf('\n number of intermediate planets=%3.0f',x(end,2))
fprintf('\nlargest planets=%3.0f\n',x(end,3))
figure(1)
plot(t,x)
xlabel('time'), ylabel('planet numbers')
grid
```

5.17. The script to solve this equation is

```
% Solution of Daisy world problem
span = 10;
[x,t] = ode45('daisyf',span,[0.2, 0.3]);
plot(x,t)
xlabel('Time'), ylabel('black and white daisy areas')
title('daisy world')
grid
```

and the function is

```
function daisyrhs = daisyf(t,x)
daisyrhs = zeros(2,1);
gamma = 0.3;
Tb = 295; Tw = 285;
betab = 1-0.003265*(295.5-Tb)^2;
betaw = 1-0.003265*(295.5-Tw)^2;
barbit = 1-x(1)-x(2);
daisyrhs(1) = x(1).*(barbit.*betab-gamma);
daisyrhs(2) = x(2).*(barbit.*betaw-gamma);
```

Chapter 6

6.1. **(a)** hyperbolic; **(b)** parabolic; **(c)** $f(x,y) > 0$, hyperbolic; $f(x,y) < 0$, elliptic.

6.2. Initial slope $= -1.6714$. The shooting and FD methods give good results.

6.3. This is an example of a stiff equation. (a) The actual slope when $x = 0$ is 1.0158×10^{-24}. Because we cannot determine this slope accurately, the shooting method gives a very inaccurate solution. (b) In this case the shooting method provides a good result because the initial slope is -120. In both cases the FD method requires a large number of divisions to give an accurate result.

6.5. The finite difference method gives $\lambda_1 = 2.4623$. Exact $\lambda_1 = (\pi/L)^2 = 2.4674$.

6.6. At $t = 0.5$ the variation of z is almost linear between the boundaries at 0 and 10.

6.7. The exact and FD approximations are very similar.

6.8. The exact and FD approximations are similar with a maximum error of 0.0479.

6.9. $\lambda = 5.8870, 14.0418, 19.6215, 27.8876, 29.8780$.

6.10. [0.7703 1.0813 1.5548 1.583 1.1943 1.5548 1.583 1.194 1.0813 0.7703].

Chapter 7

7.1. Using the `aitken` function, $E(2^\circ) = 1.5703$, $E(13^\circ) = 1.5507$, and $E(27^\circ) = 1.4864$. These are accurate to the places given.

7.2. The root is 27.8235.

7.3. (a) $p(x) = 0.9814x^2 + 0.1529$, and $p(x) = -1.2083x^4 + 2.1897x^2 + 0.0137$. The fourth-degree polynomial gives a good fit.

7.4. Interpolation gives 0.9284 (linear), 0.9463 (spline), and 0.9429 (cubic polynomial). The MATLAB function `aitken` gives 0.9455. This is the exact value to four decimal places.

7.5. $p(x) = -0.3238x^5 + 3.2x^4 - 6.9905x^3 - 12.8x^2 + 31.1429x$. Note that the polynomial oscillates between data points. The spline does not exhibit this characteristic, suggesting that it better represents any underlying function from which the data might have been taken.

7.6. **(a)** $f(x) = 3.1276 + 1.9811e^x + e^{2x}$.
(b) $f(x) = 685.1 - 2072.2/(1+x) + 1443.8/\ (1+x)^2$.
(c) $f(x) = 47.3747x^3 - 128.3479x^2 + 103.4153x - 5.2803$. Plotting these functions shows that the best fit is given by (a). The polynomial fit is a reasonable one.

7.7. The plot should diplay an airfoil section.

7.8. The product of primes less than P is given by $0.3679 + 1.0182 \log_e P$ approximately.

7.9. $a_0 = 1$, $a_1 = -0.5740$, $a_2 = 0.9456$, $a_3 = -0.6865$, $a_4 = 0.4115$, $a_5 = -0.0966$.

7.10. Exact: -78.3323. Interpolation gives -78.3340 (cubic) or -77.9876 (linear).

7.11. The minimum values of E are approximately -14.95 and -6.45 at points 40 and 170. The maximum values of E are 3.68 and 16.47 at points 110 and 252.

7.12. The data is sampled from $y = \sin(2\pi f_1 t) + 2\cos(2\pi f_2 t)$ where $f_1 = 1.25$ Hz and $f_2 = 3.4375$ Hz. At 1.25 Hz, DFT = -15.9999ι and at 3.4375 Hz, DFT = 32.0001. The negative complex coefficient is related to the positive size of the coefficient of the sine function, and the positive real component is related to the cosine function. To relate the size of the DFT components to the frequency components in the data we divide the DFT by the number of samples (32) and multiply by 2.

7.13. Algebraically,

$$32\sin^5(30t) = 20\sin(30t) - 10\sin(90t) + 2\sin(150t)$$

and

$$32\sin^6(30t) = 10 - 15\cos(60t) + 6\cos(120t) - \cos(180t)$$

To verify these results from the DFT it is necessary to divide it by n and multiply by 2. The real components are the values of the cosine coefficients. The imaginary components in the DFT are the negative of the values of the sine coefficients. Note also that the coefficient at zero frequency is 20, not 10. This is a consequence of the definition of the DFT; see Section 7.4.

7.14. Components in the spectrum at 30 Hz and 112 Hz. The reason for the large component at 112 Hz is that the component in the data at 400 Hz is above the Nyquist frequency and is folded back to give a spurious component—that is, 400 Hz is 144 Hz above the Nyquist frequency of 256 Hz; 112 Hz is 144 Hz below it.

7.16. With 32 points the frequency increment is 16 Hz and the significant components are at 96 Hz and 112 Hz (the largest amplitude). With 512 points the frequency increment is reduced to 1 Hz and the significant components are at 106, 107, and 108 Hz with the largest amplitude at 107 Hz. With 1024 points the frequency increment is reduced to 0.5 Hz and the component with the largest amplitude is at 107.5 Hz. The original data had a frequency component of 107.5 Hz.

7.17. The estimated production cost in year 6 is $31.80 using cubic extrapolation and $20.88 using quadratic extrapolation. Using the revised data the estimated costs are $24.30 and $21.57, respectively. These widely varying results, some of which are barely credible, show the dangers of trying to estimate future costs from insufficient data.

7.18. $x = 0.5304$.

7.19. $I = 1.5713$, $\alpha = 9.0038$.

7.20. $f_n = \frac{n}{6}\left(n^2 + 3n + 2\right)$.

7.23. The values are 22.70, 22.42, and 22.42.

7.24. The script for this problem is

```
load sunspot.dat
year = sunspot(:,1);
sunact = sunspot(:,2);
figure(1)
plot(year,sunact)
xlabel('Year'), ylabel('Sunspots')
title('Sunspot activity by year')
Y = fft(sunact);
N = length(Y);
Power = abs(Y(1:N/2)).^2;
freq = (1:N/2)/(N/2)*0.5;
figure(2)
plot(freq,Power)
xlabel('freq'), ylabel('Power')
```

Chapter 8

8.1. The objective is 21.6667. The solution is $x_1 = 3.6667$, $x_3 = 0.3333$; the other variables are zero.

8.2. The objective is -21.6667. The solution is $x_1 = 3.3333$, $x_2 = 1.6667$, $x_4 = 0.3333$; the other variables are zero. Thus this problem and the previous one have objective functions of equal magnitude.

8.3. The objective is -100. The solution is $x_1 = 10$, $x_3 = 20$, $x_5 = 22$; the other variables are zero.

8.4. This is a difficult function for the conjugate gradient method and this is why the accuracy of the line search for the built-in MATLAB function `fminsearch` was changed to produce more accurate results. The solution is $[1.0007\ 1.0014]$ with gradient $[0.33860.5226] \times 10^{-3}$.

8.5. The exact and computed solutions are both $[-2.9035 - 2.9035111]$.

8.6. The solution is $[-0.46000.54000.32000.8200]^\top$. `norm(b-Ax)`$= 1.3131 \times 10^{-14}$.

8.7. `[xval,maxf] = optga('p807',[0 2],8,12,20,.005,.6)` where p807 is a MATLAB function defining the problem. A test run gave the following answers:

`xval = 0.9098, maxf = 0.4980.`

8.8. The major modification is to the fitness function as follows:

```
function [fit,fitot] = fitness2d(criteria,chrom,a,b)
% calculate fitness of a set of chromosomes for a two variable
% function assuming each variable is defined in the range
% a to b using a two variable function given by criteria
[pop bitl] = size(chrom); vlength = floor(bitl/2);
for k = 1:pop
    v = [ ]; v1 = [ ]; v2 = [ ]; partchrom1 = chrom(k,1:vlength);
    partchrom2 = chrom(k,vlength+1:2*vlength);
    v1 = binvreal(partchrom1,a,b); v2 = binvreal(partchrom2,a,b);
    v = [v1 v2]; fit(k) = feval(criteria,v);
end
fitot = sum(fit);
```

A call of the modified algorithm is `optga2d('f808',[1 2],24,40,100,.005,.6)`, for example, where f808 defines $z = x^2 + y^2$. This gives the sample results `maxf=7.9795` and `xval=[1.9956 1.9993]`.

8.11. The obtained values are 0.1605 and 0.4931.

8.12. The script to solve this question is

```
clf
[x,y] = meshgrid(-4:0.1:4,-4:0.1:4);
p = x.^2+y.^2;
z = (1-x).^2.*exp(-p)- p.*exp(-p) - exp(-(x+1).^2 - y.^2);
figure(1)
surf(x,y,z)
xlabel('x-axis'), ylabel('y-axis'), zlabel('z-axis')
title('mexhat plot')
figure(2)
contour(x,y,z,20)
xlabel('x-axis'), ylabel('y-axis')
title('contour plot')
optp = ginput(3);
x = optp(:,1); y = optp(:,2);
p = x.^2+y.^2;
z = (1-x).^2.*exp(-p)- p.*exp(-p) - exp(-(x+1).^2 - y.^2)
fprintf('maximum value= %6.2f\n',max(z))
fprintf('minimum value= %6.2f\n',min(z))
x
y
```

```
P=x(1).^2+x(2).^2;
fopt=@ (x)(1-x(1)).^2 .*exp(-(x(1).^2+x(2).^2))...
    - (x(1).^2+x(2).^2).*exp(-(x(1).^2+x(2).^2))...
    - exp(-(x(1)+1).^2 - x(2).^2) ;
[x,fval] = fminsearch(fopt,[-4;4])
fprintf('\nNon global solution= %8.6f\n',fval)
```

8.13. Note that continuous GA gives good agreement with Problem 8.12. Optimum $= -0.3877$

8.14. The minimum is achieved at 63.8157.

8.15. The minimum is achieved at 63.8160, a very similar result to Problem 8.14.

Chapter 9

9.1. Use

```
>>collect((x-1/a-1/b)*(x-1/b-1/c)*(x-1/c-1/a))
```

9.2. Use

```
>>y = x^4+4*x^3-17*x^2+27*x-19; z = x^2+12*x-13;
>>horner(collect(z*y))
```

9.3. Use

```
>>expand(tan(4*x))
>>expand(cos(x+y))
>>expand(cos(3*x))
>>expand(cos(6*x))
```

9.4. Use

```
expand(cos(x+y+z))
```

9.5. Use

```
>>taylor(asin(x),8)
>>taylor(acos(x),8)
>>taylor(atan(x),8)
```

9.6. Use

```
taylor(log(cos(x)),13)
```

9.7. Use

```
>>[solution, how] = simple(symsum((r+3)/(r*(r+1)*(r+2))*(1/3)^r,1,n))
```

9.8. Use

```
symsum(k^10,1,100)
```

9.9. Use
```
symsum(k^(-4),1,inf)
```

9.10. Use
```
a = [1 a a^2;1 b b^2;1 c c^2]; factor(inv(a))
```

9.11. Set
```
a = [a1 a2 a3 a4;1 0 0 0;0 1 0 0;0 0 1 0]
```
and use
```
ev = a-lam*eye(4)
```
and
```
det(ev)
```

9.12. Set
```
trans = [cos(a1) sin(a1);-sin(a1) cos(a1)];
```
and use
```
>>[solution,how] = simple(trans^2)
>>[solution,how] = simple(trans^4)
```

9.13. Set
```
r = solve('x^3+3*h*x+g=0')
```
and use
```
[solution,s] = subexpr(r,'s')
```

9.14. Use
```
>>solve('x^3-9*x+28 = 0')
```

9.15. Use
```
>>p = solve('z^6 = 4*sqrt(2)+i*4*sqrt(2)');
>>res = double(p)
```

9.16. Use
```
>>f5 = log((1-x)*(1+x^3)/(1+x^2)); p = diff(f5);
>>factor(p)
```
Then use `pretty(ans)` to help interpret this result.

9.17. Use
```
>>f = log(x^2+y^2);
>>d2x = diff(f,x,2)
>>d2y = diff(f,y,2)
>>factor(d2x+d2y)
>>f1 = exp(-2*y)*cos(2*x);
>>r = diff(f1,'x',2)+diff(f1,'y',2)
```

9.18. Use
```
>>z = x^3*sin(y);
>>dyx = diff(diff(z,'y'),'x')
```

```
>>dxy = diff(diff(z,'x'),'y')
>>dxy = diff(diff(z,'x',4),'y',6)
>>dxy = diff(diff(z,'y',6),'x',4)
```

9.19. **(a)** Use

```
>>p = int(1/((a+f*x)*(c+g*x)));
>>[solution,how] = simple(p)
>>[solution,how] = simple(diff(solution))
```

(b) Use

```
>>solution = int((1-x^2)/(1+x^2))
>>p = diff(solution); factor(p)
```

9.20. Use

```
>>int(1/(1+cos(x)+sin(x)))
```

and

```
>>int(1/(a^4+x^4))
```

9.21. Use

```
>>int(x^3/(exp(x)-1),0,inf)
```

9.22. Use

```
>>int(1/(1+x^6),0,inf)
```

and

```
>>int(1/(1+x^10),0,inf)
```

9.23. Use

```
>>taylor(exp(-x*x),7)
>>p = int(ans,0,1); vpa(p,10)
>>taylor(exp(-x*x),15)
>>p = int(ans,0,1); vpa(p,10)
```

9.24. Use

```
>>int(sin(x^2)/x,0,inf)
```

9.25. Use

```
>>taylor(log(1+cos(x)),5)
>>int(ans,0,1)
```

9.26. Use

```
>>dint = 1/(1-x*y)
>>int(int(dint,x,0,1),y,0,1)
```

9.27. Use

```
>>[solution,s] = subexpr(dsolve('D2y+(b*p+a*q)*Dy+a*b*(p*q-1)*...
y = c*A ', 'y(0)=0', 'Dy(0)=0','t'),'s')
```

Using the subs function

```
>>subs(solution,{p,q,a,b,c,A},{1,2,2,1,1,20})
```

we obtain the solution for the given values as

```
ans =
10-5*s(2)/s(1)^(1/2)*exp(-1/2*s(3)*t)+5*s(3)/s(1)^(1/2)*exp(-1/2*s(2)*t)
```

In addition, since we also require the values of `s(1)`, `s(2)`, and `s(3)`, we again use `subs` as follows:

```
>>s = subs(s,{p,q,a,b,c},{1,2,2,1,1})

s =
[           17]
[ 5+17^(1/2)]
[ 5-17^(1/2)]
```

9.28. Use

```
>>sol = dsolve('2*Dx+4*Dy = cos(t),4*Dx-3*Dy = sin(t)','t')
```

This gives the solution in the form

```
sol =
    x: [1x1 sym]
    y: [1x1 sym]
```

To see the specific elements of the solution, use

```
>>sol.x

ans =
C1+3/22*sin(t)-2/11*cos(t)
```

and

```
>>sol.y

ans =
C2+2/11*sin(t)+1/11*cos(t)
```

9.29. Use

```
>>dsolve('(1-x^2)*D2y-2*x*Dy+2*y = 0','x')
```

9.30. **(a)** Use

```
>>laplace(cos(2*t))
```

and then

```
>>p = solve('s^2*Y+2*s+2*Y = s/(s^2+4)','Y');
>>ilaplace(p)
```

(b) Use

```
>>laplace(t)
```

and then

```
>>p = solve('s*Y-2*Y = 1/s^2','Y');
>>ilaplace(p)
```

(c) Use

```
>>laplace(exp(-2*t))
```

and then

```
>>p = solve('s^2*Y+3*s-3*(s*Y+3)+Y = 1/(s+2)','Y');]
>>ilaplace(p)
```

(d) The Laplace transform of zero is zero. Thus take the Laplace transform of the equation and then use

```
>>p = solve('(s*Y-V)+Y/c=0','Y');
>>ilaplace(p)
```

9.31. (a) The Z-transform of zero is zero. Thus take the Z-transform of the equation and then use

```
>>p = solve('Y=-2*(Y/z+4)','Y');
>>iztrans(p)
```

(b) Use

```
>>ztrans(n)
```

and then use

```
>>p = solve('Y+(Y/z+10) = z/(z-1)^2','Y');
>>iztrans(p)
```

(c) Use

```
>>ztrans(3*heaviside(n))
```

and then use

```
>>p = solve('Y-2*(Y/z+1)=3*z/(z-1)','Y');
>>iztrans(p)
```

(d) Use

```
>>ztrans(3*4^n)
```

and then use

```
>>p = solve('Y-3*(Y/z-3)+2*(Y/z^2+5-3/z) = 3*z/(z-4)','Y');
>>iztrans(p)
```

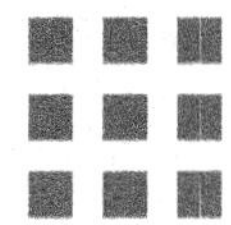

Bibliography

Abramowitz, M., and Stegun, I.A. (1965). *Handbook of Mathematical Functions*, 9th ed. Dover, New York.

Adby, P.R., and Dempster, M.A.H. (1974). *Introduction to Optimisation Methods.* Chapman and Hall, London.

Anderson, D.R., Sweeney, D.J., and Williams, T.A. (1993). *Statistics for Business and Economics.* West Publishing Co., Minneapolis.

Armstrong, R., and Kulesza, B.L.J. (1981). "An approximate solution to the equation $x = \exp(-x/c)$." *Bulletin of the Institute of Mathematics and Its Applications*, **17**(2-3), 56.

Bailey, D.H. (1988). "The computation of π to 29,360,000 decimal digits using Borweins' quadratically convergent algorithm." *Mathematics of Computation*, **50**, 283–296.

Barnes, E.R. (1986). "Affine transform method." *Mathematical Programming*, **36**, 174–182.

Beltrami, E.J. (1987). *Mathematics for Dynamic Modelling.* Academic Press, Boston.

Bracewell, R.N. (1978). *The Fourier Transform and Its Applications.* McGraw-Hill, New York.

Brent, R.P. (1971). "An algorithm with guaranteed convergence for finding the zero of a function." *Computer Journal*, **14**, 422–425.

Brigham, E.O. (1974). *The Fast Fourier Transform.* Prentice Hall, Englewood Cliffs, NJ.

Butcher, J.C. (1964). "On Runge Kutta processes of high order." *Journal of the Australian Mathematical Society*, **4**, 179–194.

Caruana, R.A., and Schaffer, J.D. (1988). "Representation and hidden bias: Grey vs. binary coding for genetic algorithms." *Proceedings of the 5th International Conference on Machine Learning*, Los Altos, CA, pp. 153–161.

Chelouah, R., and Siarry, P. (2000). "A continuous genetic algorithm design for the global optimisation of multimodal functions." *Journal of Heuristics*, **6**(2), 191–213

Cooley, P.M., and Tukey, J.W. (1965). "An algorithm for the machine calculation of complex Fourier series." *Mathematics of Computation*, **19**, 297–301.

Dantzig, G.B. (1963). *Linear Programming and Extensions.* Princeton University Press, Princeton, NJ.

Dekker, T.J. (1969). "Finding a zero by means of successive linear interpolation" in Dejon, B. and Henrici, P. (eds.). *Constructive Aspects of the Fundamental Theorem of Algebra.* Wiley-Interscience, New York.

Dongarra, J.J., Bunch, J., Moler, C.B., and Stewart, G. (1979). *LINPACK User's Guide.* SIAM, Philadelphia.

Dowell, M., and Jarrett, P. (1971). "A modified *regula falsi* method for computing the root of an equation." *BIT*, **11**, 168–174.

Draper, N.R., and Smith H. (1998). *Applied Regression Analysis*, 3rd ed. Wiley, New York.

Fiacco, A.V., and McCormick, G. (1968). *Nonlinear Programming: Sequential Unconstrained Minimization Techniques.* Wiley, New York.

Fiacco A.V. and McCormick, G. (1990). *Nonlinear Programming: Sequential Unconstrained Minimization Techniques.* SIAM Classics in Mathematics, SIAM, Philadelphia (reissue).

Fletcher, R., and Reeves, C.M. (1964). "Function minimisation by conjugate gradients." *Computer Journal*, **7**, 149–154.

Fox, L., and Mayers, D.F. (1968). *Computing Methods for Scientists and Engineers.* Oxford University Press, Oxford, UK.

Froberg, C.-E. (1969). *Introduction to Numerical Analysis,* 2nd ed. Addison-Wesley, Reading, MA.

Garbow, B.S., Boyle, J.M., Dongarra, J.J., and Moler, C.B. (1977). *Matrix Eigensystem Routines: EISPACK Guide Extension.* Lecture Notes in Computer Science, **51**. Springer-Verlag, Berlin.

Gear, C.W. (1971). *Numerical Initial Value Problems in Ordinary Differential Equations.* Prentice Hall, Englewood Cliffs, NJ.

Gilbert, J.R., Moler, C.B., and Schreiber, R. (1992). "Sparse matrices in MATLAB: Design and implementation." *SIAM Journal of Matrix Analysis and Application,* **13**(1), 333–356.

Gill, S. (1951). "Process for the step by step integration of differential equations in an automatic digital computing machine." *Proceedings of the Cambridge Philosophical Society,* **47**, 96–108.

Goldberg, D.E. (1989). *Genetic Algorithms in Search, Optimization and Machine Learning.* Addison-Wesley, Reading, MA.

Golub, G.H., and Van Loan, C.F. (1989). *Matrix Computations,* 2nd ed. John Hopkins University Press, Baltimore.

Gragg, W.B. (1965). "On extrapolation algorithms for ordinary initial value problems." *SIAM Journal of Numerical Analysis,* **2**, 384–403.

Guyan, R.J. (1965). "Reduction of stiffness and mass matrices." *AIAA Journal,* **3**(2), 380.

Hamming, R.W. (1959). "Stable predictor–corrector methods for ordinary differential equations." *Journal of the ACM,* **6**, 37–47.

Higham, D.J., and Higham, N.J. (2005). MATLAB *Guide,* 2nd ed. SIAM, Philadelphia.

Hopfield, J.J., and Tank, D.W. (1985). "Neural computation of decisions in optimisation problems." *Biological Cybernetics,* **52**(3), 141–152.

Hopfield, J.J., and Tank, D.W. (1986). "Computing with neural circuits: A model." *Science,* **233**, 625–633.

Ingber, L. (1993). "Very fast simulated annealing." *Journal of Mathematical Computer Modelling,* **18**, 29–57.

Jeffrey, A. (1979). *Mathematics for Engineers and Scientists.* Nelson, Sunburyon-Thames, UK.

Karmarkar, N.K. (1984). "A new polynomial time algorithm for linear programming." AT&T Bell Laboratories, Murray Hill, NJ.

Karmarkar, N.K., and Ramakrishnan, K.G. (1991). "Computational results of an interior point algorithm for large scale linear programming." *Mathematical Programming,* **52**(3), 555–586.

Kirkpatrick, S., Gellat, C.D., and Vecchi, M.P. (1983). "Optimisation by simulated annealing." *Science,* **220**, 206–212.

Kronrod, A.S. (1965). *Nodes and Weights of Quadrature Formulas: Sixteen Place Tables.* Consultants' Bureau, New York.

Lambert, J.D. (1973). *Computational Methods in Ordinary Differential Equations.* John Wiley & Sons, London.

Lasdon, L., Plummer, J., and Warren, A. (1996). "Nonlinear programming" in Avriel, M. and Golany, B. (eds.). *Mathematical Programming for Industrial Engineers,* Chapter 6, 385–485, Marcel Dekker, New York.

Lindfield, G.R., and Penny, J.E.T. (1989). *Microcomputers in Numerical Analysis.* Ellis Horwood, Chichester, UK.

MATLAB *User's Guide.* (1989). The MathWorks, Inc., Natick, MA. [This describes an earlier version of MATLAB.]

Merson, R.H. (1957). "An operational method for the study of integration processes." *Proceedings of the Conference on Data Processing and Automatic Computing Machines.* Weapons Research Establishment. Salisbury, South Australia.

Michalewicz, Z. (1996). *Genetic Algorithms + Data Structures = Evolution Programs*, 3rd Edition. Springer-Verlag, Berlin.

Moller M.F. (1993). "A scaled conjugate gradient algorithm for fast supervised learning." *Neural Networks*, **6**(4), 525–533.

Olver, F.W.J., Lozier, D.W., Boisvert. R.F., and Clark, C.W. (2010). *NIST Handbook of Mathematical Functions*. National Institute of Standards and Cambridge University Press, New York. See also *NIST Digital Library of Mathematical Functions*. `http://dlmf.nist.gov/`.

Percy, D.F. (2011). "Prior elicitation: A compromise between idealism and pragmatism," *Mathematics Today*, **47**(3), 142–147.

Press, W.H., Flannery, B.P., Teukolsky, S.A., and Vetterling, W.T. (1990). *Numerical Recipes: The Art of Scientific Computing in Pascal*. Cambridge University Press, Cambridge, UK.

Ralston, A. (1962). "Runge Kutta methods with minimum error bounds." *Mathematics of Computation*, **16**, 431–437.

Ralston, A., and Rabinowitz, P. (1978). *A First Course in Numerical Analysis*. McGraw-Hill, New York.

Ramirez, R.W. (1985). *The FFT, Fundamentals and Concepts*. Prentice Hall, Englewood Cliffs, NJ.

Salvadori, M.G., and Baron, M.L. (1961). *Numerical Methods in Engineering*. Prentice Hall, London.

Stakhov, A., and Rozin, B. (2005). "The golden shofar." *Chaos, Solitons and Fractals*, **26**, 677–684.

Stakhov, A., and Rozin, B. (2007). "The golden hyperbolic models of the universe." *Chaos, Solitons and Fractals*, **34**, 159–171.

Sultan, A. (1993). *Linear Programming—An Introduction with Applications*. Academic Press, San Diego.

Short, L. (1992). "Simple iteration behaving chaotically." *Bulletin of the Institute of Mathematics and its Applications*, **28**(6-8), 118–119.

Simmons, G.F. (1972). *Differential Equations with Applications and Historical Notes*. McGraw-Hill, New York.

Smith, B.T., Boyle, J.M., Dongarra, J.J., Garbow, B.S., Ikebe, Y., Kleme, V.C., and Moler, C. (1976). *Matrix Eigensystem Routines: EISPACK Guide*. Lecture Notes in Computer Science, **6**, 2nd Ed. Springer-Verlag, Berlin.

Styblinski, M.A., and Tang, T.-S. (1990). "Experiments in nonconvex optimisation: Stochastic approximation with function smoothing and simulated annealing." *Neural Networks*, **3**(4), 467–483.

Swift, A. (1977). *Course Notes*, Mathematics Department, Massey University, Wellington, New Zealand.

Thompson, I. (2010). "From Simpson to Kronrod: An elementary approach to quadrature formulae." *Mathematics Today*, **46**(6), 308–313.

Walpole, R.E., and Myers, R.H. (1993). *Probability and Statistics for Engineers and Scientists*. Macmillan, New York.

Index

M

T

U

V

W

Z